Numerical and Computer
Methods in Structural Mechanics

CONTRIBUTORS

D. J. Ayres

M. L. Baron

W. J. Batdorf

George Bugliarello

David Bushnell

S. L. Chu

G. R. Cowper

John Dainora

S. P. Desjardins

W. P. Doherty

Steven J. Fenves

Richard H. Gallagher

J. Ghaboussi

William R. Graham

Ervin D. Herness

Bruce M. Irons

H. A. Kamel

David K. Y. Kan

S. S. Kapur

Samuel W. Key

Raymond D. Krieg

D. Liu

J. M. McCormick

John F. McNamara

Pedro V. Marcal

P. Meijers

Johannes Moe

D. A. Pecknold

N. Perrone

Theodore H. H. Pian

J. R. Rice

W. W. Sable

W. C. Schnobrich

Ernst Schrem

R. L. Taylor

James L. Tocher

D. M. Tracey

T. J. Vinson

E. I. White

E. L. Wilson

Andrew Ka-Ching Wong

J. P. Wright

D. N. Yates

O. C. Zienkiewicz

Numerical and Computer Methods in Structural Mechanics

Edited by

STEVEN J. FENVES

Department of Civil Engineering
Carnegie-Mellon University
Pittsburgh, Pennsylvania

NICHOLAS PERRONE

Office of Naval Research
Department of the Navy
Arlington, Virginia

ARTHUR R. ROBINSON

Department of Civil Engineering
University of Illinois
Urbana, Illinois

WILLIAM C. SCHNOBRICH

Department of Civil Engineering
University of Illinois
Urbana, Illinois

ACADEMIC PRESS New York and London   1973

A Subsidiary of Harcourt Brace Jovanovich, Publishers

ACADEMIC PRESS, INC.
111 Fifth Avenue, New York, New York 10003

United Kingdom Edition published by
ACADEMIC PRESS, INC. (LONDON) LTD.
24/28 Oval Road, London NW1

LIBRARY OF CONGRESS CATALOG CARD NUMBER: 72-88339

PRINTED IN THE UNITED STATES OF AMERICA

Contents

Hybrid Models

Theodore H. H. Pian

Computer Implementation of the Finite–Element Procedure

Ernst Schrem

Part II. CRITICAL REVIEW OF GENERAL-PURPOSE
STRUCTURAL MECHANICS PROGRAMS

Review of the ASKA Program

P. Meijers

A Critical View of NASTRAN

James L. Tocher and Ervin D. Herness

The DAISY Code

D. N. Yates, W. W. Sable, and T. J. Vinson

An Evaluation of the STARDYNE System

John Dainora

Analysis and Design Capabilities of STRUDL Program

S. L. Chu

Elastic–Plastic and Creep Analysis via the MARC Finite-Element Computer Program

D. J. Ayres

Part III.   FINITE DIFFERENCE/FINITE ELEMENTS—A MERGING OF FORCES

A Survey of Finite-Difference Methods for Partial Differential Equations

J. P. Wright and M. L. Baron

Finite-Difference Energy Models versus Finite-Element Models: Two Variational Approaches in One Computer Program

David Bushnell

Comparison of Finite-Element and Finite-Difference Methods

Samuel W. Key and Raymond D. Krieg

Incremental Stiffness Method for Finite Element Analysis of the Nonlinear Dynamic Problem

John F. McNamara and Pedro V. Marcal

The Lumped-Parameter or Bar–Node Model Approach to Thin-Shell Analysis

W. C. Schnobrich and D. A. Pecknold

Part IV. LARGE INTERACTIVE DATA BASES

Design Philosophy of Large Interactive Systems

Steven J. Fenves

Integrated Design of Tanker Structures

Johannes Moe

The STORE Project (The Structures Oriented Exchange)

J. M. McCormick, M. L. Baron, and N. Perrone

Part V. NEW CAPABILITIES FOR COMPUTER-BASED ANALYSIS

Symbolic Computing

Andrew Ka-Ching Wong

A Review of the Capabilities and Limitations of Parallel and Pipeline Computers

William R. Graham

Equation-Solving Algorithms for the Finite-Element Method

Bruce M. Irons and David K. Y. Kan

FLING—A FORTRAN Language for Interactive Graphics

W. J. Batdorf and S. S. Kapur

Part VI. NUMERICAL METHODS FOR A CHANGING TECHNOLOGY

Trends and Directions in the Applications of Numerical Analysis

Richard H. Gallagher

Vehicle Crashworthiness

S. P. Desjardins

Computational Fracture Mechanics

J. R. Rice and D. M. Tracey

Biomechanics

George Bugliarello

The Computer in Ship Structure Design

H. A. Kamel, D. Liu, and E. I. White

List of Contributors

Numbers in parentheses indicate the pages on which the authors' contributions begin.

D. J. AYRES (247), Combustion Engineering, Inc., Windsor, Connecticut

M. L. BARON (265, 439), Paul Weidlinger, Consulting Engineer, New York, New York

W. J. BATDORF (513), Lockheed-Georgia Company, Marietta, Georgia

GEORGE BUGLIARELLO (625), College of Engineering, University of Illinois at Chicago Circle, Chicago, Illinois

DAVID BUSHNELL (291), Lockheed Missiles and Space Co., Palo Alto, California

S. L. CHU (229), Sargent and Lundy Engineers, Chicago, Illinois

G. R. COWPER (1), National Aeronautical Establishment, National Research Council of Canada, Ottawa, Canada

JOHN DAINORA (211), Stone and Webster Engineering Corporation, Boston, Massachusetts

S. P. DESJARDINS (557), Dynamic Science Engineering Operations, A Division of Marshall Industries, Phoenix, Arizona

W. P. DOHERTY (43), Department of Civil Engineering, University of California, Berkeley, California

STEVEN J. FENVES (403), Department of Civil Engineering, University of Illinois, Urbana, Illinois *

RICHARD H. GALLAGHER (543), Department of Structural Engineering, Cornell University, Ithaca, New York

J. GHABOUSSI (43), Department of Civil Engineering, University of California, Berkeley, California

WILLIAM R. GRAHAM (479), R & D Associates, Santa Monica, California

ERVIN D. HERNESS (151), Boeing Computer Services, Seattle, Washington

BRUCE M. IRONS (479), University of Wales, Swansea, Wales, U.K.

H. A. KAMEL (643), University of Arizona, Tucson, Arizona

DAVID K. Y. KAN (497), University of Wales, Swansea, Wales, U.K.

S. S. KAPUR (513), Lockheed-Georgia Company, Marietta, Georgia

SAMUEL W. KEY (337), Sandia Laboratories, Albuquerque, New Mexico

RAYMOND D. KRIEG (337), Sandia Laboratories, Albuquerque, New Mexico

D. LIU (643), American Bureau of Shipping, New York, New York

J. M. McCORMICK (439), Paul Weidlinger, Consulting Engineer, New York, New York

JOHN F. McNAMARA (353), Division of Engineering, Brown University, Providence, Rhode Island

PEDRO V. MARCAL (353), Division of Engineering, Brown University, Providence, Rhode Island

P. MEIJERS (123), Institute TNO For Mechanical Constructions, Delft, The Netherlands

JOHANNES MOE (415), Technical University, Trondheim, Norway

D. A. PECKNOLD (377), Department of Civil Engineering, University of Illinois, Urbana, Illinois

N. PERRONE (439), Office of Naval Research, Department of the Navy, Arlington, Virginia

THEODORE H. H. PIAN (59), Massachusetts Institute of Technology, Cambridge, Massachusetts

J. R. RICE (585), Brown University, Providence, Rhode Island

W. W. SABLE (175), Lockheed Missile and Space Company, Sunnyvale, California

W. C. SCHNOBRICH (377), Department of Civil Engineering, University of Illinois, Urbana, Illinois

ERNST SCHREM (79), Institut für Statik und Dynamik der Luft- und Raumfahrtkonstruktionen, University of Stuttgart, Germany

R. L. TAYLOR (43), Department of Civil Engineering, University of California, Berkeley, California

JAMES L. TOCHER (151), Boeing Computer Services, Seattle, Washington

D. M. TRACEY (585), Brown University, Providence, Rhode Island

T. J. VINSON (175), Lockheed Missile and Space Company, Sunnyvale, California

E. I. WHITE (643), Structural Engineering, Standard Oil of California

E. L. WILSON (43), Department of Civil Engineering, University of California, Berkeley, California

ANDREW KA-CHING WONG (459), Carnegie-Mellon University, Pittsburgh, Pennsylvania

J. P. WRIGHT (265), Paul Weidlinger, Consulting Engineer, New York, New York

D. N. YATES (175), Lockheed Missile and Space Company, Sunnyvale, California

O. C. ZIENKIEWICZ (1), University of Wales, Swansea, Wales, U.K.

---

* Present address: Department of Civil Engineering, Carnegie-Mellon University, Pittsburgh, Pennsylvania.

Preface

The impact of modern digital computers on the field of structural mechanics has consisted of far more than mere application of existing numerical techniques to problems of ever increasing complexity. Indeed, a far-ranging, yet often subtle, interaction has grown up between the methods employed for the static and dynamic analysis of structures and structural elements on the one hand and the capabilities of computer hardware and software on the other. It is important for the progress of structural mechanics that the implications of this interaction be recognized by practitioners of structural mechanics and by computer software designers.

The objective of this Office of Naval Research Symposium was to summarize the present and probable future status of numerical methods in structural mechanics and of related computer techniques and computer capabilities. The concern for exploration of these separate areas and their interaction is reflected in seven aspects which can be distinguished in these Proceedings.

First, the analytical basis of the finite element method—the computer technique most widely implemented in software—is examined broadly in four papers. Cowper presents a general introduction to the finite element procedure and a discussion of convergence in terms of variational principles. Isoparametric and related elements are treated by Zienkiewicz; incompatible displacement models by Wilson, Taylor, Doherty, and Ghaboussi; and hybrid models by Pian.

Second, the paper by Schrem and the one by Irons and Kan explore in depth two fundamental and interrelated aspects of the numerical and computer implementation of finite element procedures, namely the storage and retrieval of data and the selection of equation-solving algorithms.

Third, an entire session, consisting of the papers by Meijers; Tocher and Herness; Yates, Sable, and Vinson; Dainora; Chu; and Ayres, was devoted to a critical review of some of the significant general-purpose structural mechanics programs. In selecting the papers, the organizers of the Symposium were attempting to concentrate on those programs which have gained a measure of practical use outside the organization which originated the program. While novel, this attempt to obtain constructive critical information from knowledgeable users must be viewed as a token one. To provide much-needed program information exchange among users, greatly expanded efforts in this direction are warranted.

As a fourth aspect, a session concentrated on the presentation of alternatives to and extensions of the usual finite element approaches and on relations and comparisons among various analytical outlooks. The paper by Wright and Baron presents a comprehensive survey of finite difference methods. The contributions by Key and Krieg and by Bushnell provide useful comparisons and combinations of finite difference and finite element methods. The difficulties of nonlinear, dynamic finite element problems are explored by McNamara and Marcal. The paper by Schnobrich and Pecknold describes a direct physical interpretation to aid in the selection of finite difference models.

Fifth, the Symposium addressed itself to a major future trend in structural mechanics computing, namely the abandonment of individual, unconnected programs in favor of large, integrated data bases. The paper by Fenves is devoted to general questions of large interactive systems. The role of computer graphics, an important tool in such information systems, is discussed by Batdorf and Kapur. Two outstanding accomplishments in the actual implementation of integrated data bases are the subjects of papers by Moe and by McCormick, Baron, and Perrone.

Sixth, the organizers of the Symposium felt that it was their obligation to bring to the attention of workers in structural mechanics new software and hardware capabilities which have a major potential impact on computer use, both in the nature of problems to be tackled and the magnitude of problems that can be handled. The papers by Wong and by Graham discuss such outstanding new capabilities, as well as the challenges they pose.

The final session of the Symposium paralleled the last part in that it dealt specifically with new applications that seem likely to affect the content of the discipline of structural mechanics itself in the next decade. The paper by Kamel, Liu, and White extrapolates the problems which relate to ship design. Gallagher's paper points out the trends which can be perceived in numerical analysis, while the contribution of Rice and Tracey indicates the analytical problem-solving capability needed for an effective approach to problems of fracture mechanics. Technically important new areas which apply structural mechanics, biomechanics and crash safety, are discussed by Bugliarello and Desjardins.

It is hoped that the exposition of problems and interests in this Symposium will illustrate the degree of interaction existing now between numerical and computer methods in structural mechanics and will foster an even broader symbiosis between these areas in the future.

PART I

Finite Elements—Fundamentals

Variational Procedures and Convergence of Finite-Element Methods

G. R. Cowper

NATIONAL AERONAUTICAL ESTABLISHMENT
NATIONAL RESEARCH COUNCIL OF CANADA
OTTAWA, CANADA

When the finite-element method first appeared on the scene in the mid-1950s, stiffness matrices were derived on what might be termed an "intuitive" or "direct" basis. Continuous structures were regarded in the same light as those structures such as trusses and frames, which were actually made up of physically discrete elements. Thus a continuous structure was regarded as an assembly of elements which were joined only at nodes, and one spoke of the forces applied at nodes and of the equilibrium of nodes. Interelement continuity considerations did not go beyond the matching of displacements at nodes. As the method developed it was realized that the method could be regarded as an application of the variational principles of structural mechanics, especially of the principle of minimum potential energy. This realization greatly stimulated the development of the method. Certainly, variational principles are now generally used as the foundation of the finite element method, as is evident from a perusal of the proceedings of several recent symposia on the subject.

There are several advantages to basing the finite-element method on variational principles. One is that the method is thereby put on a sound theoretical foundation. Another is that the requirements for interelement continuity are clarified, as these requirements are quite explicit in the statement of the variational principles. Furthermore, greater flexibility in the formulation of elements is possible because generalized displacements need not be restricted to quantities that are conjugate to physical forces and moments. Indeed the entire concept of nodal forces as actual physical forces can be dispensed with. Higher derivatives of displacement, for example curvatures and twists in the case of a bent plate, become acceptable as generalized displacements. Greater flexibility in the formulation of elements also comes about because a variety of variational principles are available. In addition to the classical principles of minimum potential energy and minimum complementary energy there is the more general principle of Reissner, and also a number of modified principles which permit the relaxation of various requirements, particularly requirements of interelement continuity of displacements or stresses. A final advantage is that the scope of the finite element method is broadened to include nonstructural problems which can be expressed in terms of a variational principle. The method has been applied to problems of heat flow, potential and viscous fluid flow, seepage of ground water, and others.

The classical variational principles of linear structural mechanics are the principles of minimum potential energy and of minimum complementary energy. The former principle may be stated thus: The potential energy is stationary with regard to all kinematically admissible variations of displacements from the state of equilibrium. For stable equilibrium the stationary value of the potential energy is an absolute minimum. In symbols,

$$\delta\Pi = 0, \qquad \Pi = U - V \tag{1}$$

where Π is the potential energy, U is the strain energy of the body, and V is the virtual work of the applied loads. An alternative statement is: Among all kinematically admissible displacements, those satisfying the equilibrium conditions make the potential energy an absolute minimum. Note that the displacements must be kinematically admissible. This means that they must satisfy sufficient continuity conditions within the structure and must satisfy the kinematic boundary conditions. There is, however, no requirement that the stress boundary conditions be satisfied.

The principle of minimum complementary energy is concerned with stress fields that satisfy the conditions of equilibrium but not necessarily the requirements of compatibility. It may be stated thus: Among all statically admissible stress fields, the one which satisfies the stress-strain relations in the interior of the structure and the displacement boundary conditions makes the complementary energy an absolute minimum.

Some analysts have found that the requirements of kinematic admissibility or static admissibility are rather troublesome to achieve with finite elements, especially in plate and shell problems. As a result, a number of modified variational principles have been proposed which allow relaxed continuity conditions. These principles have been used effectively, chiefly by Pian and

Tong [2], as a basis for so-called hybrid elements. Reissner's variational principle has also been used as a basis for finite elements. Reissner's principle is more general than either of the principles of minimum potential or complementary energy, in that it allows simultaneous variations in both stresses and displacements. This permits independent assumptions as to the distributions of displacements and stresses, and is claimed to thus lead to more accurate approximations for the stresses. The hybrid principles and Reissner's principle, unlike the principles of minimum potential and complementary energy, are not minimum principles but state only that certain functionals are stationary.

The variational principles and their application to the finite element method have been discussed and classified by a number of writers. Mention may be made of the excellent surveys by Pian [1] and Pian and Tong [2] and the conference papers of Hansteen [3] and Tottenham [4]. Mention should also be made of the earlier papers of de Veubeke [5], [6], which point out the dual nature of the principles of minimum potential energy and minimum complementary energy and emphasize how these two principles can be used to set bounds on stiffness coefficients. The survey of Pian is particularly useful in illustrating and classifying the many different possibilities for formulating finite elements based on the various variational principles. In view of these surveys it would be superfluous to go further into the details of variational principles, and I would like to turn instead to the question of convergence.

Does an approximate solution, obtained by means of finite elements, converge to the correct solution as the mesh of finite elements is uniformly refined? For compatible displacement elements and for equilibrium elements a relatively simple proof of convergence can be given, based on the minimum property of the potential or complementary energies. A convergence proof is of more than academic interest. For one thing, it contributes to the confidence with which finite elements can be used, because the user has a guarantee that his results must approach the correct answer as the mesh of elements is refined. This has not always been the case, and the early days of the method provided examples of ill-chosen elements which did not converge to the right answer. In addition the convergence proof points up the conditions necessary for convergence and good accuracy, and thus provides useful guidance in constructing elements. The essential points of the proof have been given in a number of papers [7]–[10], of which the paper by McLay is particularly notable.

We consider the convergence as it applies to compatible displacement elements which are formulated on the basis of the principle of minimum potential energy. Using this principle, approximate solutions to a structural analysis problem can be constructed by the Rayleigh–Ritz procedure. The

deflection of the structure is represented by an assumed displacement field
which depends on a number of free parameters. The "best" approximation
is obtained by choosing the free parameters so as to minimize the potential
energy. The essential feature of the finite element method is the local nature of
the assumed displacement field. In general, the displacements within an
element are of the form

$$u(x_i) = N(x_i)q \tag{2}$$

where q is the vector of generalized displacements of an element and $N(x_i)$ is a
matrix of chosen shape functions. Usually q consists of the values of displacements at specific points on the boundary of the element. The components of q are thus common to adjacent elements and the choice of shape
functions is usually such that equality of q for adjacent elements assures
compatibility of displacements between elements.

Once the shape functions are specified, it is a relatively straightforward
matter to calculate the potential energy of each element, and to obtain the
potential energy of the entire structure by summation. The total potential
energy depends quadratically and linearly on the generalized displacements
q. The q are then chosen so as to minimize the potential energy.

The convergence proof begins by calculating the difference in potential
energies of the exact field of displacements u and of some varied field $u + \delta u$.
For linear elastic structures the potential energy is a quadratic functional
and hence

$$\Pi(u + \delta u) = \Pi(u) + \delta\Pi + \delta^2\Pi \tag{3}$$

But $\delta\Pi = 0$ since the potential energy is stationary for kinematically
admissible variations of displacement. Furthermore,

$$\delta^2\Pi = \delta^2(U - V) = \delta^2 U = U(\delta u) \tag{4}$$

since V, the virtual work of the applied loads, is linear in u, while the strain
energy U is quadratic in u. Hence it follows from Eqs. (3) and (4)

$$\Pi(u + \delta u) - \Pi(u) = U(\delta u) \tag{5}$$

To express Eq. (5) in words, the error in the potential energy is equal to the
strain energy of the variation in displacement.

If $\bar{u}$ is the finite element solution of the problem, then from Eq. (5)

$$\Pi(\bar{u}) = \Pi(u) + U(\bar{u} - u) \tag{6}$$

Now let $\tilde{u}$ be the displacement field that is obtained from the finite element
shape functions when the generalized displacements q are set equal to their
exact values. For this field

$$\Pi(\tilde{u}) = \Pi(u) + U(\tilde{u} - u) \tag{7}$$

But in the finite element solution the generalized displacements q are chosen to minimize the potential energy, and hence the setting of q equal to their exact values must result in a larger value of the potential energy. That is,

$$\Pi(\tilde{u}) \geqslant \Pi(\bar{u}) \tag{8}$$

Hence it follows from Eqs. (6)–(8) that

$$U(\tilde{u} - u) \geqslant U(\bar{u} - u) \geqslant 0 \tag{9}$$

Convergence of $U(\bar{u} - u)$ to zero will be proved if it can be shown that $U(\tilde{u} - \bar{u})$ converges to zero as the mesh is refined. If $U(\bar{u} - u)$ converges to zero then $\bar{u}$ must converge to a displacement field which differs from the correct one by no more than a rigid-body motion. The investigation of convergence is thus reduced to examining how $U(\tilde{u} - u)$ tends to zero, which is tantamount to examining how closely a finite element field $\tilde{u}$ can approximate the energy of the exact field u.

As McLay has pointed out, Taylor's theorem provides a powerful tool for examining how well $\tilde{u}$ approximates u. Let us suppose, for simplicity, that u is one-dimensional and depends only on the coordinate x. According to Taylor's theorem the exact solution u can be expanded, within each element, in a power series with remainder thus

$$u(x) = a_0 + a_1 x + \cdots + a_n x^n + R_{n+1} \tag{10}$$

where

$$|R_{n+1}| \leqslant M h^{n+1} \tag{11}$$

The remainder R_{n+1} is of order h^{n+1}, where h is a distance not larger than the length of the element, and the constant M in the bound on R_{n+1} depends on the maximum value of the $n + 1$st derivative of u within the element.

If the finite-element approximation $\tilde{u}$ consists of a polynomial of degree n it is usually not difficult to show that $\tilde{u}$ matches the power series in Eq. (10), at least to within an error of the same order as R_{n+1}. The difference $\tilde{u} - u$ then is of the same order as R_{n+1}, that is of order h^{n+1}. Since h tends to zero as the finite-element mesh is refined, it follows that $\tilde{u} - u$ also tends to zero. A similar statement can be made about the derivatives of u and $\tilde{u}$, except that the order of accuracy diminishes as the order of the derivative increases, again according to Taylor's theorem. The first derivative of $\tilde{u} - u$ is of order h^n, the second derivative of order h^{n-1}, and in general the pth derivative is of order h^{n+1-p}. The pth derivative of $\tilde{u} - u$ therefore tends to zero with h provided $n \geqslant p$. If the strain energy expression contains only first derivatives of the displacement then $U(\tilde{u} - u)$ will likewise tend to zero provided $n \geqslant 1$. Thus the approximating polynomial must be at least linear. If the strain energy expression contains second derivatives of displacement, then $n \geqslant 2$ for convergence, and so the approximating polynomial must be at least quadratic.

The extension of the argument to two or more dimensions is straightforward, and identical results for the minimum degree of the approximating polynomial are obtained. McLay [8] and Cowper et al. [11] contain examples of two-dimensional convergence investigations.

Suppose that the strain energy expression contains only first derivatives of displacement. If the displacements are approximated by a polynomial of degree n then their first derivatives have an error of order h^n, and since these derivatives appear quadratically in the strain energy the error in U is of order h^{2n}. Hence if a linear polynomial is used the error in U is of order h^2; with a quadratic polynomial it is h^4, and so on. Similarly, if the strain energy involves second derivatives of displacement the error in U is of order h^{2n-2}. These theoretical rates of convergence have been confirmed numerically by Lindberg, Olson, and co-workers [11, 12].

It is clear that the order of the error can be reduced and convergence made more rapid by using higher-degree polynomials. Although this leads to more complicated elements the gain in accuracy is often worth the effort. Consider the example shown in Fig. 1 which is taken from a paper by

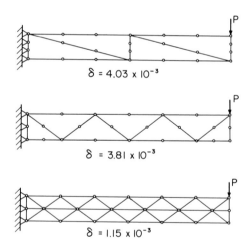

Fig. 1. Deflection of a cantilever. (After Iversen [13].)

Iversen [13]. A cantilever beam is analyzed in order to find the deflection under a point load. Three types of elements have been used which employ, respectively, linear, quadratic, and cubic polynomials as shape functions. The number of elements has been varied in such a manner that the total number of nodes and degrees of freedom is approximately constant. All elements are compatible so the deflection is underestimated and the largest answer is therefore nearest the truth. In this example the cubic element gives the best result, as far as deflection is concerned. The linear element, which is the constant strain triangle, is rather unsatisfactory.

Another fact which emerges from the comparison of shape functions with Taylor series is the importance of having complete polynomials in the shape functions for two- and three-dimensional elements. The remainder in the Taylor expansion is of order h^{n+1} only if the series contains all terms of degree n and lower. For example, a two dimensional cubic shape function must contain all of x^3, x^2y, xy^2, y^3, and all lower terms in order to have an error of order h^4. Omission of any one of these terms worsens the order of the error. This is illustrated by some results of Poppelwell and McDonald [14] which are shown in Fig. 2. Poppelwell and McDonald calculate the natural

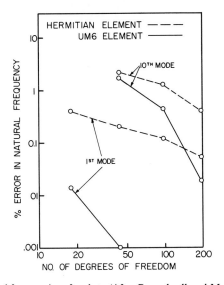

Fig. 2. Natural frequencies of a plate. (After Poppelwell and McDonald [14].)

frequencies of a simply-supported rectangular plate. Two different elements are used both of which are rectangular, fully conforming, and have the same 24 generalized coordinates. The first element, due to Bogner *et al* [15] and also to Mason [16] uses products of Hermitian interpolation polynomials for the shape functions. Although the shape functions satisfy the minimum conditions for convergence, they are not able to represent displacements of the type x^2y or xy^2 and their accuracy is inherently similar to that of a quadratic polynomial. Such deficiencies of Hermitian interpolation polynomials were pointed out by Bogner *et al.* [15], who also suggested remedial measures. The second element, due to Poppelwell and McDonald [14], uses shape functions which are able to represent an arbitrary fourth-degree polynomial. As can be seen, the second element gives much better results even though both elements have exactly the same degrees of freedom. The more rapid rate of convergence may also be noted.

The Taylor series analysis has shown that the minimum requirement for convergence is that the interpolating polynomials should be of degree p, where p is the order of the highest derivative in the expression for strain energy. It is interesting to compare this requirement with the claim that the interpolating polynomials must include all rigid-body movements and all states of uniform strain. For beams and flat plates the two requirements coincide, since the rigid-body movements of a flat plate in bending are represented by constant and linear functions of x and y, while the states of uniform bending and twisting are represented by x^2, y^2, and xy. In the case of curved beams and shells, however, the two requirements are no longer equivalent. The requirement on the minimum degree of the interpolating polynomial is the more fundamental one, and, in the writer's opinion, the requirement of exact rigid body modes has been overemphasized.

Some recent results of Fleischer and Petyt [17] are pertinent here. They examine three finite elements for the calculation of the radial vibrations of a circular arch. The first element uses a linear polynomial for the tangential displacement and a cubic for the transverse displacement. The second element uses a combination of polynomials and trigonometric functions such that all three rigid-body motions are included exactly. The third element uses cubic polynomials for both tangential and transverse displacements. These shape functions are detailed in Fig. 3 while Fleischer and Petyt's results are shown in Fig. 4. It is seen that the first element gives poor results, the second element gives reasonably good results, while the third element gives the best results. Two conclusions might be drawn. First, it is not necessary to include exact rigid-body motions in the shape functions; polynomials of sufficient

(1) $\quad v = A_1 + A_2 s$
$\quad\quad w = A_3 + A_4 s + A_5 s^2 + A_6 s^3$

(2) $\quad v = A_1 s + A_2 R(\cos\beta \cos\phi - 1) - A_3 \sin\phi + A_4 \cos\phi$
$\quad\quad w = A_2 R \cos\beta \sin\phi + A_3 \cos\phi + A_4 \sin\phi + A_5 s^2 + A_6 s^3$

(3) $\quad v = A_1 + A_2 s + A_3 s^2 + A_4 s^3$
$\quad\quad w = A_5 + A_6 s + A_7 s^2 + A_8 s^3$

Fig. 3. Shape functions for a circular arch. (After Fleischer and Petyt [17].)

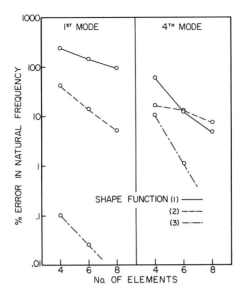

Fig. 4. Natural frequencies of an arch. (After Fleischer and Petyt [17].)

degree can give just as good, if not better, accuracy as shape functions which include rigid-body motions. Second, polynomials which satisfy only the minimum requirements for convergence, such as the polynomial for the tangential displacement in the first element, are likely to give poor accuracy and slow convergence. This was also evident in the example of Fig. 1.

All of the foregoing discussion is based on the tacit assumption that the exact solution can be expanded in a Taylor's series within each element. The argument breaks down for those elements which contain a singularity of the exact solution, such as may occur at concentrated loads or at reentrant corners. To investigate theoretically the convergence near a singularity would be a complicated business, and so far as the writer is aware has not been attempted yet. It is probably safe to say that singularities have a detrimental effect on convergence and accuracy, and this is supported by some numerical evidence. Figure 5, for example, shows the results of the following exercise carried out by the author and his colleagues [11]. A simply supported square plate is analysed for the cases of a uniformly distributed load and a central point load. The same elements and mesh are used in each case, but the error is far larger and the rate of convergence slower in the case of the point load. This can be attributed to the logarithmic singularity in the stresses at the point of application of the load.

An important situation involving singular stress distributions is the analysis of stresses near the tip of a crack. Polynomials are probably inherently

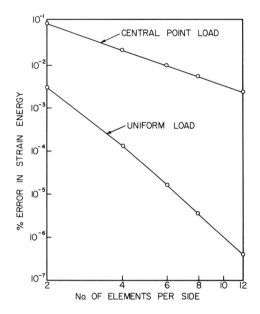

Fig. 5. Strain energy of a plate. (After Cowper *et al.* [11].)

unsuitable for representing such distributions. The efficient analysis of stresses near tips of cracks may well require special purpose elements which exploit the known form of the singular stress field.

In closing, it should be pointed out that the foregoing discussion of accuracy and convergence dealt only with the simplest situations. Many complicating factors were ignored which deserve, and in some cases have received, attention. Among these might be mentioned the questions of incompatible or non-conforming elements and of elements which are formulated on the basis of nondefinite variational principles. A proof of convergence for non-conforming elements in contained in the work of Oliveira [10, 18], while Pian and Tong have supplied a convergence proof for their hybrid elements [19]. Apart from the fundamental approximation errors the standard finite element techniques introduce additional perturbation errors. Boundary conditions are not satisfied exactly, the original domain is perturbed by triangularization, various coefficients of the differential equations are approximated, and exact integration is often replaced by numerical quadrature. These errors, among others, are examined in recent works by Fried [20] and McLay [21]. Several workers are examining the question of convergence in dynamic problems. The thesis of Fried [20] may be mentioned in this connection.

To conclude, the main point that I have tried to present in this paper is that convergence theorems for the finite element method can give useful guidance in the construction of elements. For compatible displacement elements, the theory predicts the benefits of higher degree polynomials as shape functions, shows the importance of complete polynomials, suggests the deleterious effects of singularities and puts the question of rigid-body motions in proper perspective. These points were illustrated by examples, which I hope were convincing.

References

1. Pian, T. H. H., "Formulations of Finite Element Methods for Solid Continua," in *Recent Advances in Matrix Methods of Structural Analysis and Design* (Gallagher, R. H., Oden, J. T., and Yamada, Y., eds.). Univ. of Alabama Press, 1971.
2. Pian, T. H. H., and Tong, P., "Basis of Finite Element Methods for Solid Continua," *Int. J. Numer. Methods Eng.* **1**, 3–28 (1969).
3. Hansteen, O. E., "Finite Element Methods as Applications of Variational Principles," in *Finite Element Methods in Stress Analysis* (Holand, I., and Bell, K., eds.). Tapir, Trondheim, 1969.
4. Tottenham, H., "Basic Principles," in *Finite Element Techniques in Structural Mechanics* (Tottenham, H., and Brebbia, C., eds.). Southampton Univ. Press, Southampton, 1970.
5. Fraeijs de Veubeke, B. M., "Upper and Lower Bounds in Matrix Structural Analysis," in *Matrix Methods of Structural Analysis* (de Veubeke, F., ed.). Pergamon, Oxford, 1964.
6. Fraeijs de Veubeke, B. M., "Displacement and Equilibrium Models in the Finite Element Method," in *Stress Analysis: Recent Developments in Numerical and Experimental Methods* (Zienkiewicz, O. C., and Holister, G. S., eds.). Wiley, New York, 1965.
7. Johnson, M. W., Jr., and McLay, R. W., "Convergence of the Finite Element Method in the Theory of Elasticity," *J. Appl. Mech.* **35**, 274–278 (1968).
8. McLay, R. W., "Completeness and Convergence Properties of Finite Element Displacement Functions—A General Treatment," AIAA Paper No. 67-143, AIAA 5th Aerospace Sciences Meeting, New York (January 1967).
9. Tong, Pin, and Pian, T. H. H., "The Convergence of the Finite Element Method in Solving Linear Elastic Problems," *Internat. J. Solids Structures* **3**, 865–879 (1967).
10. de Arantes e Oliviera, E. R., "Theoretical Foundations of the Finite Element Method," *Internat. J. Solids Structures* **4**, 929–952 (1968).
11. Cowper, G. R., Kosko, E., Lindberg, G. M., and Olson, M. D., "A High Precision Triangular Plate-Bending Element," Nat. Res. Council of Canada, Aeronautical Rep. LR-514 (December 1969).
12. Lindberg, G. M., and Olson, M. D., "Convergence Studies of Eigenvalue Solutions Using Two Finite Plate Bending Elements," *Internat. J. Numerical Methods Eng.* **2**, 99–116 (1970).
13. Iversen, P. A., "Some Aspects of the Finite Element Method in Two-Dimensional Problems," in *Finite Element Methods in Stress Analysis* (Holand, I., and Bell, K., eds.). Tapir, Trondheim, 1969.
14. Poppelwell, N., and McDonald, D., "Conforming Rectangular and Triangular Plate-Bending Elements," Dept of Mech. Eng. Publ. Univ. of Manitoba, Winnipeg (May 1971).
15. Bogner, F. K., Fox, R. L., and Schmit, L. A., Jr., "The Generation of Interelement Compatible Stiffness and Mass Matrices by the Use of Interpolation Formulas," *Proc. Conf.*

Matrix Methods Structural Mech., *1st*, Wright–Patterson AFB, Ohio, AFFDL TR 66-80, (November 1965).

16. Mason, V., "Rectangular Finite Elements for Analysis of Plate Vibrations," *J. Sound Vibrat.* **7**, 437–448 (1968).

17. Fleischer, C. C., and Petyt, M., "Free Vibrations of a Curved Beam," *Symp. Finite Element Techniques Structural Vibrat.* Inst. of Sound and Vibrat. Res., Univ. of Southampton (March 1971).

18. de Arantes e Oliveira, E. R., "Completeness and Convergence in the Finite Element Method," *Proc. Conf. Matrix Methods Structural Mech.*, *2nd*, Wright–Patterson AFB, Ohio, AFFDL TR 68-150 (October 1968).

19. Tong, Pin, and Pian, T. H. H., "A Variational Principle and the Convergence of a Finite Element Method Based on Assumed Stress Distribution," *Internat. J. Solids Structures* **5**, 463–472 (1969).

20. Fried, I., "Discretization and Round-off Errors in the Finite Element Analysis of Elliptic Boundary Value Problems and Eigenvalue Problems," Ph.D. Thesis, Massachusetts Inst. of Technol. (June 1971).

21. McLay, R. W., "On Certain Approximations in the Finite Element Method." *J. Appl. Mech.* **38**, Ser. E, 58–61 (1971).

Isoparametric and Allied Numerically Integrated Elements—A Review

O. C. Zienkiewicz

UNIVERSITY OF WALES
SWANSEA, WALES U.K.

1. Introduction

In the early days of finite-element analysis, simple element shapes with the minimum degrees of freedom were used almost exclusively. While such elements have much merit—not the least being a simple physical visualization of the problem by analogy with discrete networks—it was soon discovered that with elements possessing higher degrees of freedom and using more polynomial terms in the shape function expansions considerable advantages accrued. Now not only did the accuracy of representation increase for a given number of elements (as would obviously be expected), but this also improved considerably for a given number of degrees of freedom. Furthermore, the convergence rates could be shown to increase with element complexity. From the practical view-point this meant that for a given desired accuracy the engineer could reduce the number of elements and degrees of freedom drastically, and thus save effort both in data preparation and in the equation solution process which even today represents the major computer cost of finite-element analysis.

With simple element shapes (for which higher order expansions can be written simply) a difficulty arises immediately. As only a small number of elements is now required for an adequate solution—this is now only possible for simple geometrical shapes and thus appears to negate the essential advantage of the finite-element process.

13

To overcome this difficulty a distortion of the simple element shapes appears necessary. This can be accomplished by a suitable mapping procedure. It will be shown that an exceedingly simple mapping can be achieved by the use of the basic element shape functions simultaneously used for representing the variation of the unknown variables. Such mapping termed "isoparametric" was first used for deriving the properties of a linear quadrilateral by Taig [1] and later generalized by Irons [2, 3] who introduced also the concepts of numerical integration as the only practical method of establishing the general element properties economically.

Since those early days, much development and application has occurred, and today in such fields as three-dimensional analysis the use of isoparametric elements is standard [4–22]. Indeed recently their extension to the nonlinear domain has underlined the advantages of their basic formulation [23, 24].

The mapping concepts associated with formulation of isoparametric elements have certain advantages in their own right for representation of curvilinear surfaces. Thus they can form a basis for the generation of simple element meshes [25] or simply for the description of curved surfaces. Indeed it is of interest to note here that the parallel developments in computer graphics and in particular the description of so called Coons surfaces follow a similar pattern [26, 27].

Doubtless use of this fact will be soon made in the simultaneous description of the design surface and its analysis, possibly leading to developments in optimal design.

In this paper we shall concentrate on finite-element problems requiring a C_0 continuity between elements and satisfying the completeness requirement of constant first derivative type [28]. It is for such problems that the isoparametric formulation possesses essential advantages and its extension to higher-order variational principles is not easy. As the restricted class embraces all plane and three-dimensional elasticity problems—and as extension to the plate/shell domain made on this basis is useful—the limitation does not appear to be too severe.

2. Basic Principles of Shape Function (Interpolation) Mapping

Let an element be defined in the one-, two-, or multidimensional space of coordinates $(\xi, \eta, \ldots)$ by a series of nodes i and associated shape functions N_i' which interpolate an unknown ϕ (for instance, displacements) as

$$\phi = \sum N_i' \phi_i \tag{1}$$

in a manner which assures C_0 continuity between elements and, by the presence of complete linear expansion, ensures that the constant derivative criterion is satisfied.

If we now desire to map the element into another space $(x, y, \ldots)$ so that a typical node i moves to a position $(x_i, y_i \ldots)$ a relation of the form

$$x = \sum N_i' x_i, \qquad y = \sum N_i' y_i \tag{2}$$

can be used to define the mapping as

$$N_i' \equiv N_i'(\xi, \eta, \ldots) \tag{3}$$

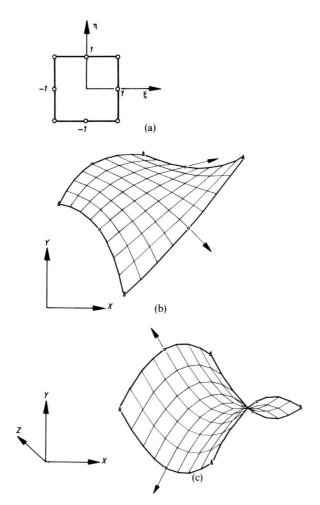

Fig. 1. Isoparametric mapping of a parabolic square element. (a) "Parent" element; (b) two-dimensional map; (c) three-dimensional map.

and the shape functions have the property that N_i' is unity at node i and zero at all other nodes.

Thus for any set of coordinates $(\xi, \eta, \ldots)$ there corresponds a set of $(x, y, \ldots)$, and desired nodal coordinates are achieved.

Such mapping allows, for instance, a two-dimensional elementary square in ξ, η coordinates ranging from -1 to 1 to be mapped into a surface in two or three Cartesian dimensions as shown in Fig. 1.

In a similar manner, any other two- or three-dimensional element may be mapped as shown in Fig. 2.

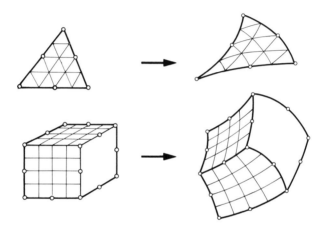

Fig. 2. Further examples of isoparametric mapping.

3. Uniqueness of Mapping

If the number of coordinates $(\xi, \eta, \ldots)$ and $(x, y, \ldots)$ are identical, then by the theorem of preservation of regions, elements will be preserved on mapping if this is of a one-to-one, regular kind. The requirement for this to be achieved if N_i' are simple, nonsingular polynomials is that the functional determinant or Jacobian, defined as

$$\frac{\partial(x, y, \ldots)}{\partial(\xi, \eta, \ldots)} = \begin{vmatrix} \dfrac{\partial x}{\partial \xi}, \dfrac{\partial x}{\partial \eta}, \cdots \\[2mm] \dfrac{\partial y}{\partial \xi}, \dfrac{\partial y}{\partial \eta}, \cdots \end{vmatrix} = |J| \tag{4}$$

does not change sign in the domain. This important check must be instituted with all element computations and what happens if this condition is violated

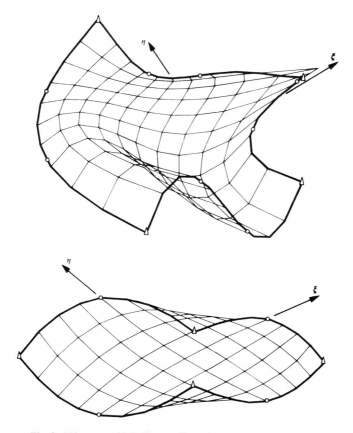

Fig. 3. "Unreasonable" element distortions. Nonunique mapping.

is illustrated in Fig. 3. Jordan [22] discusses some of the limitations on distortion of two-dimensional elements with a single midside node and formulates rules which, though given in more detail are summarized as "To stay out of trouble keep the side nodes close to the centers of these sides and corner angles well under 180°." While general rules cannot be given so simply the check on the Jacobian will point out nonuniqueness immediately.

If a two-dimensional region is mapped into a three-dimensional space the Jacobian no longer exists, and the concern regarding region preservation requires a more detailed topological investigation. Discussion of this point is, however, not relevant here.

One special property of the mapping is of great practical importance. If a series of contiguous elements is mapped the new elements remain contiguous, i.e., the space in Cartesian coordinates is derived by the curvilinear elements without interelement gap.

This property follows directly from the requirement that N_i' functions are chosen to be of such a type as to ensure C_0 continuity between elements. If this is not the case obviously the mapping will be of little practical use.

4, Iso-, Sub-, and Superparametric Elements

If the curvilinear elements are used for finite element analysis we can interpolate the unknown (or unknowns) u, by shape functions associated with the curvilinear coordinates $(\xi, \eta, \ldots)$ in the usual manner given by Eq. (1). This can be written as

$$u = \sum N_i u_i \tag{5}$$

where $N_i(\xi, \eta, \ldots)$ is a set of suitable shape functions ensuring the usual convergence cretria in $(\xi, \eta, \ldots)$ space.

If $N_i' = N_i$ the elements are called *isoparametric*. If N_i' is such that it is of a lower order than N_i and that we can express it as a linear combination

$$N_i' = \sum C_{ij} N_j$$

then the element is *subparametric*. If N_i' is of a higher order than N_i, then the element is called *superparametric*.

As we are generally concerned with limiting the physical distortion of elements, sub- and isoparametric elements are of major practical use. The following two properties apply to those elements.

1. If C_0 continuity is ensured by the N_i functions this will apply in the original or the curvilinear coordinates to the variable u.

2. If the shape functions are such that $\sum N_i = 1$ (a condition necessary simply for no straining under "rigid body" modes to occur) then both iso- and subparametric elements satisfy the constant derivative criterion.

The proof of this statement is as follows. We require that a linear expansion

$$u = \sum N_i u_i = \alpha_1 + \alpha_2 x + \alpha_3 y + \cdots \tag{6}$$

when

$$u_i = \alpha_1 + \alpha_2 x_i + \alpha_3 y_i + \cdots \tag{7}$$

be satisfied, where $\alpha_1, \ldots$ are arbitrary constants. Substituting Eq. (7) into Eq. (6)

$$\alpha_1 \sum N_i + \alpha_2 \sum N_i x_i + \cdots \equiv \alpha_1 + \alpha_2 x + \cdots \tag{8}$$

for all linear terms. This identity is true if

$$\sum N_i = 1, \ \sum N_i x_i = x, \ldots \text{etc.} \tag{9}$$

Since for both iso- and subparametric elements this is true the property is proved.

5. Evaluation of Element Properties in Curvilinear Coordinates

Stiffness and other properties of elements require evaluation of integrals such as [28]

$$\int [B]^{\mathrm{T}}[D][B]\, d\mathrm{vol} \tag{10}$$

or generally

$$\int [H]\, dv \tag{11}$$

where $[H]$ is a function of the variable u or their first derivatives with respect to an (x, y) coordinate system.

As these variables are given in terms of the curvilinear system $(\xi, \eta, \ldots)$ some transformations are necessary. In particular, derivatives $\partial N_i/\partial x$ etc. have to be determined in terms of $\partial N_i/\partial \xi$, etc.

Writing the Jacobian matrix of Eq. (4) as

$$\left[\frac{\partial(x, y, \ldots)}{\partial(\xi, \eta, \ldots)}\right] = [J] \tag{12}$$

and using following matrix definition

$$\left\{\frac{\partial N_i'}{\partial \xi}, \ldots\right\}_i^{\mathrm{T}} = \left[\frac{\partial N_i'}{\partial \xi}, \frac{\partial N_i'}{\partial \eta}, \ldots\right] \quad \text{etc.} \tag{13}$$

we have

$$\left\{\frac{\partial N_i'}{\partial x}, \ldots\right\}_i = [J^{\mathrm{T}}]^{-1}\left\{\frac{\partial N_i'}{\partial \xi}, \ldots\right\}_i \tag{14}$$

The coordinate Jacobian matrix Eq. (14) is conveniently evaluated using the relationship

$$[J] = \left[\frac{\partial(N_i'x_i, \ldots)}{\partial(\xi, \eta, \ldots)}\right] = \begin{bmatrix} \{G'\}_\xi^{\mathrm{T}}\{x\}, & \{G'\}_\eta^{\mathrm{T}}\{x\} \\ \{G'\}_\xi^{\mathrm{T}}\{y\}, & \cdots \end{bmatrix} \tag{15}$$

where

$$\{G'\}_\xi^{\mathrm{T}} = \left\{\frac{\partial N_1}{\partial \xi}, \ldots \frac{\partial N_i}{\partial \xi} \ldots\right\} \quad \text{etc.} \qquad \{x\}^{\mathrm{T}} = [x_1, \ldots \quad x_i, \ldots] \quad \text{etc.} \tag{16}$$

This is a useful presentation as it will be observed that the Jacobian matrix immediately becomes a matrix equal in size to the number of co-ordinates and only vector multiplications are required.

The volume element in Eqs. (10) and (11) has to be transformed before integration using

$$d\text{vol} = dx\, dy \cdots = |J|\, d\xi\, d\eta \cdots \tag{17}$$

Clearly, exact integration of element properties will in general be a tedious if not impossible matter and numerical integration is a necessary part of the process.

6. Required Accuracy of Numerical Integration

Before attempting numerical integration it is desirable to know the degree of accuracy required to ensure convergence to the correct result. Heuristic arguments introduced by Irons [29] are convincing in the context of continuum solid mechanics and carry over to the whole range of mathematical problems to which the finite element process is applicable.

In the context of stress analysis the interelement forces due to internal element stresses are

$$\{F\} = \left(\int [B]^{\text{T}}[D][B]\, dv \right) \{\delta\} = \int [B]^{\text{T}}\{\sigma\}\, dv \tag{18}$$

If such forces can be determined exactly by numerical integration for a constant stress state within each element, to which the exact results must tend with decreasing element size, then results obtained by exact and numerical integration must in the limit be identical.

Thus integration must be capable of determining exactly such integrals as

$$\int \frac{\partial N_i}{\partial x}|J|\, d\xi\, d\xi \cdots \tag{19}$$

and from Eqs. (14) and (15) such integrals as

$$\int \frac{\partial N_i}{\partial \xi} \left(\frac{\partial N_i'}{\partial \eta} \cdot \frac{\partial N_j'}{\partial \zeta} + \cdots \right) d\xi\, d\xi \cdots \tag{20}$$

must therefore be capable of exact evaluation.

For isoparametric elements when $N_i = N_i'$ this is equivalent to being able to determine

$$\int |J|\, d\xi\, d\eta \cdots \tag{21}$$

exactly, i.e., to find the element volume exactly in the distorted coordinates.

While the requirements are sufficient to ensure that convergence in the limit is achieved, bounding theorems are no longer valid with approximate integration and we must also ensure that integration order is not reduced so low that resulting stiffness matrices become singular after assembly.

The determination of the economic and practical limits to which the numerical integration has to be pushed is a matter of concern in much current

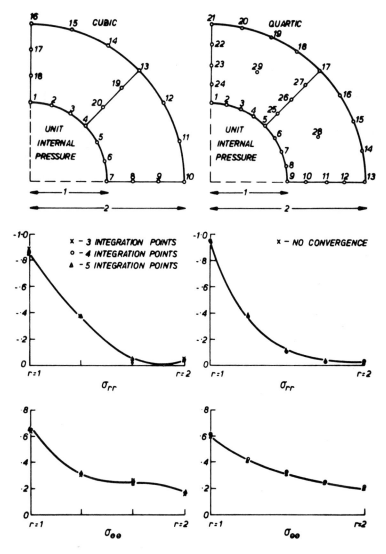

Fig. 4. Effect of varying the number of integration points.

research. Approximate integration by omitting strain energy due to the higher modes of deformation always underestimates the strain energy stored in an element for given nodal displacements and thus reduces effectively the structural stiffness. As displacement formulation always overestimates this, the error introduced is in the right direction and compensates to some extent the discretization errors. While there is no guarantee that over compensation does not occur, in many cases extreme representation improvements have been recorded [20, 21].

Figure 4 shows the analysis of a simple problem using elements of varying order of expansion and different integration orders.

7. Some Useful Elements for Two- and Three-Dimensional Analysis

It is an observable fact that improvement of element properties occurs in quanta dependent on the presence of complete polynomial expressions. While on occasion a very considerable improvement occurs due to the presence of an additional term* of one order higher than the complete expansion, isolated terms of yet higher orders contribute little to the performance [30]. Certainly the rate of convergence is governed by the highest order of complete polynomials present.

Three basic classes of elements have been developed for two- and three-dimensional elements [7, 23, 31].

1. Triangular (tetrahedron) element family where complete polynomial expansions are used if nodes are spaced on a pattern shown for a two dimensional example, Fig. 5a.

2. Lagrangian, rectangular ("brick") elements with nodes placed on a grid with shape functions established by multiplication of appropriate Lagrange interpolations. Such elements, shown in Fig. 5b in a two-dimensional context, will contain many terms surplus to the requirements of completeness as shown [11, 13].

3. "Serendipity" rectangular ("brick") elements with nodes on external edges with shape functions derived by multiplication of higher-order Lagrange interpolation in one direction, by linear terms in the others (see Fig. 5c).

Hermitian interpolations with derivatives specified at corner nodal points forms apparently a yet different group—but in fact can in general be identified with classes 2 or 3.

---

* The well known improvement of a complete linear triangle by addition of the xy term in a rectangle falls in this category.

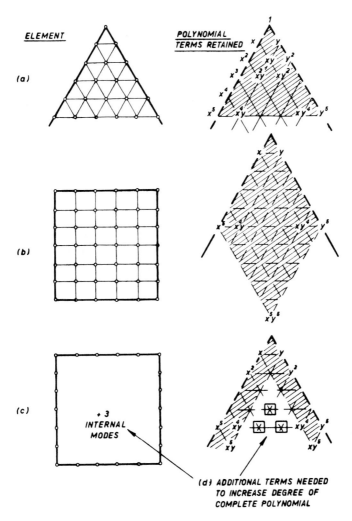

Fig. 5. Basic "parent" element types (a) triangles; (b) Lagrangian; (c) serendipity; (d) augmented serendipity.

On the face of it, it appears that elements of the first category are optimal due to no superfluity of terms while elements of the second category should be rejected a priori as inefficient. While the second proposition is universally true it should be noted that the "serendipity" elements (at least up to the cubic type) carry but few surplus terms. Furthermore, as fewer rectangular (brick) elements are needed for subdivision of a given volume of space the advantage of the serendipity type of element appears overwhelming. It is for these reasons that this particular type is widely used in iso(sub)parametric formulations.

It is worthwhile mentioning here that the simple "serendipity" region can be made complete up to any order (always leaving two surplus terms) by the addition of suitable internal degrees of freedom. This has been accomplished by, for instance, Scott [7] for a quartic element and can easily be extended to others. It should be borne in mind that with such additive terms a subparametric formulation is always desirable as the additional degrees of freedom will not influence the curvilinear boundary shapes, and if used will only add to the distortion of the curvilinear coordinates internally.

We shall therefore be concerned here with elements of the "serendipity" type alone but before proceeding further will show how shape functions can be readily developed for elements with different degrees of freedom along their sides. Such a formulation will permit elements of different order to be used in various regions of the same problem thus adding one more possible economy.

Consider, for instance, the generation of shape functions for a two-dimensional square (with coordinates $-1 < \xi < \cdots < 1$ conveniently chosen as standard) in which one side is "parabolic," i.e. contains three nodes, while the other sides are linear, Fig. 6. It is clear that this shape function

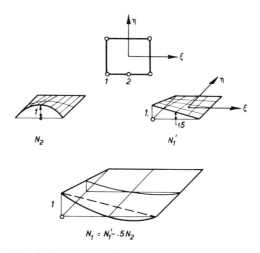

Fig. 6. Derivation of shape functions for "serendipity" elements.

for the center node (N_{i2}) is simply obtained by multiplication of a suitable parabola in ξ by a linear function in the η direction. The corner shape function is formed by a suitable combination of the above with a product of two linear functions.

This process can easily be extended to any degrees of freedom [31]. Best [32] has written simple algorithms for this as well as for triangular elements of similar kind. Extension to three dimensions is obvious.

An alternative, which avoids the need for linear combination of the various products, is to use a *hierarchical formulation* [8] in which the degrees of freedom are, respectively, the value of the variable of the corner and at the center points of each side, departure for linearity of the function, a parameter identifying the departure from parabolic variation of a cubic etc.

Functions of the form

$$P_{0,1} = (1 \pm \xi)/2, \qquad P_2 = (\xi^2 - 1), \qquad P_3 = 2\xi(\xi^2 - 1)$$
$$P_4 = (15\xi^4 - 18\xi^2 + 3)/4, \qquad P_5 = 7\xi^5 - 10\xi^3 + 3\xi \quad \text{etc.}$$

$$(22)$$

are all that is required to formulate this shape function for elements up to the fifth order (Fig. 7), in two or three dimensions. Figure 8a shows some of such elements.

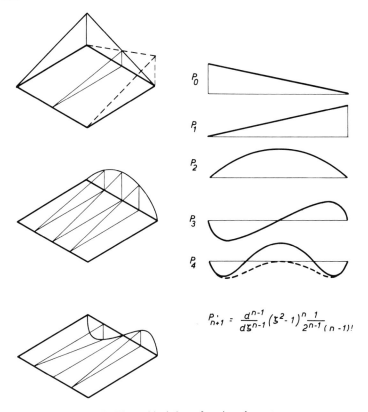

$$P_{n+1} = \frac{d^{n-1}}{d\xi^{n-1}} (\xi^2 - 1)^n \frac{1}{2^{n-1}(n-1)!}$$

Fig. 7. Hierarchical shape function elements.

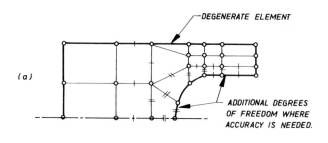

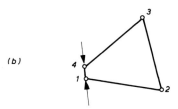

Fig. 8. Hierarchical and "degenerate" elements. (b) Triangle as degeneration of quadrilateral (1 and 4 merge).

A certain disadvantage occurs now if functions of an order higher than parabolic are used for the coordinate distortion as above that order it is difficult to identify geometrically the association of these with the co-ordinates. As a practice it is not recommended that a higher order of distortion be used—a program based on a subparametric form has been developed in which up to quartic function variation is permissible but shape distortions are parabolic.

8. Degeneration of Quadrilateral or Brick Elements

While in two dimensions the quadrilateral (or in three the brick) forms the most advantageous space division as mentioned earlier, it is occasionally necessary to use triangles (or other shapes such as wedges and tetrahedra) to complete the subdivision near boundaries (see Fig. 8a). While such elements can be generated directly it is of interest to consider their derivation as a possible case of degeneracy, for, if this is practicable, a simpler computer program can be used.

For instance it appears possible to arrive at the triangle of Fig. 8a by considering it as a case of a quadrilateral in which two corners coalesce as shown in Fig. 8b. In Fig. 9 we consider first the case of a linear quadrilateral degenerating to a triangle. The "mapping" of coordinates is still one to one

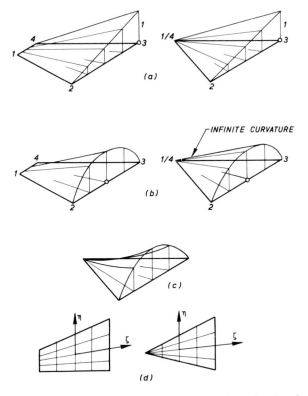

Fig. 9. Degeneration of quadrilateral to triangle. (a) Correct shape functions for linear term; (b) incorrect shape functions reached for parabolic term; (c) corrected shape functions; (d) coordinate map.

and the Jacobian has a constant value (see Fig. 9a). Furthermore, the shape function, now arrived at by superposition of the two responses of the coalesced points, whose coordinates and displacements are assumed to be the same degenerates to the linear response shown. As this in fact is the correct response for a triangle no problems arise (Fig. 9a).

With a parabolic type of quadrilateral similar remarks can be made about mapping but the shape functions corresponding to the quadratic departure (see hierarchical elements) degenerate in the manner shown in Fig. 9b presenting a linear variation along any line radiating from the coalesced corner (this gives a nonpolynomial surface with infinite curvature at the corner). The correct response of triangle involves shape function in which this variation should be parabolic along the radials as shown in Fig. 9c.

To correct for this affect is simple—by introducing a multiplying factor $(1 + \xi)$ in the shape function so degenerated.

Similar effects occur with cubic and higher distortions but the correction shown is more involved.

It is thus apparent that if the program is so written as to introduce the appropriate correction factors for degenerate elements the response of the elements will be always improved if complete (rather than near) degeneration occurs.

Irons [8] discusses in detail the concepts of degeneration of three dimensional brick elements and problems which may arise as a consequence. Some permissible degenerations of brick type elements are indicated in Fig. 10 and the practical value of these is obvious.

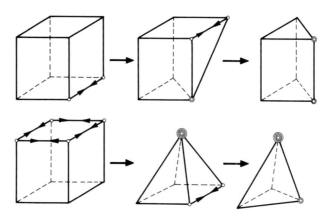

Fig. 10. Some typical degenerations of brick elements in three dimensions.

9. Computation Efficiency of Numerical Integration

For quadrilateral or brick-shaped elements the numerical integration is performed in every region in which each variable ranges from -1 to $+1$. The simplest performance of this is to use Gaussian points distributed on a regular basis, along the $\xi, \eta, \ldots$ directions.

The minimum integration requirements on elements of the serendipity type are given in Table 1 (according to the rules of Section 6).

Recently several new formulas have been developed for brick-type regions and these promise to reduce further the cost of numerical integration [33]. For instance a rule involving only 14 sampling points integrates exactly to quartic terms, i.e., with the same accuracy as 27 sampling points on a $3 \times 3 \times 3$ Gauss mesh. Application of such rules in practice is already contributing substantially to efficiency of computation.

For triangular and tetrahedral regions symmetrically placed integrating points are advantageous [28, 34] but if degenerate elements are used it is simpler to continue with the source rule as used for quadrilateral and brick. This perhaps is a slight disadvantage, cf. the new elements introduced in Section 8.

The organization of the computation of stiffness and other matrices can also contribute much to the efficiency of the operation. Some schemes of computation are obtained in Irons [35].

10. Practical Examples and Stress Computation

Figures 11–13 show some subdivisions used in three-dimensional analysis with isoparametric parabolic elements. The small number of elements used in each case—and the subsequent reduction of the labor of data preparation are manifest.

With a relatively small number of elements used attention has to be focused on the best use of the results and in particular on the problems of stress computation in structural analysis. While the stresses can be calculated at any point of an element the choice of the position giving the optimal accuracy is still a matter of debate. While in early calculations stresses were

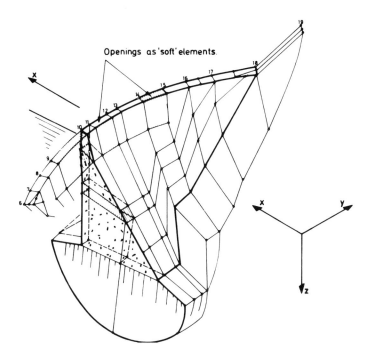

Openings as 'soft' elements.

Fig. 11. Subdivision of a gravity-arch dam.

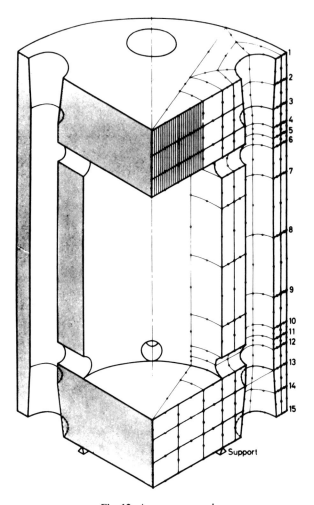

Fig. 12. A pressure vessel.

invariably output at nodes—and averaged—these using adjacent elements of the material properties did not introduce a discontinuity—it is now found that better results are obtained if the calculation is made at the integrating Gauss points.

More recently arguments have been introduced for computing stresses at midsides (or faces) of elements only. While it is probable that again better values will thus be obtained (as interpolation always gives better values at midrange than at its ends) an incidental improvement is the considerable reduction in the volume of output. This is a problem of considerable import-ance in three-dimensional situations, where the digestion and presentation

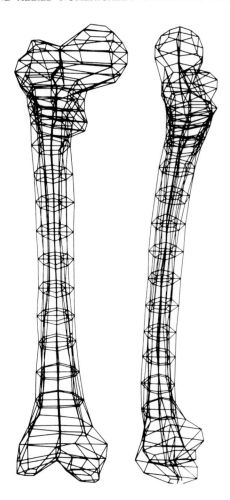

Fig. 13. A problem of biomechanics. Plot of linear element form only; curvature of elements omitted. Note degenerate element shapes.

of the information presents a prohibitive cost. More experience on this aspect will be available shortly.

The same remarks apply to other indirect output, for example velocities of fluxes in nonstructural problems, where isoparametric elements are used.

11. Shells and Plates as Limiting Cases of Three-Dimensional Analysis

It appears that a three-dimensional "brick" on progressive reduction of its thickness should function as a plate or shell element capable of representing adequately shear distortion affects as well as the usual bending and axial forces. Two possible difficulties are presented immediately.

TABLE 1
TERMS WHICH HAVE TO BE INTEGRATED EXACTLY FOR CONVERGENCE[a]

Interpolation N

[a] Gauss point requirements are in brackets.

MINIMUM INTEGRATION ORDERS FOR SERENDIPITY TYPE ELEMENTS (ISO-SUB-PARAMETRIC)[b]

N	N'	Terms to be integrated	Minimum Gauss points
		Two dimensions	
Linear	Linear	$\xi \quad \eta \quad \xi\eta$	1
Quadratic	Linear	Up to ξ^2, η^2	2×2
Quadratic	Quadratic	Up to $\xi^2, \eta^2 + \xi^2\eta^2$	2×2
Cubic	Linear	Up to ξ^2, η^2	2×2
Cubic	Quadratic	Up to $\xi^2, \eta^2 + \xi^4, \eta^4, \xi^2\eta^2$	3×3
Cubic	Cubic	Up to $\xi^2, \eta^2 + \xi^4, \eta^4, \xi^2\eta^2$	3×3
		Three dimensions	
Linear	Linear	Up to ξ^2, etc., $\xi^2\eta^2$ etc.	$2 \times 2 \times 2$
Quadratic	Linear	Up to ξ^2, etc., $\xi^2\eta^2\zeta^2$	$2 \times 2 \times 2$
Quadratic	Quadratic	Up to ξ^2, etc., $\xi^4\eta^2\zeta^2$ etc.	$3 \times 3 \times 3$
Cubic	Linear	Up to ξ^2, etc., $\xi^4\eta^2\zeta^2$, etc.	$3 \times 3 \times 3$
Cubic	Quadratic	Up to ξ^2, etc., $\xi^4\eta^4\zeta^2$	$3 \times 3 \times 3$
Cubic	Cubic	Up to ξ^2, etc., $\xi^8\eta^2$, etc.	$5 \times 5 \times 5$

[b] Only simplest regular Gauss rule quoted; more economical alternatives exist. After D. Veryard, M.Sc. Thesis, Swansea (1971).

The first is that ill-conditioning may occur due to the relatively large stiffness in the mode involving straining normal to the "middle surface."

The second, that too large a number of variables have to be included compared with more conventional analysis. Even if a linear variation of displacement is prescribed in the direction normal to middle surfaces of the shell six rather then five displacement components arise increasing the computation effort in the ratio (6/5) [3].

To overcome these difficulties Ahmad *et al.* [17, 18] suppress the effects of lateral straining and reduce suitably the degrees of freedom introducing several necessary coordinate transformations. Good results are achieved for axisymmetric shells and in general for shells where membrane stresses are predominant. For plates and other situations where bending effects are important the results are however poor when thickness/length ratios of an element become small. Here an old difficulty of the introduction of shear modes which do not exist in practice is rediscovered. A solution to this problem is suggested by the work of Wilson *et al.* [36] who succeeded in elimination of such spurious shears by using a reduced order of integration for certain stress components in a linear plane quadrilateral. Now simply by reducing the order of integration from 3 × 3 Gauss to 2 × 2 Gauss for a parabolic element a dramatic improvement in element properties is achieved [20, 21]. Figure 14 shows the improvement in deformation of a cylindrical

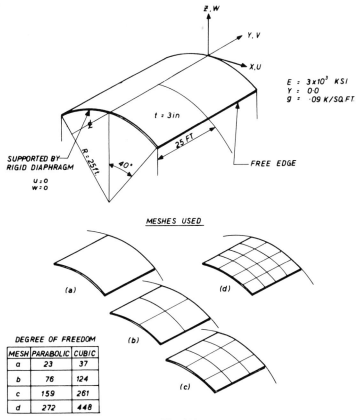

MESH	PARABOLIC	CUBIC
a	23	37
b	76	124
c	159	261
d	272	448

DEGREE OF FREEDOM

Fig. 14A

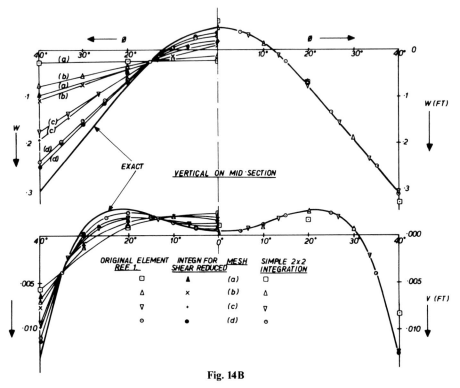

Fig. 14B

Fig. 14. Cylindrical shell under self-weight. (A) Subdivision into parabolic elements; (B) convergence of displacements for 3 × 3 and 2 × 2 point integration.

shell with such reduced integration order, while in Fig. 15, obtained by J. Too, another complex problem of a deep shell is presented. Indeed now this particular element appears to be superior to any other shell element currently available.

The effects can be illustrated on a simpler, two-dimensional problem of Fig. 16. Here four, simple, parabolic, two-dimensional elements are used to represent a cantilever of different length to depth ratios. Once again the remarkable results of reduced integration order are observed (and indeed the previously mentioned ill-conditioning effects found where the thickness is too small).

12. Applications to Nonlinear Analysis

In the application of the finite element methods to the analysis of plasticity, nonlinear geometry and other allied problems almost invariably the simplest

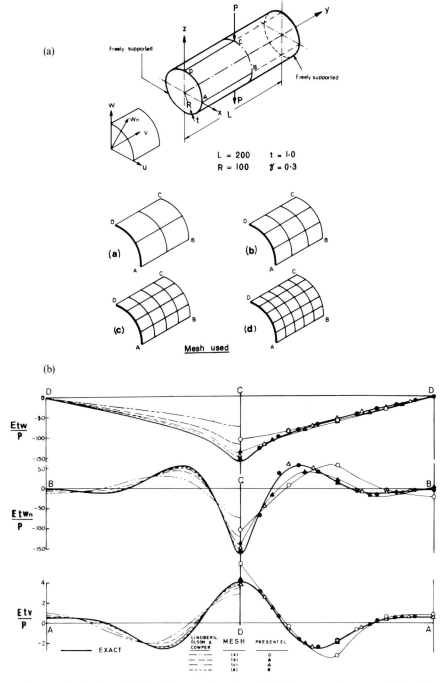

Fig. 15. (a) Pinched cylindrical shell configuration. (b) Displacement distortion of pinched cylindrical shell (solution by J. Too).

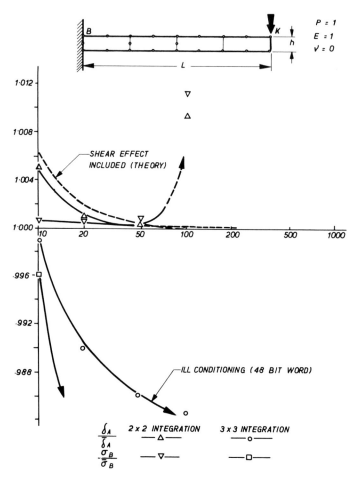

Fig. 16. Effect of reduced integration on parabolic elements in a cantilever beam of varying length/thickness ratios ($\bar{\delta}$, $\bar{\sigma}$ arc beam theory valves).

types of elements were used. Indeed it was presumed in early days that the discontinuous transition from elastic to plastic regimes could only be accomplished efficiently if essentially "constant stress" elements were used. It has now been realized that the isoparametric and allied elements are ideally suited for such analysis and indeed the economies achievable in linear analysis are emphasized in nonlinear applications [23, 24].

As all nonlinear solutions in solid mechanics involve a solution of a set of equations which can be represented as

$$\int |\bar{B}|^{\mathrm{T}}\{\sigma\}\, dv - \{F\} = 0$$

where $\{\sigma\}$ is a function of the strains developed and $|\bar{B}|$ is the matrix linking strain to displacement variations, the terms can be evaluated by numerical integration with appropriate stresses $\{\sigma\}$ found at the integration points. Indeed numerical integration is necessary for any element form once an element other than one of constant strain is used.

Furthermore, if large strain or displacement nonlinearities develop, the expressions that arise in their determination are of the same form and type as arise in transformation of coordinates. Denoting $\bar{x}$ as the deformed co-ordinate $x + u$, etc. an expression for the deformation Jacobian matrix can be written as

$$[J_d] = \left[\frac{\partial \bar{x}, \bar{v}, \ldots}{\partial x, y, \ldots} \right] \tag{23}$$

which on expansion takes a form similar to that of Eq. (15), i.e.,

$$[J_d] = \begin{bmatrix} \{G\}_x^T\{\bar{x}\}, & \{G\}_y^T\{\bar{y}\} \\ \{G\}_x^T\{\bar{y}\}, & \cdots \\ \cdots & \cdots \end{bmatrix} \tag{24}$$

with

$$\{G\}_x^T = \left[\frac{\partial N_i}{\partial x}, \ldots, \frac{\partial N_i}{\partial x}, \ldots \right] \quad \text{etc.} \tag{25}$$

As this Jacobian is related to Green's strains matrix as

$$[\varepsilon] = \tfrac{1}{2}[[J_d]^T[J_d] - I] \text{ (when } I \text{ is identity matrix)} \tag{26}$$

an easy path exists for the deformation of large strain components.

Details of nonlinear solutions are beyond the scope of this paper, and can be found in the references cited. Figure 17 shows a solution of a plasticity problem in which high-order isoparametric elements are compared with simple triangular ones. Not only is the accuracy considerably improved for a given degree of freedom but the spread of the plastic zones through the integrating points shows a much smoother pattern through the jagged development of these zones when simple triangular elements are used.

13. Concluding Remarks—Other Uses of Mapping

The relative simplicity of programming even complex numerically integrated elements and their versatility have today led to their adoption in a wide variety of problems. Indeed the simple triangles and tetrahedron can always be considered uneconomic as compared with the performances of higher order elements.

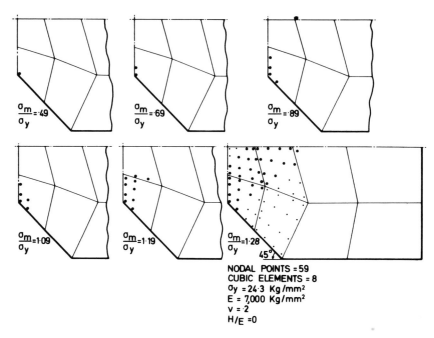

Fig. 17. Notched specimens: plastic zones for various load increments. Note smooth spread through integrating points (4 × 4).

Where the optimum lies is a problem that does not possess a unique answer. Much depends on the nature of the problem, its size, and on the relative efficiency of various components of the program. For a large majority of problems it appears that parabolic elements are sufficient but if their use can be combined (say via hierarchical formulation) with an occasional use of a cubic the best program versatility may be achieved.

Many questions raised in the paper still require detailed study—but the basic precepts can be considered as established.

In the present paper little attention was given to applications other than those requiring C_0 continuity to the context of axisymmetric thin shells isoparametric concepts have been put to good use by Delpak [28, 37]. More recently attempts have been made to extend the mapping to such problems as thin plates, where C_1 continuity and constancy of second derivatives are required. Unfortunately neither of these criteria are easy to achieve and although some attempts have been recorded they are not very satisfactory. Again here an area of possible research is open.

Finally it was already mentioned that isoparametric mapping may be put to purely geometric uses. One of such application is in automatic generation of mesh subdivisions for simple triangular networks after representing the surface by a small number of curved elements [39]. Figure 18 shows some meshes so created which have already proved to provide one of the most powerful tools for the purpose. Doubtless other similar applications will arise.

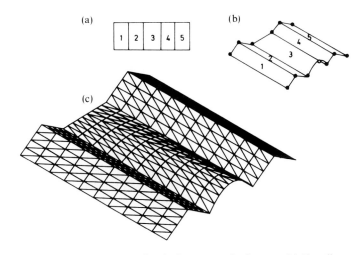

Fig. 18. Automatic mesh generation by isoparametric elements. (a) Key diagram; (b) zone diagrams (● corner points, ○ midside points), (c) final mesh-isometric view.

In the introduction we have mentioned the possibility of combining the surface description methods of computer-aided design with analysis subdivisions of isoparametric elements. The elements so far described do not lead to surfaces with slope continuity and as such do not describe smooth surfaces with complete accuracy. An obvious way out of this dilemma is the use of Hermitian-type functions of cubic type for the elements. Here the nodal variables include now not only the coordinates but the derivatives of these [7]. In the purely geometrical aspects this as is well known introduces certain difficulties of scaling—for the analysis point of view it necessitates the use of cubic polynomial expansion. Thus it is possible to forsee a larger use of such functions purely as a means of mesh generation for a start of automatic stress analysis. Interesting possibilities certainly lie ahead.

References

1. Taig, I. C., "Structural Analysis by the Matrix Displacement Method," Engl. Elec. Aviation Rep. No. SO17 (1961).
2. Irons, B. M., "Numerical Integration Applied to Finite Element Methods," *Conf. Use Digital Comput. Structure Eng.* Univ. of Newcastle (1966).

3. Irons, B. M., "Engineering Application of Numerical Integration in Stiffness Method," *A.I.A.A. J.* **14**, 2035–2037 (1966).
4. Ergatoudis, J., Irons, B. M., and Zienkiewicz, O. C., "Three Dimensional Stress Analysis of Arch Dams." Rep. to the Arch Dams Committee of the Inst. Civil Eng. AD/173S (C/R/58/66) (1966).
5. Ergatoudis, J., Irons, B. M., and Zienkiewicz, O. C., "Curved, Isoparametric, Quadrilateral Elements for Finite Element Analysis," *Int. J. Solids Struct.* **4**, 31–42 (1968).
6. Ergatoudis, J., Irons, B. M., and Zienkiewicz, O. C., "Three Dimensional Analysis of Arch Dams and Their Foundations," *Symp. Arch Dams* Inst. Civil Eng. London (1968).
7. Zienkiewicz, O. C., Irons, B. M., Ergatoudis, J., Ahmad, S., and Scott, F. C., "Isoparametric and Associated Element Families for Two- and Three-Dimensional Analysis," *Proc. Finite Element Methods Stress Analy.* (Holand, I., and Bell, K., eds.). Trondheim.
8. Zienkiewicz, O. C., Irons, B. M., Campbell, J., and Scott, F., "Three Dimensional Stress Analysis," *Internat. Un. Th. Appl. Mech. High Speed Comput. Elastic Structures.* Liège (1970).
9. Zienkiewicz, O. C., and Parekh, C. J., "Transient Field Problems; Two- and Three-Dimensional Analysis by Isoparametric Finite Elements," *Internat. J. Num. Methods Eng.* **2**, 61–71 (1970).
10. Argyris, J. H., Buck, K. E., Scharpf, D. W., Hilber, H. M., and Mareczek, G., "Some New Elements for the Matrix Displacement Method," *Proc. Conf. Matrix Methods Struct. Mech.*, *2nd*, Wright–Patterson AFFDL-TR-68-150, pp. 333–398 (1968).
11. Ergatoudis, J., "Quadrilateral Elements in Plane Analysis," M.Sc. thesis, Univ. of Wales, Swansea (1966).
12. Field, S. A., "Three Dimensional Theory of Elasticity," *Proc. Finite El. Methods Stress Anal.* (Holand, I., and Bell, K., eds.). Trondheim Tech. Univ. (1969).
13. Argyris, J. H., "The Lumina Element for Matrix Displacement Method," *Aeronaut. J. Roy. Aeronaut. Soc.* **72**, 514–517 (1968).
14. Argyris, J. H., Fried, I., and Scharpf, D. W., "The Hermes 8 Element for the Matrix Displ. Method," *Aeronaut. J. Roy. Aeronaut Soc.* **72**, 613–617 (1968).
15. Clough, R. W., "Comparison of Three Dimensional Elements," *Symp. on Appl. Finite Element Methods in Civil Eng.*, pp. 1–26. Vanderbilt Univ. (1969).
16. Hellen, T. K., and Money, H. A., "The Application of Three Dimensional Elements to a Cylinder–Cylinder Intersection," *Internat. J. Num. Methods Eng.* **2**, 415–418 (1970).
17. Ahmad, S., Irons, B. M., and Zienkiewicz, O. C., "Curved Thick Shell and Membrane Elements with Particular Reference to Axisymmetric Problems," *Proc. Conf. Matrix Methods Struct. Mech. 2nd*, Wright–Patterson AFFDL-TR-68-150, pp. 539–572 (1968).
18. Ahmad, S., Irons, B. M., and Zienkiewicz, O. C., "Analysis of Thick and Thin Shell Structures by Curved Finite Elements," *Internat. J. Num. Methods Eng.* **2**, 419–451 (1970).
19. Ahmad, S., Anderson, R. G., Zienkiewicz, O. C., "Vibration of Thick Curved Shells with Particular Reference to Turbine Blades," *Internat. J. Strain Anal.* **5**, 200–206 (1970).
20. Zienkiewicz, O. C., Taylor, R. L., Too, J. M., "Reduced Integration Technique in General Analysis of Plates and Shells," *Internat. J. Num. Methods Eng.* **3**, 275–290 (1971).
21. Pawsley, S. F., and Clough, R. W., Improved Numerical Integration of Thick Shell Finite Elements," *Internat. J. Num. Methods Eng.* **3**, 575–586 (1971).
22. Jordan, W. B., "The Plane Isoparametric Structural Element," General Electric Co. Rep. KAPL-M-7112, Schenectady, New York (February 1970).
23. Nayak, G. C., "Plasticity and Large Deformation Problems by Finite Element Method," Ph.D. Thesis. Univ. of Wales, Swansea (1971).
24. Zienkiewicz, O. C., and Nayak, G. C., "A General Approach to Problems of Plasticity and Large Deformation using Isoparametric Elements." (To be published.)

25. Zienkiewicz, O. C., and Phillips, D. V., "An Automatic Mesh Generation Scheme for Plane and Curved Surfaces by Iso-parametric Coordinates." *Internat. J. Num. Methods Eng.* **3**, 519–528 (1971).
26. Coons, S. A., "Surfaces for Computer Aided Design of Space Forms," M.I.T. Project MAC. M.A.C.-TR-41 (1967).
27. Forrest, A. R., "Curves and surfaces for computer aided design," CAD Group, Cambridge, England (1968).
28. Zienkiewicz, O. C., *The Finite Element Method in Eng. Science*, 2nd ed., McGraw-Hill, New York, 1971.
29. Irons, B. M., *Finite Element Techniques in Structure Methods* (Tottenham, H., and Brebbia, C., eds.), pp. 328–330. Southampton Univ. Press, 1970.
30. Dunne, P. C., "Complete Polynomial Displacement Fields for Finite Element Methods," *Trans. Roy Aeronaut. Soc.* **72**, 245 (1968).
31. Taylor, R. L., "On Completeness of Shape Functions for Finite Element Analysis," *Internat. J. Num. Methods Eng.* (To be published.)
32. Best, B., "An Investigation into the Use of Finite Element Methods for Analysing Stress Distributions in Block Jointed Masses," Ph.D. thesis, James Cook Univ. Australia (1971).
33. Irons, B. M., "Quadrature Rules for Brick Based Finite Elements," *Internat. J. Num. Methods Eng.* **3**, 293–294 (1971).
34. Stroud, A. H., and Secrest, D., *Gaussian Quadrature Formulae.* Prentice-Hall, Englewood Cliffs, New Jersey, 1966.
35. Irons, B. M., "Economical Computer Techniques for Numerically Integrated Elements." *Internat. J. Num. Methods Eng.* **1**, 201–203 (1969).
36. Doherty, W. P., Wilson, E. L., and Taylor, R. L., "Stress Analysis of Axisymmetric Solids Utilizing Higher Order Quadrilateral Finite Elements." Struct. Eng. Lab. Univ. of Calif. Berkeley (1969).
37. Delpak, R., "Axisymmetric Vibration of Shells of Revolution by the Finite Element Method." M.Sc. Thesis, Univ. of Wales, Swansea (1967).

Incompatible Displacement Models

E. L. Wilson, R. L. Taylor, W. P. Doherty, and J. Ghaboussi

DEPARTMENT OF CIVIL ENGINEERING
UNIVERSITY OF CALIFORNIA
BERKELEY, CALIFORNIA

1. Introduction

One of the most significant developments in the numerical solution of solid structures was the introduction of isoparametric finite elements [1]. As a result, many elements with a high degree of accuracy have been developed [2]. In addition, the technique has been extended to curved-shell elements [3]. The purpose of this paper is to present a modification of this approach which results in a further improvement in accuracy.

Other attempts have been made to improve the basic accuracy of these elements. In general they involve the use of approximate integration techniques which disregard part of the shear strain energy associated with pure bending modes [4–6]. However, these methods have been limited to idealized geometries and isotropic materials. Also, convergence may not be assured.

The method presented in this paper formally introduces incompatible displacement modes at the element level in order to improve the element accuracy. These unknowns are eliminated by requiring that the total strain energy within the element is minimum. Convergence of the solution is assured. Examples are presented for two- and three-dimensional solids and for thick shells.

43

2. Source of Errors

One of the main causes of inaccuracies in lower-order finite elements is their inability to represent certain simple stress gradients. This is clearly illustrated by subjecting a simple rectangular element to a pure bending stress as shown in Fig. 1a. The exact displacements for this type of loading are

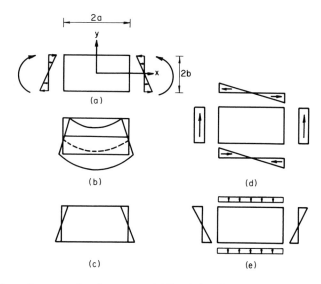

Fig. 1. Errors due to pure bending stresses. (a) Simple bending stress; (b) exact displacements; (c) finite element displacements; (d) shear stresses; (e) y stresses.

illustrated in Fig. 1b and are given by the following equations:

$$u_x = \alpha_1 xy \tag{1}$$

$$u_y = \tfrac{1}{2}\alpha_1(a^2 - x^2) + \alpha_2(b^2 - y^2) \tag{2}$$

It is clear that these displacements satisfy the pure bending condition of zero shear strain; or

$$\varepsilon_{xy} = \partial u_x/\partial y + \partial u_y/\partial x = 0 \tag{3}$$

The constant α_2 is a function of the material properties: For Poisson's ratio equal to zero $\alpha_2 = 0$.

For a finite element displacement model the only displacement activated for this type of loading is shown in Fig. 1c and is given by

$$u_x = \beta_1 xy \tag{4}$$

Therefore, the form of the error in the solution is

$$u_y = \beta_2(a^2 - x^2) + \beta_3(b^2 - y^2) \tag{5}$$

The errors in stresses associated with Eq. (5) are shown in Figs. 1d and 1e.

Previous attempts to reduce these errors have involved selecting integration formulas which disregard the strain energy associated with the stresses shown in Fig. 1d. One point integration applied at the center of the element will accomplish this. However, this technique can produce an invariant stiffness matrix which has directional properties. For shell-type structures, the normal stresses shown in Fig. 1d have been disregarded by making the assumption of plane stress in the stress–strain equations. It is clear that these approximate methods are difficult to apply in the general case of curved anisotropic elements.

In this paper, the approach adopted to minimize these errors is to add extra displacement modes to the elements which have the same form as the errors in the simple displacement approximation. In general these extra displacement modes violate interelement compatibility. The magnitudes of the modes are selected by requiring that the total strain energy of the element be a minimum. In the following sections of this paper, the method will be presented as an addition to the basic isoparametric method. Specific examples to various types of elements will be given.

3. Addition of Incompatible Modes for Two-Dimensional Isoparametric Elements

A two dimensional isoparametric element will be used to illustrate the method in detail. For a general quadrilateral element, as shown in Fig. 2, the local and global coordinate systems are related by

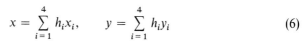

$$x = \sum_{i=1}^{4} h_i x_i, \qquad y = \sum_{i=1}^{4} h_i y_i \tag{6}$$

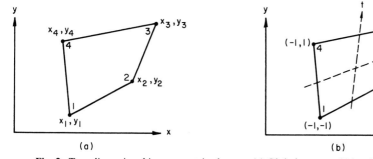

Fig. 2. Two-dimensional isoparametric element. (a) Global system; (b) local system.

where the interpolation functions are given by

$$h_1 = \tfrac{1}{4}(1 - s)(1 - t), \qquad h_2 = \tfrac{1}{4}(1 + s)(1 - t)$$
$$h_3 = \tfrac{1}{4}(1 + s)(1 + t), \qquad h_4 = \tfrac{1}{4}(1 - s)(1 + t) \tag{7}$$

3.1. STRAIN-DISPLACEMENT EQUATIONS

In order to ensure rigid-body displacement modes the same interpolation functions are used in the displacement approximation.

$$u_x(s, t) = \sum h_i u_{xi}, \qquad u_y(s, t) = \sum h_i u_{yi} \tag{8}$$

For two dimensional analysis the strain-displacement equations are

$$\varepsilon_{xx} = \partial u_x/\partial x = \sum h_{i,x} u_{xi}$$
$$\varepsilon_{yy} = \partial u_y/\partial y = \sum h_{i,y} u_{yi} \tag{9}$$
$$\varepsilon_{xy} = \partial u_x/\partial y + \partial u_y/\partial x = \sum h_{i,y} u_{xi} + \sum h_{i,x} u_{yi}$$

Or Eq. (9) can be written in matrix form as

$$\varepsilon = \mathbf{a}(s, t)\mathbf{U} = \begin{bmatrix} \mathbf{H}, x & 0 \\ 0 & \mathbf{H}, y \\ \mathbf{H}, y & \mathbf{H}, x \end{bmatrix} \begin{bmatrix} \mathbf{U}x \\ \mathbf{U}y \end{bmatrix} \tag{10}$$

In this case the three strains are related to the eight nodal point displacements by a 3×8 matrix. The submatrices in Eq. (10) are given by

$$\mathbf{H}, x = [h_{1,x} h_{2,x} h_{3,x} h_{4,x}], \qquad \mathbf{H}, y = [h_{1,y} h_{2,y} h_{3,y} h_{4,y}] \tag{11}$$

Since the functions h_i are in terms of s and t the chain rule is applied in order to compute the derivatives with respect to the global x–y system.

$$h_{i,x} = h_{i,s} s_{,x} + h_{i,t} t_{,x}, \qquad h_{i,y} = h_{i,s} s_{,y} + h_{i,t} t_{,y} \tag{12}$$

In general, the chain rule can be written as

$$\begin{bmatrix} \partial/\partial s \\ \partial/\partial t \end{bmatrix} = \begin{bmatrix} x_{,s} & y_{,s} \\ x_{,t} & y_{,t} \end{bmatrix} \begin{bmatrix} \partial/\partial x \\ \partial/\partial y \end{bmatrix} \tag{13}$$

or, inverted, as

$$\begin{bmatrix} \partial/\partial x \\ \partial/\partial y \end{bmatrix} = \begin{bmatrix} s_{,x} & t_{,x} \\ s_{,y} & t_{,y} \end{bmatrix} \begin{bmatrix} \partial/\partial s \\ \partial/\partial t \end{bmatrix} \tag{14}$$

Therefore, the derivative required in Eq. (12) is given by

$$\begin{bmatrix} s_{,x} & t_{,x} \\ s_{,y} & t_{,y} \end{bmatrix} = \frac{1}{J} \begin{bmatrix} y_{,t} & -y_{,s} \\ -x_{,t} & x_{,s} \end{bmatrix} \tag{15}$$

where the Jacobian J is

$$J = x_{,s}y_{,t} - x_{,t}y_{,s} \tag{16}$$

From Eqs. (6a) and (6b)

$$x_{,s} = \sum h_{i,s}x_i, \quad x_{,t} = \sum h_{i,t}x_i, \quad y_{,s} = \sum h_{i,s}y_i, \quad y_{,t} = \sum h_{i,t}y_i \tag{17}$$

For given numerical values of s and t the derivatives of the interpolating functions can be evaluated. Then from Eqs. (17) and (15) all required derivatives for the numerical evaluation of the strain displacement matrix, Eq. (10), can be calculated.

3.2. ELEMENT STIFFNESS AND NUMERICAL INTEGRATION

For unit thickness the element stiffness matrix is given by

$$\mathbf{K} = \int_{\text{Area}} \mathbf{a}^T \mathbf{c} \mathbf{a} \, dA \tag{18}$$

in which $\mathbf{c}$ is the stress–strain matrix and the integration is carried out over the area of the element.

For the purpose of numerical integration, Eq. (18) is written in the s and t system as

$$\mathbf{K} = \int_{-1}^{1} \int_{-1}^{1} \mathbf{a}^T \mathbf{c} \mathbf{a} J \, ds \, dt \tag{19}$$

The direct application of one-dimensional numerical integration formulas [7] yields

$$\mathbf{K} = \sum_j \sum_k W_j W_k J \mathbf{a}^T(s_j, t_k) \mathbf{c} \mathbf{a}(s_j, t_k) \tag{20}$$

in which s_j and t_k are integration points and W_j and W_k are the appropriate weight functions.

3.3. ADDITION OF INCOMPATIBLE MODES

The basic method is the same when internal degrees of freedom are added at the element level. For the quadrilateral element the displacement approximation may be of the following form:

$$u_x = \sum h_i u_{xi} + h_5 \alpha_1 + h_6 \alpha_2, \quad u_y = \sum h_i u_{yi} + h_5 \alpha_3 + h_6 \alpha_4$$

The functions h_5 and h_6 must be zero at the four nodes. The displacement amplitudes α_i are additional degrees of freedom; therefore, the resulting element stiffness will be 12×12. However, if the strain energy within the element is minimized with respect to α_i, four additional equations can be generated and the additional displacements can be eliminated and a reduced 8×8 stiffness matrix developed. This is identical to the standard static condensation procedure. An alternate approach is to consider α_i as Lagrange multipliers.

The functions h_5 and h_6 can be selected to be of the same form as the errors in the bending deformation which is given by Eq. (5). Or

$$h_5 = (1 - s^2), \qquad h_6 = (1 - t^2)$$

If the element is rectangular only the y displacements would need to be modified. However, for the general quadrilateral these modes must be added to both components of displacements.

These incompatible modes are plotted in Fig. 3. It is apparent that the energy associated with these modes is large compared with the constant strain modes. It is also clear that they will not participate significantly in areas of low stress gradients since they are added to the basic constant strain

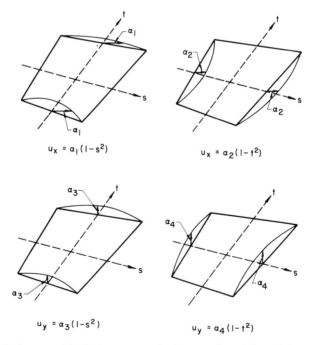

$$u_x = \alpha_1(1-s^2)$$

$$u_x = \alpha_2(1-t^2)$$

$$u_y = \alpha_3(1-s^2)$$

$$u_y = \alpha_4(1-t^2)$$

Fig. 3. Incompatible displacement modes for general quadrilateral element.

modes. The net result of the addition of the incompatible modes is that microscopic equilibrium is better satisfied within the element.

3.4. Two-Dimensional Examples

(a) *Cantilever Beam*

A cantilever beam with the dimensions shown in Fig. 4 will illustrate the accuracy of the element for plane stress structures. Results for two different loading conditions and for two different meshes are shown in Table 1. They are compared with exact solution and with a finite element solution without incompatible modes. For this case the improvement is significant.

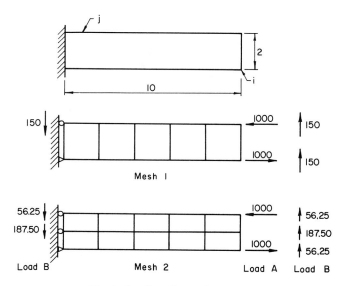

Fig. 4. Cantilever beam-plane stress.

TABLE 1
Results of Cantilever Beam Analysis

	Displacement at i		Bending stress at j	
	Load A	Load B	Load A	Load B
Beam theory	10.00	103.0	300.0	4050
Q4 Mesh 1	6.81	70.1	218.2	2945
Q4 Mesh 2	7.06	72.3	218.8	2954
Q6 Mesh 1	10.00	101.5	300.0	4050
Q6 Mesh 2	10.00	101.3	300.0	4050

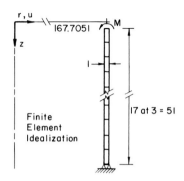

Fig. 5. Cylindrical shell analysis. $M = 2000$ per unit of arc length; $E_n = E_s = E_\theta = 11{,}250$; $V_{ns} = V_{n\theta} = V_{s\theta} = 0.25$; $G_{ns} = 4500.0$. Theoretical results based on infinitely long shell.

TABLE 2
RESULTS OF CYLINDRICAL SHELL ANALYSIS

Z	Theory	Q4CST	QM5	Q4	Q6
Lateral displacements u					
0	100.00	39.97	98.56	46.47	100.01
3	48.88	26.04	47.87	29.17	48.98
6	14.31	14.98	13.49	15.69	14.19
9	−6.57	6.56	−7.29	5.69	−6.54
12	−17.16	0.47	−17.77	−1.31	−17.15
15	−20.68	−3.65	−21.17	−5.82	−20.70
18	−19.85	−6.16	−20.21	−8.35	−19.88
21	−16.75	−7.40	−16.97	−9.39	−16.83
24	−12.82	−7.68	−12.92	−9.33	−12.85
27	−8.95	−7.27	−8.98	−8.55	−9.00
30	−5.63	−6.40	−5.65	−7.32	−5.72
33	−3.06	−5.27	−3.12	−5.87	−3.23
Hoop stresses					
1.5	4868	2210	4903	2536	4837
4.5	1991	1369	2050	1507	1986
7.5	159	716	201	718	154
10.5	−868	230	−846	147	−873
13.5	−1316	−120	−1311	−238	−1320
16.5	−1386	−334	−1392	−475	−1390
19.5	−1240	−459	−1250	−595	−1243
22.5	−994	−510	−1006	−627	−996
25.5	−727	−505	−738	−599	−729
28.5	−483	−462	−493	−532	−487
31.5	−285	−395	−297	−442	−293
34.5	−138	−315	−155	−344	153

(b) *Axisymmetric Cylindrical Shell*

The same basic expansion is used for the analysis of axisymmetric solids. An infinite cylindrical shell is idealized by 17 axisymmetric elements as shown in Fig. 5. In Table 2 results are compared with a theoretical solution and to two other types of finite elements. The Q4CST is composed of four constant strain triangles and the QM5 is quadrilateral with a restricted integration on the shear strains. The Q6 is the element presented in this paper; the Q4 is the standard isoparametric quadrilateral element.

4. Three-Dimensional Elements

The same basic method of introducing incompatible displacement modes in order to improve the bending properties can be used in three dimensions.

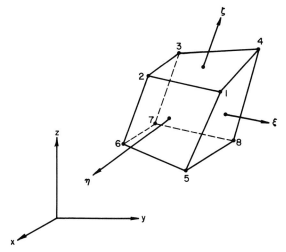

Fig. 6. Eight point three-dimensional element.

For an arbitrary eight-point brick element shown in Fig. 6 the appropriate displacement approximations are

$$u_x = \sum_{i=1}^{8} u_{xi} + h_9\alpha_{x1} + h_{10}\alpha_{x2} + h_{11}\alpha_{x3}$$

$$u_y = \sum_{i=1}^{8} u_{yi} + h_9\alpha_{y1} + h_{10}\alpha_{y2} + h_{11}\alpha_{y3}$$

$$u_z = \sum_{i=1}^{8} u_{zi} + h_9\alpha_{z1} + h_{10}\alpha_{z2} + h_{11}\alpha_{z3}$$

where

$$h_1 = \tfrac{1}{8}(1 + \xi)(1 + \eta)(1 + \zeta), \qquad h_2 = \tfrac{1}{8}(1 - \xi)(1 + \eta)(1 + \zeta)$$

$$h_3 = \tfrac{1}{8}(1 - \xi)(1 - \eta)(1 + \zeta), \qquad h_4 = \tfrac{1}{8}(1 + \xi)(1 - \eta)(1 + \zeta)$$

$$h_5 = \tfrac{1}{8}(1 + \xi)(1 + \eta)(1 - \zeta), \qquad h_6 = \tfrac{1}{8}(1 - \xi)(1 + \eta)(1 - \zeta)$$

$$h_7 = \tfrac{1}{8}(1 - \xi)(1 - \eta)(1 - \zeta), \qquad h_8 = \tfrac{1}{8}(1 + \xi)(1 - \eta)(1 - \zeta)$$

$$h_9 = (1 - \xi^2), \qquad h_{10} = (1 - \eta^2), \qquad h_{11} = (1 - \zeta^2)$$

The first eight are the standard compatible interpolation functions. The last three are incompatible and are associated with linear shear and normal strains. The nine incompatible modes are eliminated at the element stiffness level by static condensation.

Since the three-dimensional element degenerates to the same approximation as in the two-dimensional element the same general improvement in accuracy is obtained. Therefore, additional examples of its behavior will not be given here.

This element has been found to be extremely effective in the analysis of massive three-dimensional structures subjected to bending. One element in the thickness direction of arch dams or thick pipe joints have been found to be adequate. Hence, the three-dimensional analysis of this class of structure involves a reasonable amount of computer time.

4.1. EXTENSION TO 20-NODE ELEMENTS

The modified eight-node three-dimensional element has been found to be comparable in accuracy with the standard 20-node element. This is because the stresses associated with constant moment are included in the normal 20-node approximation. It is apparent that the addition of incompatible modes to the 20-node elements will improve its behavior. Three interpolation functions which are associated with linearly varying moments are

$$h_{21} = \xi(1 - \xi^2), \qquad h_{22} = \eta(1 - \eta^2), \qquad h_{23} = \zeta(1 - \zeta^2)$$

At this time the authors do not have experience with this type of modification. It is possible that the increase in computational effort to form a 69×69 matrix and to reduce it to a 60×60 matrix will not be justified.

5. Thick Shell Element

It is possible to use standard three-dimensional elements for the analysis of shell-type structures. Practically, this has not been possible because of the following three problems:

1. Most three-dimensional solid elements have not had the ability to represent bending moments. (Elements with four nodes along each edge have this property).
2. Errors in the shear and normal strains cause the element to be very stiff.
3. Because of the relatively large stiffness coefficients in the thickness direction numerical problems are introduced for thin shells.

The first two problems can be overcome by the introduction of incompatible modes. The third problem can be minimized by the use of a computer with high precision or by restricting the application of the element to thick shells.

The shell-element presented in this paper is a 16-node curved solid element shown in Fig. 7. Each node has three unknown displacements. Therefore,

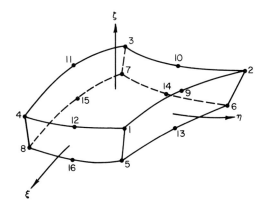

Fig. 7. Three-dimensional thick-shell element.

if the shell is considered as a two-dimensional surface there are six unknowns per point. It is apparent that this type of formulation avoids the problems associated with the sixth degree of freedom—the normal rotation is set to zero when certain finite elements are used in the idealization of shells.

The locations of the nodes are defined by the orthogonal, right-handed coordinate system (x, y, z) which is referred to as a global system. Within the element a local coordinate system (ξ, η, ζ) has been chosen such that ξ, η, ζ vary from -1 to $+1$; $(0, 0, 0)$ is located at the centroid of the element.

The local and global coordinate systems are related through a set of interpolating functions:

$$x = \sum_{i=1}^{16} h_i x_i, \qquad y = \sum_{i=1}^{16} h_i y_i, \qquad z = \sum_{i=1}^{16} h_i z_i$$

where

$$h_1 = \tfrac{1}{8}(1 + \xi)(1 + \eta)(1 + \zeta)(\xi + \eta - 1)$$

$$h_2 = \tfrac{1}{8}(1 - \xi)(1 + \eta)(1 + \zeta)(-\xi + \eta - 1)$$

$$h_3 = \tfrac{1}{8}(1 - \xi)(1 - \eta)(1 + \zeta)(-\xi - \eta - 1)$$

$$h_4 = \tfrac{1}{8}(1 + \xi)(1 - \eta)(1 + \zeta)(\xi - \eta - 1)$$

$$h_5 = \tfrac{1}{8}(1 + \xi)(1 + \eta)(1 - \zeta)(\xi + \eta - 1)$$

$$h_6 = \tfrac{1}{8}(1 - \xi)(1 + \eta)(1 - \zeta)(-\xi + \eta - 1)$$

$$h_7 = \tfrac{1}{8}(1 - \xi)(1 - \eta)(1 - \zeta)(-\xi - \eta - 1)$$

$$h_8 = \tfrac{1}{8}(1 + \xi)(1 - \eta)(1 - \zeta)(\xi - \eta - 1)$$

$$h_9 = \tfrac{1}{4}(1 - \xi^2)(1 + \eta)(1 + \zeta)$$

$$h_{10} = \tfrac{1}{4}(1 - \xi)(1 - \eta^2)(1 + \zeta)$$

$$h_{11} = \tfrac{1}{4}(1 - \xi^2)(1 - \eta)(1 + \zeta)$$

$$h_{12} = \tfrac{1}{4}(1 + \xi)(1 - \eta^2)(1 + \zeta)$$

$$h_{13} = \tfrac{1}{4}(1 - \xi^2)(1 + \eta)(1 - \zeta)$$

$$h_{14} = \tfrac{1}{4}(1 - \xi)(1 - \eta^2)(1 - \zeta)$$

$$h_{15} = \tfrac{1}{4}(1 - \xi^2)(1 - \eta)(1 - \zeta)$$

$$h_{16} - \tfrac{1}{4}(1 + \xi)(1 - \eta^2)(1 - \zeta)$$

The displacements within the element are assumed to be of the following form:

$$u_x = \sum_{i=1}^{16} h_i u_{xi} + h_{17}\alpha_{x1} + h_{18}\alpha_{x2} + h_{19}\alpha_{x3} + h_{20}\alpha_{x4} + h_{21}\alpha_{x5}$$

$$u_y = \sum_{i=1}^{16} h_i u_{yi} + h_{17}\alpha_{y1} + h_{18}\alpha_{y2} + h_{19}\alpha_{y3} + h_{20}\alpha_{y4} + h_{21}\alpha_{y5}$$

$$u_z = \sum_{i=1}^{16} h_i u_{zi} + h_{17}\alpha_{z1} + h_{18}\alpha_{z2} + h_{19}\alpha_{z3} + h_{20}\alpha_{z4} + h_{21}\alpha_{z5}$$

where;

$$h_{17} = \xi(1 - \xi^2), \qquad h_{18} = \eta(1 - \eta^2), \qquad h_{19} = (1 - \zeta^2)$$

$$h_{20} = \xi\eta(1 - \xi^2), \qquad h_{21} = \eta\xi(1 - \eta^2)$$

The motivation for addition of the interpolation functions h_{17} to h_{21} is to increase the capability of the element in producing closer approximations of the exact displacements under simple loadings, thereby increasing the convergence to exact solution. The incompatible interpolation functions h_{17} to h_{21} have zero values at the nodes and produce incompatibilities in displacement field along the interelement boundaries.

6. Thick Shell Examples

The following examples are intended to demonstrate the range of applicability of this element. Two parameters were recognized to have significant effect on the behavior of the element, namely, the ratio of thickness to the length along the surface (t/a) and the ratio of the length along the surface to the radius of curvature $(\phi = a/R)$. The effect of the first parameter is studied by the example of a square, simply supported plate and the effect of the second parameter is studied by a series of curved cantilever beam examples.

(a) Square Simply Supported Plate

The center deflection of a square, simply supported plate for three different meshes and two values of thickness to length ratio is shown in Fig. 8. It is seen

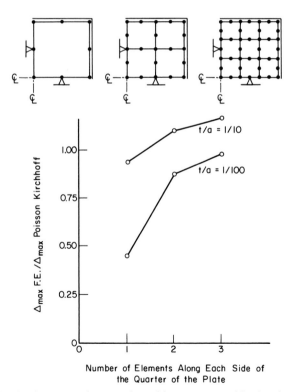

Fig. 8. Simply supported square plate with a concentrated load at the center.

that this element is more appropriate for moderately thick to thick shell problems with thickness to length ratio of more than 1/20, although final convergence is also assured for thin shell problems.

It is worthwhile to mention that the finite-element result in Fig. 8 exceeds that of Poisson–Kirchhoff which ignores the shear deformation. This is due to the fact that the thick-shell element is capable of undergoing shear deformation.

(b) Curved Cantilever Beam

The results of these examples are shown in Fig. 9. It is seen from these examples that the accuracy is only slightly affected by curvature for moderately thick to thick shell problems, whereas the high curvature affects the accuracy of the thin shell problems rather drastically.

The effect of the curvature can be explained by the fact that the number of terms of polynomial expansion of the displacement field included in the displacement expansion of the element is limited, whereas the significance of

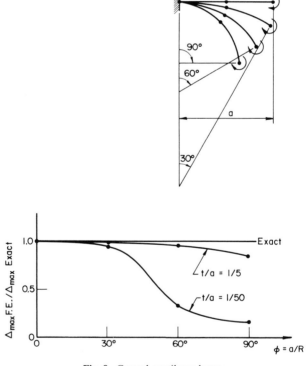

Fig. 9. Curved cantilever beam.

the higher-order terms in the polynomial expansion of the exact displacement increases with curvature. This effect is also reduced to zero at the limit as the mesh is refined, thereby assuring convergence.

References

1. Irons, B., "Stress Analysis by Stiffnesses Using Numerical Integration," Rolls Royce Co. (Internal Report) (June 1963).
2. Zienkiewicz, O. C., Irons, B. M., Ergatoudis, J., Ahmad, S., and Scott, F. C., "Isoparametric and Associated Element Families for Two and Three Dimensional Analysis," in *Finite Element Method in Stress Analysis* (Holand, I., and Bell, K., eds.), Chapter 13. Tapir, Trondheim, 1969.
3. Ahmad, S., Irons, B. M., and Zienkiewicz, O. C., "Curved Thick Shell and Membrane Elements with Particular Reference to Axisymmetric Problems," *Proc. Conf. Methods Struct. Mech.*, 2nd, Wright–Patterson AFB, Ohio (1968).
4. Doherty, W. P., Wilson, E. L., and Taylor, R. L., "Stress Analysis of Axisymmetric Solids Utilizing Higher Order Quadrilateral Finite Elements," SESM Rep. 69-3, Struct. Eng. Lab. Univ. of California, Berkeley, California (1969).
5. Zienkiewicz, O. C., Taylor, R. L., and Too, J. M., "Reduced Integration Techniques in General Analysis of Plates and Shells," *Int. J. Numer. Methods Eng.* **3**, 275–290 (1971).
6. Pawsey, S. F., "The Analysis of Moderately Thick to Thin Shells by the Finite Element Method," SESM Rep. 70-12, p. 102. Struct. Eng. Laboratory, Univ. of California, Berkeley, California (1970).
7. Zienkiewicz, O. C., *The Finite Element Method in Structural and Continuum Mechanics.* McGraw-Hill, New York, 1967.

Hybrid Models

Theodore H. H. Pian

MASSACHUSETTS INSTITUTE OF TECHNOLOGY
CAMBRIDGE, MASSACHUSETTS

1. Introduction

The finite-element methods for analyzing the stress and displacement distributions of an elastic continuum have long been interpreted as approximate methods associated with different variational principles in elasticity. In such methods, a solid continuum is first subdivided into an assemblage of discrete elements which are connected along continuous interelement boundaries, piecewise continuous displacement and/or stress fields are then assumed in each element and the resulting equations from the application of the variational principles are simultaneous algebraic equations which may have either (*i*) generalized displacements, (*ii*) generalized internal forces or stresses or (*iii*) both displacements and forces at the nodal points as unknowns to be evaluated. The nature of the final matrix equations has been used as the basis for one type of classification of the finite element methods. The three categories given above thus are often referred to as (*i*) displacement method, (*ii*) force method, and (*iii*) mixed method. The last one has been referred to by some authors as a hybrid method.

The term "hybrid" in this paper is used in a different connotation. The present classification [1, 2] is a reflection of the nature of the assumed field variables in the finite-element formulation. The conventional approach of assuming displacements which satisfy the continuity condition in each element and along the interelement boundary is termed the "compatible (or conforming displacement) model." The approach involving assumed stresses that satisfy the equilibrium conditions everywhere is termed the "equilibrium model." The approach involving simultaneously assumed stresses and displacement in each element is named a "mixed model." The hybrid model, for instance, reflects a formulation that involves either (1) assumed equilibrating stresses only within each element and compatible displacements along the interelement boundary or (2) assumed continuous displacement distribution within each element and equilibrating surface tractions along the interelement boundary. There is no definite relation between the finite-element models and the types of the resulting matrix equations. For example, the equilibrium model may lead to both a displacement method and a force method and it turns out that all four of the models just described may be formulated such that only the nodal displacements remain as the unknowns of the final matrix equations.

This paper first presents the idea behind the formulation of the element stiffness matrix based on the assumed stress hybrid model using the principle of minimum complementary energy. The general formulation of the hybrid stress model is then introduced based on a modified complementary (variational) principle which permits discontinuities of the stresses along the interelement boundaries. Various features of this hybrid method are discussed. Finally, a brief introduction is made of the various hybrid models based on assumed displacements.

2. Formulation of Hybrid Stress Model

In formulating the element stiffness matrix by the compatible model, displacement functions within each element are interpolated in terms of nodal displacements in such a manner that complete compatibility with the neighboring elements is enforced. In such a formulation the corresponding stress distribution within each element, in general, does not satisfy the equilibrium equations unless a very large number of displacement modes are introduced. A difficulty that is often experienced in the formulation of a compatible model is that for elements of arbitrary geometry it is difficult, if not impossible, to construct interpolation functions for the displacements in an element to fulfill the interelement compatibility conditions.

The motivation of the original formulation of the hybrid stress model [3] was to circumvent the difficulty described above. It was recognized that the derivation of an element stiffness matrix involves essentially the evaluation of the strain energy content in the element with prescribed boundary displacements which are interpolated in terms of a finite number of nodal displacements and are compatible with those of the neighboring elements. An approximate solution to such a problem can be accomplished by the use of the principle of minimum complementary energy, i.e., by assuming a number of stress modes which satisfy the equilibrium conditions only within each element. Clearly, it is a relatively easier task to construct displacement functions only along the element boundaries than to develop a continuous displacement function over the entire element to maintain the interelement compatibility requirements. Such an assumed stress approach was used originally to derive element stiffness matrices for plane stress and flat plate elements and to apply to problems for which no distributed body forces were involved.

A more rational and more general interpretation of the hybrid stress model was presented later [4, 5] when the entire solid continuum was considered. A solid continuum is subdivided into finite elements V_n and instead of considering stresses to be in equilibrium over the entire solid, the equilibrium conditions of the surface tractions T_i both along the interelement boundaries and along the prescribed stress boundaries are introduced as conditions of constraint. The corresponding Lagrangian multipliers can be interpreted as the displacements u_i along the element boundary. The variational functional π_{mc} which is a modified version of the conventional complementary energy functional can be written as

$$\pi_{mc} = \sum_n \left(\int_{V_n} \tfrac{1}{2} C_{ijkl} \sigma_{ij} \sigma_{kl} \, dV \right.$$
$$\left. - \int_{\partial V_n} T_i u_i \, dS + \int_{S_{\sigma_n}} \overline{T}_i u_i \, dS \right) \qquad (1)$$

where c_{ijkl} is the elastic compliance tensor, σ_{ij} is the stress tensor, u_i is the boundary displacement, ∂V_n is the nth element boundary which includes S_n, the interelement boundary, S_{u_n}, the portion over which the surface displacements are prescribed, and S_{σ_n}, the portion over which the surface tractions are prescribed. The component of the element boundary traction T_i is related to the stress components by

$$T_i = \sigma_{ij} n_j \qquad (2)$$

which n_j is the direction cosine of the normal to the boundary. Thus the original complementary energy principle involves only the stress field as the variables while in the modified principle the interelement boundary displacements u_i are the additional variables in the variational problem.

In the finite-element formulation the assumed approximate functions for the stresses $\boldsymbol{\sigma}$ in each element are divided into two parts. The first part, which consists of a finite number of parameters $\boldsymbol{\beta}$, should satisfy the homogeneous equations of equilibrium while the second part consists of a particular solution of the equations of equilibrium with the prescribed body forces. In matrix form, the stress components are expressed as

$$\boldsymbol{\sigma} = \mathbf{P}\boldsymbol{\beta} + \mathbf{P}_F\boldsymbol{\beta}_F \tag{3}$$

where $\boldsymbol{\beta}$ is unknown and $\mathbf{P}_F\boldsymbol{\beta}_F$ is known. The coefficients of $\mathbf{P}$ are functions of the coordinates. In case there exist any prescribed nonzero tractions along the boundary of the element, appropriate terms of known quantities should also be added to $\mathbf{P}_F\boldsymbol{\beta}_F$. The boundary tractions can be obtained from Eq. (2) and written in the form

$$\mathbf{T} = \mathbf{R}\boldsymbol{\beta} + \mathbf{R}_F\boldsymbol{\beta}_F \tag{4}$$

where $\mathbf{R}$ and $\mathbf{R}_F$ are obtained from $\mathbf{P}$ and $\mathbf{P}_F$, respectively.

The approximate displacements along the interelement boundaries are then interpolated in terms of the generalized nodal displacements $\mathbf{q}$ in the form of

$$\mathbf{u}_B = \mathbf{L}\mathbf{q} \tag{5}$$

Substituting Eqs. (3), (4), and (5) into Eq. (1) yields the following expression:

$$\pi_{mc} = \sum_n (\tfrac{1}{2}\boldsymbol{\beta}^T\mathbf{H}\boldsymbol{\beta} + \boldsymbol{\beta}^T\mathbf{H}_F\boldsymbol{\beta}_F - \boldsymbol{\beta}^T\mathbf{G}\mathbf{q} + \mathbf{S}^T\mathbf{q} + \text{const.}) \tag{6}$$

where

$$H = \int_{V_n} \mathbf{P}^T\mathbf{C}\mathbf{P} \, dV, \qquad \mathbf{H}_F = \int_{V_n} \mathbf{P}^T\mathbf{C}\mathbf{P}_F \, dV \tag{7}$$

$$\mathbf{C} = \text{elastic compliance matrix}$$

$$G = \int_{\partial V_n} \mathbf{R}^T\mathbf{L} \, dS, \qquad \mathbf{G}_F = \int_{\partial V_n} \mathbf{R}_F{}^T\mathbf{L} \, dS$$

$$\mathbf{S}^T = -\boldsymbol{\beta}_F\mathbf{G}_F + \int_{S\sigma_n} \bar{\mathbf{T}}\mathbf{L} \, dS$$

The stationary conditions of the functional given by Eq. (6) with respect to variations of $\boldsymbol{\beta}$ then yield

$$\mathbf{H}\boldsymbol{\beta} + \mathbf{H}_F\boldsymbol{\beta}_F - \mathbf{G}\mathbf{q} = 0 \tag{8}$$

By solving for $\boldsymbol{\beta}$ and substituting back into Eq. (6) the functional π_{mc} then contains the generalized displacements $\mathbf{q}$ only, i.e.,

$$\pi_{mc} = -\sum_n (\tfrac{1}{2}\mathbf{q}^T\mathbf{k}\mathbf{q} - \bar{\mathbf{Q}}^T\mathbf{q} + \text{const.}) \tag{9}$$

where $\bar{\mathbf{Q}}$ is the prescribed generalized nodal force defined by

$$\bar{\mathbf{Q}} = \mathbf{G}^T\mathbf{H}^{-1}\mathbf{H}_F\boldsymbol{\beta}_F + \mathbf{S} \tag{10}$$

and $\mathbf{k}$ is the element stiffness matrix defined by

$$\mathbf{k} = \mathbf{G}^T \mathbf{H}^{-1} \mathbf{G} \qquad (11)$$

It is seen that the element stiffness matrix $\mathbf{k}$ is derived from the matrices $\mathbf{G}$ and $\mathbf{H}$, which are in turn constructed from the assumed functions $\mathbf{P}$ for the stresses within the element and the interpolation functions $\mathbf{L}$ for the displacements along the element boundary. This expression is identical to that in the original derivation of the element stiffness matrix by the hybrid model [3].

In Eq. (9) when a transformation is introduced to relate the element nodal displacements $\mathbf{q}$ to a column of independent global displacement $\mathbf{q}^*$ the expression for π_{mc} becomes

$$\pi_{mc} = -(\tfrac{1}{2}\mathbf{q}^{*T}\mathbf{K}\mathbf{q}^* - \mathbf{q}^{*T}\mathbf{Q}^*) \qquad (12)$$

where $\mathbf{K}$ is the stiffness matrix of the assembled structure and $\mathbf{Q}^*$ is the matrix of applied generalized nodal forces.

Then the application of the minimum principle, $\delta\pi_{mc} = 0$, will yield the matrix equation

$$\mathbf{K}\mathbf{q}^* = \mathbf{Q}^* \qquad (13)$$

The resulting finite-element method is thus based on the element stiffness matrices and is a matrix displacement method.

3. Features of Hybrid Stress Model

Since the initial suggestion of the hybrid stress model and the application to plane stress problems by Pian in 1964 [3], the method has been extended to plate bending problems [5–9], St. Venant torsion problems [10], and shell problems [11]. The application to plate and shell problems clearly demonstrated the versatility of the hybrid stress model for deriving the stiffness matrices of elements of arbitrary geometry.

In applying the hybrid stress model in finite-element analysis it is important to make an appropriate choice of the assumed stresses in the element and the assumed element boundary displacements. For example, the problem of the central deflection of a simply supported square plate under a central load has been analyzed by the finite element method using square elements [5]. Two types of boundary displacements are used: (a) the lateral displacement w which varies as cubic function along the edge while the normal slope $w_{,n}$ varies as a linear function, and the number of degrees of freedom for each element is 12, with w, $w_{,x}$, and $w_{,y}$ at each corner as the generalized displacements, (b) both w and $w_{,n}$ varying as cubic functions, and the number of degrees of freedom for each element being 16, with w, $w_{,x}$, $w_{,y}$, and $w_{,xy}$ at each corner as the generalized displacements. In each of these two elements three different assumed functions—linear, quadratic, and cubic—are used for the bending moment distribution. In Fig. 1 the central deflections obtained by these six combinations of stresses and boundary displacements are plotted versus the number of elements per half span. It is seen that for most accurate solutions there is an appropriate combination of interior stresses and boundary displacements. For example, in this case, the linear–moment and linear–normal-slope combination appear to be a good choice.

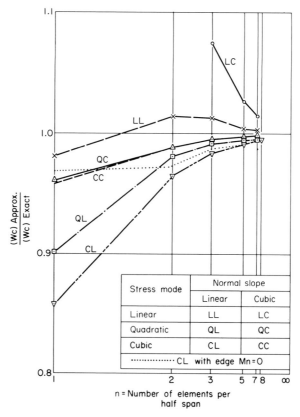

Fig. 1. Center deflection of simply supported square plate under center load (assumed stress hybrid method—rectangular elements).

The use of more boundary displacement modes with the same linear moment distribution can only yield more inaccurate solutions corresponding to more flexible structures. Indeed, as shown by Tong and Pian [4, 5], the over-abundance of the boundary displacement modes with limited number of stress parameters will cause the structure to be kinematically unstable. It is seen, however, that if the approximations for the interior stresses and boundary displacements are improved simultaneously the accuracy of the solution can be improved.

Tong and Pian [12] have proved that the direct flexibility influence coefficient obtained by the hybrid stress finite-element model is bounded from above by that of the equilibrium model, provided that they have the same type of stress distribution within each element, and is bounded from below by that of the compatible displacement model, provided that they have the same type of displacement distribution along the element boundary. The direct flexibility influence coefficient is the displacement at the point where a unit load is applied, hence it is also equal to the strain energy content in the structure.

Comparison of the compatible, equilibrium, and hybrid models for plate bending analyses have been made by several authors [6, 13]. The hybrid stress model often yields a more accurate solution than the corresponding compatible or equilibrium models, although it does not assure uniform convergence (in terms of strain energy) to the exact solution as the element pattern is refined. Figures 2, 3, and 4 are the comparisons made by Wolf [13] for the analyses of a simply supported rhombic plate under uniform loading. The results by compatible and equilibrium models were obtained by Sander [14, 15] while the results by the hybrid model were obtained by Wolf using a

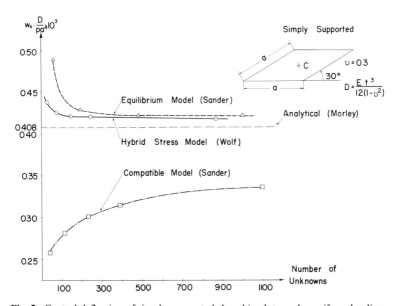

Fig. 2. Central deflection of simply supported rhombic plate under uniform loading.

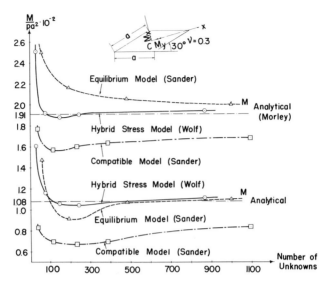

Fig. 3. Bending moments at center of simply supported rhombic plate under uniform loading.

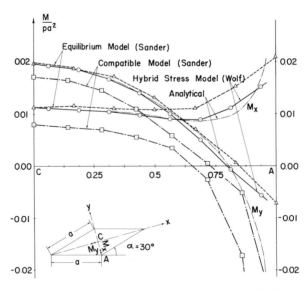

Fig. 4. Bending moment distributions of simply supported rhombic plate under uniform loading (using 16×16 mesh).

computer analysis system called STRIP. The analytical results were obtained by Morley [16]. In all these three approaches rhombic-shaped quadrilateral elements were used. Figures 2 and 3 are, respectively, plots of the deflection and bending moments at the center of the plate versus the number of unknowns in the resulting matrix equations. Figure 4 is the plot of the distributions of the bending moments M_x and M_y, respectively, along the y and x axes. It is seen that the hybrid stress model yields most accurate results for the central deflection of the plate and the bending moment distributions. Both the hybrid stress model and the equilibrium model are based on assumed stresses hence they both yield a more accurate stress solution than that obtained by the compatible model. Wolf [13] also presented excellent results in shell analysis by using hybrid stress flat plate elements in both bending and membrane actions.

Whetstone and Yen [17] evaluated several quadrilateral plane stress elements by numerical experimentations. From the study of several compatible elements and hybrid stress elements of several different element aspect ratios it is found that a hybrid stress model with a very simple stress approximation is most satisfactory from the standpoints of the accuracy in stress and displacement solutions and of the insensitivity of the solutions to element aspect ratios.

The hybrid stress model also provides several advantageous features which cannot be provided by the conventional assumed displacement compatible model. These features are discussed separately in the following paragraphs.

(a) *Convenience of Setting up Stiffness Matrices for Thick Plates and Shells*

In order to take the shear deformation effect into account in the analyses of thick plates and shells, the approach in the conventional compatible model is to add the rotations of the surface normals as additional nodal displacements which are independent of the normal displacement w and the in-plane displacements u and v. This means that under this approach there will be an increase in the number of degrees of freedom in each element when compared to one formulated under the Kirchhoff assumption. For example, when Clough and Felippa [18] modified their 12 degrees of freedom LCCT-12 plate bending element to include the transverse shear effect they had to include six nodal rotations as additional nodal displacements. Such a scheme has a more serious difficulty in that it will fail in the limiting case of a very thin shell where the shear deformation effect is negligible. A successful formulation of quadrilateral thick plate and shell elements has been made by using the limiting case of an isoparametric three-dimensional element [19]. Such an element has also been shown to lead to severe numerical difficulties for the limiting case of thin plates and thin shells unless adjustment is made by reducing the order of numerical integration applied to certain terms [20]. Nevertheless, a quadrilateral plate element formulated in this manner will contain 24 degrees of freedom. When the hybrid stress model is used for the thick plate or thick shell element, the nodal displacements may be described in exactly the same manner as that under Kirchhoff's hypothesis, and hence no additional nodal coordinates are needed. A 12-degrees-of-freedom rectangular plate element, for example, has been used successfully in analyzing a sandwich plate problem [5]. Severn [21] has also discussed recently the advantage of using the hybrid stress model for the inclusion of shear deflection in the stiffness matrix for a beam element.

(b) *Improvement of Finite Element Solution by the Introduction of Special Elements for Prescribed Stress Boundaries*

Computations using the element stiffness matrices formulated by the conventional assumed displacement approach will not generally produce the correct stress components at the prescribed stress boundaries. For example, along the stress-free boundary, spurious stress values will result, in general, by using the conventional method. The appearance of these spurious stresses can be prevented by the use of special free-boundary hybrid elements for which appropriate choice of the **Pβ** matrix in Eq. (3) can be made so as to give zero values of the generalized tractions on appropriate edges. Pian [6] and Yamada *et al.* [10] have demonstrated that the use of the special free-boundary element can make considerable improvement on the finite-element solution. In Fig. 1, for example, the six curves already presented were based on element stiffness matrices for which the free edge bending moment conditions are not considered. The seventh curve, obtained by incorporating the stress boundary conditions for the appropriate elements, certainly indicates significant improvement in the solution, particularly when the element sizes are large. Dungar and Severn [8] also demonstrated the advantage of using the special free boundary elements in an analysis of a cylindrical arch dam.

(c) *Improvement of Finite Element Solution for Problems Involving Stress Singularity*

In the conventional finite-element method based on an assumed displacement compatible model, the stress distribution within each element is finite and the method has been proved to converge to the exact solution if the corresponding stress distribution remains finite everywhere in the domain. However, for a problem involving a stress singularity such as the elastic solution at the tip of a sharp crack, one cannot prove that the finite-element method using conventional compatible elements will necessarily converge to the exact solution. When the hybrid stress model is employed, it is possible to include special stress terms which represent the correct stress singularity behavior. By extracting the singular part of the solution in its correct analytical form, the nodal displacements in the finite element analysis correspond to a solution without singularity. Thus the convergence of the finite element solution is assured.

Pian et al. [22] presents some numerical solutions for a finite plate with symmetric edge cracks under in-plane tension shown in Fig. 5. The stress singularity at the crack tip is of the form

$$\boldsymbol{\sigma} = \begin{bmatrix} \sigma_x \\ \sigma_y \\ \sigma_{xy} \end{bmatrix} = k_I/(2r)^{1/2} \begin{bmatrix} \cos(\theta/2)[1 - \sin(\theta/2)\sin(3\theta/2)] \\ \cos(\theta/2)[1 + \sin(\theta/2)\sin(3\theta/2)] \\ \sin(\theta/2)\cos(\theta/2)\cos(3\theta/2) \end{bmatrix} \quad (14)$$

where r and θ are the polar coordinates with the crack tip as the origin and k_I is the so-called elastic stress intensity factor for the mode-I type crack [23].

To apply to the crack tip stress distribution problem the stiffness matrices for elements away from the tip of the crack are constructed in the usual manner by either the compatible model or the hybrid model, while for the elements in the vicinity of the crack tip, the assumed stresses are divided into two parts, one of which contains no singularity while the other contains the proper singular behavior. In matrix form the assumed stresses are expressed as

$$\boldsymbol{\sigma} = \boldsymbol{\sigma}_{NS} + \boldsymbol{\sigma}_S = \mathbf{P}\boldsymbol{\beta} + \mathbf{P}_S k_I \quad (15)$$

where $\mathbf{P}\boldsymbol{\beta}$ may be simply polynomials which satisfy the homogeneous stress equilibrium equations while $\mathbf{P}_S k_I$ corresponds to Eq. (14). In a similar fashion to the procedure used in formulating the regular hybrid stress method the element boundary displacements are interpolated in terms of the nodal displacements $\mathbf{q}$. The resulting functional π_{mc} in Eq. (1) is thus expressed in terms of $\boldsymbol{\beta}$, k_I and $\mathbf{q}$. Since the values of $\boldsymbol{\beta}$ are independent from one element to the other, the stationary condition of π_{mc} with respect to the variation of $\boldsymbol{\beta}$ thus yields equations which can be used to solve $\boldsymbol{\beta}$ in terms of $\mathbf{q}$ and k_I. The resulting π_{mc} thus is a function of only k_I and $\mathbf{q}$ and the final matrix equations will have only $\mathbf{q}$ and k_I as unknowns.

For the finite-element solutions of the preceding problem of crack tip stress intensity, the elements used are eight-node rectangular elements. The nonsingular stress terms are assumed to vary as cubic functions in x and y. For the elements that contain the crack surface, the stress-free boundary conditions are observed. Quadratic displacements are assumed for all interelement boundaries except in the vicinity of the crack tip where the displacements are assumed to be proportional to $a + b\sqrt{r} + cr$. The $\sqrt{r}$ term is included here because it corresponds to the singular stress term.

The values of the stress intensity factor k_1 have been obtained using four different element sizes and arrangements as shown in Fig. 5. In each solution the singular stress term is included only in the two elements at the crack tip.

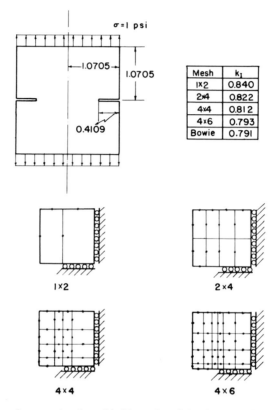

Mesh	k_1
1×2	0.840
2×4	0.822
4×4	0.812
4×6	0.793
Bowie	0.791

Fig. 5. Tension of rectangular plate with side cracks—finite element divisions and corresponding stress-intensity factor k_1 by hybrid stress model.

It is realized that to obtain a converging solution when the element size becomes smaller and smaller the singular stress term must be included in a fixed region and not in only a certain fixed number of elements. The present results, however, indicate that the stress intensity factor obtained by using the finest mesh is off by only 0.25% when compared with the analytical solution obtained by the complex variables method [24]. In fact, by using only 1 × 2 mesh for a quarter of the plate the resulting value for k_1 is still only 6% off.

In certain aspects of the finite-element formulation the hybrid stress model is handicapped when compared with a formulation based on assumed element displacements. The dynamic and static analyses by the compatible model are based, respectively, on the Hamilton's principle and the principle of minimum potential energy. The generalized nodal forces due to body forces and the generalized masses can all be derived consistently with the assumed displacement functions based on which the stiffness matrices are derived. Similarly in a buckling analysis by the compatible finite elements the derivation of the geometric stiffness matrix is based on the displacements of the structure hence is also consistent with the corresponding element stiffness matrix.

In the finite-element formulation by the hybrid stress model the element stiffness matrices are based only on the assumed element boundary displacements. Here, the consistent nodal forces due to the body forces are derived from the $P_F\beta_F$ term which is a particular solution of the nonhomogeneous differential equation and the rank of which cannot be higher than the assumed stress distribution in the element. For simple elements such as membrane and flat-plate elements the generalized nodal force can be derived rather easily. However, for a general shell element with complicated body force distribution, it has been discovered that it is not an easy task to construct the consistent nodal forces. In practice, the body force distribution is approximated by simple functions and the corresponding nodal forces can only be considered as approximately lumped ones.

In deriving the mass matrices and geometric stiffness matrices by the hybrid stress model, some assumed element interior displacement distribution functions must be introduced. These assumed functions cannot be obtained in a unique way by interpolating from the assumed element boundary displacements, thus the resulting mass matrix and geometric stiffness matrix again cannot be considered as consistent with the stiffness matrix. In practice, however, satisfactory analyses for vibration of plate and shell structures [25] and buckling of flat plate panels [26] have all been made by using element stiffness matrices derived by the hybrid stress model using, respectively, the mass matrices and geometric stiffness matrices derived by assuming element interior distributions in the manner indicated above. Henshell et al. [27] used hybrid cylindrical shell finite elements for static and vibration analyses. They started by using two different element interior displacement functions, one completely compatible but with only four rigid body modes explicitly included in the displacement function, the other noncompatible except in the limiting case of flat plate element but with all six rigid body modes included. Two hybrid stress elements were formulated by taking the edge displacements corresponding to the above-chosen interior displacement functions. The two formulations were found to yield approximately the same order of accuracy for a static problem involving a free–free circular cylinder under a pair of concentrated point loads. The vibration analyses were made by using only the noncompatible displacements. The predicted natural frequencies in two different cylindrical shell problems were indeed very accurate.

For the free-vibration problem there exists a consistent finite-element formulation based on the hybrid stress model. This can be derived by modifying the principle of stationary complementary energy used by Washizu [28] for free vibration problems to take into account the independent assumption of the element stress distributions. By adding the kinetic energy term and removing the term corresponding to prescribed boundary traction the complementary energy functional of Eq. (1) can be written as

$$\pi_{mc} = \sum_n \left[\int_{V_n} (\tfrac{1}{2}c_{ijkl}\sigma_{ij}\sigma_{kl} - \tfrac{1}{2}\omega^2\rho u_i u_i)\,dV - \int_{\partial V_n} T_i u_i\,dS \right] \tag{16}$$

where ω is the natural frequency and ρ, the density of the solid. By expressing the displacements u_i in the interior of each element in terms of the stress components by solving the equilibrium equation

$$\sigma_{ij,j} + \omega^2 \rho u_i = 0 \tag{17}$$

Eq. (16) can be written as

$$\pi_{mc} = \sum_n \left[\int_{V_n} (\tfrac{1}{2}c_{ijkl}\sigma_{ij}\sigma_{kl} - \tfrac{1}{2}(\omega^2\rho)^{-1}\sigma_{ij,j}\sigma_{ik,k})\,dV - \int_{\partial V_n} T_i u_i\,dS \right] \tag{18}$$

It is noted that in the finite-element formulation the stresses may be arbitrarily assumed as

$$\boldsymbol{\sigma} = \mathbf{P}\boldsymbol{\beta} \tag{19}$$

since the dynamic equilibrium equations are enforced a posteriori through Eq. (17). The derivative of the stress components can be written as

$$\boldsymbol{\sigma}' = \mathbf{P}'\boldsymbol{\beta} \tag{20}$$

Substituting Eqs. (19), (20), (4), and (5) into Eq. (18) yields an expression

$$\pi_{mc} = \sum_n (\tfrac{1}{2}\boldsymbol{\beta}^T\mathbf{D}\boldsymbol{\beta} - \boldsymbol{\beta}^T\mathbf{G}\mathbf{q}) \tag{21}$$

where

$$\mathbf{D} = \mathbf{H} - (\omega^2\rho)^{-1}\int_{V_n} \mathbf{P}'^T\mathbf{P}'\,dV \tag{22}$$

Note that $\mathbf{D}$ is a function of ω, the unknown natural frequency.

In the same manner as the previous formulation of the hybrid stress model, $\boldsymbol{\beta}$ can be expressed in terms of $\mathbf{q}$ and the functional π_{mc} can be expressed in terms of the generalized displacement only, i.e.,

$$\pi_{mc} = -\sum_n \tfrac{1}{2}\mathbf{q}^T\mathbf{G}^T\mathbf{D}^{-1}\mathbf{G}^T\mathbf{q} \qquad (23)$$

When a transformation is introduced to relate the element displacements $\mathbf{q}$ to the independent global displacements $\mathbf{q}^*$, Eq. (23) becomes

$$\pi_{mc} = -\tfrac{1}{2}\mathbf{q}^{*T}\mathbf{\Delta}\mathbf{q}^* \qquad (24)$$

where $\mathbf{\Delta}$ is obtained by assembling the matrices $\mathbf{G}^T\mathbf{D}^{-1}\mathbf{G}^T$ of the individual elements. The matrix $\mathbf{\Delta}$ is a function of the unknown natural frequency.

The condition for the natural vibration of the system is given by

$$|\mathbf{\Delta}(\omega)| = 0 \qquad (25)$$

Obviously, this formulation leads to a determinantal equation and not an eigenvalue matrix equation of the conventional vibration analysis. It is also obvious that the mass and stiffness matrices do not appear in the final equation. Thus, this formulation is for vibration analysis only and not for dynamic transient response analysis. Tabarrok and Sakaguchi [29, 30] first suggested this approach except they derived their equations based on Toupin's variational principle [31] with the impulse tensor and velocities as the field variables. The resulting governing equation from their formulation is, however, equivalent to Eq. (25). They have obtained satisfactory results for natural frequencies and mode shapes of flat plates both in in-plane motions and in lateral motions. Figure 6 presents the finite-element solution

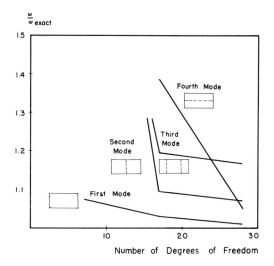

Fig. 6. Natural frequencies of simply supported rectangular plate ($a/b = 2$) by hybrid stress method.

of natural frequencies of a simply supported rectangular flat plate (aspect ratio = 2) obtained by Sakaguchi using rectangular elements. The percentage of errors of the finite-element solutions are plotted versus the number of degrees of freedom. It is seen that in all cases the finite element solutions are the upper bound solutions.

4. Hybrid Displacement Models

In the conventional principle of minimum potential energy the functional to be varied involves only the displacement field which is to be continuous over the entire solid continuum. In the finite element application, however, it is permissible to introduce the displacement compatibility conditions along the interelement boundaries as conditions of constraints and the corresponding Lagrangian multipliers are the interelement boundary tractions T_i. The modified variational functional is

$$
\pi_{\mathrm{mp_1}} = \sum_n \left[\int_{V_n} (\tfrac{1}{2} E_{ijkl} \varepsilon_{ij} \varepsilon_{kl} - \bar{F}_i u_i)\, dV - \int_{S_n} T_i u_i\, dS \right.
$$
$$
\left. - \int_{S_{u_n}} T_i(u_i - \bar{u}_i)\, dS - \int_{S_{\sigma_n}} \bar{T}_i u_i\, dS \right]
$$
(24)

where E_{ijkl} is the elastic constant tensor, and the strain components ε_{ij} is expressed in terms of the displacements. This variational principle, thus, has the displacement field and the element boundary tractions as variables. The use of this principle was first suggested by Jones [32] and has been used by Yamamoto [33] and Greene et al. [34] for finite element formulations.

In formulating the finite-element model using this variational principle, the element displacements **u** are approximated by a finite number of terms with unknown parameters **α** and the boundary tractions **T** are interpolated in terms of generalized internal forces **R**. This model is named a hybrid displacement model to indicate that the assumed displacements are continuous only within each element while the interelement forces are in equilibrium. The expression for $\pi_{\mathrm{mp_1}}$, is thus, in terms of **α** and **R**. By taking the variation of $\pi_{\mathrm{mp_1}}$, with respect to **α** and **R** a system of equations can be obtained with **α** and **R** as unknowns.

Intuitively this assumed displacement hybrid model should yield a matrix force method with a certain set of redundant forces **X** as unknowns. Such matrix equations, however, cannot be obtained simply by eliminating **α** from the system of equations obtained above by solving **α** in terms of **R** first. It is clear that not all the internal forces **R** are redundant forces. In general, in the formulation of this hybrid model both **α** and **R** are left as the unknowns as was done in Yamamoto [33] and Greene et al. [34].

Another way of applying this modified principle is to interpolate the element displacement **u** in terms of nodal displacements **q**, hence to guarantee the continuity with the neighboring element at the nodes. The complete compatibility condition along the interelement boundaries are then introduced by Lagrangian multipliers which are the generalized internal forces **R**. The final expression for π_{mp_1}, thus, is in the terms of both **q** and **R**, hence the corresponding finite element method is again a matrix mixed method. Harvey and Kelsey applied this technique in solving plate bending problems using triangular elements [35].

Another modification of the potential energy principle is due to Tong [36] and is based on the use of separate variables for the interior displacement field and the interelement boundary displacements. The compatibility condition of these two variables at the element boundary can be introduced by means of Lagrangian multipliers, which can again be recognized as the boundary tractions which are, in this case, independent for the two neighboring elements. The functional to be varied under this principle is

$$\pi_{mp_2} = \sum_n \left[\int_{V_n} (\tfrac{1}{2}E_{ijkl}\varepsilon_{ij}\varepsilon_{kl} - \bar{F}_i u_i)\,dV - \int_{S_n} T_i(u_i - \tilde{u}_i)\,dS \right.$$
$$\left. - \int_{S_{\sigma n}} \bar{T}_i u_i\,dS - \int_{S_{u_n}} T_i(u_i - \tilde{u}_i)\,dS \right] \tag{25}$$

where $\tilde{u}_i$ is the interelement boundary displacement. The variables in this variational principle are the element displacements u_i, the interelement boundary displacements $\tilde{u}_i$ and the element boundary tractions T_i.

In applying the finite-element formulation the element displacements **u** are again approximated in terms of a finite number of parameters $\boldsymbol{\alpha}$, while boundary tractions **T** and boundary displacements $\tilde{\mathbf{u}}$ are interpolated in terms of generalized internal forces **R** and nodal displacements **q**, respectively. Here $\boldsymbol{\alpha}$ and **R** for one element are all independent of those for other elements, hence they can be solved in terms of the generalized nodal displacements **q**. The resulting expression for π_{mp_2} is of the same form as π_{mc} in Eq. (9) and the resulting finite element method is a matrix displacement method.

The hybrid displacement model by Tong was used to solve plate bending problems [36] and shell problems [37]. This model may have certain advantages over the assumed stress hybrid model. Since the formulation is based on assumed element displacements, the generalized nodal forces and the generalized masses can be evaluated, respectively, from the virtual work and kinetic energy considerations. Similar to the hybrid stress model, this hybrid displacement model is a more convenient scheme for taking into consideration the transverse shear effect for thick plates and shells.

5. Conclusion

The hybrid models in finite-element methods are formulated by the replacing of the condition of complete compatibility of the field variables at the interelement boundary by one which is satisfied in an average sense. Such models are based on rigorous variational principles in solid mechanics, and can be proved to converge as the element size is reduced. The hybrid stress model cannot only avoid the complexity of choosing displacement functions in the interior of the element that also satisfy the complete displacement compatibility along the interelement boundaries, but also has been found, through several applications, to be a more versatile and accurate scheme when compared with the conventional compatible model. Among the other features of the hybrid stress model are its capability of taking prescribed stress boundary conditions and known types of stress singularity into account and its convenience in formulating thick-plate and thick-shell problems.

Acknowledgment

The author wishes to acknowledge the support provided by the Air Force Office of Scientific Research under Contract No. F44620-67-C-0019.

References

1. Pian, T. H. H., and Tong, P., "Basis of Finite Element Methods for Solid Continua," *Int. J. Numer. Methods Eng.* **1**, 3–28 (1969).
2. Pian, T. H. H., "Formulations of Finite Element Methods for Solid Continua," *Recent Advances in Matrix Methods of Structural Analysis and Design* (Oden, J. T., Gallagher, R. H., and Yamada, Y., eds.), pp. 49–83. Univ. of Alabama Press, Tuscaloosa, Alabama, 1971.
3. Pian, T. H. H., "Derivation of Element Stiffness Matrices by Assumed Stress Distributions," *AIAA J.* **2**, 1333–1336 (1964).
4. Tong, P., and Pian, T. H. H., "A Variational Principle and the Convergence of a Finite Element Method Based on Assumed Stress Distribution," *Int. J. Solids Struct.* **5**, 463–472 (1969).
5. Pian, T. H. H., and Tong, P., "Rationalization in Deriving Element Stiffness Matrix by Assumed Stress Approach," *Proc. Conf. Matrix Methods Struct. Mech. 2nd*, AFFDL-TR-68-150, pp. 441–469. Wright Patterson AFB, Ohio (1968).
6. Pian, T. H. H., "Element Stiffness Matrices for Boundary Compatibility and for Prescribed Boundary Stresses," *Proc. Conf. Matrix Methods Struct. Mech.*, pp. 457–477. AFFDL-TR-66-80, Wright Patterson AFB, Ohio (1965).
7. Severn, R. T., and Taylor, P. R., "The Finite Element Method for Flexure of Slabs when Stress Distributions are Assumed," *Proc. Inst. Civil Eng.* **34**, 153–170 (1966).
8. Dungar, R., and Severn, R. T., "Triangular Finite Element of Variable Thickness and their Application to Plate and Shell Problems," *J. Strain Anal.* **4**, No. 1, 10–21 (1969).

9. Allwood, R. J., and Cornes, G. M. M., "A Polygonal Finite Element for Plate Bending Problems Using the Assumed Stress Approach," *Int. J. Numer. Methods Eng.* **1**, 135–149 (1969).

10. Yamada, Y., Nakagiri, S., and Takatsuka, K., "Analysis of Saint-Venant Torsion Problem by a Hybrid Stress Model," *Seisan-Kenkyu* (*Mon. J. Inst. Ind. Sci., Univ. Tokyo*) **21**, No. 11 (1969) (paper presented at Japan–U.S. Seminar on Matrix Methods of Structural Analysis and Design, Aug. 25–30, 1969, Tokyo).

11. Tanaka, M., "Development and Evaluation of a Triangular Thin Shell Element Based upon the Hybrid Assumed Stress Model." Ph.D. Thesis, Massachusetts Inst. of Technol., Dept. of Aeronaut. and Astronaut. (June 1970).

12. Tong, P., and Pian, T. H. H., "Bounds to the Influence Coefficients by the Assumed Stress Method," *Int. J. Solids Struct.* **6**, 1429–1432 (1970).

13. Wolf, J. P., "Programme STRIP pour le calcul des structures en surface porteuse," *Bull. Tech. Suisse Romande, Lausanne* 97ᵉ année, No. 17, August 21, pp. 381–397 (1971).

14. Sander, G., "Applications de la Methode des Elements Finis a la flexion des Plaques," Univ. de Liege, Faculte des Sci. Appl., Collect. des Publ., No. 15 (1969).

15. Sander, G., "Application of the Dual Analysis Principle," *High Speed Computing of Elastic Structures* (*Proc. IUTAM Symp. 1970, Congr. Colloque Univ. Liege*), pp. 167–207. Place du XX Aout, 16 B-4000, Liege, Belgium (1971).

16. Morley, L. S. D., "Bending of Simply Supported Rhombic Plate under Uniform Normal Loading," *Quart. J. Mech. Appl. Math.* **15**, 413–426 (1962).

17. Whetstone, W. D., and Yen, C. L., "Comparison of Membrane Finite Element Formulations," Lockheed Huntsville Res. and Eng. Center, LMSC/HREC D162553 (September 1970).

18. Clough, R. W., and Felippa, C. A., "A Refined Quadrilateral Element for Analysis of Plate Bending," *Proc. Conf. Matrix Methods Struct. Mech., 2nd,* AFFDL-TR-68-150, pp. 399–440. Wright Patterson AFB, Ohio (1968).

19. Ahmad, S., Irons, B. M., and Zienkiewicz, O. C., "Analysis of Thick and Thin Shell Structures by Curved Finite Elements," *Int. J. Numer. Methods Eng.* **2**, 419–451 (1970).

20. Zienkiewicz, O. C., Taylor, R. L., and Too, J. M., "Reduced Integration Technique in General Analysis of Plates and Shells," *Int. J. Numer. Methods Eng.,* **3**, 275–290 (1971).

21. Severn, R. T., "Inclusion of Shear Deflection in the Stiffness Matrix for a Beam Element," *J. Strain Anal.* **5**, No. 4, 239–241 (1970).

22. Pian, T. H. H., Tong, P., and Luk, C. H., "Elastic Crack Analysis by a Finite Element Hybrid Methods," paper presented at the *Conf. Matrix Method Struct. Mech. 3rd,* Wright Patterson AFB, Ohio (October 1971).

23. Irwin, G. R., "Fracture Mechanics," *Structural Mechanics* (*Proc. Symp. Naval Struct. Mech. 1st*) (Goodier, J. H., and Hoff, H. J., eds.), pp. 557–594. Pergamon Press, Oxford, (1960).

24. Bowie, O. L., "Rectangular Tensile Sheet with Symmetric Edge Cracks," *J. Appl. Mech., Trans. ASME* **31**, Ser. E, No. 2, 208–212 (1964).

25. Dungar, R., Severn, R. T., and Taylor, P. R., "Vibration of Plate and Shell Structures Using Triangular Finite Elements," *J. Strain Anal.* **2**, No. 1, 73–83 (1967).

26. Lundgren, H. R., "Buckling of Multilayer Plates by Finite Elements," Ph.D. Thesis, Oklahoma State Univ. (May 1967).

27. Henshell, R. D., Neale, B. K., and Warburton, G. B., "New Hybrid Cylindrical Shell Finite Element," *J. Sound Vibrat.* **16**, No. 4, 519–531 (1971).

28. Washizu, K., "Note on the Principle of Stationary Complementary Energy Applied to Free Vibration of an Elastic Body," *Int. J. Solids Struct.* **2**, 27–35 (1966).

29. Tabarrok, B., "A Variational Principle for the Vibration Analysis of Continua by the Hybrid Finite Element Method," *Int. J. Solids Struct.* **7**, No. 3, 251–268 (1971).
30. Sakaguchi, R. L., "Energy Methods in Plate Vibration Analysis," Ph.D. Thesis, Univ. of Toronto, Dept. of Mech. Eng. (1970).
31. Toupin, R. A., "A Variational Principle for the Mesh-type Analysis of a Mechanical System," *J. Appl. Mech., Trans. ASME* **74**, 151–152 (1952).
32. Jones, R. E., "A Generalization of the Direct-Stiffness Method of Structural Analysis," *AIAA J.* **2**, 821–826 (1964).
33. Yamamoto, Y., "A Formulation of Matrix Displacement Method," Dept. of Aeronaut. and Astronaut., Massachusetts Inst. of Technol. (1966).
34. Greene, B. E., Jones, R. E., McLay, R. W., and Strome, D. R., "General Variational Principles in the Finite-Element Method," *AIAA J.* **7**, 1254–1260 (1969).
35. Harvey, J. W., and Kelsey, S., "Triangular Plate Bending Element with Enforced Compatibility," *AIAA J.* **9**, No. 6, 1023–1026 (1971).
36. Tong, P., "New Displacement Hybrid Finite Element Model for Solid Continua," *Int. J. Numer. Methods Eng.* **2**, 73–83 (1970).
37. Atluri, S., "Static Analysis of Shells of Revolution Using Doubly-Curved Quadrilateral Elements Derived from Alternate Variational Models," Sc.D. Thesis, Massachusetts Inst. of Technol., Dept. of Aeronaut. and Astronaut. (June 1969).

Computer Implementation of the Finite-Element Procedure

Ernst Schrem

INSTITUT FÜR STATIK UND DYNAMIK DER LUFT- UND RAUMFAHRTKONSTRUKTIONEN
UNIVERSITY OF STUTTGART GERMANY

1. Introduction

"The matrix method of structural analysis, as developed by Argyris [1], is a powerful analytical tool. The notation is convenient and general and is applicable to a very wide range of structural problems. In application, however, the matrix notation is often mis-interpreted from the point of view of efficient numerical computation." These words written in 1964 by Tocher [4], have lost nothing of their actuality [5]. In the 1960s the development of finite-element methods exhibited an exponential growth [3], and the demand for the appropriate computer programs increased accordingly.

During that decade a large amount of effort was spent in the implementation of finite-element procedures, whereby many efficient methods well known in other disciplines had to be rediscovered. Apparently cross-fertilization with neighboring disciplines [6, 7] took place to only a limited extent. This paper, which is mainly of review nature, aims at presenting some fundamental concepts for the implementation of the finite-element method, with primary emphasis on static analysis based upon the displacement method.

The total number of larger finite element computer programs which have been developed during the past decade probably is closer to five hundred than to two hundred. Gallagher [8] recently summarized the properties of 13 large-scale general purpose programs for finite element structural analysis.

The basic properties of about 20 general purpose finite element programs will be tabulated in [C]. In Fig. 1 a classification in tabular form is attempted. Naturally, most of the existing programs fall into class A. Due to their limited capabilities they will be excluded from the subsequent considerations.

Class	Effort of development [Man year] 1	Amount of code [K] 2	Core storage for execution		Dominant data-organization 5	Limitations of problem size due to core storage 6	Variety of element types 7
			Code [K] 3	Data area [K] 4			
A	.LT.2	.LT.10	3 to 10	5 to 30	in-core	serious	poor
B	1 to 10	10 to 100	5 to 15	5 to 25	sequential	not critical for medium size problems	limited to certain classes[a]
C	10 to 100	100 to 1000	5–20–40	8–20–60	direct access, write in-place	none	large

1 [K] = 1024 central memory words (FORTRAN Type REAL).
[a] E.g.: only a fixed number of degrees of freedom per node.

Fig. 1. Classification of finite-element computer programs.

The number of program packages in class B and C having a wide range of applicability already exceeds 30, whereby a few of them [9, 10, 20, 23] result from development efforts of more than 100 man years and budgets exceeding $one million. Unfortunately only a rather small part of the large fund of experience accumulated in the past is readily accessible in the open literature. On the other hand new projects are being started literally on a weekly rate all over the world, so that a word of caution seems to be appropriate for the case of new developments starting from scratch.

2. Prerequisites

As depicted in Fig. 2, the principal scientific disciplines contributing to the computer implementation of finite-element procedures are software engineering, numerical analysis, and mechanics. The heaviest impact upon the basic design, however, results from the experience gathered during the application of finite-element methods in the solution of actual engineering problems.

The correct interpretation of information-feedback from the user's side in the course of the development of a large-scale program system should have a significant effect upon its applicability. Thus the ideal developers of computer software for the application of finite element methods would be design engineers with several years of experience in the development of large computer systems and with some basic understanding of numerical analysis. It is quite clear that the program should pass through a period of verification prior to being released for general use. On the other hand, during application the need for additional verification might also arise with a corresponding impact on program development (Fig. 2). The important question of program verification will not be discussed here since Griffin recently has devoted an excellent paper to this subject [11].

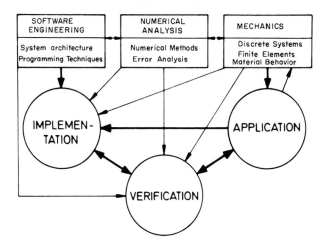

Fig. 2. Principal sources of finite-element computer programs.

As the development of large-scale computer software is rather time-consuming and expensive, a high degree of independence from a specific computer installation or operation system must be attempted [12a]. In meeting this objective, the first decision is the choice of the basic procedural language (e.g. ALGOL, COBOL, PL/1). Unfortunately, there does not remain much choice at present, since FORTRAN seems to be the only acceptable procedural language which is implemented on a large number of computer types [13] and has a common standardized subset of capabilities [14]. When supplemented by a number of assembly language subroutines enabling partial word operations, recursivity [15], and efficient data transfer, Standard FORTRAN is a rather efficient programming tool. Furthermore, many computer installations already maintain a large fund of FORTRAN subroutines for matrix manipulations [16]. Thus it is little wonder that all programs documented in [C] are based upon FORTRAN.

On the other hand, higher-level general purpose program packages for matrix operations and/or data management [e.g. 17–19] have not in their current stage of development provided an optimal basis for machine-independent large scale production programs. Due to their generality, their demands on core storage and computation time are larger than those of program packages specifically designed to satisfy the requirements of a certain production program. Nevertheless, many of their basic design concepts can be adapted to a finite element program system. In similar ways the architecture of modern computer operating systems may inspire the basic design of finite element production programs.

In the attempt to attain machine independence, another complex of problems arises in the organization and representation of the data [12] as well as data transfer between different computer installations. For the external data (input and results) the highest degree of transferability can be achieved at present by representing them in fixed length coded (BCD/EBCDIC) records (e.g. 80 column punched cards or 110/132 column printed lines), using a rather restricted character set (e.g. 47 legal ANSI-FORTRAN characters). This representation is independent of the computer word length, but subject to an inherent conversion error which may be cumulative, if the coded-binary-coded conversion cycle is repeated several times.

The internal data are handled in the internal machine representation, thus excluding transferability between computers with different word length. All internal data handling can be concentrated in a separate data retrieval package with sufficient parametric generality to enable its adaptation to a large spectrum of computer configurations [10, 21, 22]. A data retrieval system based upon fixed length records (pages) can be implemented on nearly all computer installations in an efficient way, especially if the page size is adjustable in order to suit the particular hardware or software specifications (e.g. size of a track or of a buffer).

The sequential organization of the internal data on backing storage (tape) imposes many difficulties and restrictions:

1. Random searches are very expensive.
2. Backwards read operations are often inefficient (BACKSPACE-READ-BACKSPACE cycles).
3. Updating of already written records (write-in-place) is not possible.
4. The storage capacity per reel is insufficient, even for medium size problems.
5. Data faults occur more frequently on tapes than on disks or drums.

The management of internal data in high-capacity finite element programs is thus best based upon direct access backing storage (disks or drums) with a capacity of 2 to 100 million computer words, according to problem size (Fig. 3).

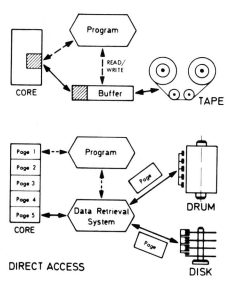

Fig. 3. Management of the internal data. (a) Sequential access; (b) direct access.

3. Solution Methods for the Load-Deflection Equations

The solution of the linear load-deflection equations is usually the most time-consuming computation step in the displacement method. As the finite-element method is ideally suited for analyzing highly complex structures, problems with several thousand unknowns occur frequently. In fact there are already a considerable number of problems in the range of 10,000 to 20,000 unknowns which have been solved successfully. Therefore, one of the most important features of a finite-element program package is its ability to solve efficiently very large sets of linear equations. This requirement has led to severe difficulties in the past, but by now there exist program systems which can handle equations of any size and population pattern quite effectively in a core working area of about 8000 to 14,000 computer words [10]. Nevertheless the various solution methods deserve our attention, as their demands on data organization will strongly influence the overall architecture of the program system. A well-designed program system should apply uniform methods of data organization throughout all its parts.

3.1. SPARSE MATRICES

The point at issue is the method used to take advantage of the sparse population of the various matrices. Basically there are two ways of handling sparsity (Fig. 4):

1. In the data structure, by excluding from storage all areas with zero matrix coefficients,
2. Within the calculations, by excluding all arithmetic operations on zeros as far as practicable.

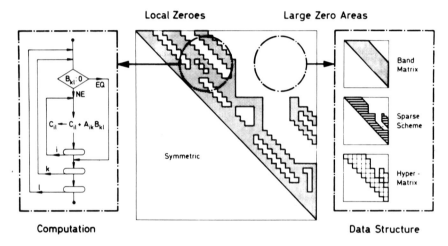

Fig. 4. Basic methods of handling sparse matrices.

The second way can often be handled by checking for zeros prior to entering the innermost loop. The positions of the nonzero coefficients can also be recorded in a binary mask which is used to control the arithmetic operations.

A data structure having the zeros squeezed out requires additional information about the location of the nonzero elements within the matrix [26–29, 37]. This can be rather uneconomical when no zeros at all are stored. In the most unfavorable case the storage requirements for the column and row indices are of the same order of magnitude as those for the matrix elements themselves. Matrix elements stored in such a way cannot be handled efficiently in a direct manner during the calculations. The elements of at least one operand have to be transferred ("unpacked") to their proper locations prior to the computations. Unless organized in an extremely professional way [29], these pack- and unpack-operations are quite expensive. An inadequate sparse scheme can lead to very inefficient programs due to a large amount of data handling.

By taking into account (a) the larger zero areas in the data structure and (b) the zeros scattered into nonzero areas in the calculations, the handling of sparse matrices is probably most economically performed (Fig. 4). One application of this principle is the hypermatrix scheme, where large matrices are subdivided both by rows and columns into submatrices (level 1, Fig. 5), which can be regarded as elements of a hypermatrix in the analogous way as numbers are the elements of a matrix [18].

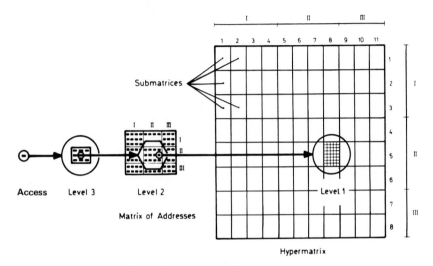

Fig. 5. Recursively partitioned matrices.

The addresses of the submatrices are arranged in a matrix of addresses (level 2, Fig. 5) which itself may be too large to fit into core as a whole. The matrix of addresses can then in turn be divided into submatrices, the addresses of which are stored in an address matrix of the next level. In that way a hypermatrix can be organized in a multilevel addressing scheme.

The size of the submatrices can be chosen in order to be best accommodated in the available space in executable storage and/or to optimize the record lengths of a given storage device. By increasing the number of levels in the addressing scheme, matrices of any size and population pattern can be processed in a given working area of executable storage.

The purpose of the address-matrices is twofold:

1. They represent the sparsity pattern of the matrix. Zero submatrices are never generated nor stored. They are merely indicated by a zero address.
2. They provide a simple means of accessing the submatrices. When the secondary storage allows for direct access, processing by rows or by columns or in any other way is equally simple.

The basic matrix operations, such as duplication, addition, multiplication, and the solution of linear equations can be organized in a fully recursive way [18, 30, 45], thus leading to rather simple modular programs. A matrix can be handled as its transpose simply by interchanging the row and column indices at all levels. Sparsity can be taken into account by merely skipping zero submatrices in the algorithm. Thus the algorithms are completely independent of the peculiar sparsity pattern of the matrices processed, and the same techniques are applied whether the overall matrix is sparse or dense.

The data management can be organized in a very simple way. The level 2 and 1 matrices of the two operands and the result are allocated within executable storage, and matrices of the same level are exchanged simultaneously. In contrast to sparse schemes, the expense for bookkeeping amounts to only a small percentage of the actual calculation costs. If n denotes the order of the level 1 submatrices $O(n^3)$ multiply–add operations can be performed in matrix multiplication until the next two data transfers become necessary. Thus the IO/CP time ratio is much smaller than in the case of row-by-row or column-by-column representation [30].

The arithmetic operations are simple manipulations of the level 1 matrices. Thus the innermost loops can be programmed very effectively in machine language [24, 31] in order to avoid inefficient compiler-generated code or to take advantage of parallel processing features of the hardware (e.g., CDC 6600). In a similar way the program logic can take into account the microstructure of sparsity. The hypermatrix scheme is not well suited to algorithms based upon the manipulation of specific rows or columns (e.g., elimination with pivoting). The same holds for a number of algorithms customary for the solution of dynamic problems (e.g., tridiagonalization methods), although these algorithms can often be generalized to permit hypermatrix operation [25].

3.2. DIRECT METHODS

Direct (closed) methods as distinct from iterative methods yield the solution by performing a fixed number of arithmetic operations. There is no direct method possible which solves a general system of linear algebraic equations by a smaller total number of arithmetic operations than that which is required by Gaussian elimination [32], although the proportion of multiplications and additions often can be modified to speed solution [33, 34]. A number of variants of the Gaussian elimination was developed for desk calculators (Gauss, 1826; Doolittle, 1879; Cholesky, 1916; Banachiewicz, 1938; Crout, 1941). There are also more recent methods, tailored for the digital computer (e.g. wave front method [41, 42], bifactorization method [35], optimally ordered triangular factorization [37, 53]). Although all these variants require about the same number of arithmetic operations, they differ in the required data organization and in the arrangement of the inter-mediate calculations.

Given the load-deflection equations

$$Kr = R, \tag{1}$$

the Gaussian elimination process can be divided into three steps:

1. Decomposition of the matrix of coefficients K into the product of a lower L and upper U triangular matrix

$$K = LU$$

2. Forward substitution by solving

$$LR' = R$$

3. Backward substitution by solving

$$Ur = R'$$

Steps 1 and 2 could be executed simultaneously. If the triangular matrices are saved, steps 2 and 3 can be applied repeatedly to yield the solutions for any given set of right-hand sides. In most program packages a limited number (5 to 3000 [C, 10]) of loading cases can be handled simultaneously in either step.

Computing the inverse in order to be able to solve Eq. (1) by post-multiplying the inverse by the load matrix,

$$r = K^{-1}R,$$

would be extremely inefficient. Not only is the number of operations con-siderably larger than that for the elimination method, but also the sparsity structure will be destroyed. The flexibility matrix K^{-1} for general structures is fully populated, whereas the triangular factors L and U (or their equiva-lents in the variants) retain the same sparse population pattern as K, apart from the additional "fill-in" terms in the active columns (Fig. 7a,b). Besides taking advantage of the sparse structure of the operand matrices, a well-designed procedure should also try to conserve sparsity in the result as far as possible.

Due to energy considerations the structural stiffness matrix is always symmetric and positive definite (omitting some exceptional cases of mixed-type finite elements). In this case Gaussian elimination is guaranteed to be numerically stable, irrespective of the order in which the equations are eliminated. Therefore, a pivot-search is not necessary. Furthermore, there exists a real nonsingular upper triangular matrix U so that

$$U^T U = K$$

This symmetric decomposition due to Choleski can also be formulated in the following way
$$\bar{U}^T D \bar{U} = K$$

with a unit upper triangular matrix $\bar{U}$ and a diagonal matrix D. This second formulation avoids taking N square roots in the decomposition process, but does not have essential advantages in comparison to the first [36].

Among all the variants of the Gaussian elimination the Choleski method seems to have some computational advantages. If direct access backing storage is available, the IO-operations can be organized most economically. Only the upper triangle of the structural stiffness matrix needs to be stored, and this is normally progressively overwritten by the triangularized result matrix. The basic arithmetic operations are scalar products which are readily written in machine language to speed up the calculations. It is also possible to utilize to a large extent parallel processing features in the arithmetic section of the computer hardware. The results of a very large number of arithmetic operations are accumulated into the one result area.

In order to illustrate the basic procedures proposed to handle sparse matrices, four of the most powerful methods of organizing the elimination process will be considered in detail.

(a) *Band-Algorithm*

In the field of structural analysis band algorithms date back to the application of finite difference methods. The stiffness matrix of a net of nodal points generated by uniform division of a simply connected ("rectangular") region can always be arranged as a symmetric band matrix:

$$K_{ij} = 0 \ \forall \ |i - j| \geqslant s, \qquad 1 \leqslant i, j \leqslant N$$

In many cases the (maximum) semibandwidth s is much smaller than the order N of the stiffness matrix. In this case a special band algorithm can be used to advantage [38, 39].

In Gaussian elimination, the arithmetic operations affect only a triangular area below the row currently being reduced. After one reduction cycle the reduced row leaves the working area, and a new column is picked up (Fig. 6). By contrast, in the Choleski decomposition only the coefficients in a triangle above the row currently being reduced are required. After one reduction cycle a new row enters the working area, and a column of reduced coefficients can be output. In both cases the triangular working area is shifted down the diagonal during the decomposition process (Fig. 6).

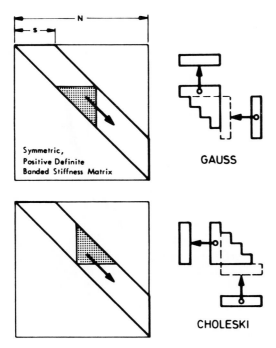

$Ns - (s^2 - s)/2$ storage locations in band
$(s^2 - s)/2$ storage locations in working area
$O(Ns^2/2)$ multiply–add operations in decomposition
$O(2Ns)$ multiply–add operations per right-hand vector

Fig. 6. Schematic diagram for (a) Gauss and (b) Choleski decomposition of band matrices.

The complete band matrix requires about Ns storage locations, whereas the decomposition can be carried out in a working area of about $\frac{1}{2}s^2 + O(s)$ storage locations. The number of multiply–add operations is about $\frac{1}{2}Ns^2$ in the decomposition step and about Ns for each loading case in either forward or backward substitution.

Meshes generated over multiply connected or branched regions lead to a population pattern which does not readily fit into a band matrix scheme (Fig. 7a). Despite high sparsity the maximum bandwidth can be very large so that the area actually occupied by nonzero elements (Fig. 7b) is in great disproportion to the area needed for the band calculations (Fig. 7c). This can be improved by modifications of the band algorithm to account for a variable band width. Although in this case the numerical operations are not much more complex, the data organization has to provide for variable length records (matrix rows). Nevertheless, there remains a rather large area of zeros within the stepped band envelope (Fig. 7d).

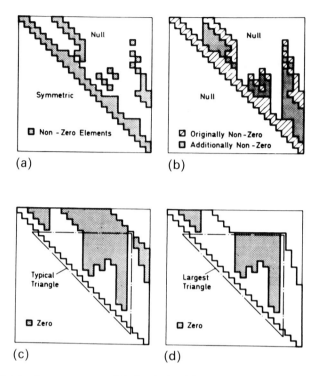

Fig. 7. Matrix elements processed in band algorithms (schematic). (a) structural stiffness matrix before decomposition; (b) upper triangular matrix after decomposition; (c) constant band algorithm; (d) stepped band algorithm.

(b) *Wavefront Method*

One elegant method of excluding the zeros within the band from the calculations is the wavefront method of Gaussian elimination [41, 42].

The elimination of a row s leads to a modification of the elements in the remaining rows according to

$$k_{ij}^* = k_{ij} - (k_{si}k_{sj}/k_{ss}) \qquad (2)$$

Of course the modifications have to be carried out only for all admissible pairs (i, j) where both k_{si} and k_{sj} are nonzero. The affected elements k_{ij} can be collected into a dense triangle (Fig. 8) which often is considerably smaller than the triangle required in the band algorithm (Fig. 7c, d), particularly when the population pattern cannot be regarded as a dense band.

This fact leads to one of the basic ideas of the wavefront method: All degrees of freedom corresponding to the columns which have to be modified according to (2) ("active columns"), constitute the so-called "wavefront,"

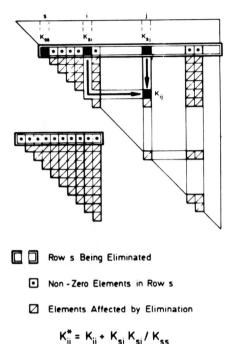

Row s Being Eliminated

Non - Zero Elements in Row s

Elements Affected by Elimination

$$K_{ij}^* = K_{ij} + K_{si} K_{sj} / K_{ss}$$

Fig. 8. Matrix elements directly affected by elimination of row s.

and are retained in executable storage. A column is added to the wavefront as soon as its first nonzero element appears during the elimination process. The successive elimination of rows causes other columns to be removed from the wavefront. In any case the number of columns in the wavefront never can exceed the maximum semibandwidth s.

Figure 9 illustrates the advantages of the wavefront method over the band method in the most extreme case of a bordered diagonal matrix. The associated graph (Fig. 9a)—which is closely related to the actual graph of the structure—explains the name "wavefront." Nodes 8, 9, 10, 22, 23, and 24 currently constitute the wavefront. In the next step, node 8 will be eliminated, and node 11 will enter the wavefront which is thus progressing across the structure, leaving behind all nodes already reduced.

Irons [41] made full use of this characteristic feature by controlling the order of elimination according to the sequence of finite elements crossed by the wavefront (Fig. 10). As indicated by (2), the original value k_{ij} can be assembled from the elemental stiffness matrices independently from the rows s being eliminated. The elimination and the assembly of the original k_{ij} can be performed alternately in the same program, thus avoiding the retrieval

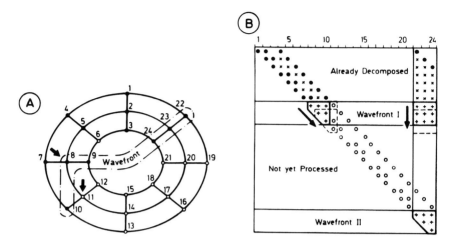

Fig. 9. Wavefront method (simple example).

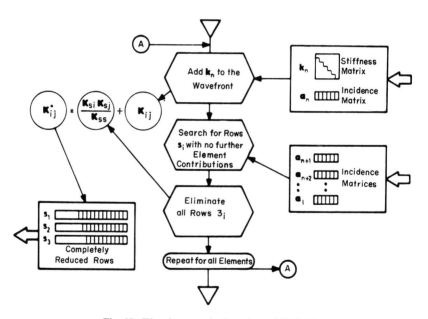

Fig. 10. Wavefront method, variant of IRONS.

of the coefficients k_{ij} from an already assembled matrix which would be rather cumbersome in the wavefront method. As soon as there are no further elemental stiffness matrices contributing to a given degree of freedom, the corresponding row can be immediately eliminated. The sequence of rows being eliminated, and thus the size of the wavefront as well as the efficiency (number of operations) of the elimination process, are determined by the order of the elements. The node numbers are merely unique identifiers, relating elemental and structural degrees of freedom. The numbering scheme of the nodes has no influence upon the order of the eliminations.

In the same way the contributions of elemental loads to the right-hand sides can be summed during the forward substitution process. The connectivity information required in the elimination can also be utilized in the backward substitution to distribute directly the resulting displacements into the elemental displacement matrices. The resulting violation of the principle of modularity could be carried so far that in the same program complex even the elemental stresses could be calculated and averaged at the approximate nodal points.

On the other hand, modularity will be retained to a very high degree, if the process is divided into two parts executed consecutively:

Step 1: The kinematic connection information is processed element by element, yielding all information necessary to control Step 2, i.e., the locations at which the elemental stiffness coefficients have to be added, the columns which will be added to the front and the rows which will be eliminated. This control information is output element by element onto a sequential data set. Due to the large core storage available, the program could be designed in a quite comprehensive way, taking into account all kinds of degrees of freedom for any types of elements. Even resequencing the elements in order to increase the efficiency of Step 2 could be taken into account.

Step 2: The control information generated in Step 1 is processed together with the elemental stiffness matrices. The program logic is extremely compact since it is completely controlled by optimally prepared data. More than 80 % of the available core storage could be used to hold the wavefront.

(c) *Choleski Method for Sparse Schemes*

Another method of avoiding unnecessary operations upon the zeros within the band (Fig. 7) is the *sparse scheme method* of Choleski decomposition, as documented by Jennings and Tuff [40]. In every column only the elements between the first nonzero and the diagonal are stored in a column sparse scheme (Fig. 11b). The zeros in this interval normally become non-zero during the decomposition process. A certain number of columns are arranged into a segment to be transferred as a whole to and from backing storage. The method is quite effective and simple to program. However, in every computation step a considerable portion of data within the segments is superfluous because it is lacking a counterpart in the complimentary segment (Fig. 11a). This does not affect the CPU-time, but storage allocation and data

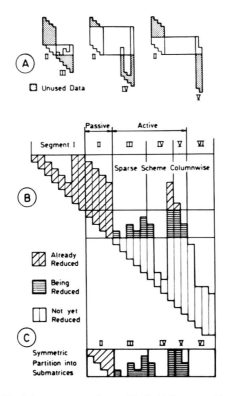

Fig. 11. Column sparse scheme. Choleski decomposition.

transfers increase unnecessarily. As indicated in Fig. 11c this deficiency can be easily avoided if the matrix is divided not only by columns, but also by rows. This leads to the

(3) Hypermatrix Algorithm

Every direct method for solution of linear equations is required to work both by rows and by columns. This demand is met in the most natural way by partitioning the matrices in a symmetric way both by rows and by columns. Furthermore, the multilevel hypermatrix scheme allows the handling of matrices of any size in a given area of executable storage, in contrast to any method based upon partitioning by rows only (or columns only).

If a matrix operation, defined by an algebraic relation between the matrix elements, also can be defined by an algebraic relation of the same form between the submatrices, then this matrix operation can be organized in a recursive way as a multilevel hypermatrix operation, the total number of arithmetic operations remaining unaltered. For example, matrix multi-

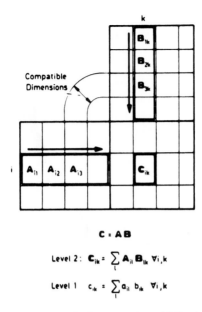

$$C = AB$$

Level 2 : $C_{ik} = \sum_{l} A_{il} B_{lk} \quad \forall i, k$

Level 1 $c_{ik} = \sum_{l} a_{il} b_{lk} \quad \forall i, k$

Fig. 12. Recursive hypermatrix multiplication.

plication (Fig. 12)

$$c_{ik} = \sum_{i} a_{ij} b_{jk}$$

could also be defined on the submatrix level as

$$C_{ik} = \sum_{i} A_{ij} B_{jk}$$

Similarly the basic relation of Gaussian elimination (2) could be expressed on submatrix level by the equation:

$$K_{ij}^* = K_{ij} - K_{si}^{T} K_{ss}^{-1} K_{sj} \tag{3}$$

The deviations in the form of the two analogous equations are due to the noncommutativity of the matrix product and the special treatment of matrix "division." The transcription of scalar equations into matrix equations is only possible because the laws of associativity and distributivity are valid for both matrix and scalar operations. Equation (3) also reveals the close connection between the elimination methods for linear equations and the methods of substructure analysis [43, 44].

Von Fuchs and Roy [45] have demonstrated that Choleski decomposition can be organized as a recursive multilevel hypermatrix algorithm (Fig. 13).

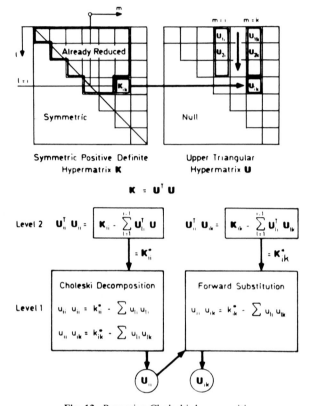

Fig. 13. Recursive Choleski decomposition.

The basic equation

$$u_{ii}u_{ik} = k_{ik} - \sum_{l=1}^{i-1} u_{il}u_{lk} \qquad (4)$$

has an analogous counterpart on the submatrix level:

$$U_{ii}^{\mathrm{T}}U_{ik} = K_{ik} - \sum_{l=1}^{i-1} U_{li}^{\mathrm{T}}U_{lk} \qquad (5)$$

The right-hand side K_{ik} can be expressed for the submatrices on the diagonal $(i = k)$

$$U_{ii}^{\mathrm{T}}U_{ii} = K_{ii}^{*}$$

yielding U_{ii} by an ordinary Choleski decomposition of K_{ii}^{*}. In the next step the products in Eq. (5) are accumulated into the off-diagonal submatrices $K_{ik}^{*} (i \neq k)$, leading to a lower triangular set of linear equations

$$U_{ii}^{\mathrm{T}}U_{ik} = K_{ik}^{*}$$

which can be solved for U_{ik} by a forward substitution operation for each value of k between $(i + 1)$ and N. Forward- and backward-substitution can be organized as recursive hypermatrix algorithms in a similar way [45]. In all these cases simple and efficient means have been devised to skip the majority of operations on zeros within the various submatrices [10].

The various direct methods for solving the load-deflection equations are difficult to compare. The obvious criteria are as follows:

1. numerical accuracy
2. efficiency and simplicity of the data organization and arithmetic operations
3. intrinsic limitations of problem size (e.g., number of equations, maximum bandwidth)
4. suitability for modular design
5. ability for implementation on a large number of different hardware/ software configurations.

The usefulness of a given method also depends upon the degree of optimization of the underlying programming techniques. A proper design might improve the efficiency by a factor of 4 to 20 (Fig. 14). Unfortunately, the band

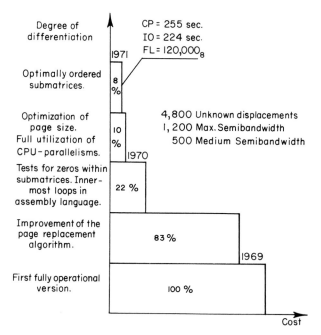

Fig. 14. Typical reduction of computation time for Choleski decomposition (CDC 6600).

algorithms and wavefront methods are limited in the manageable problem size, although they could be extended to handle larger problems [42] at the expense of reduced efficiency and simplicity. Furthermore, the procedure for the solution of the structural equations is always embedded into program parts which generate the stiffness and load matrices and postprocess the deflection matrix. Thus the influence of a given solution method upon the overall architecture of the program system must be taken into consideration. Coining firm statements about the superiority or inferiority of a single method [27, 42] seems at least to be imprudent, as also emphasized in [46].

3.3. OPTIMAL ORDERING OF THE ELIMINATIONS

The number of arithmetic operations required in the decomposition process, and hence the extent to which roundoff errors may become significant, depends heavily upon the order in which the equations are eliminated. Several structural software packages provide more-or-less automated algorithms for resequencing the linear equations in a way which is equivalent to renumbering the mesh nodes. Such a feature is particularly useful when automatic mesh generators are used [47] or when the connectivity of the structure is so complex that it is almost impossible to determine the optimal numbering scheme by inspection.

Reordering of the linear equations can be attempted according to three different criteria :

1. to reduce the maximum bandwidth [47–51]
2. to minimize the storage requirements for the triangularized matrix (or its equivalent) [37, 53, 54]
3. to reduce the number of elementary multiply–add operations required for the decomposition process [37, 52–54]

It is very important to note that the reduction of the maximum bandwidth is not equivalent to the reduction of the number of elementary operations. If there are isolated nonzero terms far away from the diagonal, it is always possible to reduce the maximum bandwidth at the expense of an overall increase in the median bandwidth. In general this will also increase the final number of nonzero elements in the triangularized matrix. The only justification for deliberately reducing the bandwidth can be found in cases where the equations are solved by a band-algorithm. Here the allowable bandwidth provides in general the most crucial limitation of the permissible problem size. In extreme cases more computational effort can be spent in cutting down the bandwidth to a feasible size than for the actual solution of the equations.

The second and the third objectives are closely related. In the former, the number of additional "fill-in" terms is minimized. Generally this will also result in a smaller number of elementary operations. The effect of sparsity on the generation of new nonzero submatrices is demonstrated by Eq. (5). If, in that equation, for every l either U_{li} or U_{lk} is zero, no arithmetic operations at all will be carried out. If K_{ik} also is zero, no additional submatrix U_{ik} will be generated.

There are $N!$ possible permutations of the N equations. Even if the large number of equivalent permutations could be excluded from consideration there remain an astronomically large number of possible configurations. Today there is no general algorithm available which can economically perform the interchange of single equations necessary to significantly reduce the storage requirements or the total computational effort.

The situation is more hopeful if entire groups of equations are reordered. Particularly in the hypermatrix scheme a dummy execution of the decomposition algorithm is very fast on the higher-level address matrices and yields a quite close estimate of the number of operations and storage locations which will be required in the actual decomposition. This provides an inexpensive means to control the decisions between the optimization steps.

In Fig. 15a the structural stiffness matrix is plotted, as obtained from the analysis of a thick-walled pressure vessel having four large attached tubes. The total number of unknowns was about 12,000, and the size of the level 1 submatrices was 75 by 75. Apparently the analyst did a good job with the nodal point numbering, as indicated by the rather narrow band structure with a really aesthetic appearance (maximum half-bandwidth $s = 1350$, $s/N = 11.2\%$).

Surprisingly, the optimal ordering program developed at ISD by Bernhardt came to a different conclusion, yielding the pattern of Fig. 15b, where the narrow band structure of Fig. 15a is heavily disturbed. In this rearranged form the CPU time required on CDC 6600 for the Choleski decomposition was reduced from 2486 to 1676 sec, i.e., by 32.6%. Similarly, the IO-time was reduced from 1970 to 1280 sec, i.e., by 35%. The number of fill-in matrices generated during the decomposition process was reduced from 562 to 320, resulting in total storage savings on disk of about 15%.

A slightly more sophisticated optimization algorithm yielded the pattern of Fig. 15c. It is interesting that although this pattern leads to practically the same savings as the pattern of Fig. 15b, the banded population structure is retained. The additional computation time required by the optimization program (60/25 CPU sec) is in almost all practical applications heavily outweighed by the corresponding savings in the expensive Choleski decomposition process.

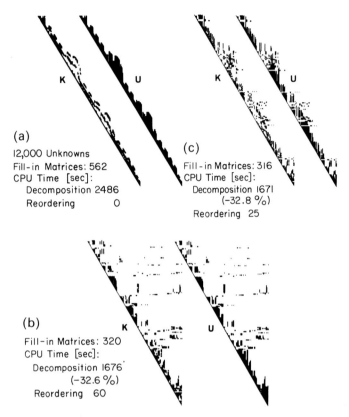

Fig. 15. (a) Pattern of original stiffness matrix ; (b) pattern of optimally reordered matrix ; (c) equivalently reordered matrix.

3.4. Iterative Methods

It appears that iterative methods for the solution of the load-deflection equations [55–58, 81] are seldom provided as a standard facility in the larger finite element program packages. They usually require less storage than the direct methods, and are more simply programmed. The population pattern of the structural stiffness matrix is less important. The conjugate gradient method can be even organized so that only the elemental stiffness, displacement, and load matrices are processed, without ever assembling the structural stiffness matrix explicitly [56, 57].

It is true that iterative methods converge rapidly for strongly diagonal dominant matrices. However, in the case of stiffness matrices of general structures they occasionally display a rather slow convergence. In cases where the conjugate gradient method requires nearly N iterations until the required accuracy is achieved (which was occasionally observed in practice) direct methods are clearly superior. Also the impossibility of repeating the solution process for an additional set of right-hand sides without paying a good part of the cost of the preceding solution does not support arguments in favor of iterative methods.

3.5. ERROR ANALYSIS

Apart from the discretization error, which is already introduced when the mesh size and the element types are specified by the user, three major sources of error can be distinguished [45, 62]:

1. *inaccuracies of the basic input parameters* (e.g., coordinates, geometrical dimensions, material data and loads)
2. *initial truncation errors* (representation errors) arising when a matrix is originally represented in the computer to a fixed number of digits. Important information may be lost due to truncation of some of the lower digits
3. *rounding errors* (process errors) which occur during the arithmetic operations upon the computer representation of the matrices.

The first sort of error is present in all calculations arising from engineering problems. If the underlying mathematical model is adequately chosen and if the input parameters are not interdependent, the results will just describe a neighboring physical behavior which, in general, can be tolerated.

The other two types of error result from the computer representation of real numbers by a fixed number of digits [59]. Rounding errors always require special consideration due to their statistical nature [60]. Often they can be reduced by using an optimally ordered calculation process or by applying additional features such as double precision accumulation of scalar products. Initial truncation errors impose the largest difficulties since they are already present at the very beginning of a computation step. The symmetric $N \times N$ structural stiffness matrix can be represented as the sum of the N dyadic products $V_i V_i^\mathrm{T}$ of its normalized eigenvectors which are scaled by the corresponding eigenvalues λ_i:

$$K = \sum_{i=1}^{N} \lambda_i V_i V_i^\mathrm{T} \tag{6}$$

Since the eigenvectors are orthonormal

$$V_i^\mathrm{T} V_j = \delta_{ij}$$

the inverse of the structural stiffness matrix K can be written in the form

$$K^{-1} = \sum_{i=1}^{N} \lambda_i^{-1} V_i V_i^{\mathrm{T}} \tag{7}$$

Equations (6) and (7) reveal the nature of the initial truncation error [61]. The elements of the structural stiffness matrix K are represented in the computer to a fixed number of digits p. Thus the contributions of the smaller eigenvalues to the elements of the stiffness matrix in (6) will be truncated, provided that the nonnegligible elements of all dyads $V_i V_i^{\mathrm{T}}$ are distributed in a rather uniform way. The lowest eigenvalue will be represented in the structural stiffness matrix by about q decimal digits less than the largest one, if the ratio of the largest eigenvalue λ_1 to the smallest one λ_N is about

$$\lambda_1/\lambda_N \approx 10^q$$

On the other hand, the smallest eigenvalue is predominant in the inverse (7), which can only be about as accurate as the representation of the smallest eigenvalue in K.

The most important result of these considerations is that the magnitude of the initial truncation error is completely independent of the precision and the method of the solution process, be it direct or iterative. Once the structural stiffness matrix is built, there is no possibility of recovering the information lost due to initial truncation. For poorly conditioned problems the various structural matrices should be represented and calculated from the start in double precision. In particular, the programs for calculating the elemental stiffness matrices should already use double precision on all arithmetical operations on the basic single precision input data [62].

The spectral condition number

$$\mathrm{cond}(K) = \lambda_1/\lambda_N \tag{8}$$

provides an upper bound for the loss of accuracy which has to be expected. The stiffness matrix K can be scaled by replacing the basis $\{u_i\}$ by a basis $\{d_i u_i\}$, where d_i are the elements of a diagonal matrix D. In contrast to iterative methods, scaling does not have any significant influence on the accuracy of the direct elimination methods, in fact it has none at all in the case of binary scaling [66, p. 38]. The minimum condition number [63]

$$\mathrm{cond}(K) = \inf_{D \in \mathcal{M}} \frac{\lambda_{\max}(DKD)}{\lambda_{\min}(DKD)} \tag{9}$$

thus will provide a closer upper bound than (8), as indicated by Rosanoff *et al.* [61]. $\lambda_{\max}(DKD)$ and $\lambda_{\min}(DKD)$ are, respectively, the maximal and the minimal eigenvalues of the scaled stiffness matrix DKD, and $\mathcal{M}$ is the set of all diagonal matrices with positive diagonal elements. It is true that there is no economical method available to obtain the optimal matrix of scale factors D. Simple scaling, however, according to

$$d_i = (K_{ii})^{-1/2}$$

will usually reduce the condition number to a realistic value [62].

The condition number according to (9) can be used to estimate the loss of accuracy due to rounding errors in the decomposition process [36, 39, 64]. As the rounding errors are distributed statistically, such bounds are usually extremely pessimistic. In practice the relative error will be essentially smaller than

$$m2^{-t} \operatorname{cond}(K) \qquad (10)$$

where m is the average half bandwidth, and t is the number of mantissa bits in the single precision floating point numbers [65]. If inner products are accumulated in double precision (Choleski), m should be replaced by $m^{1/2}$. As pointed out by Rosanoff et al. [61], equation (9) also provides a most useful upper bound for the relative error arising from initial truncation.

Combining (9) and (10), the number s of correct decimal places in the solution of (1) can be estimated by

$$s \geqslant p - (1 \pm \log 2^{-t}m) \log \operatorname{cond}(K)$$

where p is the number of decimal places to which K is represented in the computer.

Experience with practical applications indicates that there are cases where the roundoff error and the initial truncation error work in opposite directions [62], and that the number of correct digits can be estimated quite satisfactorily by the simpler formula

$$s \geqslant p - \log \operatorname{cond}(K) \qquad (11)$$

There are two cases where even this estimate can be too pessimistic:

1. Special loading vectors: From (1) and (7) the solution vector r can be represented as

$$r = \sum_{i=1}^{N} \lambda_i^{-1} V_i V_i^T R$$

Thus the lowest eigenvalue does not contribute to the solution at all if the loading vector R happens to be orthogonal to the eigenvector V_N:

$$V_N^T R = 0$$

This demonstrates that the displacements can be obtained to a higher accuracy when the loading vector is closely orthogonal to the eigenvectors associated with the lowest eigenvalues [62].

2. Artificial ill-conditioning: If the nonnegligible elements of the dyads $V_i V_i^T$ are not distributed in a uniform way for all i, the lowest eigenvalue may be represented in the elements of K to a higher accuracy. The important role of scaling in this case is discussed in [61].

The calculation of the condition number of the scaled stiffness matrix requires only a small percentage of the total computation time. As the condition number provides a useful means to confirm the degree of accuracy of the solution, it is a real deficiency that this feature is lacking in most of the finite-element program packages.

Practically all program packages allow for checking the validity of the solution by performing the equilibrium test

$$\Delta R = K r - R$$

where r are the displacements resulting from the solution of the load-deflection equations. If the residual forces ΔR are large, the solution is definitely incorrect. On the other hand small residuals do not imply sufficient accuracy. As demonstrated by Roy [62], the formulation of the a posteriori bound on r [66]

$$\frac{\|r - K^{-1}R\|}{\|r\|} \leqslant \frac{\|\Delta R\|\, \|K^{-1}\|}{\|r\|}$$

is inadmissible in cases where a large initial truncation error is present. Using the residual loads R for iterative improvement of the solution [67] is also wasted effort, even when calculated in double precision, because thereby only the rounding error could be reduced. The initial truncation error remains unchanged.

4. Incorporation of Different Finite-Element Types

The various types of finite elements available in structural mechanics can be divided into two major classes:

1. Element types suitable for the idealization of the basic domains over which the fields of the theory of elasticity are defined:
   (*a*) One dimensional continua (flange elements)
   (*b*) Two-dimensional continua
       (*i*) Plane stress/strain (triangles and quadrilaterals)
       (*ii*) Axisymmetric domains (ring elements with triangular or quadrilateral cross section)
   (*c*) Three-dimensional continua (tetrahedrons, pentahedrons, hexahedrons)

2. Element types derived from structural components which are themselves idealizations based upon St. Venant's principle. All elements with bending effects fall into this class (beams, plates, shells). Within this category, the possibility of including or neglecting various special effects (e.g. torsion-bending, curvature, shear strain effects) leads to a large variety of specific element models.

If the elements of group (a) and (i) are permitted to be arbitrarily situated in three-space, they also can be combined to discretize skin–skeleton structures. Together with beam, plate and shell elements they can be combined to cover a large variety of structures whereby the program requirements for the description of the idealized structure and for the representation of the results are much more complex than in the case of the analysis of continua.

For all element types of class 1 there exist displacement models which behave excellently in practical applications. In principle the development of displacement models for this class can be considered to be completed. In addition their computer implementation does not impose severe difficulties. This is probably the reason why this class of element types is available in practically all well-designed, large-scale, finite-element program packages.

The situation is much worse for the element types of class 2. In this area the development is far from being concluded. Obviously there is no model which ideally suits all applications. The most sophisticated displacement models for shell elements have displacements, and their first and second-order derivatives as nodal degrees of freedom. They have proved to be very useful for shell structures with continuously varying tangential plane. The elements of class 2 comprise the domain where equilibrium, hybrid and mixed models are frequently used. Especially the mixed models having displacements and moments as nodal parameters can be used to advantage for plate and shell structures with discontinuous variation of the tangential plane. Mixed models, however, have a severe numerical disadvantage: In general they yield a nondefinite matrix of coefficients for the linear system equations, thus rendering inapplicable all basic results of Section 3.

The most important intrinsic limitations of a program system with respect to the incorporation of new element types are as follows:

1. the maximum number of element nodes (30)
2. the maximum number of degrees of freedom per node (32)
3. the maximum dimensions of the elemental stiffness matrix (120 by 120).

The numbers in parentheses indicate a reasonable capacity which is sufficient for most element types. Furthermore, the program system should be capable of handling the following:

1. any number of elements attached to a structural node,

2. any combination of degrees of freedom at the various elemental and structural nodes, and
3. any kind of geometrical and material data which will be required for the elements.

The necessity of handling a great variety of material and geometrical data for the various element types also leads to severe difficulties in planning the basic architecture of a general program package. The same holds for the representation of the results, especially for membrane and bending elements situated arbitrarily in three-space. In this case the stresses (or stress resultants) have to be referred to a suitably chosen local coordinate system.

5. Modular Design

In the development of any large scale program system the following design objectives are worthy of attainment:

1. Versatility: The program system should be able to respond as far as possible to a broad spectrum of applications implied by the underlying mathematical model. The intrinsic limitations of the problem size should define margins which could never be exceeded by realistic problems, provided that the corresponding computer hardware is available. In modern finite-element programs this goal can be achieved by selecting an appropriate organization of the internal data.

2. Controllability: The amount and type of output as well as the extent of error checking should be controllable by the user. Also the user should be able to specify the sequence of the various procedures to be used in the solution process. While the computer often is occupied for a period of hours with regard to the solution of large problems, the user should have the option of saving intermediate results and allowing for interruption and later restart of the solution process.

3. Efficiency: Efficient utilization of main storage, backup storage and of the central processing unit(s) is a basic requirement. Any kind of parallelism allowed by hardware/software should be utilized as far as possible. Especially when operating in the multiprogramming mode the resources of the operating system should be engaged only to the extent for which they are actually needed. For future development the restriction of executable storage to 25,000–30,000 CM words seems to be realistic.

4. Expandability: The extension to unforeseen applications always necessitates further development. The addition of new or more efficient capabilities or new finite element types should be anticipated in the basic design.

5. Integrability: Large finite-element program packages should be able to be integrated into the environment of other programs which serve for the solution of thermal, dynamic, aerodynamic, geometric, economic, and even (e.g., in architecture) sociological problems.

6. Adaptability: The basic economical aspects as well as the rapid obsolescence of hardware and software require that the program is general enough to be installed on a large variety of machine configurations with different speed, capacity and operating modes. Especially the management of the internal data should be insensitive to the actual complement of I/O devices.

7. Availability: The frequency and diversity of the applications of a large program package is an important factor in estimating the level of confidence which can be credited to the correctness of its operation [11]. Of course, a high degree of adaptability is a prerequisite for widespread distribution of a program system.

These far-reaching objectives can only be approximated by a program system which is highly modular at all its levels. The possibility of substituting alternate program modules and of treating each functional capability as a separable option will also facilitate program development, testing and maintenance.

Obviously, the success of integrating various modules into a complex system will depend primarily on the degree of systemization and creativity underlying their specifications. The breakdown for the modules can be carried out in a systematic way by the famous outside-in method as described by Ross [68]. Knowledge of the natural boundaries between the various functions of the system will help to specify the highest level modules which in turn can be repeatedly decomposed into lower level modules. In this way modularization can be carried on to a quite fine degree, whereby all aspects of the modularity are ordered into hierarchical structures.

Intermodule communication must be routed exclusively via externally defined interfaces, such as subroutine arguments, control tables, control blocks, lists and self-descriptive data. Any private communication based upon special knowledge of the internal structure of the modules must be regarded as a severe violation of the principle of modularity [69].

5.1. MODULAR STRUCTURE OF THE INTERNAL DATA

Internal data comprise both control information and numerical data which are to be communicated among the program modules. All information which is actually interchanged among the highest level modules resides on direct access secondary storage and is transferred to executable storage only when needed. Figure 16 demonstrates a simple example of organizing the data transfers to and from secondary storage [10, 22]: Data arrays of

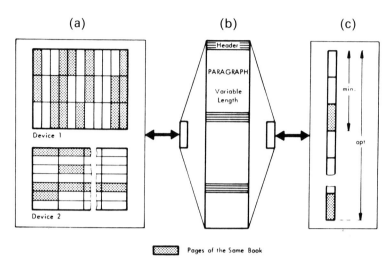

Fig. 16. Modularization of the internal data on physical (storage) level. (a) Backing storage, direct access device(s), permanent home; (b) page, fixed length record; (c) working area, one-dimensional array in executable memory.

variable length (paragraphs) are packed into records of fixed length (pages).

A book is a named collection of pages which contain data belonging to the same logical entity (e.g., hypermatrix, control list). Being quite similar to "files" or "data sets" in operating systems, books are handled as individual data packages:

1. Books are identified by alphanumeric names.
2. Books can be saved onto magnetic tape and restored in a later job.
3. All space on backing storage occupied by a book can be released.

All information necessary for properly processing the contents of a book (e.g., status, disposition, type of contents, storage format, structure of partitions) is collected in a data descriptor block (Fig. 17). Whenever a book is opened for processing, its data descriptor is brought into a specific control area in executable memory.

5.2. MODULAR STRUCTURE OF THE PROGRAM

Books serve as the basic interfaces among the highest level procedure/process-oriented modules called processors. Apart from the specific input/output processors which are required to generate books or to display their contents, a processor performs within the overall solution procedure a

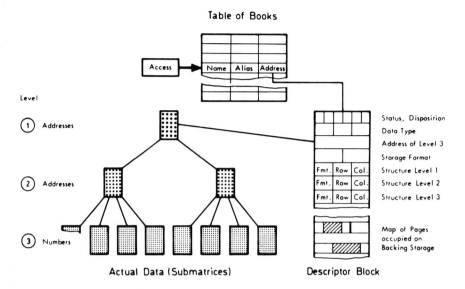

Fig. 17. Modularization of the internal data on logical level.

well-defined distinct computation step, such as generation of the elemental stiffness matrices, assembly of the structural stiffness matrix or solution of a set of linear equations (Fig. 18).

Since books are self-descriptive data structures, it is sufficient to provide the names of all the books involved when activating a processor. By analyzing the data descriptor blocks of the input books the processor can detect inconsistent or illegal input prior to entering the actual computations. New output books can be generated automatically by using the information in the data descriptor blocks of the input books. In that way it is possible to develop general processors which can be used in a very flexible way, independently of the specific overall problem. For instance a processor for matrix multiplication will multiply any two matrices submitted as input books, irrespective of their size and population pattern, so long as they are organized in a consistent (compatible) format.

In the extreme case, the processors could be regarded as self-contained programs (or even computers) which are linked together via books contained in a data base and executed in separate job steps (or jobs). More frequently, however, the processors reside in overlay segments and are activated by simple subroutine calls from the root segment.

From the programming viewpoint the processors themselves are highly modularized. Logical branches of reasonable size outside the innermost loops are put into separate overlay segments in order to reduce the core

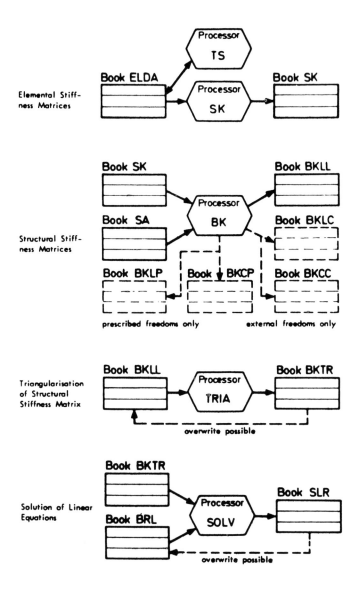

Fig. 18. Typical processors, relational level of the internal data.

storage requirements. Standard operations, such as data retrieval and book management, are performed by low level ("slave") utility modules which are either integrated into every separate processor or are commonly available through residing in the root segment.

5.3. Modularization of the Problem

The more sophisticated finite-element program packages perform extensive data checking prior to entering the actual computations. However, the number of potential input errors which thereby cannot be detected is still large enough so that their occurrence is very likely in larger problems. The preparation of the input data for complex applications of the finite element method is by no means a trivial task. On the other hand, even on the fastest third-generation computers a considerable amount of computer time is required for the solution of large problems.

In order to get as much benefit as possible out of the considerable expense of manpower and money, it must be possible to extend the solution to new loading and boundary conditions or to closely related structures without repeating the analysis from scratch. Similarly the occurrence of errors during a late stage of computation should not automatically necessitate the annulling of all intermediate results and the repetition of the analysis from the start.

These goals can be reached simultaneously in the most elegant way by modularizing the problem itself (Fig. 19): The structure to be analyzed is broken into a number of individual components called substructures. In

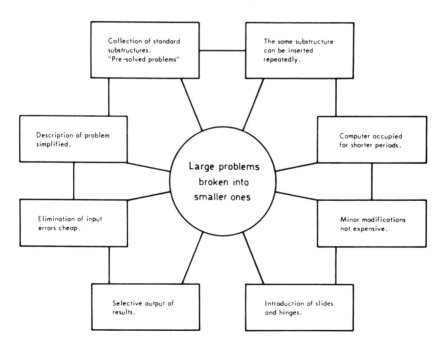

Fig. 19. Modularization of the problem by substructure analysis.

every substructure the unknown degrees of freedom which are not connected to other substructures are eliminated first. The remaining degrees of freedom are handled by inserting the substructures into a hypernet of nodal points in quite an analogous way to that in which elements are inserted into a net of nodal points. The underlying theory is very simple [1, 2, 43, 44, 82]. Substructure analysis can be regarded as a two-level Gaussian or Choleski elimination process and has the same theoretical basis as the hypermatrix method described in Section 3.2.

Conclusions about the merits of substructure analysis should not be drawn from the way it is used occasionally in finite element programs in order to overcome severe limitations of problem size imposed by poor programming techniques [70]. The real benefits of substructure analysis (Fig. 19) materialize only in program systems where suitable modularity of both program and data is provided. Substructures then can be handled as individual components which are stored as permanent data sets which can later be combined with other substructures to produce a variety of new hyperstructures. Some hyperstructures can be assembled by inserting the same substructure repeatedly, resulting in drastic savings of computational effort [10].

6. Problem Description and Representation of the Results

In the past, most of the development effort for structural analysis programs was devoted to the analytical and numerical aspects (e.g., better finite elements, more efficient equation solvers, and improved techniques for dynamic and nonlinear analysis), despite the fact that the current capabilities for generating the input parameters and evaluating the results are far from providing the optimum convenience and simplicity required from the production viewpoint [72]. This problem was already recognized in the early 1960s, and a number of attempts have been made to develop problem-oriented languages which were regarded as the most powerful means of providing more user-comfort in describing the idealized structure. However, problem-oriented languages cover only one aspect of facilitating data preparation. For instance no advantage is apparent, when users decide to write their own FORTRAN programs in order to generate the statements of a problem-oriented language, as was done occasionally in the past.

From the viewpoint of program architecture as well as of production requirements the problems of user–program interaction are very complex. Especially in large-scale general program packages, an integrated optimal solution seems not to be feasible because too many differing requirements have to be satisfied simultaneously. Also the advent of interactive computation techniques and the development of various devices for graphical display have added very powerful tools, which until now have not been utilized to their full potentiality.

6.1. PREPARATION OF THE INPUT DATA

There are three major methods of generating the input data:

(a) *Data Records Processed in Batch Mode*

The data are contained in fixed-length records (punched cards, magnetic tape), arranged in tabular form, and identified by preceding header records. On cards, free-field format and continuation cards are normally allowed. Omission of data which remain zero (e.g., certain components of load vectors) and repeated generation of regular data are desirable options. Interactive or batch mode manipulation of the data (e.g., updating, modifications, editing) can be performed externally with standard software.

(b) *Automatic Data Generators*

By their very nature program packages for computer-assisted data generation (e.g. automatic mesh generators [47]) have to be restricted to specific types of structures. When properly used within their domain of application only a minimum of parameters has to be specified by the user, allowing the bulk of data to be calculated automatically. In general further postprocessing of the automatically generated data is required in order to improve the idealization of the structure or the efficiency of the analysis (e.g. renumbering of nodal points).

(c) *User-Oriented Input Languages*

Many concepts and methods which have evolved from the development of compilers and compiler-compilers are also of great value for the basic design of problem-oriented languages [73–75]. They are either based upon a precompiler or an interpretive driver (i.e. control program). A precompiler translates the commands of the problem-oriented language into a procedural language program (e.g., FORTRAN or ALGOL). This program is processed subsequently by the appropriate compiler, linked and finally executed as a whole. All features of the underlying procedural language (e.g. arithmetic expressions, loops, branches, recursivity, data management) can be easily extended to the problem-oriented language (POL). The precompiler method, however, is not well suited for interactive operation mode on the POL-level. An interpretive driver normally decodes one statement of the user-oriented language at a time and activates immediately the appropriate sub-routines, thus allowing fast interactive operation.

In addition to the generation of the input data, another important task is data verification which has to be simplified using similar methods as those being applied for the representation of the results.

6.2. REPRESENTATION OF THE RESULTS

The sophisticated finite element program packages are a most powerful analytical tool as they permit the inclusion of a large variety of structural details into the analysis. As a consequence, the corresponding results (e.g. elemental stresses, nodal point deflections) can be extremely voluminous and have to be reduced to a concentrated form comprising only the effects which are of real interest to the engineer. There are two basic ways of displaying the results:

(a) Alphanumeric Representation

The data items are allocated in fixed length records (line printer, magnetic tape), arranged in tabular form and preceded by header records. Important features include special flagging of the dominant results as well as various options for selective output, either predefined by the user or automatically determined by the magnitude of the numbers to be printed.

(b) Graphic Representation

Generating a pattern of characters by a line printer [72] is one of the oldest "tricks" used to produce graphic representations. This method is fast and a large variety of software packages are available for that purpose. Off-line plotters are employed to generate very accurate large-scale drawings. As their operation is rather slow they serve primarily to generate plots containing a large number of details and for the final documentation of the results. The most flexible tool for handling graphic information both in batch and interactive operation mode is the cathode ray tube (scope) [76–78, 83]. Due to the limited size of the screen, the displays cannot comprise too many details. This disadvantage is outweighed by the possibility of selectively displaying information of immediate interest:

1. Plotting only selected regions of a structure, e.g., substructures, element groups, elements.
2. Removing portions of the structure ("peel-off effect"), e.g., the skin of an airplane wing.
3. Cutting the structure into separately displayed parts ("explosion views").
4. Isometric/perspective views with arbitrary selection of viewing angle/ center of projection or vanishing point ("walk-around effect").
5. Selective magnification of a region ("zoom effect").

All these options are best controlled interactively by the user, sitting in front of the scope. The user should then have the option of requesting large-scale hard-copy plots in order to document the most useful displays. The human hand is much better suited for pointing than for working on a typewriter. Thus the light pen can be used quite effectively for selecting commands by pointing to an item on a "menu" on the CRT (cathode ray tube) or for generating raw data by moving a tracking cross over the screen and optionally adjusting the less significant digits of a data item by pointing to an element of a character string.

6.3. Consequences for the General Architecture of Large-Scale Finite-Element Program Systems

The analytical sections (processors) in a finite-element program system are expensive and time-consuming to program if adequate capacity and efficient performance are to be achieved. On the other hand, considerable costs are incurred in training hundreds (or even thousands) of users to become accustomed to the necessary input and output conventions. The mathematical processors can be essentially the same in very general program packages as in highly efficient special-purpose programs, whereas special-purpose programs for data preparation and representation of results can normally provide much more user convenience than ever could be achieved by general purpose programs. Consequently the complete separation of the analysis modules from the operations which primarily process input data and results (Fig. 20) seems to be a most useful design principle.

On the computation level there is a kernel of procedural modules (processors) for solution methods and finite-element calculations, based upon a computation-oriented internal data structure. A high degree of independence from the specific kind of problems to be solved as well as from the actual hardware and software to be used can be achieved for these modules through proper design. Hence there is a payoff in devoting a great deal of development effort in order to achieve a high life expectancy (normally programs are obsolete after a few years) and to tune the various processors to maximum efficiency.

The internal data in computation-oriented format cannot be easily accessed from outside, because detailed knowledge of their structure and interrelation cannot be presumed to be possessed by the user. Thus there has to be at least one data conversion module which translates the internal data into an easily accessible, process-oriented format (Fig. 20), and vice versa [79]. All communication between the data processing satellite programs and the analysis program complex is routed exclusively via the data converter(s). The process-oriented intermediate data format is ideally suited for

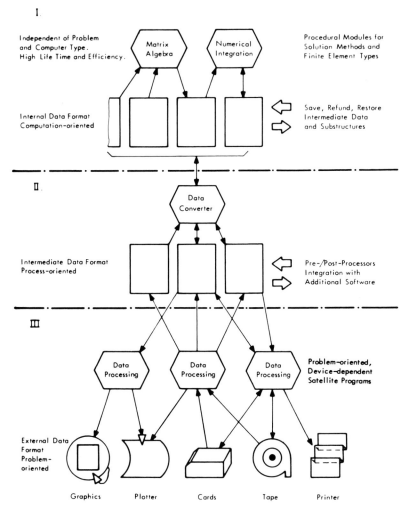

Fig. 20. Schematic layout of a large-scale finite-element program system. (I) Computation; (II) adaptation; (III) problem orientation.

linking the procedural modules to data pre- and postprocessors as well as to additional program complexes (e.g., to calculate loads, temperatures, fluid motion, or to add dynamic analysis capabilities [80]).

All the various possibilities of input creation and output evaluation will be implemented in separate data processing software packages which can be designed to specifically satisfy the requirements and needs of a given user environment. Simple problem-oriented, device-dependent satellite programs are written very rapidly. On the other hand, highly sophisticated general purpose programs for interactive graphics or input languages could be developed independently from the specific procedural modules in the analysis complex. Obsolete analysis programs could even be replaced by redesigned ones without changing the external specifications of the data processors. This allows continuous application of the same user conventions over a relatively long period of time, thus saving a remarkable amount of expense in user education. In addition, the sudden increase of user errors which inevitably occurs upon changing the external data specifications is avoided.

Acknowledgments

This paper contains many principles and ideas which evolved during the development and implementation of the ASKA program package under the direction of Prof. Dr. Dr. h.c. J. H. Argyris who, during the past decade, has always succeeded in providing the optimum environment and resources necessary for the creative work of a large team. Among the members of the ASKA team H. A. Balmer, K. Bernhardt, G. v. Fuchs, R. E. Hammer, H. M. Hilber, H. Knapp, M. König, J. R. Roy, and P. Streiner have to be mentioned explicitly for their fruitful cooperation. Also G. Mareczek provided a permanent challenge to the ASKA team in concurrently developing a program package of his own. J. R. Roy contributed to the preparation of this paper with a large number of valuable suggestions. K. Bernhardt made available Figs. 15a, b, c from his own unpublished work.

References

A. Proceedings of Conferences

A. *Symp. Sparse Matrices and Their Applications*, IBM Watson Research Center, Sept. 9–10, 1968 (Willoughby, R. A., ed.). *Sparse Matrix Proceedings*. RA-1, IBM Research Center, Yorktown Heights, New York, 1969.

B. *Conf. Inst. Math. and Its Applications*, St. Catherine's College, Oxford, April 5–8, 1970 (Reid, J. K., ed.), *Large Sparse Sets of Linear Equations*. Academic Press, New York, 1971.

C. *Sem. Meth. finiten Elemente*, Stuttgart, April 27–28 (Buck, Scharpf, Stein, and Wunderlich, eds.), *Finite Elemente in der Statik*. W. Ernst und Sohn, Munich, 1973.

D. *IUTAM Symp. High Speed Computing of Elastic Structures*, Liège, August 24–28, 1970. *Congrès et Colloques de l'Université de Liège*, Vol. 69. Place du XX Aout 16, Liège, Belgium, 1971.

E. *ASME/ACM 1969 Comput. Conf.* IIT Chicago, June 19–20, 1969. (Sevin, E., ed.), *Computational Approaches in Applied Mechanics*, ASME 345. East 47th Street, New York, 1969.

F. *ASME 91st Winter Ann. Meeting, New York* November 30, 1970, *Pressure Vessels and Piping Division*. (Marcal, P. V., ed.). On General Purpose Finite Element Computer Programs, ASME, 345 East 47th Street, New York, 1970.

G. *ASME 2nd Pressure Vessels Piping Conf., Denver* September 16, 1970 (Berman, I., ed.), *Computer Software in Structural Analysis*. ASME, 345 East 47th Street, New York, 1970.

H. *AFIPS Spring Joint Comput. Conf.*, *Boston* May 14–16, 1969. *AFIPS Conf. Proc.* **34**. AFIPS Press, 210 Summit Avenue, Montvale, New Jersey 07645, 1969.

I. *AFIPS Fall Joint Comput. Conf. Las Vegas*, November 18–20, 1969. *AFIPS Conf. Proc.* **35**. AFIPS Press Montvale, 1969.

J. COINS-69, *Int. Symp. Comput. Information Sci. 3rd, Miami Beach*, Dec. 18–20, 1969. (Tou, J. T., ed.). *Software Eng.*, Academic Press, New York, 1970.

B. Books and Articles

1. Argyris, J. H., "Energy Theorems and Structural Analysis," *Aircraft Eng.* **26** (1954)/**27** (1956). (Also as book: Butterworths, London, 1960.)

2. Argyris, J. H., *Recent Advances in Matrix Methods of Structural Analysis*. Pergamon Press, London, 1964.

3. Argyris, J. H., "The Impact of the Digital Computer on Engineering Sciences, 12th Lancaster Memorial Lecture," *Aeronaut. J. Roy. Aeronaut. Soc.* **74**, 13–41, 111–127 (1970).

4. Tocher, J. L., "Selective Inversion of Stiffness Matrices," *ASCE J. Struct. Div.* **92**, 75–88 (1966).

5. Rosanoff, R., "A Survey of Modern Nonsense as Applied to Matrix Computations," *Proc. AIAA/ASME Struct., Struct. Dynam. Mater. Conf.* (April 1969).

6. Branin, F. H., Jr., "Computer Methods of Network Analysis," *Proc. IEEE* **55**, 1787–1801 (1967).

7. Fenves, S. J., and Branin, F. H., Jr., "A Network-Topological Formulation of Structural Analysis," *ASCE J. Struct. Div.* **89**, 483–514 (1963).

8. Gallagher, R. H., "Large-Scale Computer Programs for Structural Analysis," [F], pp. 1–14.

9. MacNeal, R. H., and McCormick, C. W., "The NASTRAN Computer Program for Structural Analysis," SAE Natl. Aeronaut. Space Engn. Manufacturing Meeting, Los Angeles (October 1969).

10. Schrem, E., and Roy, J. R., "An Automatic System for Kinematic Analysis," ASKA Part I [D], pp. 477–507.

11. Griffin, D. S., "The Verification and Acceptance of Computer Programs for Design Analysis" [F], pp. 143–150.

12. Ward, J. A., Bemer, R. W., Gosden, J. A., Hopper, G. M., Morenoff, E., and Sable, J. D., "Software Transferability" [H], pp. 605–612.

12a. Gosden, J. A., "Software Compatibility: What was Promised, What We Have, What We Need," *AFIPS Conf. Proc.* **33**, 81–87 (1968).

13. Spencer, D. D., *Programming with USA Standard FORTRAN and FORTRAN IV*. Ginn (Blaisdell), Boston, Massachusetts, 1969.

14. ANSI, "FORTRAN Standard X 3.9-1966," Amer. Nat. Std. Inst., New York, 1966.

15. Rechenberg, P., and Seyferth, A., "Rekursive Unterprogramme in FORTRAN," *Elektron. Datenverarbeitung* **12**, 208–215 (1970).

16. Almond, J. C. (ed.), "FORMAT, a general purpose matrix analysis system in FORTRAN," RRZ Rep. 7, Regionales Rechenzentrum am Inst. f. Statik u. Dynamik d. Luft- und Raumfahrtkonstr., Univ. Stuttgart, Rev. A (1971).

17. "System 360 Matrix Language MATLAN (360A-CM-05X)," Program Description Manual, IBM Publ. H20-0564-0 (1968).

18. Skagestein, G. M., "Rekursiv unterteilte Matrizen und Methoden zur Erstellung von Rechenprogrammen für ihre Verarbeitung," Dr.-Ing. thesis, Univ. Stuttgart (1972).

19. Bayer, R., and Witzgall, C., "Some Complete Calculi for Matrices," *Comm. ACM* **13**, 223–237 (1970).

20. Roos, D., *ICES System Design*. MIT Press, Cambridge, Massachusetts, 1966.
21. Hammer, R. E., "DRS Manual," ISD Rep. 90, Inst. f. Statik u. Dynamik d. Luft- u. Raumfahrtkonstr., Univ. Stuttgart (1970).
22. Schrem, E., Die Konzipierung eines allgemeinen Rechenprogramms für die Anwendung der Methode der finiten Elemente, [C], pp. 302–320.
23. Klotz, L. H., "On the Application of ICES and Universal Program Software Systems," *ASCE* Civil Eng, pp. 72–77 (1969).
24. Melliar-Smith, P. M., "A Design for a Fast Computer for Scientific Calculations" [1], pp. 201–208.
25. Brönlund, O. E., "Eigenvalues of Large Matrices," *Int. Symp. Finite Element Techniques Int. Ship Struct. Congr.*, *4th ISD Stuttgart*, June 10–12, 1969. (Soerensen, M., ed.), Finite Element Techniques, Inst. für Statik und Dynamik der Luft- und Raumfahrtkonstr., Univ. Stuttgart (1969).
26. Jennings, A., "A Sparse Matrix Scheme for the Computer Analysis of Structures," *Int. J. Comput. Math.* **2**, 1–21 (1968).
27. Larcombe, M. H. E., "A List Processing Approach to the Solution of Large Sparse Sets of Matrix Equations and the Factorization of the Overall Matrix" [B], pp. 25–40.
28. McCormick, C. W., "Application of Partially Banded Matrix Methods to Structural Analysis" [A].
29. McCormick, C. W., "Sparse Matrix Methods in Structural Analysis," *ASME 91st Winter Ann. Meeting*, Nov. 29–Dec. 3, 1970, New York.
30. McKellar, A. C., and Coffmann, E. G., "Organising Matrices and Matrix Operations for Paged Memory Systems," *Comm. ACM* **12**, 153–165 (1969).
31. Forsythe, G. E., "Today's Computational Methods of Linear Algebra," *SIAM Rev.* **9**, 489–515 (1967).
32. Klyuyev, V. V., and Kokovkin-Shcherbak, N. I., "On the Minimization of the Number of Arithmetic Operations for the Solution of Linear Algebraic Systems of Equations" (Tee, G. I., transl.), Tech. Rep. CS 24, Comput. Sci. Dept., Stanford Univ., Stanford California (1965).
33. Winograd, S., "A New Algorithm for Inner Product," *IEEE Trans.* **C-17**, 693–694 (1968).
34. Brent, R. P., "Error Analysis of Algorithms for Matrix Multiplication and Triangular Decomposition using Winograd's Identity," *Numer. Math.* **16**, 145–156 (1970).
35. Zollenkopf, K., "Bifactorisation—Basic Computational Algorithm and Programming Techniques" [B], pp. 75–96.
36. Martin, R. S., Peters, G., and Wilkinson, J. H., "Symmetric Decomposition of a Positive Definite Matrix, *Numer. Math.* **7**, 362–383 (1965).
37. Jensen, H. G., and Parks, G. A., "Efficient Solutions for Linear Matrix Equations," *ASCE J. Struct. Div.* **96**, 49–64 (1970).
38. Tezcan, S. S., "Computer Analysis of Plane and Space Structure," *ASCE J. Struct. Div.* **92**, 143–173 (1966).
39. Martin, R. S., and Wilkinson, J. H., "Symmetric Decomposition of Positive Definite Band Matrices," *Numer. Math.* **7**, 355–361 (1965).
40. Jennings, A., and Tuff, A. D., "A Direct Method for the Solution of Large Sparse Symmetric Simultaneous Equations" [B], pp. 97–104.
41. Irons, B. M., "A Frontal Solution Program for Finite Element Analysis," *Int. J. Numer. Methods Eng.* **2**, 5–32 (1970). Discussion by Hellen, T. K., *ibid.* **2**, 149 (1971).
42. Melosh, R. J., and Bamford, R. M., "Efficient Solution of Load-Deflection Equations," *ASCE J. Struct. Div.* **95**, 661–676 (1969). Discussions by J. S. Campbell, J. R. Roy, and R. D. Cook in ST 12 (1969), by V. B. Venkayya, and B. M. Irons in ST 1 (1970), by S. S. Tezcan, G. Kostro, C. A. Felippa, and J. L. Tocher in ST 2 (1970), by S. Klein in ST 5 (1970), Closure in ST 2 (1971).

43. Przemieniecki, J. S., "Matrix Structural Analysis of Substructures," *AIAA J.* **1**, 138–147 (1963).
44. Rosen, R., and Rubinstein, M. F., "Substructure Analysis by Matrix Decomposition," *ASCE J. Struct. Div.* **96**, 663–670 (1970). Discussions by T. S. Tarpy and W. H. Rowan, *ibid.* **96**, 2249–2251 (1970). N.-E. Wiberg, *ibid.* **97**, 724–726 (1971).
45. von Fuchs, G., and Roy, J. R., "Solution of the Stiffness Matrix Equations in ASKA," Rep. No. 50, Inst. für Statik und Dynamik der Luft- und Raumfahrtkonstruktionen, Univ. Stuttgart (1968).
46. Irons, B. M., "Discussion to 42," *ASCE J. Struct. Div.* **96**, 152–153 (1970).
47. Kamel, H. A., and Eisenstein, H. K., "Automatic Mesh Generation in Two and Three Dimensional Inter-connected Domains [D], pp. 455–475.
48. Alway, G. G., and Martin, O. W., "An Algorithm for Reducing the Bandwidth of a Matrix of Symmetric Configuration," *Computer J.* **8**, 264–272 (1965).
49. Rosen, R., "Matrix Bandwidth Minimization," *Proc. Nat. Conf. ACM, 23rd,* Publ. P-68, pp. 585–595. Brandon Systems Press, Princeton, New Jersey, 1968.
50. Barlow, J., and Marples, C. G., Comment on "Automatic Node-Relabelling Scheme for Bandwidth Minimization of Stiffness Matrices," *AIAA J.* **7**, 380–381 (1969).
51. Tewarson, R. P., "Sorting and Ordering Sparse Linear Systems" [B], pp. 151–167.
52. Spillers, W. R., and Hickerson, N., "Optimal Elimination for Sparse Symmetric Systems as a Graph Problem," *Quart. Appl. Math.* **26**, 425–432 (1968).
53. Tinney, W. F., and Walker, J. W., "Solutions of Sparse Network Equations by Optimally Ordered Triangular Factorization," *Proc. IEEE* **55**, 1801–1809 (1967).
54. Bree, D., Jr., "Some Remarks on the Application of Graph Theory to the Solution of Sparse Systems of Linear Equations." Thesis, Dept. of Math., Princeton Univ., Princeton, New Jersey (May 1965).
55. Varga, R. S., *Matrix Iterative Analysis.* Prentice Hall, Englewood Cliffs, New Jersey, 1962.
56. Fried, I., "A Gradient Computational Procedure for the Solution of Large Problems Arising from the Finite Element Discretization Method," *Int. J. Numer. Methods Eng.* **2**, 477–494 (1970).
57. Fox, R. L., and Stanton, E. L., "Developments in Structural Analysis by Direct Energy Minimization," *AIAA J.* **6**, 1063–1042 (1968).
58. Reid, J. K., "On the Method of Conjugate Gradients for the Solution of Large Sparse Systems of Linear Equations" [B], pp. 231–254.
59. Matula, D. W., "Towards an Abstract Mathematical Theory of Floating-Point Arithmetic" [H], pp. 765–772.
60. Wilkinson, J. H., *Rounding Errors in Algebraic Processes.* H.M. Stationery Office, London, 1963.
61. Rosanoff, R. A., Gloudeman, J. F., and Levi, S., "Numerical Conditions of Stiffness Matrix Formulations for Frame Structures," *Proc. Conf. Matrix Methods Struct. Mech.* AFFDL-TR-68-140, Wright–Patterson Air Force Base, Ohio (1968).
62. Roy, J. R., "Numerical Error in Structural Solutions," *ASCE J. Struct. Div.* **97**, 1039–1054 (1971).
63. Bauer, F. L., "Optimally Scaled Matrices," *Numer. Math.* **5**, 73–87 (1963).
64. Wilkinson, J. H., *The Algebraic Eigenvalue Problem.* Oxford Univ. Press (Clarendon) London and New York, 1965.
65. Felippa, C. A., and Tocher, J. L., "Discussion to 42," *ASCE J. Struct. Div.* **96**, 422–426 (1970).
66. Forsythe, G., and Moler, C. B., *Computer Solution of Linear Algebraic Systems.* Prentice-Hall, Englewood Cliffs, New Jersey, 1967.
67. Martin, R. S., Peters, G., and Wilkinson, J. H., "Iterative Refinement of the Solution of a Positive Definite System of Equations," *Numer. Math.* **8**, 203–216 (1966).

68. Ross, D. T., "Uniform Referents: An Essential Property for a Software Engineering Language" [J], pp. 91–101.
69. Trapnell, F. M., "A Systematic Approach to the Development of System Programs" [H], pp. 411–418.
70. Meissner, C. J., "A Multiple Coupling Algorithm for the Stiffness Method of Structural Analysis," *AIAA J.* **6**, pp. 2184–2185 (1968).
71. Bremer, R. W., "Manageable Software Engineering" [J], pp. 121–138.
72. Tocher, J. L., and Felippa, C. A., "Computer Graphics Applied to Production Structural Analysis" [D], pp. 521–545.
73. Fenves, S. J., "Problem-Oriented Languages" [E], pp. 1–9.
74. Newman, W. M., "A System for Interactive Graphical Programming," *AFIPS Conf. Proc.* **32**, 47–54 (1968).
75. Murphree, E. L., and Fenves, S. J., "A Technique for Generating Interpretive Translators for Problem-Oriented Languages," *BIT* **10**, 310–323 (1970).
76. Batdorf, W. J., Kapur, S. S., and Sayer, R. B., "The Role of Computer Graphics in the Structural Design Process," *AIAA Ann. Meeting Tech. Display, 5th, Philadelphia*, Oct., 21–24, 1968, AIAA Paper No. 68-1012.
77. Butlin, G. A., "Man-Machine Interactive Structural Analysis as a Preliminary Design Aid," *29th Meeting Struct. Mater. Panel AGARD, Istanbul* Sept. 28–Oct. 8, 1969.
78. Grieger, I., "Über den Einsatz von Bildschirmgeräten bei der Tragwerksberechnung," Dr.-Ing. Univ. of Stuttgart (1972).
79. Bernhardt, K., and Streiner, P., "GIO—a Set of ASKA Processors Providing Direct Access to Internal Data," ASKA PM 111, Inst. für Statik und Dynamik der Luft- und Raumfahrtkonstruktionen, Univ. Stuttgart (1971).
80. Brönlung, O. E., Bühlmeier, J., Dietrich, G., Frik, G., Johnsen, T. L., Kiesbauer, H. T., Malejannakis, G. A., and Straub, K., "Dynan User's Reference Manual, ISD-Rep. 97, Inst. für Statik und Dynamik der Luft- und Raumfahrtkonstruktionen, Univ. of Stuttgart (July 1971).
81. Galligani, I., "Numerical Considerations on the Finite Element Method," *Conf. Finite Element Methods Stress Anal. Problems Nucl. Eng.*, Ispra (Italy) (June 30–July 1, 1971) (Proceedings forthcoming).
82. Argyris, J. H., and Kelsey, S., "Modern Fuselage Analysis and the Elastic Aircraft," *Aircraft Eng.* **31** (1959); **33** (1961); also as book: Butterworths, London and Washington, D.C., 1963.
83. Argyris, J. H., Grieger, I., and Schrem, E., "Structural Analysis by Problem-Oriented Languages," *Aeronaut. Soc. India Ann. General Meeting, 23rd*, Kanpur-16 (India), February 26–28, 1971.

PART II

Critical Review of General-Purpose Structural Mechanics Programs

Review of the ASKA Program

P. Meijers

INSTITUTE TNO FOR MECHANICAL CONSTRUCTIONS
DELFT, THE NETHERLANDS

ASKA is a software package for structural analysis. The first part is for static, linear, finite-element analysis; it is connected with a package for dynamic analysis DYNAN. The element types available in the static ASKA package are flange, membrane, plate, shell, ring, three-dimensional, and beam. The input is a topological description which fixes, for example, the net of nodal points and the element types, followed by blocks of physical data. The programming is reduced to calling a set of highest-order subroutines, each carrying out a set of operations on all elements involved. The output either provides us with stresses and displacements, or a set of matrices for dynamic analysis. DYNAN operates on data from the static ASKA package or on matrices put in directly. It contains eigenvalue routines and programs for response calculations. For dynamic analysis, the program to be written consists of FORTRAN subroutine and function calls. A few examples give information on computation time and accuracy, and finally an overall user viewpoint is given.

1. Introduction

ASKA (Automatic System for Kinematic Analysis) is a software package for structural analysis based on the finite-element displacement method. It is being developed by Argyris and his co-workers at the Institut für Statik und Dynamik der Luft- und Raumfahrtkonstruktionen, University of

Stuttgart (ISD). The final version of ASKA will contain a static, a dynamic and a nonlinear part. The almost machine-independent version of the static and the dynamic parts are now available, the nonlinear part is expected to be available in 1972.

In this paper we intend to give an objective review on ASKA from a user's viewpoint, but no details on data-retrieval system, etc. For a general elucidation on this subject, including the system applied in ASKA, we refer to Schrem [1].

Sections 2–5 are devoted to ASKA version 4.1, which is the version without the routines for dynamic analysis recently built in. First, we shall give some general information on the ASKA program followed by a review of the element library currently built-in. This element library gives a good indication as to whether or not a certain problem can be solved efficiently with the ASKA program. In subsequent paragraphs, we shall discuss how ASKA jobs should be prepared, as well as some special features of ASKA: substructuring, saving and error messages, etc.

The dynamic part, which has been available only since July 1971, and with which we therefore have little experience, is discussed in Section 6. It is not known to the present author precisely what the nonlinear part of ASKA will contain. In ASKA elastic–plastic analysis will be possible for membrane, ring, and three-dimensional elements, but certainly not for all element types. Large displacement and strain analysis will be built-in for a few elements, but no details are available as yet. It is not to be expected that, in the near future, ASKA will become an important program for nonlinear analysis.

The final section is devoted to an overall discussion on the pros and cons of the ASKA program.

2. General Information

Most of the information given here on ASKA is already available in papers by Argyris and his co-workers [2–4]; we shall summarize this matter to make the present paper self-contained.

Apart from a very small number of subroutines ASKA 4.1 is machine-independent. It can efficiently be implemented only on large computer systems such as IBM 360–65/75, CDC 6600 and Univac 1108. The core size for IBM 360-65 should preferably be not less than 512 kbytes, and direct access backing store for large problems should be about 10,000,000 computer words.

ASKA aims [2] to be an automatic system for practical engineers to solve their stress and vibration problems without using additional programs.

Moreover, due to its generality and modular design, ASKA aims to be a research tool that can be used in combination with user programs and is suited to support development of new elements and new solution techniques.

The ASKA program consists of so-called segments, which can be loaded into core during execution. These segments are divided into a root segment, which always remains in core, primary and secondary segments, which when necessary are loaded into core, and overlay other segments.

The data blocks are packed into so-called pages which are data blocks of a fixed length. The pages are directly accessible; when needed they are brought into a core working area. The rest of the working area is available for dynamic allocation of temporary data blocks. The data transfer and the dynamic data allocation of temporary data is performed by a set of subroutines called data retrieval system.

For large problems the structural matrices do not completely fit in core. Therefore a row- and column-wise partitioning of hypermatrices is built in.

For the calculations one can choose between single and double precision. When one chooses single precision, all the calculations are carried out in single precision. When a double precision level is chosen, all the operations for which it may be important are carried out in double precision. Since the computation time for a double precision run on IBM computers is mostly in the order of 15% more than for single precision, it is advisable to do all the calculations on these computers in double precision.

3. Element Library

The total number of element types currently installed is 41. They can be subdivided into flange elements (2), membrane elements (7), plate bending elements (3), shell elements (3), ring elements (3), beam elements (9), and three-dimensional elements (14). For the names of the elements we have adopted the ones used by Argyris et al. [2–4].

For a detailed description of the elements developed by Argyris and his co-workers, we refer to the numerous publications of ISD, Stuttgart. The currently installed elements [2] are given in Figs. 1–3.

3.1. FLANGE ELEMENTS

The degrees of freedom of both FLA elements are the displacements at the nodal points. The displacements vary linearly along the flange for a FLA 2 element and quadratically along a FLA 3 element. The elements may be tapered. The standard stress output of ASKA gives the normal force at the endpoints of the elements.

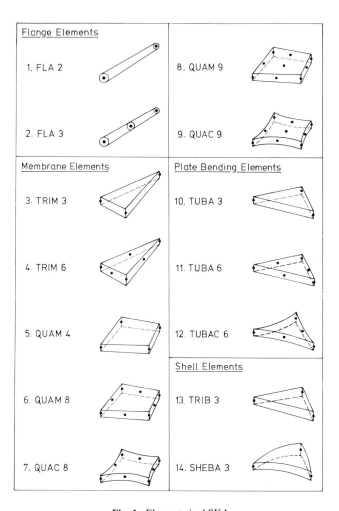

Fig. 1. Elements in ASKA.

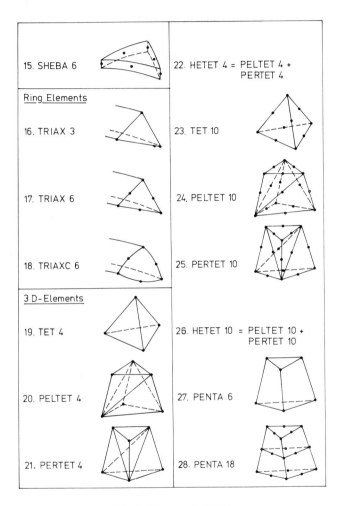

Fig. 2. Elements in ASKA.

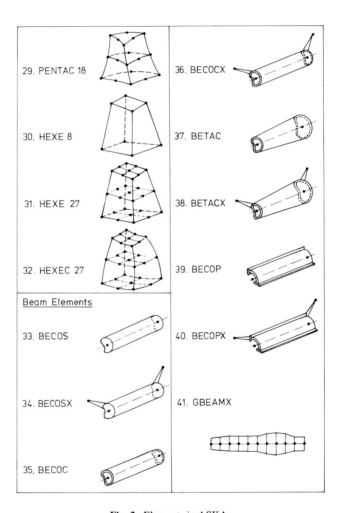

Fig. 3. Elements in ASKA.

3.2. MEMBRANE ELEMENTS

The triangular membrane elements are TRIM 3, for which the strains are constant, and TRIM 6 with linearly varying strains. Quadrilateral element QUAM 4 has eight independent in-plane freedoms, and QUAM 9 is comparable with two TRIM 6 elements. QUAM 8 is obtained from QUAM 9 by eliminating the internal node. Curved elements QUAC 8 and QUAC 9 have parabolically curved edges, but are otherwise similar to QUAM 8 and QUAM 9, respectively. The plate thickness may be variable for all elements. The computer interpolates between the thickness at the corner points. In ASKA many element types can be combined, but when combining elements FLA 3, QUAM 8, and QUAM 9 the displacements are fully compatible. The same holds for FLA 2 and TRIM 3. A comparison of the different membrane elements is given in Section 7. The standard stress output gives the stresses at the corner points of the elements in a local, element dependent coordinate system.

3.3. PLATE-BENDING ELEMENTS

Three plate-bending elements are available in ASKA. TUBA 6 is the well-known, fully compatible element [5–7] for which the displacement field is a complete fifth-order polynomial and the degrees of freedom are the normal displacement, the first- and second-derivatives at the corner points and the normal derivatives at the midpoints of the sides. TUBAC 6 is similar to TUBA 6 but it has parabolically curved sides. The displacement field for TUBA 3 is an incomplete fifth-order polynomial, the degrees of freedom at the corner points are the same as for TUBA 6. The plate may be solid or a sandwich. The thickness of the plate may vary linearly over the element. The standard stress output shows, for each element, the moments per unit length at the corner points.

3.4. SHELL ELEMENTS

TRIB 3 is a flat shell element. The degrees of freedom are the displacements and rotations at the corner points. The membrane part of the displacements is equal to that of TRIM 3 and hence a fairly fine grid will be necessary for accurate shell analysis. The standard stress output shows, for each element, the stress components at the center of gravity of the top and bottom surface in a local coordinate system [8].

SHEBA 6 has complete fifth-order polynomials for the three translations and hence 63 degrees of freedom. For the corner points, the degrees of freedom are translations and the first- and second-order derivatives. For the midpoints of the sides, the degrees of freedom are the normal derivatives of the displacements. SHEBA 3 has incomplete fifth-order polynomials for the displacements. The degrees of freedom of the corner points are the same as for SHEBA 6. The standard stress output gives, for each element, the in-plane forces and the moments per unit length at the corner nodes in a local coordinate system.

The shell elements may have variable thickness and are, in general, sandwich elements.

3.5. Axisymmetric Elements

For axisymmetric structures, loaded axisymmetrically, three ring elements are available. TRIAX 3 has a linearly varying radial and axial displacement field, whereas for TRIAX 6 these displacement functions are complete second-order polynomials. TRIAXC 6 is similar to TRIAX 6, but the cross section has parabolically curved sides. The standard stress output gives the four stress components at the corner points of a cross section.

3.6. Three-Dimensional Elements

TET 4 is a constant stress element and TET 10 can give an exact description of linearly varying stress distributions. PERTET and PELTET elements are pentahedronal macroelements composed of three TET elements, whereas HETET elements are hexahedronal ones composed of six TET elements. Another group of pentahedronal and hexahedronal elements using Lagrangian interpolation polynomials for the displacements between the nodes are the PENTA and HEXE elements. These elements may have plane surfaces or parabolically curved surfaces (PENTAC 18, HEXEC 27).

For a discussion on the accuracy in relation to computation time see Section 7. For all three-dimensional elements the standard stress output gives the stress components at the corner points.

3.7. Beam Elements

BECOS and BECOSX elements are solid beam elements of constant cross section for which shear deformation is neglected. The degrees of freedom are the displacements and the rotations at the nodal points. For BECOS elements the nodes lie at the centroid of the end cross sections; these nodes are placed excentrically for BECOSX BECOC(X) elements are closed, thin-walled elements of constant cross section for which the shear deformation is taken into account. BETAC(X) is similar to BECOC(X), but tapered. BECOP(X) elements are to be used for the analysis of open, thin-walled beams of constant cross section; warping is taken into account (Vlassov-theory). Finally, the general beam element GBEAMX is a closed, thin-walled element of variable cross section connected to eccentric nodal points.

The cross section properties are given at 10 equally spaced stations. This element is a generalization of the BETACX element. For all beam elements, the standard stress output gives the normal forces, the shear forces and the moments at the centroid of the end cross sections and, for BECOP(X), moreover the bimoment at these points.

4. Preparation of an ASKA Job

For every ASKA job the user has to prepare a topological description, physical data, and a steering program called the ASKA processor control program. To all this, a number of installation dependent cards have to be added for compilation and link editing of selected load modules from the ASKA library.

4.1. TOPOLOGICAL DESCRIPTION

The topological description specifies the following:

1. nets of nodal points;
2. groups of elements, and net nodal points connected to each element;
3. freedoms suppressed, prescribed, and, in case of substructures, the freedoms connected to the main net;
4. points for which a local coordinate system is prescribed and points that are renumbered to obtain a smaller bandwidth in the structural stiffness matrix;
5. for substructures, the relation between the substructure nodes and the main net nodes.

Loops in the topological description of a regular structure can be made with help of a so-called topological variable and a repetition feature. A constant, a linear and a quadratic topological variable have been defined [8]. Constant topological variable (i) specifies integer i, linear topological variable (i, j)

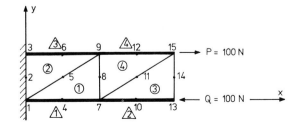

Fig. 4. Simple example. $\triangle$ = group 1, $\bigcirc$ = group 2.

specifies integers $i, i + j, i + 2j, \ldots$ and quadratic topological variable (i, j, k) specifies integers $i, i + j, i + 2j + k, i + 3j + 3k, i + 4j + 6k, \ldots$.

The first level of repetition R causes repetition of a certain statement in the topological description with modified starting values. The second level of repetition RR causes repetition of the basic statement and that of the first level of repetition.

The whole can perhaps be most easily understood with reference to a simple example. The structure in Fig. 4 is divided into a group of four FLA 3 elements and a group of four TRIM 6 elements. The topological description for ASKA could here be as follows:

```
TOPOLOGICAL DESCRIPTION
NET (1) (15)                      —net number 1, 15 nodal points
FLA 3(1) (2) (1,6) (4,6) (7,6)    —group number 1, 2 elements, nodes 1, 4, 7 and 7, 10, 13
R (2) (0) (2) (2) (2)             —2 loops, number of elements equal for both loops, starting
                                     values increased by 2
TRIM 6 (2) (2) (1,8) (4,2) (7,–4) (8,–6) (9,–8) (5,0)
R (2) (0) (6) (6) (6) (6) (6) (6)
SUPPRESS (1,2) (3) (1,1)          —suppress displacements in x and y directions at nodes 1, 2
                                     and 3
SUPPRESS (3) (15) (1,1)           —suppress displacements normal to plate at all nodes
END NET
END TOPOLOGICAL DESCRIPTION
```

The mainstream of the topological description is by cards, but it is also possible to read information on the topological description from tape, disk, or drum.

4.2. DATA INPUT

The input data consist of all the necessary information except the topological description: nodal point coordinates, presecribed loads, prescribed displacements, element properties (e.g., plate thickness for membrane elements), material properties, etc.

The input is either by cards or by tape. The data tape may be a tape of input data generated in a prerun by the user, or a tape generated by processor DATEX. The data input for the simple example (Fig. 4) could be:

```
$ DATA
$ NPCO  N = 1  C = 2
          1     0.     0.     ⎫
          3     0.    10.     ⎪
          7    20.     0.     ⎪
          9    20.    10.     ⎬  coordinates of nodal points (x, y)
         13    40.     0.     ⎪
         15    40.    10.     ⎭
```

```
$ NPBR N = 1 C = 1 L = 1
        13   − 100.           ⎫
        15   + 100.           ⎬ nodal point forces
                              ⎭
$ GEDA N = 1 G = 1 E = A C = 3    geometrical data FLA elements
         1  10.  10.  10.          (cross-section at nodal points)
$ GEDA N = 1 G = 2 E = A C = 3    geometrical data TRIM elements
         1   1.   1.   1.          (plate thickness at corner points)
$ EMOD N = 1 C = 2 G = A
         1  2.1E6  0.3           elastic constants
$ FIN                            end of data input.
```

The data are divided into blocks, each containing data of a certain type; each block starts with a header card giving the information necessary for identification. The first block for the example shows the nodal point coordinates for the only net, number 1. The header card shows there are two columns. The row description of the data block gives the nodal point numbers, and the block gives, respectively, the x- and y-coordinate. All data that are not specified are assumed to be zero. When required, the data can be divided into files and subfiles by delimiter cards.

4.3. ASKA-Processor Control Program

The programming is reduced to calling a series of highest-order subroutines (processors). This program is called the ASKA processor control program. Each processor carries out a set of operations on data or hypermatrices. The input and output of the processors is a book or a number of books; here books may be defined as: logical units of data blocks.

For example, processor SK, which produces all the elemental stiffness matrices, needs as input book ELDA, which contains all the element data, and it produces book SK.

For the sample problem, the processor control program could be:

CALL START (1, 1)	—start program, number of loading cases 1, all calculations in double precision
CALL SA	—evaluate topological description
CALL DATIN (0, 4HFIN ⊔)	—read data from card till delimiter FIN, ⊔ = blank
CALL INFEL	—print element information
CALL INFUNK	—print information on type of freedoms for each node
CALL ELCO	—prepare and store element coordinates
CALL TS	—check all element data that are independent of loading-case
CALL SK	—calculate elemental stiffness matrices
CALL BK	—assemble structural stiffness matrix
CALL BR	—prepare loading matrix
CALL SR	—solve set of linear equations
CALL USR	—convert displacement matrix into user format
CALL DATEX (0, 4HUSR ⊔)	—print displacement matrix

```
CALL  SP                      —rearrange displacements elementwise
CALL  BP                      —calculate element forces
CALL  BRR                     —assemble resulting forces
CALL  UBRR                    —convert resulting forces into user format
CALL  DATEX (0, 4HUBRR)       —print resulting forces
CALL  ST                      —calculate elemental stresses
CALL  SIGEX (0, 0)            —print stresses for all the loading cases
CALL  NPST                    —calculate average stresses at nodal points
CALL  DATEX (0, 4HNPST)       —print average stresses
END                           —stop calculations.
```

We see that several processors are available for testing. When the program has been fully tested, these processor-calls could be omitted.

For most static problems, this processor control program can remain unchanged. Sometimes a few calls must be added, for example, to take into account initial strains or to determine the kinematically consistent nodal point loads from a prescribed distributed load.

For more advanced use of ASKA, a large number of additional processors are available. For example, for substructuring, saving, testing for special output, etc. These features will be discussed in the next section. Although the ASKA processor control program becomes more complicated, it makes ASKA more flexible.

4.4. Data Output

Besides information on input and error messages, the standard output for a static analysis is, normally for each loading case, the displacement-vector, the elemental stresses or the average stresses at the nodal points and when required the reaction forces. All the output except the elemental stresses is generated by a call to processor DATEX; it prints all the data blocks of one type complete with header for the net currently being processed. When not all the output is required, a selection can be made.

The element stresses are outputted via a special processor, SIGEX; each element type has its own format, which is self-explanatory. It contains, for each element and each loading case, a data block with header for identification and a row descriptor being the point number.

Output is via a printer, or on tape; the output via DATEX is, in the correct format, to be read as input in a subsequent run. No plot routines for plotting of output results are available in ASKA, although for interpretation they can hardly be dispensed with.

5. Special Features in ASKA

5.1. Saving

Especially for large programs, it is important that books in ASKA can be put on tape in internal format and can be reinput afterward. In case of a computer stop, all the books saved on tape remain intact; this, moreover makes it possible to break a long running program into pieces in order to prevent blocking of the computer for a long time.

5.2. SUBSTRUCTURING

Another important feature in ASKA is the possibility of substructuring; this can have certain advantages. When the substructures are identical, several operations have to be carried out only for one substructure and the results may be copied for the rest. This can reduce computation time. Another more general advantage of substructures is that a large structure can be cut into parts, and each of these parts prepared separately. When the books prepared are saved one can correct input errors locally. Similarly a structure which is locally modified with respects to a former design can be analyzed with far less cost.

Another point which can make substructuring of interest, is that for substructures which are often applied in structures the substructure solutions may be stored in a data-library.

A point that makes substructuring absolutely necessary in the new versions of ASKA is the fact that it is the only way to uncouple certain freedoms at a node, for example: to introduce a hinge or sliding.

From the example discussed in Section 7 it follows that it is not always advisable to apply substructuring, since computation time may increase and the costs for excps (execute channel programs) will certainly rise.

5.3. NEW ELEMENTS

ISD has provided everything to make it possible for the user to build in his own elements with help of the so-called VIRGO processors.

With the documentation provided in the programmers manual this could indeed be performed. Of course, newly built-in elements have the same restrictions as have the elements currently built-in. The number of nodal points can not be higher than 30, and the number of freedoms per node is restricted to 32.

5.4. ERROR MESSAGES

Input errors are hardly avoidable for large structures; checks on input are, therefore, necessary. In the example we see that certain processors may be used for checking. Besides, the computer automatically carries out several checks during execution. When the checks give rise to a certain comment, a so-called error message is printed. Depending on the level of severity it may be:

1. Comment (C): This gives useful information on the problem for checking input data, etc.

2. Warning (W):   A condition has been encountered which may or may not lead to incorrect results. For example, the warning may be in a code: "suspicious element shape."

3. Error (E):   An error is found and the job is stopped at the end of the processor step. For example: maximum element on diagonal of elemental stiffness matrix is zero.

4. Fatal error (F): An error is found giving unpredictable results; job is stopped. For example: zero submatrix found in diagonal of structural stiffness matrix.

5. System error (X): An unrecoverable error is found; job is stopped. For example: programming error in data-retrieval system.

Besides a message, extra information is sometimes printed to locate the errors. It is also possible to perform all kinds of traces and dumps and, of course, a posteriori checks on the results are possible. The error diagnostics indeed turn out to be very useful in tracing errors. Plotting of the nodal points and the network of elements would also be very useful for debugging, but programs for that are not available in the standard ASKA package.

5.5. Pre- and Postprocessing

As mentioned before, ASKA does not have pre- and postprocessing programs and it is rather difficult, on the one hand, to make such programs as general as the ASKA system is now, and, on the other hand, to satisfy all requirements of the users.

Now, access to internal data is rather difficult for the user who has not carefully studied the ASKA system and it is, therefore, difficult to manipulate on internal data. Moreover, it is also rather difficult to put in external data in the internal data structure.

In principle the following solution is now being worked out at ISD [9]. A set of processors is being prepared that make it possible to take certain internal data and put them in a prescribed core area a "box," and also to take external data from that box and put it in the right place in the internal data structure. The box must be easily accessible by a user's program. It will be clear that certain steering parameters are necessary to know exactly where to place the data internally and to know the format in which the internal data are placed in the box.

In future, the set of processors will probably be built-in in the standard version of ASKA. It is to be expected that ISD will also develop certain packages for manipulating on output data. The user can then take the package that suits him. But at this point ISD's intention is not known to the author.

6. Dynamic Analysis

In mid-1971, the extension of the ASKA program with a machine-independent package for dynamic analysis, called DYNAN (DYNamic ANalyzer) [10], was made available for use outside ISD.

The structural stiffness and mass matrices are produced within the static ASKA package. Although not necessary, this will in general be followed by a condensation of both matrices. For computation of a few eigenvalues and eigenvectors one eigenvalue routine is now available in static ASKA, although it is essentially a part of the dynamic package. The algorithm is a simultaneous vector iteration. The matrices are not restricted in size.

For a further dynamic analysis, the required matrices are handled in the separate package, DYNAN. It is a number of FORTRAN programs for analyzing dynamic systems described by the set of second-order differential equations:

$$M\ddot{q} + C\dot{q} + Kq = r(t)$$

where M is the mass matrix, C the matrix of damping coefficients, and K the stiffness matrix. Vector $r(t)$ is a prescribed vector due to prescribed forces and/or displacements, and q is the vector of nonprescribed displacements.

In DYNAN two possibilities are built-in for eigenvalue analysis of undamped and proportionally damped systems, viz. a Householder reduction [11] to a tridiagonal form followed either by the QR-algorithm or the bisection method to obtain the eigenvalues. The computation of eigenvectors is by inverse iteration with shift and for close eigenvalues simultaneous inverse iteration with shift [12].

In case of nonproportionally damped systems there are again two possibilities. The $A–P$ algorithm [13] is applied to improve a set of approximate eigenvalues and eigenvectors or the procedure is: reduction to Hessenberg form, QR algorithms to compute eigenvalues and Wielandt algorithm to determine distinct eigenvectors. Eigenvectors for close eigenvalues of nonproportionally damped systems cannot be determined. Since the analysis for nonproportionally damped systems takes much computer time, the matrices are completely in core during the operations. This limits the problem size.

Special routines have been built-in for handling systems with rigid body modes. Apart from the condensation in the static ASKA package, which could be indicated as a static condensation, it is possible to apply a dynamic condensation with help of a number of selected natural modes of vibration of the undamped system.

For normal mode response calculations of a proportionally damped system, a vector of time-dependent displacements and/or forces must be prescribed as an analytical expression in the form of a summation of sine terms, as a table of numerical values, or as a combination of both. For separate transient, and steady-state response calculations, only a summation of sine terms is allowed. Response calculations for nonproportionally damped systems are not yet built-in. Neither have been built-in Runge–Kutta type procedures for direct numerical integration of the differential equations.

Besides input from the static ASKA package, direct input of matrices from tape or cards is possible. Important examples for which this is necessary, are damping matrices and diagonal mass matrices for lumped masses, but all other matrices may be input or modified as well.

An important difference between the static ASKA package and DYNAN is that the program consists of FORTRAN subroutine and function calls, and not a set of user-oriented higher-order subroutine calls. This makes programming more difficult for the user who is familiar with structural analysis and not with FORTRAN programming. On the other hand, the programming is more flexible.

For eigenvalue problems, the output will normally be the natural frequencies and eigenvectors, and for response calculations the displacements, velocities and accelerations at prescribed time intervals. Moreover it is possible to print all kinds of matrices, information and error messages. No plotting or other postprocessing routines are available.

In the present version, the substructuring feature cannot be applied for dynamic analysis in DYNAN.

7. Problems Solved with ASKA

7.1. PLANE-STRESS PROBLEM

To compare the different membrane elements and to experience the substructuring feature in ASKA, we have analyzed the part of a perforated plate shown in Fig. 5 with QUAM 8, QUAC 8, QUAM 9, QUAC 9, and TRIM 6 elements. Since TRIM 3 and QUAM 4 elements describe the stress-field very poorly, they cannot be recommended for a detailed stress analysis and have not been included in this comparison. The element distribution applied is indicated in Fig. 5. The structure is divided into three substructures; it was moreover analyzed without substructuring for the QUAM 8 elements only. For the latter case, the numbering was chosen such that the bandwidth in the structural stiffness matrix was as narrow as possible; for all other cases the bandwidths of the stiffness matrices of the individual substructures were made as small as possible.

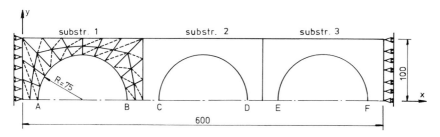

Loading case 1 : displacement u_y along upper edge 0.1 mm
Loading case 2: displacement u_y along upper edge varies linearly from −0.3 mm
for x = 0 till + 0.3 mm for x = 600 mm

Fig. 5. Loading case 1 : displacement u_y along upper edge 0.1 mm ; loading case II : displacement u_y along upper edge varies linearly from −0.3 mm for $x = 0$ to +0.3 mm for $x = 600$ mm.

The structure and the loading are symmetric with respect to the x axis. Two loading cases are considered. The first is a constant normal displacement along the upper edge, and in the second that displacement varies linearly along the edge. For the first case, the plate can be considered as a part of an infinite perforated plate with a square hole pattern. Such a plate behaves macroscopically as an orthotropic solid plate, the effective Young's modulus and the effective Poisson's ratio for tension in the x or y direction are assumed to be E^* and v^*. These elastic constants, and also the stresses along the hole boundary, can be compared with an analytical solution available [14].

TABLE 1

	QUAM 8 (substr.)	QUAM 8	QUAC 8 (substr.)	QUAM 9 (substr.)	QUAC 9 (substr.)	TRIM 6 (substr.)	Exact
E^*/E	0.3587	0.3574	0.3541	0.3586	0.3547	0.3591	0.3539
	(+1.4%)	(+1.0%)	(<0.1%)	(+1.3%)	(+0.2%)	(+1.5%)	
v^*/v	0.5885	0.5906	0.5818	0.5887	0.5800	0.5859	0.5831
	(+0.9%)	(+1.3%)	(−0.2%)	(−1.0%)	(−0.5%)	(+0.5%)	

Table 1 shows the effective elastic constants and Table 2 the stresses at the most heavily loaded points, A–C. For the second loading case, for which no analytical solution is available, the stresses at points A–F are presented in Table 3. When several elements are connected to one point (TRIM 6), the average of the values for the different elements is taken. The cpu time, the numbers of excps and the core storage used on IBM 360-65 are indicated

TABLE 2
LOADING CASE I (σ_φ IN NEWTONS PER SQUARE MILLIMETER):
CONSTANT DISPLACEMENT (FIG. 5)

Point	QUAM 8 (substr.)	QUAM 8	QUAC 8 (substr.)	QUAM 9 (substr.)	QUAC 9 (substr.)	TRIM 6 (substr.)	Exact
A	388	398	377	387	383	366	370
B	388	388	377	387	368	379	370
C	388	388	377	387	383	366	370

TABLE 3
LOADING CASE II: LINEARLY VARYING DISPLACEMENT (FIG. 5)

Point	QUAM 8 (substr.)	QUAM 8	QUAC 8 (substr.)	QUAM 9 (substr.)	QUAC 9 (substr.)	TRIM 6 (substr.)
A	−1046	−1083	−1017	−1043	−1029	−981
B	−535	−535	−514	−534	−508	−521
C	−220	−220	−220	−219	−225	−216
D	+220	+220	+220	+219	+209	+217
E	+535	+535	+514	+534	+519	+497
F	+1046	+1083	+1017	+1043	+992	+1021

TABLE 4

	QUAM 8 (substr.)	QUAM 8	QUAC 8 (substr.)	QUAM 9 (substr.)	QUAC 9 (substr.)	TRIM 6 (substr.)
cpu (seconds)	257	189	284	283	323	270
excps	18,257	7711	18,467	18,959	19,291	19,977
store (k bytes)	276	348	276	276	276	274
time SK (seconds)	90	75	113	91	132	52

in Table 4. In addition it gives the (elapse) time necessary to prepare the elemental stiffness matrices by processor SK. That processor will cause the difference in computation time between the elements with straight and curved edges.

We conclude that all the results are fairly accurate, and that the curved elements are somewhat better than the others. From the point of view of costs, substructuring cannot be recommended for a problem as indicated here. The computation time is greater and the number of excps is greatly increased whereas the savings in core storage are negligible. With the given information we can calculate the costs for each case; it is then found that the price for all substructures runs is about equal but that for the nonsubstructures run the price is far less. For the rates currently in use a drop in the order of 50% applies. In comparison with user-made ad-hoc programs for membrane and axisymmetric structures ASKA, although a large program, turned out to be efficient.

The calculations have been carried out as if the substructures were not identical and hence the conclusions are valid for that general case. For this special case of repeated substructures the cost could be reduced by just copying the books that are identical for each substructure.

For all cases we have calculated and printed information on nodal points and on elements, displacement vectors, element stresses, average nodal point stresses, and reaction forces.

7.2. PLATE-BENDING AND SHELL PROBLEMS

Figure 6 shows a part of a perforated plate with regular triangular penetration pattern. Due to the doubly periodic character of the loading to be

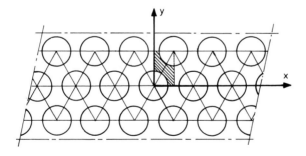

Fig. 6. Infinite plate with regular triangular pattern of holes.

considered, we only have to analyze the shaded part. The loading of that part indicated in Fig. 7 is a constant rotation along one edge and a rotation equal zero along the other edges.

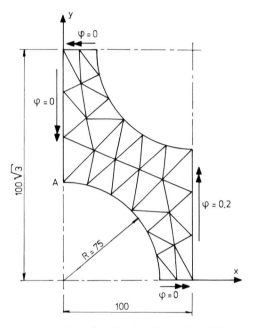

Fig. 7. Test problem for plate-bending and shell elements.

From the reaction moments obtained with ASKA we have calculated the effective Young's modulus E^* and the effective Poisson's ratio v^*, of the perforated plate in bending. In Table 5, these elastic constants, the maximum stress at point A and some information on computation time and core storage are presented for all the element types applied.

TABLE 5

	TRIB 3	TUBA 3	TUBA 6	TUBAC 6	SHEBA 3	SHEBA 6	Exact
E^*/E	0.3501	0.3533	0.3532	0.3109	0.3533	0.3532	0.3477
v^*/v	0.1148	0.1099	0.1100	0.0383	0.1099	0.1100	0.1028
$\sigma_A/E\varepsilon_{x_{av}}$ [a]	1.115	1.172	1.173	0.880	1.172	1.173	1.1259
cpu							
(seconds)	182	52	66	138	1198	1310	
store							
(k bytes)	348	348	348	348	348	348	
excps	4466	3387	4048	4128	5487	7989	
time SK							
(seconds)	109	17	20.5	96	1153	1057	

[a] $\varepsilon_{x_{av}}$ = average strain in x direction on plate surface.

TUBAC 6 gave erroneous results for this case and the same was found by Visser [15] for a circular, simply supported plate. The other elements gave accurate results. The cpu time for the SHEBA elements was extremely high; SHEBA 6 required more than seven times the cpu time necessary for the flat shell-element, TRIB 3, when a comparison is made with the same number of nodal points. The computation time is mainly absorbed in making the elemental stiffness matrices. For shell analysis with ASKA almost the only choice is the TRIB 3 element. It is far more generally applicable than the SHEBA element [16]; the input is simpler and the computation costs are far lower. On the other hand TRIB 3 is a fairly simple element since the in-plane

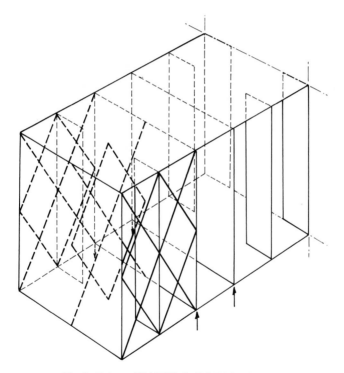

Fig. 8. Caisson (304 TRIB 3, 48 BECOS elements).

stresses are assumed to be constant. Another problem solved is shown in Fig. 8; it is a caisson analyzed with TRIB 3 and beam elements. The most severe loading case was a torsion introduced at the supports. The cpu time for the IBM 360-65 was about 12 min for one parameter combination.

7.3. SOLUTION OF THREE-DIMENSIONAL PROBLEMS

In ASKA are available 14 three-dimensional elements. Six of them are macroelements built up of either TET 4 or TET 10 elements. From a point of view of accuracy, these macroelements are entirely equal to the corresponding tetrahedronal element, and the computation time will hardly be influenced, but they are far more convenient for the user than the corresponding basic tetrahedronal element. From a simple test problem we found that the constant stress element TET 4, the corresponding macroelements, and the PENTA 6 and HEXE 8 elements were far less accurate than the other elements with the same number of nodal points for the whole structure.

There remain the TET 10 element (and the corresponding macroelements), PENTA 18, PENTAC 18, HEXE 27, and HEXEC 27. With these five element types we have determined the effective elastic constants of a perforated plate with a regular triangular hole pattern and the stress distribution along the hole boundaries for a doubly periodic loading case. We only had to analyze the part of the plate given in Fig. 9. Two layers of elements over half the plate thickness were chosen. The loading is such that plane $x = 100$ is rotated over an angle ϕ_y with respect to plane $x = 0$, whereas plane $y = 0$ and $y = 100\sqrt{3}$ are prevented from rotation about an axis parallel to the x axis. The calculations were carried out with the HETET 10 element and the pentahedronal and hexahedronal elements already mentioned. For all cases, the network of nodal points was as indicated in Fig. 9.

Table 6 shows the results obtained for the effective elastic constants, and for the maximum stress along the hole boundary. Moreover the computation time is indicated and so is the elapse time to prepare the elemental stiffness

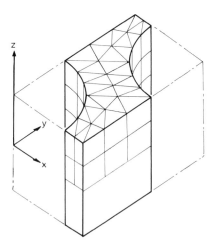

Fig. 9. Three-dimensional analysis of perforated plate.

TABLE 6

	HETET 10	PENTA 18	PENTAC 18	HEXE 27	HEXEC 27
E^*/E	0.266	0.263	0.257	0.262	0.257
v^*/v	0.909	0.932	0.942	0.937	0.946
$\dfrac{\sigma_{\phi max}}{E\varepsilon_{x_{av}}}$	1.269	1.313	1.300	1.332	1.307
cpu (seconds)	839	855	1934	1437	1478
time SK (seconds)	255	131	1324	840	952

matrices (SK) for the different element types. We conclude that the accuracy with the same number of nodal points is practically the same; the computation time differs very much, which is caused mainly by the time necessary to prepare matrices SK. The number of degrees of freedom was 1138, and the bandwidth in the system matrix can easily be obtained from the element distribution; the core storage used was 348 k bytes. The triangularization and solution of the set of equations took about 450 sec elapse time. From a point of view of efficiency, the best elements turn out to be the tetrahedron based elements and element PENTA 18. The other element types and especially PENTAC 18 take far more computer time.

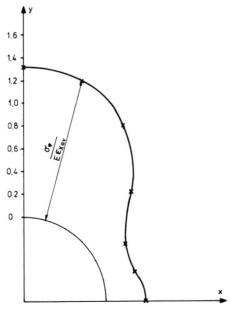

Fig. 10. Stress distribution along hole boundary at plate surface ($\varepsilon_{x_{av}}$ is the average strain in the x direction at the plate surface).

Figure 10 gives the stress distribution along the hole boundary on the plate surface obtained with PENTA 18. For a more detailed discussion of the problem of a thick perforated plate in bending we refer the reader to [17].

Another three-dimensional problem discussed in several papers of Argyris *et al.* [3] is the pipe-junction problem, Fig. 11. The structure was divided into 125 HEXEC 27 elements. The computation time on IBM 360-75 was 123 min and the core storage used approximately 400 k bytes.

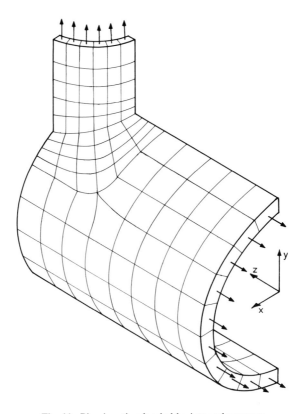

Fig. 11. Pipe junction loaded by internal pressure.

Since in the standard ASKA package it is not possible to handle directly the surface pressure and volume loads for the HEXE and PENTA elements, whereas it is possible for elements built of tetrahedrons, the latter are preferable for boundary conditions in forces.

The conclusion is that ASKA is suitable for three-dimensional analysis, but the choice of the correct element type and element distribution is very important because of the computer costs involved.

7.4. Dynamic Problem

As mentioned before, until now, there has been little experience with DYNAN. A test problem run, having 35 master freedoms, took 191 sec to solve the eigenvalue problem on an IBM 360-65.

The only nontrivial practical problem solved by the author is shown in Fig. 12. It is a multistorey deckhouse idealized with orthotropic QUAM 4

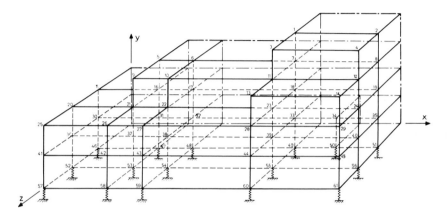

Fig. 12. Deckhouse.

elements in combination with FLA 2 elements. As a first approximation, the stiffness of the springs was assumed to be infinite. The mass was concentrated at 26 master nodes for each half of the structure. For symmetric vibrations, the lowest two natural frequencies turned out to be 21.6 and 27.8 Hz, which was in good agreement with experimentally obtained results (22 and 27 Hz).

8. Concluding Remarks

1. The static ASKA package was found to be a very efficient tool for linear static analysis.

2. All pre- and postprocessing, which is absolutely necessary, is left to the user. It is to be expected that packages for this purpose will be available in the future. The simpler interface that is being developed will make it easier to connect user programs for mesh generating, plotting, etc. with ASKA.

3. Specially for three-dimensional analysis it probably offers more than other packages now available, but one must be careful in choosing the element type.

4. For shell analysis, ASKA could be considerably improved by building-in a better shell element. The simple shell element TRIB 3 is useful and it fits well in the ASKA system since it is compatible with beam elements. But it is rather elementary because it assumes constant in-plane stresses. SHEBA shell elements are very complicated, require an extremely high amount of computer time, and are very restricted in the application for practical problems.

5. TUBA elements are not compatible with other elements in ASKA and are not very well suited for boundary conditions in the displacements along curved boundaries. The TUBAC element gave unreliable results, but this was corrected in later versions of ASKA.

6. Substructuring and saving are important features for large programs, and the possibility for the user to build-in his own elements is very practical.

7. Judging from what is built-in, the dynamic package DYNAN will be very useful for eigenvalue analysis and response calculations but until now there has been little experience with that package.

8. To prepare a static ASKA job is very simple and can be learned in a very short time (1 week). For DYNAN, experience with FORTRAN programming is necessary.

9. Error messages built-in are very useful in tracing errors.

10. We have no complaints regarding the service provided by ISD.

Acknowledgment

The author is grateful to the participants in the PROGEL association who allowed him to publish results of ASKA tests carried out for the association. Moreover, he thanks the Netherlands Ship Research Centre TNO and the Dutch Rijkswaterstaat who permitted his mentioning results of research that they had sponsored.

References

1. Schrem, E., "Implementation of the Finite-Element Procedure," *ONR Symp. Numer. Comput. Methods Struct. Mech. Sept. 8–10, 1971* Urbana, Illinois.
2. Schrem, E., and Roy, J. R., "An Automatic System for Kinematic Analysis ASKA Part I," *Proc. IUTAM Colloq. High Speed Comput. of Elastic Structures. Univ. of Liège, Belgium, Aug. 23–28, 1970.*
3. Argyris, J. H., Brönlund, O. E., and Sorensen, M., "Computer Aided Structural Analysis. The Machine-independent System ASKA." *Nord. Data-70 Conf., August 26–28, 1970* Copenhagen.
4. Argyris, J. H., Grieger, I., and Schrem, E., "Structural Analysis by Problem Oriented Languages," 23rd Ann. General Meeting of the Aeronaut. Soc. of India, Indian Inst. of Technol., Kanpur, India, February 26–28 (1971).
5. Visser, W., The Finite Element Method in Deformation and Heat Conduction Problems," Thesis, Delft (1968).

6. Bell, K., "Analysis of Thin Plates in Bending, Using Triangular Finite Elements." Thesis, Trondheim (1968).
7. Argyris, J. H., Fried, I., and Scharpf, D. W., The TUBA Family of Plate Elements for the Matrix Displacement Method," *Aeronaut. J. Roy. Aeronaut. Soc.* **72**, 701–709 (1968).
8. Schrem, E., "ASKA User's Reference Manual," ISD-Rep. No. 73, Stuttgart (1971).
9. Bernhardt, K., and Streiner, P., "Gio. A set of ASKA processors providing direct access to internal data." Institut für Statik und Dynamik der Technische Hochschule Stuttgart, Stuttgart, March 1971.
10. Brönlund, O. E., "DYNAN User's Reference Manual," ISD-Rep. No. 97, Stuttgart. No. 97, Stuttgart.
11. Wilkinson, J. H., *The Algebraic Eigenvalue Problem.* Oxford Univ. Press (Clarendon), London and New York, 1965.
12. Rutishauser, H., "Computational Aspects of F. L. Bauer's Simultaneous Iteration Method," *Numer. Math.* **13**, 4–13 (1969).
13. Patton, P. C. "Matrix Analysis of Non-proportionally Damped Dynamical Systems Employing the A-P Algorithm," *Ingenieur-Arch.* **37**, 73–80 (1968).
14. Meijers, P. "Doubly-periodic Stress Distributions in Perforated Plates," Thesis, Delft (1967).
15. Visser, W., "Discussion on TUBA-element Tests PRGL TEST M 71-2," Royal/Shell Exploration and Production Lab., Rijswijk, The Netherlands.
16. Visser, W., "Defects of the SHEBA-element in ASKA," PRGL-TEST M 70-1, Royal/Shell Exploration and Production Lab., Rijswijk, The Netherlands.
17. Meijers, P., "Three-dimensional Stress Analysis for Perforated Plates with a regular Triangular Hole Pattern." Inst. TNO for Mechanical Construction, Delft, Rep. No. 81290 (June 1971).

A Critical View of NASTRAN

James L. Tocher and Ervin D. Herness

BOEING COMPUTER SERVICES, SEATTLE, WASHINGTON

1. Introduction

In the Fall of 1970 the NASTRAN (NAsa STRuctural ANalysis) program was released for public use and is currently being distributed by COSMIC (University of Georgia) for a nominal fee. The impact of this large general-purpose structural analysis program is bound to be significant. This paper will attempt to evaluate NASTRAN from a number of viewpoints and present a (hopefully) unbiased critical review. It is felt that this review will be useful to the wide range of computer users who are considering the use and/or installation of NASTRAN.

2. History of NASTRAN

In 1964 the various NASA centers agreed that the development of a large-capacity, general-purpose structural analysis computer program should be undertaken. The specifications included the following general objectives (quoted from the introduction to the NASTRAN manuals [1–4]:

> Combine the best of the state-of-the-arts in three disciplines: analytical mechanics, numerical methods and computer programming.
>
> Incorporate both the Force and the Displacement approaches of finite elements.
>
> Organize to be general purpose.

Embody large three dimensional structural capability.
Provide for modification without cascading effects.
Build in the maximum of user convenience.
Document all aspects to gain maximum visibility.

A Request for Proposal was prepared by NASA and in mid-1965 a contract was awarded to Computer Sciences Corporation with MacNeal Schwendler, Martin Baltimore, and later Bell Aero Systems as subcontractors. NASA's project manager for the NASTRAN development was Thomas G. Butler of the Goddard Space Flight Center. The project cost for its first 5 years is estimated to be 3 to 4 million dollars if NASA labor and computer time are included. The result is a massive program (some 150,000 FORTRAN cards) with very broad capabilities.

The NASTRAN project has now moved into a new phase—maintenance and improvement. Large maintenance [5] and development [6] contracts have recently been let for this improvement phase (see Appendix A). The project is currently directed by the newly established NASTRAN System Management Office by J. Phil Raney of the Langley Research Center.

3. Boeing Evaluation Project

During the development stage, the NASA centers received various preliminary releases and in January of 1970 selected firms (including Boeing) received Version 8.1 for evaluation. Early experience with this preliminary release plus usage of the public release Version 12 provides the basis of this review. From the days of Turner *et al.* [7], The Boeing Company has been deeply involved in the development and use of finite-element structural analysis methods and computer programs. Several large-scale structural analysis computer programs (including COSMOS, ASTRA, and SAMECS) have been developed and used extensively for a wide variety of product design work. For this reason, Boeing was anxious to participate in the evaluation of the prerelease version of NASTRAN. In order that the evaluations would be related directly to specific existing analysis tasks, each division of Boeing was asked to participate in the evaluation.

The NASTRAN documentation was reviewed from the theoretical, user, and programmer viewpoints. (The four volumes amount to 3300 pages.) A series of test problems were defined and analyzed, including some from the NASTRAN demonstration problem set. During the evaluation period and subsequently, several project production requirements for NASTRAN arose and experience with NASTRAN was gained on these real-life problems. Additional experience was gained by installing NASTRAN on several different IBM 360 facilities and on a CDC 6600.

A 131 page evaluation report [8] was prepared and sent to NASA. In order to help the Boeing user learn NASTRAN, a capabilities guide [9] was prepared giving examples of sample data, deck stacking, and results plus explanations of the NASTRAN program capabilities. All of these activities have produced a substantial body of knowledge about NASTRAN. This paper is an attempt to summarize that knowledge, not from the normally enthusiastic viewpoint of a program developer, but from the viewpoint of an experienced user with broad requirements. Since users and requirements differ greatly, the sections which follow will discuss NASTRAN from a number of different viewpoints.

4. Analytic Capability

The NASTRAN system is capable of analyzing a wide range of static and dynamic structural analysis problems. This capability, integrated into a single system, was previously available to industry only in the form of non-integrated, special-purpose programs. (The relative merits of a general purpose program versus several special purpose programs will not be argued here.) The NASTRAN system is composed of independent but related functional modules that are controlled by the executive system and the DMAP (Direct Matrix Abstraction Program) language. A set of DMAP statements (control cards) define a solution procedure that is called a "rigid format." Twelve rigid formats have been provided initially and these give a good basic analytical capability. It is possible to modify rigid formats with alter cards as well as to construct new ones. This, however, is a job for an experienced NASTRAN programmer or analyst.

The rigid formats provide a convenient basis for describing the analytic capabilities of NASTRAN. The following 12 subsections are based on the rigid formats. All analysis is based on the direct stiffness-matrix displacement method. (The force method was never integrated into NASTRAN.)

4.1. STATIC ANALYSIS

This rigid format performs the static stress analysis. The usual element stiffness matrices are generated, and, if deadload effects are to be considered, element mass matrices also. This allows the computation of total weight and balance information.

Each right-hand side (loading condition) is defined by a subcase card in the Case Control desk. For each subcase, thermal loads, pressure loads, concentrated loads, gravity loads, etc., may be defined along with the desired boundary condition case (single and multipoint constraints). The boundary conditions may be different between subcases. If multiple boundary condition cases have been specified, then the stiffness matrix is partitioned, constrained, and decomposed in an overall analysis loop.

After a displacement set has been calculated, the case control is examined to determine which output is requested. Partial groups of element stresses, element forces, load columns, and reactions are some of the quantities that may be selectively requested and displayed.

The solution procedure used to solve the stiffness equations can be summarized as follows:

Form the stiffness matrix from the built-in elements and general (hand input) elements.

Apply multipoint constraints.

Apply single point constraints.

Perform matrix reduction for omitted freedoms. (This is an expensive process and should be used only when necessary.)

Decompose the reduced stiffness matrix $[K] = [L][U]$.

Perform similar operations on the load columns.

Recover stresses, reactions, forces, and displacements.

Loop back to multipoint constraints if other constraint sets are specified.

This rigid format uses 24 major functional modules and consists of 133 DMAP statements (execution control cards).

4.2. STATIC ANALYSIS WITH INERTIA RELIEF

This procedure is similar to the static analysis procedure except that inertia loads produced by the acceleration of an unsupported structure are included in the load vectors. This free-floating structure is constrained with the NASTRAN free body supports and then solved as a statics problem.

4.3. STATIC ANALYSIS WITH DIFFERENTIAL STIFFNESS

This procedure gives a first approximation to large deflection effects. An initial linear stress analysis is first performed. Then the geometric stiffness matrix is generated using the stresses just computed. The geometric stiffness matrix is then constrained and reduced to correspond to the linear stiffness matrix.

The procedure then cycles through a number of solution phases for which the user has defined a number of scalars β_i. For each cycle, a matrix is generated which is equal to β_i times the differential stiffness matrix plus the original linear stiffness matrix. This newly generated set of equations is solved for deflections and stresses. (The deflections and stresses are completely independent of the preceding cycle.)

It would not be difficult to develop a new rigid format that would perform an incremental finite deflection analysis using a geometric stiffness matrix updated with the current stresses. However, it would be a difficult task to update the nodal coordinate data of the geometric stiffness matrix.

It is not presently possible for the user to specify (by input data) the element stresses from which a geometric stiffness matrix could be computed.

4.4. BUCKLING ANALYSIS

For the buckling analysis, the stiffness matrix $[K]$ and the differential (geometric) stiffness matrix $[K^d]$ are used to formulate the eigenvalue problem

$$[K + \lambda K^d]\{u\} = \{0\}.$$

The eigenvalues, λ_i are the load level factors for various buckling modes. A stress analysis is performed initially to obtain the stress state on which the differential stiffness generation is based. For each buckling mode the load levels, eigenvectors, reduced out displacement components, and stress patterns can be computed and printed.

Eigenvalues can be extracted using either inverse power iteration or determinant tracking. This general buckling analysis is probably the most valuable new capability that NASTRAN has provided Boeing.

4.5. NONLINEAR MATERIAL ANALYSIS

The nonlinear material analysis procedure is performed using a rigid format called *piecewise linear analysis*. Solutions are obtained for structures with nonlinear, stress-dependent material properties. The load is applied in increments until its full intensity has been reached. At each increment, the stiffness matrix for the nonlinear elements is computed and then added to the basic linear stiffness matrix. The incremented solutions are added to the current solution after each increment of load is applied to the structure.

An isotropic material with an arbitrary nondecreasing stress–strain curve may be specified. Unloading effects will use the existing slope of the curve, not the original Young's modulus. Temperature and enforced element deformation loading options are not allowed. The rod, tube, and beam element consider plasticity only on the basis of the extensional stress. The modulus of the plate elements is modified on the basis of just the in-plane stress state of the element. The modulus used for the next step is based on the existing strain and an estimate of the strain at the end of the next step.

4.6. NORMAL MODES ANALYSIS

This procedure takes the reduced stiffness matrix as generated in the static analysis procedure and the mass matrix after a Guyan reduction and formulates the eigenvalue problem

$$[K - \lambda M]\{u\} = \{0\}.$$

The eigenvalues and the eigenvectors are then computed. At the user's option, the eigensolver can be selected as either the inverse power method, the determinant method, or the Given's method.

Also at the user's request, the eigenvectors can be normalized so that either the generalized masses are unity, the largest element of the vector is unity, or a selected element of the vector is unity. The eigenvalues and eigenvectors calculated by the normal modes procedure can be used by other NASTRAN dynamic analysis procedures.

4.7. DIRECT COMPLEX EIGENVALUE ANALYSIS

The eigenvalue problem solved by this procedure has the general form

$$[K + \lambda B + \lambda^2 M]\{x\} = \{0\}.$$

The matrices may be real or complex and may include both elastic and geometric stiffness, structural, and/or viscous damping and direct input matrices. Problems such as flutter and control-system feedback can be solved.

4.8. MODAL COMPLEX EIGENVALUE ANALYSIS

The problem

$$[K - \lambda M]\{x\} = \{0\}$$

for real symmetric K and M is first solved to obtain a set of modal coordinates. These are then used to formulate the modal equivalent to the matrices K, B, and M of the direct complex eigenvalue procedure. The modal matrices are normally much smaller than those formulated by the direct method and thus the eigensolution is quicker.

4.9. DIRECT TRANSIENT ANALYSIS

The coupled system of differential equations,

$$[M]\{\ddot{u}\} + [B]\{\dot{u}\} + [K]\{u\} = \{P(t)\}$$

is solved, where u are the physical (not modal) degrees of freedom. Spcial capability for describing the time varying loading is provided. Matrices may be input directly and/or generated by NASTRAN.

4.10. MODAL TRANSIENT ANALYSIS

The uncoupled differential equations,

$$m_i \ddot{\xi}_i + b_i \dot{\xi}_i + k_i \xi_i = P_i(t), \qquad i = 1, n$$

are solved for all n equations.

4.11. DIRECT FREQUENCY AND RANDOM RESPONSE

Frequency response due to a set of specified sinusoidal excitations and the subsequent random response analysis is provided by this rigid format. A set of excitation frequencies is specified by the user. NASTRAN solves the frequency equation

$$[-\omega^2 M + i\omega B + K]\{u\} = \{P\}$$

where u are the physical coordinates to be found and ω is the forcing frequency.

4.12. MODAL FREQUENCY AND RANDOM RESPONSE

This rigid format solves the same problem as above except that M, B and K are modal matrices.

(a) Special Analytical Features

A number of special features are available to all analytical paths. The structural coordinate system can be specified as either Cartesian, cylindrical, or spherical. The coordinate system may be skewed arbitrarily at different nodal points. The multipoint constraint feature allows arbitrary linear constraints on groups of nodal freedoms. Several points may be constrained to move through space on a plane, for example.

The stiffness matrix is formed according to the numbering sequence from low node number to high node number. Reordering can be specified with additional data cards, but no optimal reordering scheme is provided to minimize matrix bandwidth. NASTRAN is presently designed to treat the entire structure in one unit—no substructure analysis is possible. Matrices may be partitioned to omit (reduce out) degrees of freedom in order to manipulate smaller sets of equations in dynamic analysis.

A substantial plotting capability is available. The element layout, both original and deformed may be plotted. Node points and element numbers may be displayed. Selected sections of a structure may be plotted and symmetry can be projected across the symmetry plane. Plots of time varying functions and various other x–y plots are available.

One significant limitation for modern finite-element technology is that only 6 degrees of freedom per nodal point are allowed. This limitation is built rather firmly into the code.

5. Element Technology

The finite-element technology used in NASTRAN is one of the weaker parts of the whole system. The element technology is a mixture of some clever, advanced technology elements and some rather antiquated ones. The scalar element for example, offers a new concept with interesting potential. However, the plane stress quadrilateral performs very poorly indeed in light of today's element technology. A summary of the present element technology is given in the following pages.

1. *Beam (Bar)*. The beam element is a straight, constant cross-section member capable of resisting two-way bending, axial loads, torsion, and shear. Nodal point offsets (in three directions) are available, plus the ability to provide pins or slot releases at the ends of a beam. Temperature gradients through the depth of the beam are not available. The ability to taper the cross-section or to handle a curved axis is not presently available. Loads must be applied at the endpoints since there is no capability for handling a distributed loading. An enforced axial deformation (misfit) is available. Geometric stiffness and distributed mass matrices are provided. Bending moments, shears, torque and axial force, plus stresses at selected points on the cross-section are provided.

2. *Rod (Axial Member)*. The rod element provides both axial stiffness and torsional stiffness. A special purpose circular cross-section tube element is also available. This element is essentially a simplification of the beam element.

3. *Shear Panel*. A special-purpose quadrilateral, shear-only element using Garvey's theory [10] is provided. The panel does not resist normal forces. Lumped corner masses are generated from the total panel mass. A geometric stiffness matrix is provided. Corner loads, corner shear stresses, plus the maximum shear stress is computed.

4. *Twist Panel*. The twist panel is the bending analog to the shear panel.

5. *Plane Stress Element*. The basic element is the constant-strain triangle [9]. The quadrilateral is generated from two layers of two triangles each with the two layers having their diagonals crossing. No interior node point is used (see Fig. 1a). Presumably, programming considerations led the program designers to avoid internal nodes which must be reduced out during genera-

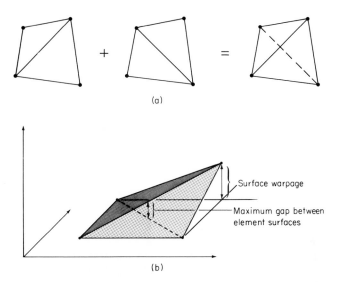

Fig. 1. (a) Formation of two-layer quadrilateral element. (b) Distortion of quadrilateral element to a hollow tetrahedron.

tion. The element performs very poorly in light of today's wide array of improved elements. Anisotropic material properties, constant-temperature thermal expansion, and geometric stiffness matrices are available. Principal stresses and directions are computed. Distributed edge loads are not incorporated.

6. *Plate Bending Element.* The HCT triangle [11] is the basic element used for bending. The quadrilateral is formed in a manner similar to the plane stress quadrilateral with four overlapping triangles. Results with quadrilaterals are considerably better than with single triangle meshes. Anistropic material properties, constant-temperature thermal expansion, and distributed mass matrices are available. Bending moments, twisting moments, and shears plus surface stresses are computed. Distributed lateral loads cannot be applied in a consistent manner to an element, nor are nodal point offsets available.

7. *General Shell Element (Facet).* The plane stress and plate bending elements above are combined to form the general shell element. Because of the two layer, double diagonal construction, the quadrilateral element will be distorted in an unusual pattern if the four nodal points are not coplanar (see Fig. 1b). If distortion is too severe, single triangular elements are recommended to cover the surface.

8. *Conical shell element.* An axisymmetric shell structure can be analyzed with either the conical or doubly curved (see next paragraph) element. Bending, membrane and shear stiffnesses are incorporated in the conical shell element. Nonaxisymmetric nodal and pressure loads are handled with Fourier series analysis. Thermal expansion and enforced strains are available, although thermal gradients through the shell thickness are not considered. A geometric stiffness matrix which is axisymmetric is available. Bending moments, twists, and shears plus stresses (including principal stresses) are calculated.

At the present time, the conical shell element cannot be used in conjunction with any other element.

9. *Doubly Curved Shell* (*Toroidal Element*). An axisymmetric toroidal ring (plus a shell cap) is provided. This doubly curved element is restricted to problems with axisymmetric loadings. Since the element incorporates 5 degrees of freedom at each node point (as opposed to the conical shell's 3), the additional degrees of freedom necessary for nonaxisymmetric analysis would exceed the NASTRAN limit of 6 degrees of freedom per node. Nodal loads, gravity loads and pressure loads may be applied as well as thermal expansion and prestrain. (Thermal gradients through the thickness are not available.) A distributed mass matrix is available. Stress results include both membrane and bending stresses.

The torodial element cannot be used in conjunction with any other element.

10. *Axisymmetric Solid Element.* The solid of revolution element is presently limited to axisymmetric loading cases. The triangular ring is based on the work of Wilson [12]. Orthotropic material properties are incorporated as are thermal expansion, prestrain and pressure, gravity, and centrifugal loading capabilities. A distributed mass matrix is also available. Stresses are computed at the centroid but principal stresses are not given. The theoretical development incorporates "exact" integration of the various volume integrals. This is a rather dubious "improvement," either from the viewpoint of numerical stability, improvement in accuracy, or the simplicity with which numerical integration can be performed.

The triangular ring elements cannot be used in conjunction with any other element.

There is also available an element with trapezoidal cross section. This element is based on shape functions which are compatible for a rectangular region, i.e., $u(r, z) = a_0 + a_1 r + a_2 z + a_3 rz$. Incompatible interelement displacements occur for the trapezoidal shape. All of the analysis capabilities of the triangular element are available. In addition, corner stresses are given.

It is not known why a general quadrilateral ring similar to the plane stress element was not developed. The trapezoidal element cannot be used in conjunction with any other element.

11. *Scalar Element.* A set of 1 degree of freedom elements are provided which connect pairs of degrees of freedom or connect a degree of freedom to the ground. Different nodal points, scalar nodal points, and (presumably) 2 degrees of freedom at the same nodal point can be connected. The elements available are the scalar spring, the scalar viscous damper and the scalar mass. Problems of electrical networks and heat transfer can be solved with these elements. These elements presently represent one of the most interesting (and unexplored) areas of NASTRAN's capability.

12. *General Element.* NASTRAN will accept a user-defined general element matrix. This element may have any number of degrees of freedom. The element is input as a flexibility matrix, inverted by NASTRAN, and augmented with a rigid body matrix (either user-supplied or internally generated).

13. *General Comments on Element Technology.* The availability of differential stiffness matrices and distributed mass matrices for most of the elements is considered to be valuable. The differential stiffness matrices do not, however, allow the user to specify the initial element prestress. The element stress must be computed from the first load increment analysis. The fact that consistent loads for most elements are not automatically generated, but rather require the user to specify the equivalent loads on the nodes, makes loads preparation awkward. Presently there are no general-purpose solid elements, no good general-purpose shell elements, and no elements with more than four nodal points.

6. Numerical Methods

The numerical analysis procedures used in the NASTRAN system are commensurate with the state-of-the-art for large scale numerical analyses. The procedures are basically sound and represent a substantial improvement over those used in many existing structural analysis programs.

Machine word size is a critical factor in any analysis. NASTRAN uses double precision in its IBM 360 and UNIVAC 1108 versions. (This is essential for short-word-length machines.) Conversely, the CDC 6600 version is presently hampered with that same double precision (which is equivalent to IBM 360 quadruple precision). It is felt that few problems at the present time require 96-bit accuracy. Double precision is not uniformly used in all routines, however, and this can cause trouble in certain applications. For example, the generation of the thermal loads columns and the stress recovery matrices is in single precision. The analysis of flexible structures subjected to high temperatures will produce misleading or inaccurate results.

6.1. LINEAR EQUATION SOLVER

The linear equation solver uses the LU decomposition method. Decomposition of either real or complex matrices is possible. Symmetry is used if it occurs. Partial row pivoting for unsymmetric matrices is utilized. The algorithm incorporates automatic spill logic for matrices which do not fit inside the core and utilizes all available core space. (This approach is used in all matrix routines.) The algorithm uses a fixed-width band plus offband "active columns." Preliminary processing determines the optimal relation between the bandwidth and the number of active columns.

An error measure of dubious value is computed and printed as the sole measure of solution quality. Using the computed solution $\{x_0\}$ of the system $[A]\{x\} = \{b\}$, an error $\{\delta b\}$ is computed from $\{\delta b\} = \{b\} - [A]\{x_0\}$. The error measure ε is then computed from

$$\varepsilon = \{x_0\}^T\{\delta b\}/\{x_0\}^T\{b\}.$$

A small value of ε is desirable; however, an unstable structure will give a small value.

6.2. MATRIX MULTIPLY

The matrix multiply module uses one of two strategies depending upon the sparseness of the two matrices to be multiplied. One strategy uses a sparse–sparse multiply with every coefficient individually subscripted. The other strategy uses a sparse–full multiply. Matrices may be either real or complex. A higher precision, vector inner-product multiply and accumulate option is not available.

6.3. EIGENSOLVERS

NASTRAN gives the user a choice of three eigensolvers. The determinant tracking method is appropriate for buckling analysis and complex eigenvalue extraction where few roots are required. The general problem

$$[A(\lambda)]\{x\} = \{0\}$$

can be solved for real or complex A. The matrix A is repeatedly decomposed into the LU form for different values of λ in order to find a zero determinant. Row interchanges are performed during decomposition. For real eigenvalue extraction, an option is provided without row interchanges.

Inverse power iteration with shifts is available and is most appropriate when applied to narrowly banded matrices for which relatively few eigenvalues are required. The most general problem solved is of the form

$$[K + \lambda B + \lambda^2 M]\{x\} = \{0\},$$

where K, B, and M may be either real or complex, and either symmetric or unsymmetric. A specialized version is available for real, symmetric K and M, i.e.,

$$[K - \lambda M]\{x\} = \{0\}.$$

The Givens tridiagonalization method is used for real symmetric matrices of the form

$$[A - \lambda I]\{x\} = \{0\}.$$

This form is usually obtained by transforming from the real symmetric dynamics problem

$$[K - \lambda M]\{w\} = \{0\}.$$

The eigenvalues of the tridiagonal system are extracted with the QR method. Eigenvectors are found using inverse iteration and Gram–Schmidt orthogonalization.

All of these methods will automatically go out of core if the problem is large.

6.4. Differential Equation Solver

The choice of numerical methods to integrate the differential equations of motion seems to be carefully made. NASTRAN is able to handle either the complete set of equilibrium equations or the modal formulation. The analysis of the equilbrium equations

$$M\ddot{x} + C\dot{x} + Kx = f(t),$$

where M, C, and K are nondiagonal, constant matrices (real or complex) is done with a second order finite difference integration method with a stability correction. No step size advice or aids are given, and one of our analyses took an inordinate amount of computer time because a very conservative step length was chosen. Restart in midstream with a new step length is not possible.

The modal formulation (uncoupled, second-order differential equations), is handled by exact integration of each equation as it is subjected to a stepwise linear loading function.

7. Ease of Use

The ability of users to easily obtain correct solutions to their problems is an important measure of any computer program. Factors which should be considered include:

Experience of the user
Design and format of the input
Design of the output results
Program diagnostics
Uncorrected errors in program logic
Organization and clarity of documentation

The NASTRAN program is large, general purpose, and comprehensive. It requires that the users have a significant amount of training and experience before its full potential can be realized. For the analyst who is inexperienced in using large-scale finite-element analysis programs, the initial encounter would undoubtedly be overwhelming. Unless additional introductory and/or tutorial documentation is developed, some personalized indoctrination courses will be required in order to install and utilize this code effectively in most engineering computer facilities.

The flexibility of the computer configurations that may be used and the complexity of the system necessitates a close working relationship between a computer specialist and the engineering user to eliminate abuses to the computing system and to the user's computer budget. An unfortunate choice is record length or region size could easily make a problem run four times longer than necessary.

NASTRAN input consists of three parts: an *executive control deck*, a *case control deck* and a *bulk data deck*. The executive control deck identifies the job and the solution procedure (rigid format) to be used and also establishes the maximum computational time allowed and the restart conditions. NASTRAN executive control data are similar to those in other general purpose programs and is easy to use and understand. The NASTRAN case control data represent an extension in generality that is not found in many other codes. They allow tremendous flexibility in the specification of the problem and the requested output, but being new they are often difficult to use correctly. The case control deck is used to select load temperature sets and boundary conditions from the bulk data deck to form solution cases. It is also used to request selective output on the printer or on any one of several plotting devices.

The bulk data deck contains all of the data necessary to describe the structure and its loading conditions. The fixed format of this deck is easy to use, but there are 150 formats to search through in the User's Manual.

The bulk data specification is very voluminous and for many problems the required bulk data deck would be so large that automatic data generators become essential. One practical limitation to production use of the system is the tremendous amount of hand-prepared bulk data required to describe a significant structure. Some short cuts must be developed if the code is to become a useful tool in the day-to-day solution of practical structural analysis problems.

The printed and plotted output results were found to be self-explanatory and easy to read. NASTRAN has an extensive list of diagnostics that define the error codes and likely causes.

As is the case in every large computer program when first released, there are a number of errors and program difficulties that have been uncovered. However, considering the size of the program and its many options, the number of errors is relatively small. Many of the errors may be avoided by different combinations of case control or bulk data cards. In general the standard options of the rigid formats are well checked-out.

The NASTRAN documentation consists of four manuals: the NASTRAN Theoretical Manual; the NASTRAN User's Manual; the NASTRAN Programmer's Manual; and the NASTRAN Demonstration Problem Manual. The documentation, an essential factor in ease of use considerations, is extensive and well written. However, it is written for the experienced NASTRAN user. An index is urgently needed to permit rapid access to sections relevant to a given topic. It is very easy for a new user to spend a long time looking for required minor details. There are 150 bulk data card formats which should be summarized by functions such as loads, elements, and boundary conditions in a single table.

An effective way to learn to use a computer code is by examples. A sample problem document which lists input and output for the available rigid formats in the code would be very helpful. The user also should know the practical limits and cost of solving his particular problem. A table of run times for variously sized problems under a particular computer configuration would be a good start toward helping the analyst define the most appropriate size of his model.

8. Problem Size

The NASTRAN stress analysis procedures are adequate for the size and type of structure encountered in the majority of analytical studies. They are not competitive with the specialized codes utilized in the analysis of very large structural systems. The dynamic analysis capabilities are more extensive than those found in the codes currently used within industry.

The major computational boxes such as the linear equation solver, the matrix multiply routine, the eigenvalue/eigenvector routines, and the differential equation routines as well as the supporting routines are essentially unlimited in design. There is a current limit on the number of records that may be output by the general input/output routines of 65,566. This means that the no matrix can have more than 65,566 rows or columns. The major limitation of NASTRAN will be excessive run times (including mean-time-to-failure problems of the computer) and excessive machine time costs. The IBM 360/65 probably is effective for static analyses up to 3000 to 5000 degrees of freedom. The IBM 360/95 probably will handle problems twice that size effectively.

In the current version, NASTRAN considers a single structure at a time. There is no substructuring capability available and, thus, the code is limited to the size of problem that can be solved in one computer run.

9. Performance

The performance of the NASTRAN code was evaluated by comparing the central processing time and the charge time of selected problems that had been previously run on Boeing Computer Services (BCS)-developed general purpose codes. The core size used and the length of the input/output records for the NASTRAN runs were chosen to be similar to those used in the runs of the BCS codes. (This is very important because NASTRAN can be made to run two to five times slower by limiting its open core and choosing very small buffers.) NASTRAN's flexibility in the DMAP statements and its case control data allow a user to increase the program's performance by eliminating unnecessary computations and/or printout.

Version 8.1 of NASTRAN on the IBM 360 was found to run somewhat slower than local, general-purpose finite-element codes for a reasonably wide range of problems. This slower running was emphasized on small problems which leads to our assumption that NASTRAN has a large, relatively fixed overhead cost.

There are obviously inefficiencies in the 8.1/360 version which may have resulted from the laudable effort to write as much as possible in machine independent FORTRAN. Examples of this are the use of the less capable "G" compiler, the general input/output routines in FORTRAN and the use of some inefficient algorithms in the computational modules.

The checkpoint restart feature is a significant capability which greatly reduces lost time in solving large problems. It permits engineering review at various stages in the computational process and minimizes the amount of repeated operations during a restart. In the solution of very large problems, however, the time between checkpoints can be excessive because there are no intramodule checkpoints.

The performance of NASTRAN on the CDC 6600 is severely penalized by the use of double precision. It is expected that NASTRAN would run three to five times faster if the computations were performed in single precision.

Appendix B summarizes performance data on two medium sized structures. The first example is the static stress analysis of a hyperbolic paraboloid. The second example is computation of the mode shapes and frequencies of a helicopter fuselage.

10. Design Criteria

NASTRAN was designed as a general-purpose computer program capable of solving significant problems in both statics and dynamics. System design decisions were based on the following guidelines:

The total framework of the program was considered to be most important.
The executive system must handle problems unbounded by core.
The system must be compact in its core requirements.
Problem restart must be available.
The system should operate efficiently on different computers.
The system should be maintainable.

The NASTRAN development activity is virtually unique in the formality with which it established its design criteria and the discipline displayed in adhering to them. The project managers are to be commended for their long range perspective in spending more money to create a superior overall executive system rather than in developing the fastest immediate problem-solving capability at the expense of future development.

A set of general input/output routines and data formats is used to accomplish all external input and output, thus providing compatability among all functional computation modules. The disadvantage of this design decision is that it increases data handling and confronts the maintenance programmer with added complexity.

DMAP provides a considerable flexibility in solution procedure. However, this is sufficiently complex in concept and application that even the experienced user will be limited to the use of the rigid formats.

One of the severest penalties paid to achieve common FORTRAN code occurs in the area of numerical precision. Double precision (64 bits) is essential for the IBM 360 series of computers; however, the concomitant double precision on the CDC 6600 (120 bits) is indefensible as a technical necessity.

NASTRAN is not designed to take advantage of cathode ray tube (CRT) graphics display hardware. Much of the current plotting capability could be used directly to produce graphic displays.

The initial NASTRAN development did not include a higher level input capability. This user–program interface is important and should not be considered a convenience which can be added locally if desired. Experience has demonstrated that it is a cost-effective necessity which can be justified by better, more accurate idealization; quicker data certification; and a choice of analysis programs which will accept the engineer's single data deck.

The most important advantage of the NASTRAN design philosophy is the ability to interchange analysis information (and perhaps even input data decks) among industrial and governmental technical organizations which are supported by computing facilities using different hardware suppliers.

11. Maintainability

A long-term development program and a disciplined, responsive configuration control activity will be essential to the widespread acceptance of NASTRAN. The size and complexity of this system demands that a formal and rigorous maintenance program be instituted [5]. In one sense, its magnitude will help protect it; that is, it is less likely that inexperienced programmers will attempt modifications. On the other hand, widespread acceptance of the code, particularly by large organizations and agencies having similar local codes, can occur only after NASA has demonstrated that NASTRAN will be assured a long, useful life.

Some essential elements of a well-conceived configuration control program that will ensure continuing viability include the following:

good maintenance documentation;
continued NASA-sponsored development;
timely dissemination of error discovery, temporary fix and permanent correction information;
coordination of privately funded enhancements to the code;
development of cost-effective source code correction methods.

Previous experience with other codes indicates that codes must be maintained by actual users rather than by uninvolved computer-code custodians. When this is not done, the code soon loses its relevancy or the users soon lose their confidence in it. Only reasonably heavy production users have the experience, motivation, perspective, and knowledge to maintain codes effectively.

12. Conclusion

NASTRAN is a massive program with extremely broad capabilities that can operate on the IBM 360, SRU 1108, and CDC 6000 series. This generality and commonality results in high computing overhead costs. The potential user must balance this higher cost against the time, resources, and cost required to develop a special purpose code to do a job. A large firm with existing codes might decide to keep NASTRAN "on the shelf," available for the solution of problems for which their existing codes are not suitable. Since NASTRAN can be obtained from COSMIC for about $2000, the cost of this strategy is very low.

The finite elements in Version 12 were all developed before 1966, but improved elements will be added in 1973. New elements can be added now to the existing code by an experienced NASTRAN programmer without disrupting the system. The solution methods are modern and capable and limited not by problem size but by excessive run times. The documentation is extensive and complete, but needs an introductory manual and an index or cross-reference list. Many organizations planning to install NASTRAN should get help with installation, orientation, and training.

The long-term success of NASTRAN will depend upon the involvement of a number of users other than the NASA centers. And finally, as computers get bigger and faster, NASTRAN may not seem so enormous.

Appendix A. Future NASTRAN Developments

A NASTRAN System Management Office (NSMO) has been established at the Langley Research Center to ensure the future usefulness of the NASTRAN computer program. A contract for the maintenance of NASTRAN [5] has been awarded. In addition to error correction and system maintenance, some of the development tasks will include the following:

1. A single precision option for NASTRAN. This will greatly improve the computer run time on the CDC 6600.
2. An improved version of the general input/output routine GINO. This development should greatly improve the efficiency of the program on all machines, and IBM 360 in particular.
3. New matrix multiply and add routines. This matrix routine is used many places throughout the program and an improved version will speed up the program.
4. A dummy element capability in NASTRAN. With this capability new elements can easily be added to the program.
5. A substructuring capability for NASTRAN. Many of the current NASTRAN size limitations will be eliminated by the addition of a good substructuring capability.

A contract for the development of new elements [6] has been awarded and should eliminate many short comings of the current NASTRAN elements.

The following is quoted from the Request for Proposal:

> These elements and capabilities are intended to give NASTRAN increased capability to analyze curved shell-type structures, solid structures, and heated structures.
> A. *Elements.* The new elements to be added can be put into three categories as follows:
> 1. Elements obtained by modifying existing elements:
>   a. Nonprismatic beam with transverse shear deformation, offset shear center, offset center of mass and distributed torsional inertia
>   b. Solid ring elements (solid of revolution) applicable to non-axi-symmetric as well as axi-symmetric motion, and compatible with other elements in the NASTRAN library of elements
>   c. Modification of the general element which will permit direct stiffness matrix input
>   d. Multilayered triangular plate element having any number of generally anisotropic layers having arbitrary principal material directions
>   e. Triangular and quadrilateral shell elements.
> 2. New elements that are compatible with existing NASTRAN code:
>   a. Rigid body element
>   b. Triangular and quadrilateral plate elements with coupling between bending and membrane deformation
>   c. Curved beam element
>   d. Axi-symmetric thin ring element for representing stiffeners on shells of revolution
>   e. Dummy elements, used for making trial runs of new or candidate elements NASTRAN
>   f. Solid elements—tetrahedron and wedge.
> 3. New elements that may possibly be incompatible with the bulk of existing NASTRAN elements because of degree-of-freedom per grid point limitation or other reasons:
>   a. Shell-of-revolution element for non-axi-symmetric as well as axi-symmetric deformation and orthotropic elastic properties
>   b. Higher order triangular and quadrilateral plate element having extra grid points (not at vertices)
>   c. Dummy elements by which new or candidate higher order elements may be used in NASTRAN.
> B. *New Capability for Heat Transfer Analysis.* The capability shall be provided to calculate the transient and steady state temperature distributions using bar, plate, shell, and shell of revolution elements. Effects of convective heat loss and applied heat flux shall be included but not radiation effects.

Appendix B. Examples of Analysis

A static stress analysis and a natural frequency analysis of two medium-sized problems are given to illustrate run times and results. The NASTRAN runs are compared to results from other BCS structural analysis programs.

B.1. Hyperbolic Paraboloid Membrane Shell

A hyperbolic paraboloid shell was solved using quadrilateral membrane plate elements. Warpage of the structure allows the membrane elements to carry the pressure load. The shell was idealized by a 10×10 grid of quadrilateral membrane elements as shown in Fig. 2. Rod elements were placed around the four edges of the shell in order to match a previous analysis done with the BCS SAMECS program. The shell was subjected to a uniform load of 1 lb/in.² on the projected surface in the x-y plane acting in the z direction.

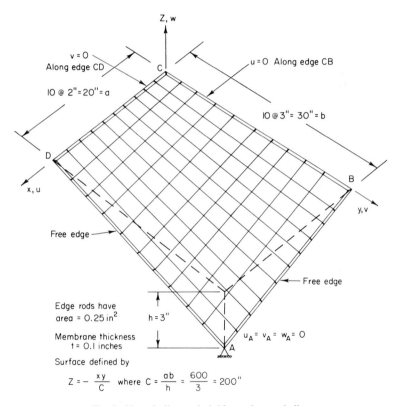

Fig. 2. Hyperbolic paraboloid membrane shell.

The displacements in the z direction along both the x and y coordinate axes were compared between the SAMECS and NASTRAN computer runs. The excessive stiffness of the warped NASTRAN quadrilateral membrane elements resulted in displacements of approximately $\frac{1}{3}$ that given by

SAMECS (whose quadrilateral element uses four constant strain triangles with a central nodal point). Although no theoretical solution is readily available, the maximum deflection of 4.25 in. given by SAMECS is expected to be below the theoretical maximum.

The NASTRAN and SAMECS computer runs both gave reasonable stress profiles. The theoretical stress for an unstiffened shell [13] gives $\tau_{xy} = 1000$ lb/in.^{2} over the whole structure. NASTRAN and SAMECS results were very close to this value in the interior (less than 2% error), but at points A and C, where concentrated loads occur, the distribution of τ_{xy} deviates substantially.

The significant details of the NASTRAN run are as follows:

Computer: IBM 360/67
Run Time: 254 CPU seconds, 766 elapsed seconds
Number of Nodal Points: 121
Number of Elements: 135

B.2. Helicopter Fuselage Vibration

The mode shapes and frequencies of a helicopter fuselage (Fig. 3) were calculated in NASTRAN and compared to the results obtained from the

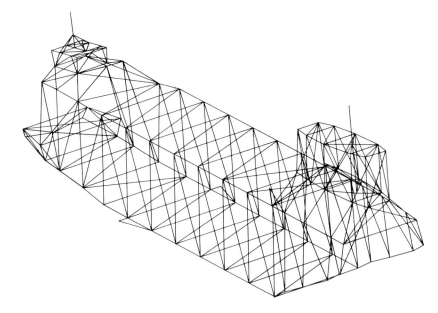

Fig. 3. Vertol helicopter fuselage, Calcomp plot.

BCS ASTRA program. The fuselage idealization contained 115 nodal points and 343 elements (membrane, beams and rods). The original 450 degrees of freedom were reduced to 123 for the eigenvalue problem.

The calculation of modes shapes and frequencies was attempted on NASTRAN Version 8.1 without success. The partial results that were obtained were erroneous.

The same problem was successfully solved on NASTRAN Version 12 using both inverse power iteration and Givens methods. The frequencies were almost identical to those calculated with ASTRA. A run time breakdown is given in Table 1.

TABLE 1

EXECUTION COMPARISONS FOR THE HELICOPTER FUSELAGE

Program	Eigenvalues extraction Method	Number of eigenvectors and eigenvalues extracted	CPU times in minutes				Computer
			Pre- and post-processing	Reduction to 123 df	Eigen-solution	Total	
ASTRA	Household, LR	(19,[a] 19)	5	4	5	14	IBM 360/65
NASTRAN	Inverse power iteration	(19, 19)	3	4	22	31	IBM 360/65
NASTRAN	Inverse power iteration	(19, 19)	2.3	2.7	12	17	CDC 6600
NASTRAN	Givens, QR	(All[a], 20)	2.3[b]	2.7[b]	1.8	6.8	CDC 6600

[a] The number of eigenvalues extracted has very little effect on the total time required for Householder or Givens methods.

[b] This run used the restart feature, skipping the initialization and reduction phases. These numbers are estimated. Actual total time for this restarted run (Givens method chosen instead of inverse power iteration) was 3.1 min.

References

1. "The NASTRAN Theoretical Manual," *NASA SP-221* (September 1970).
2. "The NASTRAN User's Manual," *NASA SP-222* (September 1970).
3. "The NASTRAN Programmer's Manual," *NASA SP-223* (September 1970).
4. "NASTRAN Demonstration Problem Manual," *NASA SP-224* (September 1970).
5. "Maintenance and Improvement of the NASTRAN System of Computer Programs," Request for Proposal L13-1603, NASA Langley Res. Center (February 1971).

6. "Addition of New Elements and Capabilities to NASA Structural Analysis Program (NASTRAN)," Request for Proposal L13-1706, NASA Langley Res. Center (March 1971).
7. Turner, M. J., Clough, R. W., Martin, H. C., and Topp, L. J., "Stiffness and Deflection Analysis of Complex Structures," *J. Aeronaut. Sci.* **23**, No. 9 (1956).
8. "A Technical Evaluation of the NASTRAN Computer Program," Corporate Engineering, The Boeing Co., Seattle, Washington (February 1971).
9. Beste, D. L., Herness, E. D., and Ice, M. W., "A Capabilities Guide to the NASTRAN Computer Code," Scientific Systems Rep. Boeing Comput. Services, Seattle, Washington (May 1971).
10. Garvey, S. J., "The Quadrilateral Shear Panel," *Aircraft Eng.*, p. 134 (May 1951).
11. Clough, R. W., and Tocher, J. L., "Finite Element Stiffness Matrices for Analysis of Plate Bending," *Proc. Conf. Matrix Methods Struct. Mech.* Wright–Patterson AFB, Ohio (October 1965).
12. Wilson, E. L., "Structural Analysis of Axi-Symmetric Solids," *AIAA J.* **3**, No. 12 (1965).
13. Timoskenko, S., and Woinowsky-Krieger, S., *Plates and Shells*, pp. 464–465. McGraw-Hill, New York, 1965.

The DAISY Code

D. N. Yates, W. W. Sable, T. J. Vinson

LOCKHEED MISSILE AND SPACE COMPANY
SUNNYVALE, CALIFORNIA

Since the original version of the DAISY code was received in March 1968 via contract with Dr. H. A. Kamel of the University of Arizona, a considerable research and development effort has been continuously conducted by the Missile Systems Division of the Lockheed Missile and Space Company to extend the utility of this code and bring it to its present highly automated production status.

By virtue of the flexibility and modularity of the basic program as structured by Dr. Kamel, significant advances have been allowed in all aspects of the program's capabilities and applicability. The basic assembly and solution routines have been altered to yield radical reductions in assembly and solution times without affecting the program's ability to operate without bandwidth restrictions or user ease of input. Mesh generation schemes have been added which greatly reduce input time spans and errors while a library of new, advanced elements has been incorporated to replace earlier formulations. Displacement boundary conditions, constrained nodes, iterative accuracy improvement, thermal gradients, and selective double precisioning and machine language routines have further added to DAISY's capabilities. Finally, the development of a complete graphics package coupled to SC4020 and FR 80 CRT plotters has enabled the program to reach a full production stress analysis status.

In this production form, the DAISY code has proved its accuracy and applicability over a wide range of advanced structural problems in both aerospace and nonaerospace applications. For example, the program has successfully solved a variety of tasks involving many thousands of unknowns ranging from submarine missile launch tube/hull intersections to tanker web frames and from civil nuclear reactor vessels to pipe tee intersections. This paper discusses the results of such analyses and presents comparisons with available test data. Experience has shown, therefore, the practicality of employing a large-scale general structural analysis computer program as a production tool. However, to attain such a status, a program's initial organization must be highly flexible and modular, while an intensive development effort must be continuously made by the using group.

1. Introduction

Lockheed Missile and Space Company has been heavily involved in large scale computerized structural analysis since 1966, when the seminal axisymmetric-plane stress finite-element program authored by Professor E. L. Wilson at the University of California at Berkeley was procured. The extended analytical capabilities afforded by this program were inspirational, leading us to obtain such other large-scale digital computer codes for structural analysis as Kalnins' program for axisymmetric shells with nonaxisymmetric loads, Carr's program with beams and membrane elements, and Bushnell's BOSOR code for buckling of shells of revolution. We began closely following the literature on the subject of finite-element analyses and, in 1966, in the Journal of the Royal Aeronautical Society, a short article by Dr. Hussein A. Kamel impressed us with its practical tone. Subsequently we asked Dr. Kamel to visit us in Sunnyvale to discuss the development of a truly general finite-element digital computer code which in turn led to the funding of the University of Arizona School of Engineering under the direction of Dr. Kamel to produce such a code for Lockheed.

Early in 1967, we received a preliminary CDC 6400 in-core version of what we have since called, the KAMEL program. The early program has grown in size and scope as a result of our own efforts and Dr. Kamel's, and today's version is still in heavy use on a wide variety of structures. In 1968, our first copy of an updated version (now called DAISY) was delivered, and this too has been greatly extended over the years both by Dr. Kamel and ourselves. In order to understand why this code has survived both obsolescence and the many attempts made to have it displaced by more recently developed computer codes, it is essential to understand our requirements as production-oriented engineers.

First, we are required to attest to the structural integrity of missile components, the failure of which could cost our employer millions of dollars. We therefore insist on understanding all analytical tools employed in our work.

It is our philosophy that no analyst should use a computer code blindly; he should understand completely the entire function of the code and know both its potential and limitations. To feed a deck of punched cards to a black box and accept the set of stresses and deflections that it prints out is not only foolhardy and dangerous, but also, by our definition, irresponsible.

Additionally, in aerospace work there is a continual problem of schedule requirements; a limited time is available to conduct an analysis before drawings must be released for procurement. We must be able to respond quickly and accurately with useful engineering data in a reasonable time during the preliminary development, production, and test phases of a contract. To fulfill this primary function, we must have tools of general applicability; if the facility we require is not in a computer program, and may not be rapidly included, the program is useless no matter how quickly it will solve the wrong problem. This may appear to be a truism, but we have, during our early years of development, been asked seriously to use structural computer codes that do not calculate stresses, have no plate elements, or that totally lack any thermal capabilities. We have also been asked to use programs coded purposely in such a fashion as to obscure the functions of the subroutines, or where a usable bandwidth is reduced to an unusable one solely to improve run times (for trivial narrow bandwidth test cases). The people making these suggestions were very proud of their products, and were gravely offended when their programs were rejected in favor of DAISY. Finally, unusual design problems are the rule rather than the exception, making the use of advanced structural tools of general applicability a necessity. All such facets of our operation were fully discussed with Dr. Kamel prior to his development of the DAISY code, and our continued very active development of his code attests to the success of Dr. Kamel in meeting our needs.

In early 1967, an engineer was dispatched to the University of Arizona just prior to the official transmittal of our first program version. The first major and, at the time, rather shocking discovery was that Dr. Kamel assumed a certain level of competence on the part of the person using his program, with only a few vestiges of the old "black box for an idiot" organization remaining. The program may, however, be used in the same fashion as the common data deck readers, but this is equivalent to hauling hay in a Ferrari. Documentation is thorough, defining the function of every routine, providing a list of variable names used in the coding, and explaining how in general a solution is achieved. The program coding itself strikes a very attractive balance between efficiency and clarity, although, of course, the two concepts are not mutually exclusive. In short, the program is ideally structured to suit our demand for comprehension.

The years subsequent to the delivery of the original program deck have seen parallel development and continuous updating of structural capability at the

University of Arizona and at Lockheed Missile and Space Company (LMSC). Dr. Kamel has been funded to visit periodically and aid in the implementation of improvements to the system. There has, of course, been some divergence between the development efforts of Dr. Kamel and ourselves; Lockheed's interest tending toward refined production analysis and the associated development of computer graphics with Dr. Kamel's interest tending toward enhanced program organization and more efficient use of machines to analyze extremely large structures, such as ships. These special interests, however, have been mutually beneficial and a close technical liaison and interchange has been continuously maintained from program inception to the present day between Dr. Kamel and ourselves.

2. Some Features of DAISY

The chief advantages of the DAISY computer code are its flexibility and level of production automation. Capability exists for efficiently analyzing a wide range of structures that are constructed from a variety of general materials and subjected to arbitrary mechanical, thermal, or displacement loadings. The program is constructed in a modular fashion so that modification of any of the various features such as loading, materials, elements, input, or output is easily accomplished This is of primary importance. The structural analysis group must be able to rapidly respond to unusual design situations. Digital computer codes must be of very broad capability, or in a form conducive to rapid alteration by the analyst. The high frequency with which projects are encountered requiring program modification has led us to place a high premium on program modularity. Two typical examples of recent Lockheed short-notice alterations to program functions are given here to illustrate analysis problems that frequently occur and require maximum program flexibility and modularity together with an intimate user knowledge of program content and structure.

1. Supplementary equations of equilibrium were included in the analysis of an oil tanker web-frame to prevent the build up of reactions at displacement boundary condition points.
2. A new preassembled quadrilateral plate element was developed and employed on a technology contract when it was discovered that the previously used triangular plate element was yielding erroneous results due to element directionality.

The term "modularity," as applied to the DAISY code, means that the program is structured so that all major program functions, and other general functions that are repetitively performed, are separated into distinct subroutines. Each subroutine is like a building block, its size and scope being determined by Dr. Kamel so that its function is obvious, logical, and reasonable. To use the program, the engineer simply writes a FORTRAN driver subroutine that generates the modeling data and ties together the various program aspects to perform the specific analysis task required for his individual job. (For the benefit of new users, a standard driver routine is provided so that the analyst may, if he so desires, supply a deck of standard data cards.) Therefore, the user has the maximum amount of flexibility to perform exactly the type of analysis that his job requires.

Subroutine SKRBM is reproduced in Fig. 1 to illustrate how a beam stiffness matrix is generated and assembled into the master stiffness matrix. The modular separation of program functions, as shown in the figure, allows a programmer–analyst to alter pertinent areas of the program with ease.

Fig. 1. Example of DAISY code modularity (demonstration of stiffness matrix assembly for a beam element).

A change in the program philosophy, element formulation, or the matrix manipulation routines can be quickly and confidently effected without fear of destroying some other phase of the program. It should also be pointed out that the clear logic demonstrated by this example is carried out throughout the entire DAISY code. Very little time is required to train new users of the code to a high level of competency.

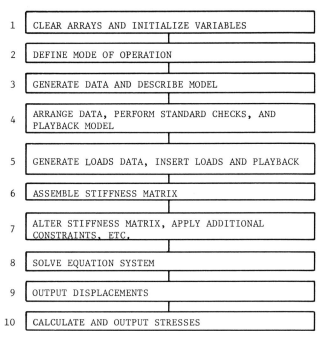

1	CLEAR ARRAYS AND INITIALIZE VARIABLES
2	DEFINE MODE OF OPERATION
3	GENERATE DATA AND DESCRIBE MODEL
4	ARRANGE DATA, PERFORM STANDARD CHECKS, AND PLAYBACK MODEL
5	GENERATE LOADS DATA, INSERT LOADS AND PLAYBACK
6	ASSEMBLE STIFFNESS MATRIX
7	ALTER STIFFNESS MATRIX, APPLY ADDITIONAL CONSTRAINTS, ETC.
8	SOLVE EQUATION SYSTEM
9	OUTPUT DISPLACEMENTS
10	CALCULATE AND OUTPUT STRESSES

Fig. 2. Block diagram of typical DAISY analysis.

The block diagram shown in Fig. 2 illustrates the overall structure of the DAISY code and the required analysis steps that must be defined and implemented for each analysis. A typical user-supplied driving routine is shown below to demonstrate the essential elements for a simple, standard-type analysis. It is obvious that as the complexity of the model and analysis increases, so does the complexity of the driver routine.

CALL CLEAR	Clears arrays, initializes
CALL THERM	Communicates to DAISY that a thermal stress analysis will be performed
CALL GEOMN	Incremental mode selected, geometric nonlinearities to be considered

GENERATE DATA   (Data card tape, disc, etc., input or user supplied data generation
NODAL POINTS   statements. May use any of mesh generation routines)
AND ELEMENTS
PLAYBACK MODEL   Printed and/or graphical output
NODES
ELEMENTS
CALL UNLOAD
APPLY LOADINGS   Data card, tape, disk, etc., or FORTRAN statement generated
CALL SOLVE   Assembles, solves, prints displacements and stresses
END

Thus, while DAISY is not, perhaps, unique as a general purpose code, it does stand somewhat apart from many other such codes in that it is deliberately designed for development and extension by the user if he so desires without reference to the author. The practicality and virtues of this approach may be illustrated by reference to Table 1 which summarizes the

TABLE 1
DAISY—LOCKHEED DEVELOPMENTS

Initial (1967) version	1971 version
1. *Size and machine*	
CDC 6400	Univac 1108 (EXEC 2 and 8)
32K core	64K core
Approximately 100 elements	Typically 2000 elements standard (can be readily increased by change in mass storage)
2. *Elements*	
Axial member	Axial member
Torsion bar	General beam with eccentric neutral axis
Single axis beam	Isoparametric linear stress quadrilateral membrane
Constant stress triangular membrane	Quadrilateral plate
Triangular plate	Isoparametric linear stress hexahedron solid
Constant stress tetrahedron solid	
3. *Solution procedure*	
Gaussian elimination of partitioned matrix	Modified Gaussian elimination, "wavefront" technique Extensive buffering
4. *Graphics*	
None	Model—arbitrary viewing angles, element and nodal numbering
	Partial displays of regions of model
	Deflected figures (any desired magnification)
	Graphs (single or multiple) of stress or strain quantities along any selected cut of elements
	Individual displays of selected element types
	Isostress and isostrain contours on surface of model
	Interactive graphics (pre- and postprocessing)

TABLE 1 (continued)

Initial (1967) version	1971 version
5. *Features*	
Modularity	Modularity
38 mechanical load cases	38 mechanical load cases
Single thermal case (membranes and links only)	Multiple thermal load cases (all elements)
No bandwidth constraints	No bandwidth constraints
	Thermal gradients (plates)
	Constrained nodes
	Double precision
	Accuracy improvement
	Displacement and skew boundaries
	Large displacements
	Pressure loadings
	Plasticity
	Parametric and standard driver versions
6. *Typical run times* (1108 *version*)	*Example problem*
Assembly: 35 min	⎧ ~4000 unknowns ⎫ Assembly: 7.5 min
	⎨ Plates, beams, membranes, links ⎬
Solution: 70 min	⎩ 1230 max semibandwidth ⎭ Solution: 10.5 min

in-house Lockheed development of DAISY which has been performed independently of Dr. Kamel since our initial receipt of the program in 1967. Furthermore, DAISY is designed to keep the analyst in control and force him to think as an engineer, not merely as a card encoder. Such features as these have led us to pursue an active development program during the past 5 years on DAISY and its various derivatives, the major aspects of which are discussed in Section 3.

3. Lockheed's Development of DAISY

Our present computer system at Lockheed is a UNIVAC 1108 with 64K core, 1 million words of high-speed FH-432 drum storage and several tape drives. Generally, on runs of less than 7 min and 200 pages, we can depend on 4 hr turnaround during the day, longer runs being executed overnight. We are charged straight wall-clock time on a dedicated machine basis of about $540 per hour, with no charge for accessing mass storage or SC-4020 (now FR-80) graphic display software. These factors must be kept in mind when, in the following discussion, the terms "fast" or "efficient" are used. There is, for example, no point in using less than full core, but it is extremely important to avoid using low speed mass storage (magnetic tapes).

Naturally, our first step after obtaining the original deck, taking a short lesson from Dr. Kamel, and reading the documentation, was to perform the necessary programming changes to allow an execution in the Lockheed computer system. Dealing with mass storage access was a major item as we are forced to use a somewhat limited-capacity extremely fast addressable drum storage, while the University of Arizona computer system employs large disks. With characteristic DAISY clarity, all mass storage input/output is handled through four routines. There are two input routines, one for two-dimensional arrays and the other for one-dimensional arrays. Output routines are handled in similar fashion. Thus the user may, with knowledge of only these four routines, easily alter external input/output processes without chasing through 250 subroutines scanning for READ and WRITE statements. In a short time the program was functioning reliably using the most helpful peculiarities of our computer such as NTRAN, a system routine allowing simultaneous input/output and Central Processing Unit activity, an especially important facility for wall-clock charge systems.

The character of our computer system demands that core be used as fully as possible to keep the Central Processing Unit busy. Buffering systems, as a result, have been used extensively at Lockheed to avoid waiting for mass storage input/output to be complete before mathematical functions may begin. Since the stiffness matrix is in partitioned form, each element may have to access as many as $N(N + 1)/2$ separate blocks, where N is the number of nodal points of the element. The original procedure of calling in the blocks as they are required, adding the contribution of each element, and outputting the blocks is extremely expensive in terms of input/output, especially since elements in sequence often contribute to the same blocks. A buffering system was installed to retain several blocks of the stiffness matrix, exchanging buffers in a cyclical fashion only when a submatrix not currently in core is required. This simple, machine-oriented bookkeeping procedure cut assembly times in half, and required only 3 weeks to install and test, chiefly due to the modular organization of assembly and input/output routines. No assistance from the program author was required, whereas our experience with other programs indicates that such a task is almost impossible without the author.

Neither was DAISY'S solution procedure held inviolate. A modified Gauss–Jordan row-by-row wavefront technique has replaced the matrix operations used in the original Gaussian elimination procedure, thereby reducing trivial arithmetic and unnecessary mass storage utilization to a minimum. This procedure cut solution time (wall-clock time from start to end of solution routine) by a factor of three from a solution time already reduced by a factor of two by use of NTRAN. Although these reductions in computation cost are extremely desirable, their value must be considered in the light of our operating environment, which emphasizes production of useful engineering data in a minimal time span. This fact, however, does not preclude the situation where in-house program development may be necessary prior to performance of the required analysis. Indeed, our experience has shown that such development is generally necessary, another reason why we favor a modular program and a group which understands all aspects of the code. Thus, our development of DAISY has emphasized reduction of the overall elaspsed time required to perform a given analysis rather than reducing just a segment of that time, such as the solution or assembly time of a computer run. Our studies have clearly demonstrated that the primary areas to reduce time are on each side of the final computer run rather than during it. There are too many examples of codes where "speed kills" and our versions of DAISY are, in reply to questions, "fast enough."

Hence, in order to efficiently accomplish the rapid analysis of complex structures, it is essential that the modeling phase be accomplished in as short a time span as possible and at the lowest costs possible. Experience has shown that preparation of the required input data along with necessary initial program changes account for up to 50% of the total analysis and cost. For this reason, it is felt that a wide range of data input options should be available in any general-purpose computer code. In DAISY we have developed a variety of input options which include the following:

- Standard point-by-point or element-by-element data input from punched cards, magnetic tape, or high-speed drum.
- Use of general FORTRAN subroutines or FORTRAN statements from the DAISY library or provided by the user.
- Use of an interactive graphics terminal with the LMSC CDC 6400 computer. Here a preprocessor program is used to generate and modify the input with the final model deck being generated via cards or tape and submitted as a regular batch run into the 1108 System.

Data card input is, of course, the most general and widely used form of input, and requires little programming knowledge or thought. However, it has major disadvantages in that it is extremely time consuming, highly prone to errors, and very inflexible. The truth is that the majority of any structure (other than pathological aberrations) is mathematically describable, and the analyst should take advantage of this fact. Hence, FORTRAN statement or FORTRAN subroutine is the input method highly favored by our group at LMSC. The analyst constructs a FORTRAN driving routine to generate the input data and direct the analysis. Available to the analyst for the driving routine construction are a number of DAISY subroutines for such functions as point, line, arc, and surface generation. Additionally, a comprehensive package of matrix and mathematical routines are present. The wing structure of Fig. 3, for example, required no data cards at all. Coordinate calculations,

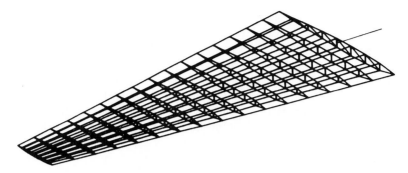

Fig. 2. Fine element wing model.

meshing, and loading are totally performed by the computer; hence little room for error exists. As a further example, the submarine structure of Fig. 4 required 600 FORTRAN statements and 50 data cards for a complete stress analysis. Finally, another advantage to this method of data generation is that a change to the geometry of the structure can usually be accomplished by the substitution of a few cards, rather than the change of a whole deck as required by data card-type input.

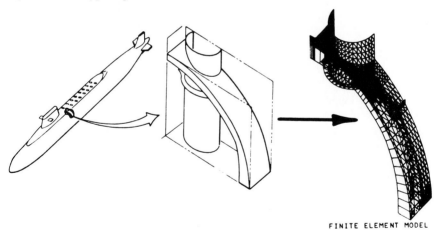

FINITE ELEMENT MODEL

Fig. 4. Submarine model.

Use of the interactive graphics terminal for data generation has several obvious advantages. The model is immediately displaced so any errors can be readily eliminated and meshing inadequacy (such as badly shaped elements) can be quickly corrected. Since the entire process is performed on a time-sharing basis, computer costs are low. Calendar time spans, rather than computer time requirements, are the biggest gains. Since the effects of corrections can be immediately seen, refinement of a model, usually taking at least 1 day and more typically requiring as much as 3 days, can be accomplished in a 1- or 2-hr session at the interactive terminal. The data for the final model are either punched on cards or stored on magnetic tape in a format consistent with the standard DAISY input. This type of data generation has become efficient for our work, largely due to the developments in graphics display and mesh generation that we had incorporated earlier. Additionally, the experience of the structures engineers in our group with FORTRAN programming has made possible efficient and rapid use of this terminal. The primary disadvantage of this method of data generation is that work levels must be stable at fairly high levels to justify the cost of this system. We have, fortunately, not found this to be a problem. It is felt that this method may soon be our prime method of data generation and display.

The advantage of finite-element analysis over other methods, namely that it provides all of the stress analysis information required for a complex structure subjected to a complex set of loading conditions, proved to be a disadvantage when compared to the more traditional "hand-solution" methods previously relied upon. Reduction, evaluation, and presentation of the typical output (which can run into hundreds of pages of computer paper) generally requires 35–50% of the total analysis time and budget. To reduce these time and cost quantities, we have developed and incorporated a complete FR 80 graphics package allowing graphic display of the model and all output quantities.

The model playback facilities that we have incorporated can be best demonstrated by the accompanying figures. Figures 5–8 show how reduction of several hundred lines of output can be reduced to four pictures which can be interpreted in only a fraction of the time that would normally be required. Errors are readily detected, and meshing adequacy becomes immediately apparent. For complex geometries, orthographic views and isometric views as shown in Fig. 9 are extremely useful. In these examples, it is next to impossible to check the model without some form of graphic display. Blown up partial views of particularly complex regions, as shown in Fig. 10, provide extremely useful modeling information in areas which are most susceptible to errors or to poor element geometries.

Graphic output of stress and deflection quantities not only greatly reduces the time required to evaluate a large amount of information, but also provides a highly effective means of reporting and presenting analysis results. Deflected figures of the loaded model as shown in Fig. 11 provide a rapid check of the validity of the model and point out potential problem areas. Figure 12 demonstrates the use of the FR80 plotter for displaying the stresses along any specified line in the structure. This feature, along with the graphic playback of the model itself, has proven extremely valuable and is used on all but the

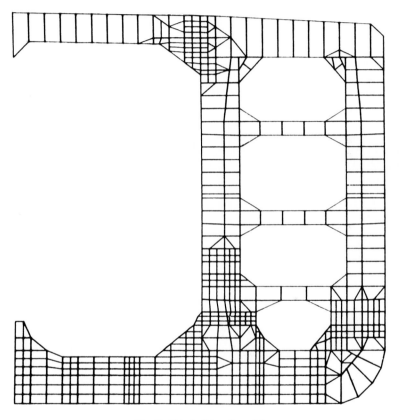

Fig. 5. Ship bulkhead model.

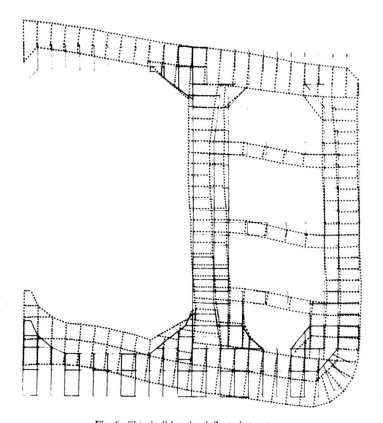

Fig. 6. Ship bulkhead—deflected structure.

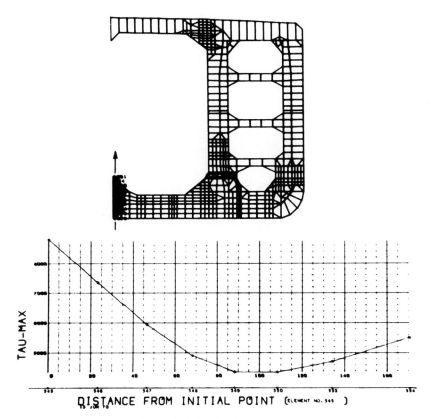

Fig. 7. Ship bulkhead—stress cut.

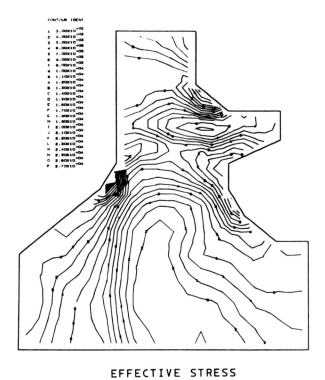

EFFECTIVE STRESS

Fig. 8. Stress contours—tanker web frame.

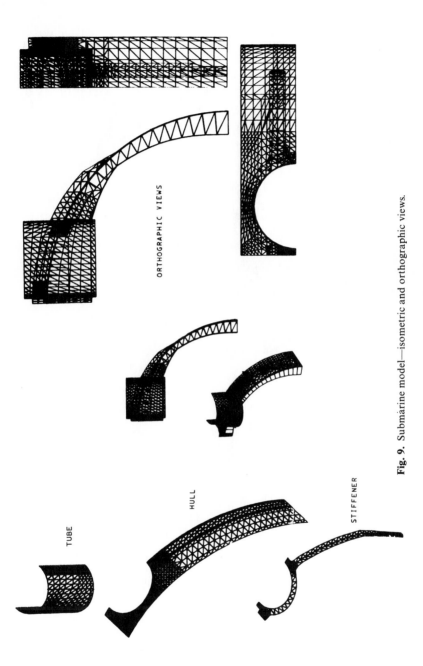

Fig. 9. Submarine model—isometric and orthographic views.

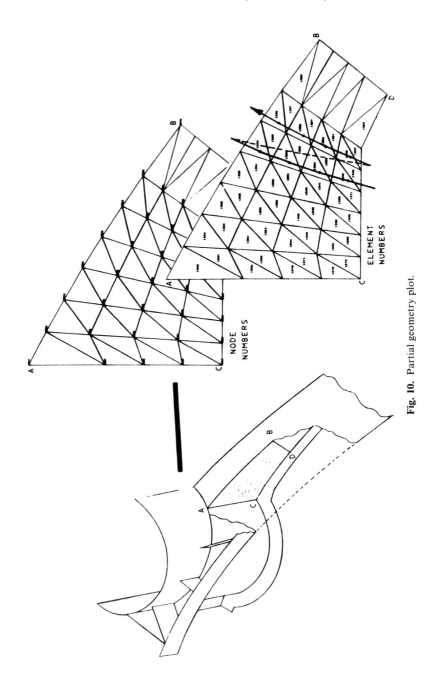

Fig. 10. Partial geometry plot.

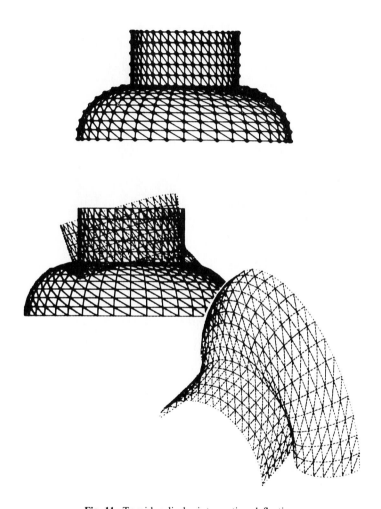

Fig. 11. Toroid–cylinder intersection deflection.

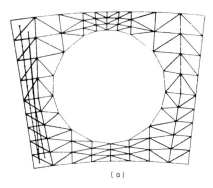

(a)

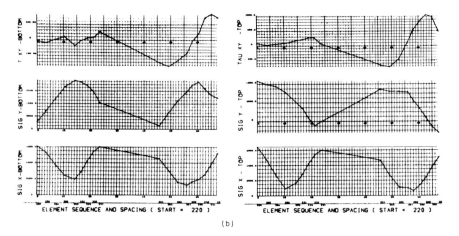

(b)

Fig. 12. Nuclear pressure vessel. (a) Stress cut model. (b) Element stress plots through stress cut.

most trivial analyses at LMSC. Another feature for the graphical display of stresses, strains, or any other quantity, is the contour plots shown in Fig. 13. Data reduction of all quantities of interest is reduced to a degree where little additional hand effort is required. It can be seen from the foregoing examples that normally time-consuming and error-prone data reduction requirements can be minimized by the extensive application of advanced graphic and pictorial output techniques.

Mesh generation and graphics routines have been developed on a steady basis since 1966 at Lockheed because, to our analysis group, they are obviously essential. A general plan was adopted to generate visual displays

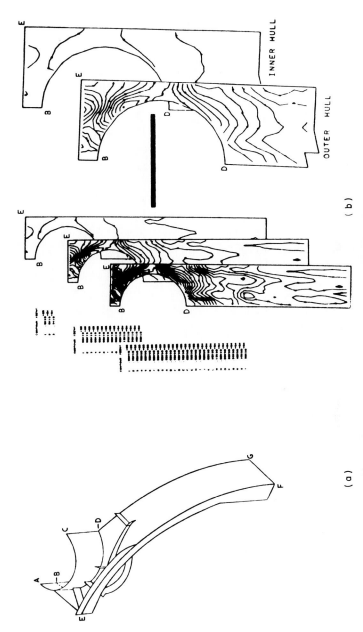

Fig. 13. Submarine model—stress contours. (a) Contour stress plots, sigma-I. (b) Unrolled hull, various densities.

to fit our needs, and a reasonable amount of man-time was set aside to accomplish a list of tasks. Many other facilities relating to functional capability, however, have been developed on a "panic" basis when a short-term need is encountered. A structural component may, for example, have thermal gradients through the thickness of plates, and if the necessary capability for such an analysis is not already included in the program, it must be coded and tested as soon as possible.

Displacement boundary conditions were developed during the performance of a structural analysis consulting contract involving a ship bulkhead (Fig. 5). In a calm sea the relative deflections at several common points on transverse bulkheads were measured for a variety of fuel tank fill levels. Although the theory behind displacement boundary conditions is well documented, the proper coding of the theory and its subsequent incorporation into a large scale digital computer program requires ready access to load, deflection, and stiffness arrays, some handy matrix manipulation routines (inverters, transposers, multipliers), and a clear, logical bookkeeping system. DAISY, in our experience, is the only program extant which would allow a user with the requisite physical understanding of the problem to promise confidently to produce results in a short time without help from the author of the program. In this same bulkhead analysis a set of influence coefficients were generated and deflection vectors stored for forces applied at boundary condition points, for external crushing pressure, and for distributed loads representing the shear transmitted from a longitudinal bulkhead. Augmenting this set of influence coefficients with a vertical equilibrium equation, a set of equilibrated boundary forces can be calculated and the proper combination of deflection vectors used to compute stresses. All the manipulations required to perform the analysis were programmed with ease and produced a considerable improvement in results. This type of machination is certainly not automated in any program of which we are aware.

During the execution of a submarine analysis contract (Fig. 4) for NAVSEC we discovered a need to attach rotational element to nonrotational elements, specifically plates to solids. A launch tube–pressure hull intersection is formed by welding thick steel plate to an extremely heavy cast section. The stress state in this region is three-dimensional with a significant amount of shear deflection in the hull plate, thereby violating the usual assumptions used in finite-element plate formulations; but the structure some distance away from the intersection region is quite adequately modeled using plates. One might, of course, use three-dimensional solid elements for the entire model and suffer under the quintupled element formation time and extreme complication of the mesh, but such an approach is very uneconomical. A much more direct approach is to use the less expensive plate element where it is adequate and the solid element where it is required, thereby necessitating that the elements be joined at some point in the structure. The common method used to make this connection is the "constrained nodes" approach where the deflections and/or rotations of some "slave" points are made independent upon a "master" point, and this is precisely what has been done. The presence of matrix manipulation routines, direction cosine routines, and simplified, direct mass storage tracking routines made it a comparatively easy task to perform the load and stiffness matrix alterations necessary to impose the constraints. This new facility was added to the program in a short time with no help from the author, and has since been used on several other contracts. Dr. Kamel has recently extended this procedure to produce his "super elements" in which the deflections along connective edges of sub-structures are related to the deflection of end points by a polynomial of order specified by the user.

These examples merely serve to illustrate how well the DAISY program has performed in our operating environment. Other analysis projects led to the development of accuracy improvement routines, some selective double precisioning, the introduction of through-thickness thermal gradients for plates, orthotropic properties, and, particularly important, the updating of the element library. Our original version of the program had six elements: a link, a torsion bar, a constant strain membrane triangle, a 3-degree-of-freedom (in local coordinates) beam, a triangular bending element, and a constant strain tetrahedron. Of these elements only the link remains. The torsion bar and beam have been used to develop a 6-degree-of-freedom beam with eccentricity, the triangular membrane has been replaced by an iso-parametric linear strain quadrilateral, the triangular bending element has been replaced by a quadrilateral, and the tetrahedron has been replaced by an isoparametric hexahedron (Fig. 14). Introducing a new element into the system requires about 10 cards plus a routine to compute the element stiffness and a routine to compute stress, given deflections. It has been necessary in the past to install new elements on short notice. A short-term wing analysis showed the extremely poor performance of constant strain triangles for modeling long thin structures where the primary members (spars) are in a state of flexure. The new linear strain quadrilateral was installed within 1 week without assistance from Dr. Kamel or the element's author, and produced excellent results.

4. Some Examples of Problems Solved with DAISY

Our experience with DAISY has given us a great deal of confidence in its accuracy. It is an extremely reliable program, having been successfully used on nearly every major analysis effort that we have undertaken in the past

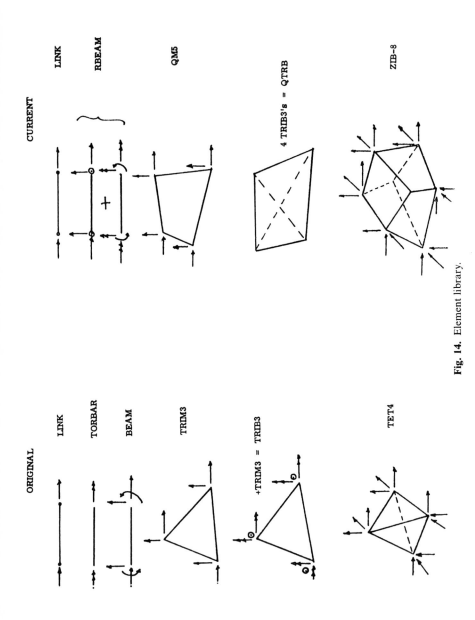

Fig. 14. Element library.

3 years. That DAISY is truly a general purpose analysis code is borne out by the following examples:

4.1. TANKER WEB FRAME

An analysis was undertaken to determine the stress distributions in a web frame of a supertanker. The web frame is constructed of numerous in-plane frame members and connecting plates. The finite element model is shown in Fig. 5. The structure was analyzed for three load cases which involved application of concentrated and distributed mechanical loadings as well as specified displacements at a number of locations. The output presented to the customer consisted of deflected shapes, stresses along critical lines in the structure, and principal stress contours in the highly loaded regions, as illustrated previously in Figs. 6–8. The analysis and reporting required approximately a month. The results, according to the customer who had already tested a complete tanker, gave excellent agreement with strain gauges.

4.2. NUCLEAR REACTOR VESSEL

A contract for the analysis of the aft jet pump region of a nuclear reactor vessel (Fig. 15) subjected to 18 mechanical and thermal load cases was recently completed using the DAISY code. The effects of thermal gradients through the thickness and over the surfaces combined with mechanical concentrated loads, pressures, and displacement design conditions. Deflected shapes, isostress contours in highly stressed areas, and graphs of stresses along every major line of interest, were graphically presented for each load case. Reduction of the vast amounts of data generated by the 18 load cases was greatly simplified and the time required considerably shortened.

4.3. SUBMARINE STRUCTURE

Figure 9 shows one of the models of the launch tube/pressure hull intersection region recently analyzed at Lockheed. The model consisted of plate, membrane, and beam elements, and was subjected to external pressure. The results were compared by the customer with previously obtained test data from a fully instrumented submarine. Agreement was excellent in nearly all regions as shown by Figs. 16 and 17 and their accompanying tables (Table 2 and 3). For the few isolated locations where agreement was poor, the customer commented that the gauges were bad or mislocated. The results transmitted to the customer relied heavily on the use of graphic output and included stresses along critical lines of the structure (Fig. 18) as well as isostress contours on the pressure hull (Fig. 13) and the launch tube. Preparation of the results into report format required only three days and largely consisted of pasting the graphs to report format paper.

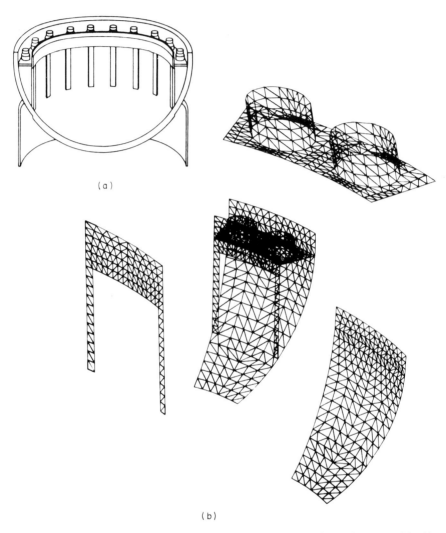

(a)

(b)

Fig. 15. Nuclear reactor. (a) Reactor cut-away. (b) Thirty-degree finite element model with exploded view.

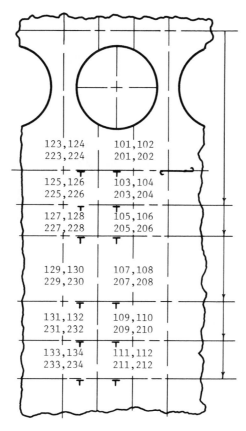

Fig. 16. Strain gauge correlation. 100 Series, outboard; 200 series, inboard; even numbers, circumferential; odd numbers, longitudinal.

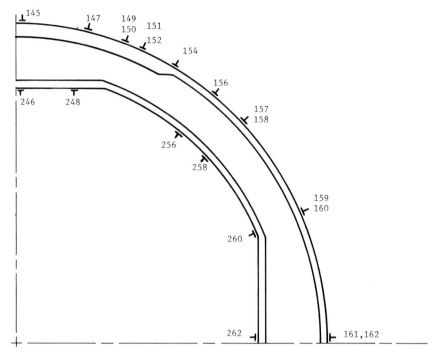

Fig. 17. Strain gauge correlation. Even numbers, circumferential; odd numbers, longitudinal.

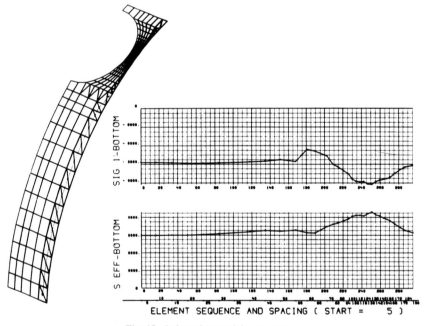

Fig. 18. Submarine model—stress cuts.

TABLE 2
STRAIN GAUGE CORRELATION

| Gauge numbers | | | | Element numbers | For descriptions see (1) and (2) | | | |
A	B	C	D		SA	SB	SC	SD
101	102	201	202	345	97.3	89.4	72.0	85.6
					(13.4)	(6.4)	(49.8)	(6.7)
123	124	223	224	354	72.1	80.3	69.4	90.0
					(33.9)	(−13.7)	(44.2)	(−16.4)
103	104	203	204	289	93.3	91.9	60.7	81.7
					(17.3)	(−0.3)	(65.4)	(7.8)
125	126	225	226	299	88.0	88.5	64.2	79.5
					(22.6)	(1.6)	(66.7)	(10.2)
105	106	205	206	241	91.6	91.8	60.6	80.8
					(20.3)	(5.7)	(97.3)	(93.3)
127	128	227	228	251	87.0	88.0	65.8	80.7
					(19.8)	(5.0)	(56.1)	(3.8)
107	108	207	208	129	90.5	90.7	62.0	83.7
					(19.7)	(0.0)	(37.6)	(3.8)
129	130	229	139	139	85.3	88.2	68.4	85.5
					(34.4)	(−5.5)	(45.3)	(1.5)
109	110	209	210	65	91.6	96.5	63.9	83.3
					(20.6)	(3.3)	(34.5)	(−2.7)
131	132	231	232	75	86.8	94.9	70.1	85.7
					(6.8)	(−7.7)	(52.8)	(3.2)
111	112	211	212	1	98.0	100.0	57.4	80.6
					(24.6)	(−11.8)		
133	134	233	234	11	88.8	96.5	66.4	85.4
					(37.6)	(2.2)	(34.8)	(−5.4)

(1) Upper number is predicted stress expressed as a percentage of maximum predicted stress. Lower number (in parentheses) is percentage difference between test gauge reading and predicted stress

$$\text{Percentage difference} = \frac{\text{Predicted stress-test readings}}{\text{Predicted stress}} \times 100$$

(2) Actual stress values (both predicted and test) can be supplied separately by the authors upon receipt of necessary clearance credentials and need-to-know justification.

TABLE 3
Strain Gauge Correlation

Gauge numbers	Predicted stress[a] expressed as a percentage of maximum predicted stress	Percentage difference[a,b] between test gauge reading and predicted stress
145	2.3	5.6
246	100.0	19.3
147	3.1	−140.4
248	94.1	−5.0
149	12.6	−6.7
150	69.7	11.5
151	23.4	−1.9
152	54.4	16.4
154	76.9	−0.8
156	79.1	16.1
256	66.3	−30.4
157	61.6	88.7*
158	82.0	18.4
258	63.2	−15.7
159	65.7	59.7*
160	76.3	−7.6
260	50.7	−7.3
161	75.1	38.1*
162	98.2	2.6
262	22.9	−2.5

[a] Actual stress values (both predicted and test) can be supplied separately by the authors upon recept of necessary clearance credentials and need-to-know justifications. Asterisked results (*) identified by customer as bad gauge.

[b] $\text{Percentage difference} = \dfrac{(\text{Predicted stress} - \text{test reading})}{\text{Predicted stress}} \times 100$

4.4. Pipe Tee Intersection

Figures 19 and 20 show two of the models that were used for the analysis of the pipe tee intersection region. The effects of internal pressure coupled with shear, axial, and bending loads on the outstanding leg were considered. The model consisted entirely of quadrilateral bending elements and gave extremely good results, corresponding within 5% to strain gauge results. A comparison between analytical and test results is presented in Fig. 21. The graphical output presented to the customer included stresses along

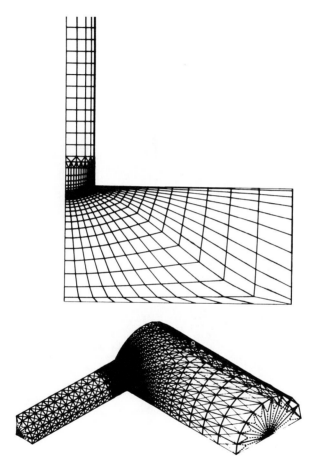

Fig. 19. Pipe tee intersection model (internal pressure case).

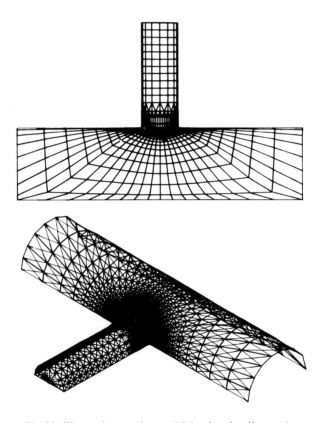

Fig. 20. Pipe tee intersection model (in-plane bending case).

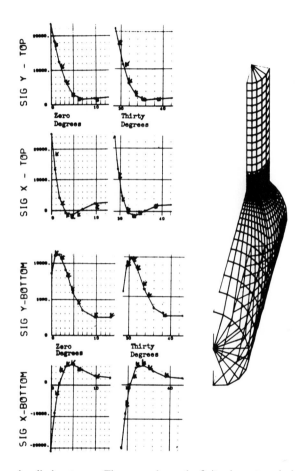

Fig. 21. Branch cylinder stresses. The curve shows the finite element analysis, arrows strain gauge results.

selected lines of elements and isostress contours over the surfaces of the pipe tee (Fig. 22).

In general, the DAISY program in use at Lockheed has proven successful on every job that we have attempted. Where strain gauge results have been available, we have achieved excellent agreement with analysis results.

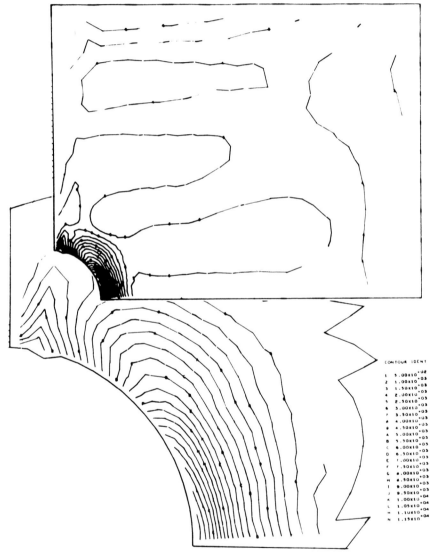

Fig. 22. Sigma-effective top.

5. Future Plans

Our future plans involve improving capability through the use of more sophisticated methods, and through more efficient use of computer facilities. Interactive terminals allow almost instantaneous checking of input, and have the potential of reducing modeling time for a complex structure from 1 week to 1 afternoon. We have recently begun to extend the facilities of interactive terminals to match those in our working programs, for output as well as input. Effort in this area is divided into the same general sections as a standard structural analysis, i.e., input, solution, output. As mentioned earlier in this paper, input preparation and output interpretation both require considerably more engineering time than the solution itself. Consequently, these areas are being attacked first. The DAISY interactive preprocessor allows the user to display all or any part of his structure at any orientation with either element or node numbers superimposed on the mesh. Numerical data may be changed through the terminal and copies made of anything appearing on the screen. A postprocessor being developed will make it possible to examine data such as stresses along lines through structures, deflected shapes, etc., and select pertinent data for permanent records. These simple facilities will be of tremendous aid in design support even before the more advanced capabilities such as automatic substructuring or accuracy improvement become available.

There are, however, some very basic areas requiring attention that are at least as important as reducing our reaction time to analysis problems. We feel that there is plenty of room for improvement in the quality of plate and solid elements. Our quadrilateral plate is simply four TRIB3 elements pre-assembled, a process we found necessary to eliminate the basic TRIB3 directionality. In trying other more sophisticated plate elements we found none that would improve accuracy without extreme penalties in formation time and reliability. Likewise, the Zienkiewicz–Irons isoparametric linear strain hexahedron represents a significant improvement over constant strain solids, but even this advanced formulation has serious shortcomings for skew shapes.

Also, we hope to select in the near future methods for performing stability and large deflection analysis, currently the forte of finite difference programs. Geometric stiffness matrices are available in the literature for several basic elements, and it should be reasonably easy to add the necessary solution routines. Dynamics, of course, require similar effort. Our examination of new formulations will be greatly aided by the experience we have gained in the past with practical structural problems. The ease with which new facilities may be inserted in DAISY will prove of inestimable value.

In conclusion, the DAISY code has proven to be our most powerful analysis tool. It has survived obsolescence through its clarity and generality, and it has allowed the solution of problems requiring special facilities without necessitating the release of study contracts to outside parties. Having a common organizational ground with Dr. Kamel ensures that developments at Lockheed and at the University of Arizona may be readily interchanged for our mutual benefit. DAISY has enabled us to react successfully to our design problems since 1966, and we have great confidence that it will continue to grow to fit our needs.

An Evaluation of the STARDYNE System

John Dainora

STONE AND WEBSTER ENGINEERING CORPORATION, BOSTON, MASSACHUSETTS

1. Introduction

The STARDYNE system is a proprietary software package which can be accessed through a network of computer service bureaus, public and private terminals. STARDYNE was developed by Mechanics Research, Inc. (MRI) and made available nationally on Control Data Corporation's Cybernet system in the latter part of 1968. A new version of the program, designated as STARDYNE2, has been available since August 1971. STARDYNE2 incorporates solid-element capabilities as well as improved static and dynamic analysis features.

STARDYNE consists of a series of compatible structural engineering programs designed to analyze linear elastic structural models. The programs which are based on the stiffness method of structural analysis can be used to solve a wide range of static, dynamic, and stability problems.

2. STARDYNE

2.1. System Capability

The physical structure to be modeled is represented as an array of nodes interconnected by finite modeling elements. The following finite elements are available in the STARDYNE system:

1. three-dimensional beam elements

2. infinitely rigid bar member
3. triangular plate element
4. rectangular plate element
5. hexahedron (cube) solid element
6. tetrahedron solid element

The three-dimensional beam element or bar member may experience loadings described by two transverse forces and two moments at each end of the member combined with an axial load and a torsional moment. The geometric and elastic properties between two connected nodes are assumed to be constant along the length of the member.

Infinitely rigid bar members are available for modeling components which are very stiff in relation to the total stiffness of the structure. One or more of these members are considered to form an Infinitely Rigid Substructure System (IRS). Only the highest node in each IRS remains a part of the elastic mathematical model, and the mass of the IRS must be transferred to this node or other nodes not associated with an IRS. Although the remaining nodes in each IRS are not represented in the stiffness matrix, any static loads applied to these dependent nodes will be correctly transferred to the elastic model.

The triangular plate element connects any three nodes of the structural model. The displacement of a triangular plate element is defined by three translations and two rotations at each corner of the plate. This element will resist in-plane shear and direct forces and transverse shear and bending forces. The triangular plate element assumes a linear displacement function.

The four-node rectangular plate element uses a higher-order displacement function and should be used instead of the triangular plate element wherever practical. However, if sandwich element capabilities are required or if in-plane shear forces are to be resisted, the triangular plate element should be used. The triangular and rectangular plate elements may connect the same three and four nodes, respectively.

The solid hexahedron and tetrahedron elements assume a linear displacement distribution (constant strain). The order of the eight node numbers that define the hexahedron is restricted. The node numbers that define the tetrahedron may be listed in any order. The average three-dimensional stress field is printed for each solid element in the global coordinate system. Principal stresses and the direction cosine matrix are also indicated.

The effect of structural components such as nonstandard elements or substructures not available to STARDYNE can be accounted for by direct alterations to the stiffness matrix. Output in the form of loads and stresses are not generated for this structural system, but the equilibrium check for each of the nodes will be equal to the internal force in this system.

Maximum dimensional capabilities of STARDYNE are indicated in Table 1.

TABLE 1
STARDYNE DIMENSIONAL CAPABILITY

Item	Maximum allowables	
	6 df/Node	3 df/Node
Degree of freedom (df)	6000	3000
Nodes	999	999
Beams	9999	9999
Triangular plates	9999	9999
Rectangular plates	9999	9999
Cubes	0	9999
Tetrahedrons	0	9999
Rigid systems	100	100
Nodes per rigid system	18	18
Rigid bar elements	300	300
Static load cases	No limit	No limit
Dynamic df for modal extraction		
Householder QR	330	330
Inverse iteration	6000	3000

2.2. STATIC ANALYSIS

The static structural analysis capabilities of STARDYNE include the following:

1. applied element and nodal loading;
2. specified displacements;
3. automated thermal analysis;
4. inertia loading;
5. combined loading cases;

The static portion of STARDYNE determines the displacements of the structural system as well as the internal forces on all components which make up the system. The pseudostatic load or displacement vectors obtained from dynamic response calculations can be processed in the static portion of the program. Pressure loading can be specified directly for the triangular and rectangular plate elements.

Boundary displacements correspond to the global coordinate system unless an individual coordinate system has been specified for the boundary node. The input for thermal analysis consists of temperatures on each face of only those elements that are being heated or cooled. Inertia loading is based on the nodal weights and is accomplished by inputting translational and rotational accelerations and rotational velocities in the global reference system. Previous load cases including dynamic response analysis results can be factored and combined with the current load case. A load case that contains thermal data should not be factored because initial element loads and stresses due to thermal loads are not scaled. However, the applied loads, displacements, and equilibrium check will be given correctly in the final output.

2.3. DYNAMIC ANALYSIS

STARDYNE will extract eigenvalues and eigenvectors for any desired frequency range. In addition, the program will compute the generalized weights, participation factors, and internal forces on the elements associated with each mode. Participation factors are computed for each of the three global translational directions. The largest component of the eigenvector for each mode is set equal to unity and the remaining components are scaled proportionately. A dynamic equilibrium check is performed in order to facilitate the evaluation of the accuracy of the solution.

Modal extraction can be performed using either the Householder–QR or inverse iteration methods. The Householder–QR method can be used for problems up to 330 dynamic degrees of freedom. The inverse iteration method can be used for solving problems when the Householder–QR maximum has been exceeded or the modal extraction is to be performed over a narrow frequency range and this method is more efficient. The inverse iteration procedure during any iteration attempts to converge simultaneously upon a single eigenvector, two close roots (quadratic procedure) and three close roots (cubic procedure).

The time history response analysis is based on the "normal mode" method. The transient analysis is preceeded by the extraction of a suitable number of eigenvalues and eigenvectors. The transient response of a linear elastic structural modal is accomplished by specifying the force–time history of loading and the time points at which the program is to compute the response. The maximum number of forcing functions which can be applied is limited to 100. One set of initial displacements and velocities may be entered for a set of load vectors. One value of damping is entered for each mode. The static option of the program is used to calculate final loads and stresses at the specified time points of interest.

The steady-state response of a structural model is determined by entering frequency dependent sinusoidal excitations and specifying the frequencies at which the structural response is to be computed. Modal extraction precedes the steady-state response analysis. The forcing function input can consist of either a base excitation, a distributed force loading, or a unit sinusoidal excitation.

The base of the structure may be excited in any of the six directions of motion by describing the excitation and the corresponding amplitude versus frequency of the excitation. The base motion may consist of accelerations, velocities, or displacements.

The distributed force loading consists of a force vector applied at the nodes of the structure. The force vector is specified in each of the six directions of motion using amplitude versus frequency relationships. A phase lag angle may be used for each amplitude versus frequency relationship.

A unit sinusoidal excitation may be applied at a specific node of the model and transfer functions computed for all of the remaining nodes. The unit excitation can consist of either an acceleration, velocity, displacement, or force.

The problem size for transient and steady-state dynamic analyses is limited to a maximum of 5000 static degrees of freedom. The largest node number in the structural model cannot be above 833.

2.4. Special Features

A significant feature of the STARDYNE system enables the program to detect ill-conditioning and/or singularities during the decomposition of the stiffness matrix. The user has several options whereby he may direct the course of the program during the error analysis. The user may select the normal option which terminates the run if ill-conditioning is detected or a singularity is encountered. Another option is to check for and solve all singularities detected but terminate the run if ill-conditioning is determined.

A unique feature of STARDYNE is the solution of a free–free system. Static load analysis may be performed on a free–free structure if the applied loading is self-equilibrating. The solution of a free–free structure is obtained by selecting the appropriate error analysis option.

Nodal numbering is reordered internally to produce a minimum bandwidth. This does not affect either the input or output data. Element numbers may have gaps in the list of numbers defined and any number from one to the maximum may be used. As a result of this feature a rapid series of parametric model studies is possible by deleting or adding nodes to the existing mathematical models and rerunning the program.

2.5. Input/Output

The input/output features of STARDYNE have been significantly improved in the STARDYNE2 version of the program. The geometric data to a large extent are now entered in the "data table" format. The input data preparation features include tables of material properties in which the user may enter up to 250 different material types. Tables of member section properties, element weight computation, automated thermal loads for all elements, member end release conditions and data generators are other new features. The coding rules for IRSs have been simplified. The program retains its modular structure although the data handling aspects of the program especially for dynamic analysis has been streamlined.

A wide variety of output options is available to the user. Initial model and deformed structure plots can be specified. Deformation patterns resulting from static loading or mode shapes can be plotted. Options are presently available to present the computer output in a format suitable for a report.

The final conditions from a time-history analysis can be saved for use as initial conditions on a subsequent dynamic analysis. The content of the output is dependent upon the options selected. For example, corner forces in the element system, corner forces in the global system, stresses in the element system, principal stresses, stress resultants in the element system, and principal forces are output items that can be selected for the triangular plate element. Stresses in the global system and principal stresses and directions are available for the two solid elements.

3. Examples of Problems Solved

3.1. CONTAINMENT STRUCTURE

Since 1968, a large variety of complex, static, and dynamic problems related to the nuclear power industry have been solved using STARDYNE. The application of STARDYNE to the solution of several structural problems is discussed to illustrate program capabilities and performance.

The validity of the solutions of structural problems using digital computers is primarily a function of the capability of the analyst to translate a real structure to a truly representative mathematical model. The formulation of the model must be consistent with the program capabilities and expected behavior of the structure. Once a program such as STARDYNE has been thoroughly checked and debugged, the responsibility for engineering judgments necessary to implement the program must be borne by the structural analyst.

The first decision that must be faced by the analyst is the extent of detail to be included in the model. It is immediately obvious from the computational procedures involved that a large, highly detailed model may not necessarily give more correct answers if problems of ill-conditioning and numerical round-off errors are considered. Program limitations on allowable stiffness and mass ratios consistent with accuracy are frequently determined by a trial-and-error procedure. Because STARDYNE performs an error analysis and an equilibrium check, the accuracy of the numerical computations can be quickly assessed.

The general tendency in modeling a structure is to overmodel by including detail that will not significantly affect the final results or that is not warranted by or consistent with the problem assumptions. On occasion an overcomplex model may result in prohibitive solution costs and preclude a timely solution. Finally, the more complex the model, the greater the possibility of coding or modeling errors which cannot be detected by an independent check of the computer results. An independent check may not be even feasible for a very complex structural model which cannot be tested under the postulated loading conditions.

Even if a simple model is initially developed, changes in design criteria, requirements for additional load path information, or external factors may result in the model becoming gradually more complex. A mathematical model was initially developed for the containment structure and components which are schematically represented in Fig. 1. The model included the reactor building and steel superstructure, drywell, reactor pressure vessel and

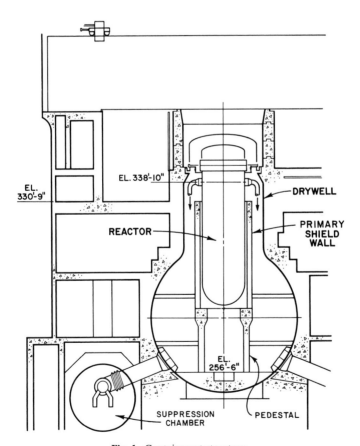

Fig. 1. Containment structure.

pedestal, primary shield wall, and a ground spring representing the ground–structure interaction. Because of the presence of expansion joints the toroidal suppression chamber was uncoupled from the containment structure model and was represented by a separate mathematical model.

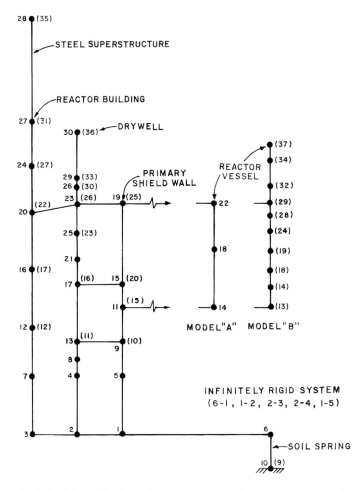

Fig. 2. Initial models of containment structure and reactor pressure vessel.

Figure 2 indicates the initial mathematical model (Model A) which was used to determine preliminary design loadings resulting from a postulated earthquake. Eigenvalues, eigenvectors, and participation factors were calculated using STARDYNE, and results were punched on cards for further processing using an in-house computer program to calculate modal accelerations and displacements and seismic nodal loads. An intermediate model (Model B) with an expanded representation of the reactor pressure vessel was used to investigate the response of the structural system to postulate pipe breaks. The dynamic transient analysis included the consideration of break locations at nodes 18 and 32 (Model B).

Figures 3 and 4 indicate the final mathematical model which was constructed as design activities progressed and as-built dimensions became available. The final model of the reactor pressure vessel (Fig. 4) which includes a detailed representation of the vessel and internals was generated by the nuclear steam supply system vendor and was incorporated into the structural

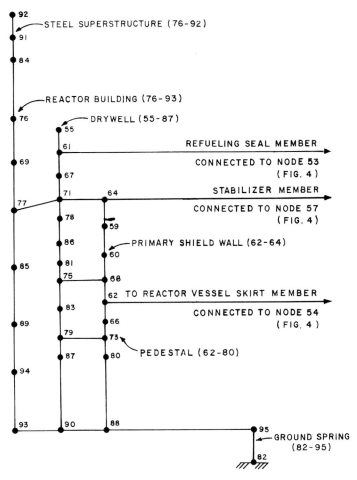

Fig. 3. Final model of containment structure.

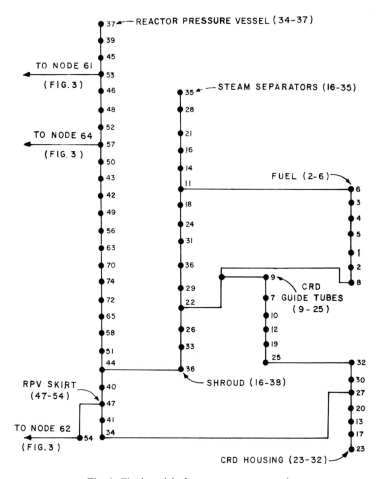

Fig. 4. Final model of reactor pressure vessel.

model represented by Fig. 3. Results from the two simpler model versions, which were checked qualitatively and quantitatively by several methods, were used to debug the final model. Although the two earlier models did not predict the localized response of the substructures, predictions of the overall structural behavior given by the three models were in close agreement.

The containment structure model parameters are presented in Table 2. The eigenvalue extraction and time-history analysis was performed using the STARDYNE version prior to the release of STARDYNE2. The nodal renumbering was performed by PREP requiring 30 system seconds of computer time. Eigenvectors, and participation factors were calculated using the Householder–QR option of the STARDYNE program. Extraction of

TABLE 2
CONTAINMENT STRUCTURE MODEL PARAMETERS

Nodes	95
Members	93
Static (degrees of freedom)	254
Dynamic (degrees of freedom)	95
Bandwidth (degrees of freedom)	30
Modes	95
Running time (system seconds)	
Prep	30
STARDYNE	281
DYNRE-I	278

95 modes required 281 system seconds. The time-history transient analysis which consisted of one forcing function and 30 significant modes was performed using the DYNRE-I program. This calculation required 278 system seconds. The present STARDYNE2 version is more efficient and these running times should be significantly less.

3.2. REACTOR VESSEL SUPPORTS

Figure 5 indicates the mathematical model used to investigate the effect of postulated earthquake and pipe-break loads on the structural system supporting the reactor vessel. The reactor vessel is supported by six sliding foot assemblies mounted on the neutron shield tank. The support feet are designed to restrain lateral and rotational movement of the reactor vessel for simultaneously applied earthquake and pipe-rupture loads, while allowing thermal expansion. The neutron shield tank is a double-walled cylindrical structure of steel which transfers the loadings to the heavily reinforced concrete mat of the containment structure. Overturning moments and horizontal forces which are induced on the tank during normal operation or accident condition are taken by the reinforced concrete primary shield wall poured around the neutron shield tank. The shield tank is fastened down by anchor bolts which resist any resulting vertical loads.

Beam elements with approximate properties were used to represent the reactor vessel and its support. Triangular plate elements were used to represent the complex neutron shield tank and primary shield wall structure. In the earlier version of STARDYNE a moment release capability was not included and a special ball joint substructure model was used to provide the capability of releasing the moment restraints about the three orthogonal axes. In the present example, the reactor vessel supports attached to each of the six reactor vessel nozzles were designed as ball supports. A typical

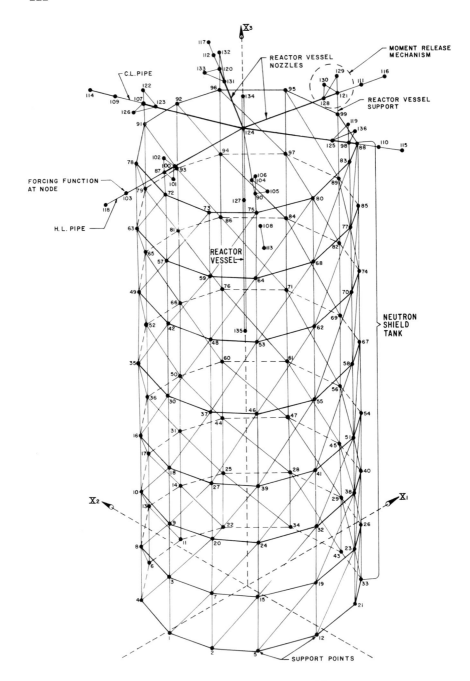

Fig. 5. Reactor vessel support model.

moment release substructure consisting of nodes 121, 128, 129, and 130 is indicated in Fig. 5.

The mathematical model given in Figure 5 consisted of approximately: 136 nodes, 57 beam elements, 168 triangular plate elements, 588 final static degrees of freedom, and 96 dynamic degrees of freedom. Numerous postulated break locations were considered and the mathematical model was modified to represent any changes in the geometrical configuration of the structural system. The Householder–QR option was employed to extract the modal data. A forcing function described by 20 to 40 time points was applied at the postulated break location. The time-history response was calculated at selected time increments specified by the user. Generally, 40 to 60 time points were considered during the first second. Each of the specified critical time points was subsequently rerun as a static loading case in order to obtain the internal member loads.

TABLE 3
MODEL PARAMETERS—REACTOR VESSEL SUPPORTS

Number of modes extracted	96
Modes used—time-history analysis	30
Number of forcing functions	2
Number of time points describing each forcing function	20
Number of selected critical time points	40
Number of static loads	40
Running time (system seconds)	
Modal extraction	1656
Time-history analysis	470
Static analysis	698
Post	260

Representative program running times are indicated in Table 3. The Householder–QR–modal extraction of 96 modes required 1656 system seconds. The time-history calculation for 40 selected critical time points was completed in 470 system seconds. Each of the selected critical time points in the transient analysis can be further processed in the static portion of STARDYNE to obtain internal member loads. The transient analysis output for 40 static load cases used 698 system seconds. Additional output options to calculate triangular plate stresses in the POST program (original STARDYNE version) were selected. This calculation required 260 system seconds. If the mass and stiffness matrices are unchanged and the modal extraction output is punched on cards or stored on tape, additional transient analysis can be performed without repeating the modal extraction. The modular construction of STARDYNE permits the user to supplement the intital data processing and retrieval as required, if the data base is unchanged.

The number of modes used in the time-history analysis affects the program running time, and an insufficient number of significant modes may also affect the program results. For the majority of problems considered, the higher modes are generally insignificant and can be excluded from the time-history analysis if an appropriate independent check of the results can be made. The investigation using the mathematical model given by Fig. 5 included the consideration of different pipe break locations. Except for one series of runs the time-history analysis based on the first 30 modes gave accurate results.

The applied loading is transmitted to the support system through a cantilevered pipe (nodes 103–87, Fig. 5). The member length (nodes 103–87) was decreased by a factor of two prior to the 30-mode time-history run (Case A), which gave inaccurate results. An evaluation of the data indicated that the higher shear modes of the cantilevered component had been neglected. The time-history analysis was rerun with the addition of nodes 51, 52, 91, and 93 (Case B). A comparison for Cases A and B of calculated cantilevered component loads are presented in Table 4. The shear and moment in the

TABLE 4
EFFECT OF MODES CONSIDERED ON TIME-HISTORY ANALYSIS RESULTS

	Case A Modes 1–30	Case B Modes 1–30, 51, 52, 91, 93
Shear—pipe (kips)	0.2	1,510
Moment—pipe (in-kips)	2.9	26,400
Shear—vessel support (kips)	599	643
Moment—vessel support (in-kips)	24,330	26,090

reactor support increased by less than 10% whereas the shear and moment in the cantilevered end of the pipe increased from 0.2 to 1510 kips and 2.9 to 26,400 in-kips, respectively. These increased loads were in close agreement with the hand calculations used as a check on the computer results.

4. Performance

A computer program is, at best, only a tool available to the engineer or analyst to assist him in the resolution of his structural problems. The end result, as with the application of any tool, is always dependent on the skill and judgment of the user and the effectiveness of the program in generating acceptable results. Because a computer program generally contains some errors and deficiences, the user's evaluation of the merits of a particular program is influenced to a large extent by the program's performance or total productivity.

The program's productivity may be assessed on the basis of the time and effort required to solve the problem once the problem is defined. Factors such as the operating environment, analytical capacity, availability of technical support, ease of modeling, input/output features, reliability, program speed, and turnaround affect the total performance of the program.

The STARDYNE system is available internationally through Control Data Corporation's Cybernet system of service bureaus and terminals. The program is maintained and supported by CDC structural analysts and Mechanical Research, Inc. technical staff. Additional program features and improvements in program efficiency are periodically implemented by MRI/CDC analysts. Programming errors are corrected quickly as they are discovered. As a result of the constant availability of technical support the program is highly reliable and the user is assured of continuing assistance to meet his individual needs. Control Data Corporation's structural analysts are available to assist the first-time user in implementing the program quickly. The experienced user receives support in applying the many data manipulation features of the program as well as in the resolution of unusual problems. Because of the combined MRI/CDC support of the program, the user is assured that the solution of a particular structural problem will be accomplished with a minimum delay resulting from program malfunction.

The program capabilities and input/output features are generally fixed, and, except for unusual circumstances, STARDYNE is available to the user as a "black box." The user must adopt his mathematical model to the capabilities of the program. The STARDYNE2 version of the program contains many new features which reduce the amount of time required to generate a suitable model. Data input has been simplified and limited element-generating capability is presently available. Many options are available for specifying the amount and format of the output. However, on occasion, the user may obtain more output than he actually desires, or find that data he require are not available in the desired format. An inexperienced user should check very carefully the output options to ensure that the desired results can be extracted from the program.

The program is designed with the user in mind. However, the program capabilities are limited to the solution of static and dynamic problems which are elastic and fall within the realm of the small displacement theory. The selection of element types is limited, but is considered to be adequate for the majority of the problems normally encountered. The user should investigate carefully the characteristic of the problem to be solved to ensure that the finite-elements available in STARDYNE will represent adequately the features of the problem.

The user's manual has been revised several times; the current version explains thoroughly the program capabilities and the procedures for preparing input data and selecting the proper output options. Input data errors are the primary cause of premature termination of program runs and additional error messages might be beneficial to the inexperienced user. A significant feature of STARDYNE is the error analysis and a detailed explanation of acceptable error bounds would benefit the user. The manual does guide the user in avoiding some of the more frequent modeling errors. It also indicates how to estimate the program running times.

The time required to solve a problem is also affected by the run turnaround time and the availability of back-up hardware support. The author's experience with STARDYNE and the CYBERNET system has been excellent in this area. The initial model runs generally contain some input errors and fast turnaround cuts down substantially the overall time required to obtain a solution. As many as 29 individual STARDYNE runs have been processed by the author and his co-workers during a 24-hr period. On rare occasions when the computer hardware was inoperative at the local data center the runs where processed at other data centers which are part of the CYBERNET system.

The question of program running cost is difficult to assess for a general class of problems because the program running time may be problem dependent. If the hardware and program characteristics are known, it is always possible to pick a bench mark to substantiate a position either for or against a software package. The results of any comparison can be misleading especially if they are taken out of context. Finally if a computer program cost comparison is made between the run charges of an in-house computer and service bureau run charges, the comparison is almost meaningless. It is not possible to store machine cycles and the in-house run charges can be assumed to vary by as much as a factor of five, again depending on the desired end result of the comparison.

It is the author's opinion that the total cost to obtain a solution to a problem using STARDYNE via the CYBERNET system is very competitive with respect to other programs available on either in-house or service bureau computers. In addition, if the problem must be solved over a short time interval the reliability and turnaround time of STARDYNE becomes a significant factor favoring the application of STARDYNE.

5. Conclusions

The STARDYNE system is a highly reliable finite-element program that can be used for static and dynamic structural analysis. The program has been fully tested and its ease and possible range of application recommend it highly to the user whose problems can be solved using small displacement theory. STARDYNE's performance when defined on the basis of total problem cost of solution is considerably better than the performance of many of the general-purpose structural programs readily available to the user.

Analysis and Design Capabilities of STRUDL Program

S. L. Chu

SARGENT & LUNDY ENGINEERS
CHICAGO, ILLINOIS

1. Introduction

The STRUctural Design Language [1], known by the acronym STRUDL, is a computer program that operates as a subsystem of the Integrated Civil Engineering System (ICES) [2] developed by M.I.T. Civil Engineering System Laboratory as a design aid and research tool for analysis and design in structural engineering.

There are two versions of STRUDL-II released by M.I.T. The use of the command-like input language and analysis/design facilities are presented in the ICES STRUDL-II Engineering User's Manual. The manual contains three volumes. *Volume 1, Frame Analysis* [3] describes the elements of command language and the use of the analysis for frame structures. *Volume 2, Additional Design and Analysis Facilities* [4] describes the use of STRUDL-II for the frame member selection, cost optimization, finite element analysis, dynamic analysis, geometrically nonlinear and buckling analysis. *Volume 3, Reinforced Concrete Structures* [5] describes the use of STRUDL-II for investigating and designing reinforced concrete beams, columns and slabs in accordance with ACI 318-63 code.

The first edition of Volume 1–3 presents the discussion on the design and analysis facilities of Version 1 of STRUDL-II. Version 2 of STRUDL-II provides a number of expanded capabilities on the finite-element analysis, dynamic analysis, and reinforced concrete structures; the second edition of Volumes 2 and 3 contains the description of those features. It should be noted that Version 1 of STRUDL-II is not STRUDL-I [6]. The latter only provides a limited portion of the analysis facilities available in STRUDL-II.

The purpose of this paper is to present an assessment of the present status of the various analysis and design capabilities of the STRUDL-II program.

2. Definition of the Problem

STRUDL utilizes Problem Oriented Language (POL) [7] to define the structure, loading and solution requirements of the problems. The flexibility and versatility inherent in POL provide greater efficiency in problem-solving than would ever be possible using conventional fixed-format input. A number of advantages in the use of STRUDL-POL are summarized as follows:

1. It directs overall flow of computation. STRUDL offers a wide variety of analysis and design facilities; POL provides the control for performing computation steps in a sequence which best suits the engineer's needs.

2. The engineer is only required to enter the data which differ from that defined by the STRUDL program; therefore, the amount of input data is greatly reduced.

3. The engineer has complete control over the amount of results to be produced at various stages of computation.

4. The free-format input of POL eliminates the use of cumbersome input data forms.

5. The order of problem input is not restricted by a rigid sequence. Rather, the data input may be entered in any convenient manner to suit the logical order of the computing procedure.

6. The structural element and nodal identifications can be alphabetic, numeric, or a mixture of both.

7. The similarity of structural element properties can be indicated by using labels and descriptive language, thus reducing the amount of repetitious input.

8. The user may choose any convenient units for his input and output at any stage of the analysis.

9. The STRUDL–POL input is in engineering language which is governed by a few basic rules. It facilitates both the input data preparation and input data checking.

10. No change of data input format is required when modifying and expanding the program facilities. Only an additional POL command need be added.

The program also offers other features that would not be possible with a conventional fixed-format type of input. For example, changes in structural configuration or loading can be made by entering only the data that are different from the original. Approximate solutions to complex structures can be found readily and with a saving in computation time. Eliminating redundant reactions or activating only one plane of elements in a space structure are but two options the engineer can utilize to obtain preliminary solutions.

One of the shortcomings of STRUDL is its lack of a mesh generation routine, which is quite essential in finite-element analysis. Although STRUDL-II does provide a grid generation for othogonal buildings, there is still a definite need for finite-element mesh generation in order to make this phase of the program more attractive to the designer.

3. Analysis Facilities

3.1. STIFFNESS ANALYSIS

STRUDL-II employs the stiffness method to solve large two- or three-dimensional linear, static, structural problems. The structure may be composed of frame members, finite plate, shell, or solid elements, or a mixture of these. It offers five types of frame members as shown in Table 1. A plane

TABLE 1
STRUDL-II MEMBER TYPES[a]

Type	Nodal degrees of freedom
Plane truss	U_1, U_2
Plane frame	U_1, U_2, U_6
Plane grid	U_3, U_4, U_5
Space truss	U_1, U_2, U_3
Space frame	$U_1, U_2, U_3, U_4, U_5, U_6$

[a] Directions shown in Fig. 1.

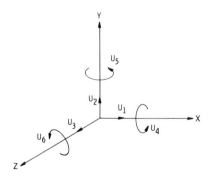

Fig. 1. Displacement directions.

TABLE 2
STRUDL-II ELEMENT TYPES (VERSION 1)

Type	Element name	Shape	Number of nodal points	Nodal degrees of freedom
Plane stress, plane strain	PSR		4	U_1, U_2
	CSTL		3	U_1, U_2
	CSTG		3	U_1, U_2
	LST		6	U_1, U_2
	PSROT		4	U_1, U_2, U_6
Plate bending	CPT		3	U_3, U_4, U_5
	BPR		4	U_3, U_4, U_5
	BPP		4	U_3, U_4, U_5
Shallow shell	SSCR		4	U_1, U_2, U_3 U_4, U_5
	SSCT		3	U_1, U_2, U_3 U_4, U_5
Plate stretching and bending	SBCT		3	U_1, U_2, U_3 U_4, U_5
Tridimensional	TETRA		4	U_1, U_2, U_3

structure may be described in any principal plane in a global coordinate system. The member properties can be input in a variety of forms as follows:

1. *Prismatic*—a straight member having a constant cross section.
2. *Variable*—a straight member having several segments with each segment having constant cross section.
3. *Tabulated*—given a designation, geometric properties of a member can be accessed from the secondary storage in conjunction with the use of ICES TABLE Subsystem [8].
4. Flexibility matrix.
5. Stiffness matrix.

The STRUDL-II Engineering User's Manual, Volume 2, First Edition, contains the description of the use of finite-element analysis for Version 1 of the STRUDL-II program. As shown in Table 2, it presents a variety of element types of plane stress, plane strain, plate bending, shallow shell, and three-dimensional stress analysis problems. Among those 12 elements, PSROT, SBCT and TETRA4 had not been released by M.I.T. as of fall 1971. Furthermore, it has also been found that the results obtained from the shell elements, SSCR and SSCT, are not reliable.

Table 3, which is taken from the STRUDL-II Engineering User's Manual, Volume 2, Second Edition, reviews the variety of finite elements, either currently operational or to be released by M.I.T. in the near future. Here a group of isoparametric elements has been added.

Both Versions 1 and 2 of STRUDL-II allow for the mixing of frame members and finite elements. Version 1 restricts the mixing to those frame members and finite-element types which have the same number of components of nodal degrees of freedom. Such a restriction is removed in Version 2 of STRUDL-II by the introduction of the use of the local-element coordinate and plane-element coordinate systems.

3.2. DETERMINATE AND PRELIMINARY ANALYSIS

As already mentioned, STRUDL allows the engineer to assume the internal member force components and joint reactions for a statically indeterminate problem for each loading condition to render the problem statically determinate. An intelligent user may use this simplified analysis procedure to approximate the real behavior of the structure for the preliminary design process.

3.3. DYNAMIC ANALYSIS

It appears that the dynamic analysis facilities in Version 1 are not operational in practice. However, the dynamic analysis facilities have been

TABLE 3
STRUDL-II ELEMENT TYPES (VERSION 2)

Type	Element name	Shape	Number of nodal points	Nodal degrees of freedom
Plane stress, plane strain	CSTL		3	U_1, U_2
	CSTG		3	U_1, U_2
	LST		6	U_1, U_2
	PSR		4	U_1, U_2
	PSROT		4	U_1, U_2, U_6
	LSR		8	U_1, U_2
	PSQ1		4	U_1, U_2
	IPLQ		8	U_1, U_2
	IPQQ		8	U_1, U_2
	IPCQ		12	U_1, U_2
	PSRCSH		4	U_1, U_2
	IPLQCSH		4	U_1, U_2
Plate bending	CPT		3	U_3, U_4, U_5
	BPR		4	U_3, U_4, U_5
	BPP		4	U_3, U_4, U_5

TABLE 3 (continued)

Type	Element name	Shape	Number of nodal points	Nodal degrees of freedom
Plate bending	PBQ1		4	U_3, U_4, U_5
Tridimensional	IPLS		8	U_1, U_2, U_3
	IPQS		20	U_1, U_2, U_3
	IPLSCSH		8	U_1, U_2, U_3
Bending	SBCT		3	U_1, U_2, U_3 U_4, U_5
Shallow shell	SSCR		4	U_1, U_2, U_3 U_4, U_5
	SSCT		3	U_1, U_2, U_3 U_4, U_5

rewritten for Version 2 with a considerable amount of expansion and were released in 1971. Version 2 of STRUDL-II performs modal analysis; it has the option of solving either "lumped" or "consistent" mass systems. The eigenvalues and eigenvectors can be extracted either by tridiagonalization or iteration processes. It performs both "time history" and "response spectra" analyses.

The technique commonly known as "kinematic condensation" (which reduces a system of equations of motion from the static degrees of freedom to the inertia degrees of freedom, and consequently, reduces the size of the eigenvalue problem) was not incorporated into Version 2 as of fall 1971.

3.4. NONLINEAR ANALYSIS AND BUCKLING ANALYSIS

It has been reported that these analysis facilities provide good results for the small, classroom-type problems. At present, no large practical problem has been reported as solved by this portion of the program.

4. Design Facilities

STRUDL-II is one of very few general-purpose programs available to the profession today offering steel and reinforced concrete member design and investigation capabilities. Employing the analysis procedures described before and aided by a recycling process, STRUDL-II provides the design facilities to accommodate a great variety of structural elements and loading conditions. Here, the users may impose the architectural constraints to a particular structural component in addition to the code and stress requirements to suit a particular structural configuration. To save computer time, the approximate analysis techniques such as determinate analysis and frame segmentation are employed extensively in the initial stress analysis to estimate the member forces and moments. Since an experienced engineer can often tell that a specific loading will only affect a particular portion of the structure, the intelligent use of these approximate analysis techniques can result in economic computer costs.

The user has complete freedom to specify the upper- and lower-bound dimensions for a particular member as well as the compatibility of dimensions among members. He may further design a group of members for the same size by equating them or by taking the most critical load carrying requirements.

4.1. SELECTION AND CHECKING OF STEEL MEMBERS

The steel member selection and checking procedure conforms to the 1963 AISC Specification. The properties of steel members are obtained from the secondary storage in conjunction with the use of ICES Table Subsystem. The steel member may be subject to the axial loads, bending moments, or a combination of both. The program does not design or investigate connections.

Experience has shown that the computation cost per member selection is quite high for Version 1. This is due to the fact that the internal member selection program algorithm and the manner in which the steel member properties are accessed from the secondary storage are not very efficient. However, the access speed for steel members has been improved in Version 2. In addition, it has been reported that McDonnell–ECI STRUDL has made program modifications on the member selection program algorithm.

The STRUDL-II User's Manual states that the STRUDL-II program is organized for easy addition of new codes and procedures, and the manual describes how the additions may be accomplished. Nevertheless, the author does not know of any users that have done so.

4.2. PROPORTIONING AND INVESTIGATION OF REINFORCED CONCRETE MEMBERS

A considerable amount of work has been done for reinforced concrete building design in STRUDL-II to provide the structural engineer with a convenient and powerful computer-aided design facility [9]. Although the reinforced concrete structure facilities released for Version 1 seem quite attractive for high-rise building design, this author knows of no such implementation as of the present.

The reinforced concrete facilities have been rewritten for Version 2 of STRUDL-II with an extensive number of additions to the program capabilities. At the time of this writing, there were no user experience reports for the reinforced concrete STRUDL facilities in Version 2, which had only just been released.

Version 2 of STRUDL-II provides reinforced concrete member proportioning and investigation in accordance with ACI 318-63 Building Code. All the member proportioning is based on the ultimate strength method, and the member investigations can be performed by either the working stress method or the ultimate strength method. It provides the proportioning and investigation of the following structural elements.

1. *Beam*—rectangular, T or L shaped beams.
2. *Column*—rectangular, square or round, tied and spiral columns.
3. *Flat slab*—with or without capitals and drop panels.
4. *One way slab system*—solid slab or joist floors.

In addition to the cross section dimensions, STRUDL-II also produces output of the primary and secondary reinforcement, the location and length of all bars, the critical loading condition and the quantity take-off for the concrete and reinforcement.

Version 2 of STRUDL-II allows a simplified method to define the structural configuration and its loading for typical orthogonal RC buildings. Here, the geometry of a building can be defined through the use of grid generation and the wind load acting upon the wall and the floor loadings can be obtained automatically.

5. Nondestructive Save/Restore and Graphic/Output

STRUDL-II is intended to be used as an information system to provide the facility to store and retrieve information about a structure. Such a capability is not available on versions released by M.I.T. However, some of these provisions are available in the proprietary McDonnell–ECI Version.

Although STRUDL-II provides commands to present graphic data on plotter and scope, such facilities are not operational in M.I.T. versions because the graphic equipment varies from one installation to another. It is reported that a few users have modified their STRUDL to provide the plotting capability. The printer graphic presentation is presently partially operational in the M.I.T. version.

6. Maintenance, Improvement, and Implementation

Although M.I.T. has made a continuous effort to improve STRUDL-II, their resources for system maintenance are limited. As early as 1968, M.I.T. announced that in view of the rapidly increasing number of ICES users and computing facilities, it had become extremely difficult for them to perform all the maintenance and improvement functions adequately.

In response to this, an ICES User's Group was formed to exchange experiences regarding the use of ICES, as well as to discuss maintenance and improvement of the ICES package. The User's Group is composed of members from consulting engineering firms, academic institutions, government agencies, computer service bureaus, and computer manufacturers. It holds meetings twice a year and publishes a newsletter three times a year to disseminate the information of the latest activity, use, and development of ICES. It also publishes the program "bugs" and remedies reported by users.

Because there is no organization specifically charged with the maintenance of STRUDL, many disappointed users have found that the problem solving speed is rather low or that some STRUDL commands are not operational.

A number of STRUDL users, favoring its scope and versatility, have devoted a great deal of time to its improvement and have reported their experiences at the ICES User's Group Conference. Among those are:

1. McDonnell–Douglas Automation Company, which helped to bring STRUDL-I to completion, has since been working on the verification and improvement of STRUDL-II. In order to service their customers better, they have been continuously upgrading their proprietary program package, known as McDonnell–ECI STRUDL, with the assistance of Engineering Computing International, Inc. They have made improvements on the STRUDL module and secondary memory usage, added new capabilities, endorsed the part of STRUDL facilities which they have tested, and periodically issued a STRUDL Advisory to their customers.

2. Pennsylvania State University, in the recent 7th ICES User's Group Conference, presented the underlying concepts of their PSU-ICES implementation. They developed a new ICETRAN precompiler which generates FORTRAN IV levels G and H language and devises a new method of addressing a common data area as improvements in primary and secondary memory usage and ICES subsystem generation.

3. A number of users have presented their experience on the implementation and modification of STRUDL to solve large buildings, steam piping systems and dynamic analysis, as well as problems used for instruction [10–15].

7. Machine Configuration

M.I.T. STRUDL, as a subsystem of ICES, was initially implemented on the IBM System 360. It requires, as a minimum machine configuration, the IBM System 360, Model 40 with 128,000 bytes and two 2311 disk drives. It also has been processed as part of ICES on the IBM System 360, Models 50, 65, 75, 85 and 91 and the IBM System 370, Models 145 and 155.

STRUDL is a machine dependent program because it is written in the machine oriented ICETRAN Language and Command Definition Language and processed under ICES EXECUTIVE Program. In the recent 7th Semi-annual ICES User's Group Conference held in Cherry Hill, New Jersey (June 1971), RCA announced their release of objective modules for STRUDL-I along with ICES EXECUTIVE, ICETRAN Precompiler, Command Definition Language and other ICES subsystems for their computing system. They also announced their computing service on their SPECTRA 70/40 at the South Jersey Information Systems Center in Cherry Hill, New Jersey.

At the same meeting, ACI-TEK Computer, Inc. discussed their development work on the conversion of ICES to the Univac 1108 system. It was stated that the conversion is 80% complete and they expect it to be in operation by September 1971.

8. Computer Cost

A number of problems, varying in size, are discussed here, comparing the actual cost of solving by STRUDL-II with that of solving with other programs. The actual execution time is not reported here in view of the fact that most of these problems are processed in the multiprogram environment, where the so-called input–output time is merely a number resulting from a particular accounting procedure for a particular operating system rather than real time. Also, some of these programs have overhead charges added to the standard computer rate, which varies from one to another. The STRUDL-II problems presented here were processed on the IBM System 360, Model 85 at McDonnell-Douglas Automation Company. It is believed that McDonnell–ECI STRUDL is more efficient than the original one released by M.I.T. On the other hand, there is a 30% overhead charge for this. The first three problems were also processed on the IBM System 370, Model 155 at the nearby IBM Data Center by M.I.T. STRUDL-II. The System 370 generally has a more favorable rate structure than the System 360 but the commercial rate is not readily available at the present time. It is estimated that the costs of STRUDL problem solving by the two computers would be approximately the same.

The plane truss shown in Fig. 2 has six joints, ten members, and one loading condition. It costs $5 to process it by STRUDL-II The same problem

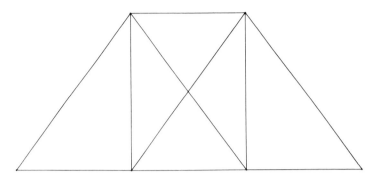

Fig. 2. Plane truss.

was solved by UCC STRESS-II [16, 17] on the Univac 1108 computing system at a cost of 67¢.

The space frame shown in Fig. 3 represents a nuclear steam supply system containing steam generator, pump, primary piping, and supporting structure. It has 91 joints, 96 members, and one loading condition. It costs $14 for STRUDL-II and $16 for UCC STRESS-II.

Fig. 3. Nuclear steam supply system.

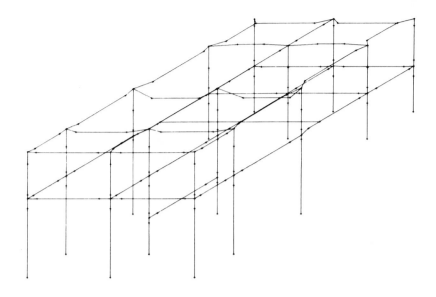

Fig. 4. Turbine foundation.

The space frame shown in Fig. 4 is an analytical model for a turbine-generator foundation. It has 179 joints, 207 members, and 11 loading conditions. The cost for processing it by STRUDL-II is $182. This same problem was solved by the SPACE Program [18] on the Univac 1108 system at the cost of $62.

Finite-element representations of a reinforced concrete mat foundation are shown in Figs. 5 and 6. The problem solved here consists of the mixing of finite plate elements and frame members. There are 298 joints, 142 beams, and two loading conditions. The problem was solved by STRUDL-II, EASE [19], and SAP [20]. Their respective finite elements and costs are shown in Table 4.

Experience indicates that it is more economical to process small analysis problems on UCC STRESS-II than on STRUDL. However, as the problem size increases the STRUDL-II program gains more advantage. For large analysis problems, the user finds that there are considerable savings in processing the problem by SAP, EASE, SPACE, or other large-scale computer programs, reducing costs by a factor of two or more. The excessive cost in

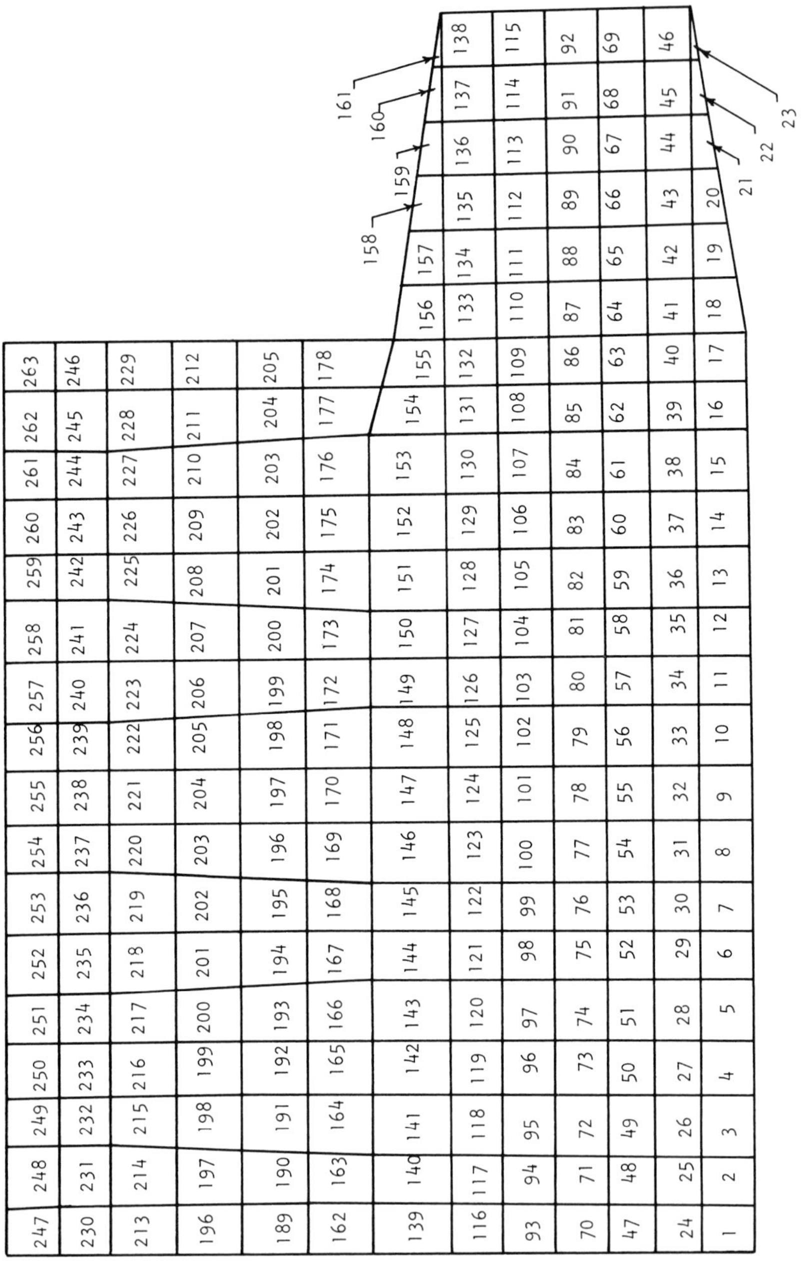

Fig. 5. Mat foundation (quadrilateral element).

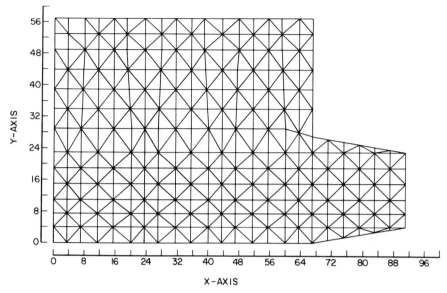

Fig. 6. Mat foundation (triangular element).

TABLE 4
FINITE ELEMENTS AND COSTS FOR THE SAME PROBLEM ON THREE PROGRAMS

Program name	STRUDL-II	STRUDL-II	EASE	SAP
Computing system	IBM 360/85	IBM 360/85	CDC 6600	CDC 6600
Type of element	Triangular	Rectangular	Triangular	Quadrilateral
No. of elements	524	263	524	263
Cost	$166	$157	$73	$55

processing problems by STRUDL-II is partially attributed to its IBM operating system; one often finds that it costs twice as much or more to solve a problem on an IBM system than it does to solve the same problem using the same program on the UNIVAC 1108 or CED 6600.

The excessive computer cost in processing STRUDL analysis problems has overshadowed the many versatile design and analysis facilities of STRUDL. In a check with several large engineering firms, it was found that they seldom use STRUDL to solve problems, primarily because of its high computer cost.

9. Conclusion

There is no doubt that the availability of ICES STRUDL has made a significant advance in the man–machine communication environment. Its conceptual design and implementation of a problem-oriented language, the command modular program generation, program linkage mechanisms, and dynamic data management have furthered the process of development of large scale integrated programs.

To a potential ICES STRUDL program user, the flexibility of the STRUDL command input data preparation and the versatility of analysis and design facilities are very inviting. However, sophisticated use of the problem-oriented language, problem-solving and data input facilities requires a great deal of experience. Because of the lack of confidence in the workability and reliability of a certain number of STRUDL program facilities, and in view of its computer costs as compared to other computer programs in problem solving, it is often difficult to justify the use of STRUDL from an economical point of view.

Some practicing engineers have questioned the value of the design facilities provided by STRUDL. The proportioning and selection of many structural members are controlled by the constraints of architecture, construction, human comfort, or economics, rather than the load-carrying capability.

Some structures, it is pointed out, have a great variety of loading conditions, and a particular loading is only relevant to a particular portion of a structure. Frequently, a number of those loadings are subjected to change as the design proceeds. For such structures, it would be beyond any engineer's experience and comprehension to provide an efficient programming algorithm that would account for all the design constraints. Thus, it is argued, an experienced engineer, confident of his judgment and aided by simple calculations, often can relate a particular loading or its change to a specific portion of a structure to accomplish his design purpose more effectively than the rigorous repetitive design process used by STRUDL for an entire structure.

The cost of optimization, nonlinear analysis and buckling analysis facilities of STRUDL have limited their use in practical application. Since excessive calculation is required for this type of analysis and since a great deal of uncertainty exists with respect to the reliability of the results, there has been little application in this area. For these reasons, engineers have turned to other programs such as STRESS, FRAN, EASE, SPACE, STARDYNE, or in-house programs, which show considerable savings on the computer cost and whose results they have learned to trust.

The success of a large-scale computer program system depends on both program developers and program users. Qualified programmers must provide the users with their needed computing facilities, efficient computing algorithms and availability to continuously upgrade its operation. The interested users, on the other hand, utilize these program facilities and, at the same time willingly assist the programmers in debugging possible program errors. The perfection of a large program relies on the interaction of both groups. As of today, such interaction in STRUDL seems lacking.

The efficiency of a program operation vitally depends on its operating computer system. As long as STRUDL remains operating only on the IBM System 360, with its current rate structure, the expectation for wide acceptance of STRUDL for solving large problems appears dim.

References

1. Logcher, R. D., and Sturman, G. M., "STRUDL—A Computer System for Structural Design," *J. Structural Div. Proc. ASCE* **92**, No. ST6, p. 191.
2. "ICES System: General Description," Report R67-94. Massachusetts Inst. of Technology, September, 1967.
3. "ICES STRUDL-II, Engineering User's Manual, Vol. 1, Frame Analysis," 1st ed., Report R68-91. Massachusetts Inst. of Technology, September, 1967.
4. "ICES STRUDL-II, Engineering User's Manual, Vol. 2, Additional Design and Analysis Facilities," 1st ed., Report R68-92, June, 1969; 2nd ed., Report R70-71, June, 1970. Massachusetts Inst. of Technology.
5. "ICES STRUDL-II, Engineering User's Manual, Vol. 3, Reinforced Concrete Structure," 1st ed., Report R68-93, December, 1968; 2nd ed., Report R70-35, June, 1970. Massachusetts Inst. of Technology.
6. "ICES STRUDL-I, Engineering User's Manual," 1st ed., Report R68-91, Massachusetts Inst. of Technology, September, 1967.
7. Miller, C. L., "Man-Machine Communication in Civil Engineering," *J. Structural Div., Proc. ASCE* **89**, No. ST4, p. 9.
8. "ICES TABLE-I, Engineering User's Manual," 1st ed., Report R67-58, Massachusetts Inst. of Technology, September, 1967.
9. Biggs, J. M., Pahl, P. J., and Wenke, H. N., "Integrated System for RC Building Design," *J. Structural Div., Proc. ASCE* **97**, No. ST1, p. 13.
10. Gilkey, C. H., "STRUDL, As An Effective Piping Analysis Program." Paper presented at the Sixth ICES Users Group Conference, January, 1971.
11. Wells, R. A., "Efficient Use of STRUDL for the Solution of Large Problems." Paper presented at the Sixth ICES Users Group Conference, January 1971.
12. Ringe, B. C., "Design of Christ Lutheran Church Frame." Paper presented at Fifth ICES Users Group Conference, June 1970.
13. Logcher, R. D., Patel, K., and Harmon, T., "Analysis and Design of the 87-Storey Standard Oil of Indiana Building, Chicago, Illinois." Paper presented at the Fifth ICES Users Group Conference, June 1970.
14. Carver, B. R., "Use of STRUDL-II to Analyze Free Vibrations of Structures." Paper presented at Fifth ICES Users Group Conference, June 1970.
15. Chalobi, A. F., "Users of ICES STRUDL in Education at WPI." Paper presented at the Fifth ICES Users Group Conference, June 1970.
16. Fenves, S. J., Logcher, R. D., Mauch, S. P., and Reinschmidt, K. F., *STRESS User's Manual*, M.I.T. Press, Cambridge, Massachusetts, 1964.

17. *STRESS-II*, *User's Guide*, University Computing Company.
18. *SPACE—Structural Preprogrammed Analysis Capability for Engineers*, *User's Manual*. Digital Analysis Consultants, Inc., La Jolla, California.
19. *EASE—Elastic Analysis for Structural Engineering*, *User's Manual*. Engineering/Analysis Corporation, Redondo Beach, California, March 1970.
20. Wilson, E. L., "SAP—Structural Analysis Program." Univ. of California, Berkeley, California, September 1970.

Elastic–Plastic and Creep Analysis via the MARC Finite-Element Computer Program

D. J. Ayres

COMBUSTION ENGINEERING, INC.
WINDSOR, CONNECTICUT

The MARC general-purpose, nonlinear finite-element computer program is a powerful tool for nonlinear stress analysis. The program is oriented toward flexibility of application and user modification. It is presently being used to solve elastic–plastic and creep problems associated with the design of high-temperature pressure containing components. In order to illustrate some of the capabilities of the program, two example problems are discussed in detail. The first is the elastic–plastic analysis of the effect of a temperature transient on a nozzle in a thick vessel. The second is an elastic–primary creep analysis of a thick, high-temperature boiler stop valve. Several other types of problems are discussed briefly to illustrate the variety of problems that have been solved. The agreement of these solutions with experimental results and experience indicate that a high degree of confidence can be placed in an analysis using the MARC program.

1. Introduction

Structures and pressure vessels which are subjected to conditions that cause plastic flow and creep have been designed, built, and operated for many years. These have traditionally been analyzed by the techniques of linear elasticity. The results of these elastic analyses have been interpreted in the light of experience, and "experience factors" have been applied so the

247

components would operate safely and reliably. A good design required a substantial back-up of experience and test data to assure that these "experience factors" were adequate but not excessive. This design process was acceptable so long as designs changed slowly so that extrapolations from experience were not too great.

Today, this design process is often not acceptable because designs are changing faster than previous ones can be constructed, and totally new designs, materials, and situations are encountered. In these cases, the "experience factors" are not known, and an analysis of the plastic and creep behavior of the component may be required.

At Combustion Engineering, the search for a way to analyze a wide variety of elastic–plastic and creep problems led to the MARC program developed by Prof. P. V. Marcal of Brown University. This program was selected because of its orientation toward flexibility of application and user modification, and because it computed acceptable answers to a series of test problems.

MARC is a general-purpose finite-element program for the nonlinear static analysis of structures with large displacement. The element library contains two- and three-dimensional elements and plate and shell elements. New elements can be added easily by the user. A node-tying option is available which allows arbitrary linear constraints between degrees of freedom and the connection of elements with different degrees of freedom. Plasticity behavior is based on the theory of isotropic, elastic–plastic, time-independent materials with a Mises yield criterion, isotropic or kinematic strain hardening, and temperature-dependent elastic properties, and equivalent yield stress. Creep behavior is based on a Mises flow criterion with isotropic behavior described by an equivalent creep rate law specified by the user. The program uses the tangent modulus method for plasticity and an iterated initial strain method for creep calculations [1].

The capabilities of the program can best be illustrated by some example problems. The first is the elastic–plastic analysis of the effect of a temperature transient on a nozzle in a thick vessel. The second is an elastic–primary creep analysis of a thick, high-temperature boiler stop valve. Other examples include plastic bending of a thin tube and the collapse of a nearly rectangular pressure vessel onto a rigid mass. Once the capabilities are exposed, the merits of the program can be discussed.

2. Example 1—Plasticity Analysis

One type of problem which occurs fairly frequently is the determination of thermal stresses in a complex component such as a nozzle. The geometry of an example nozzle is shown in Fig. 1. In this case, hot fluid flows through the

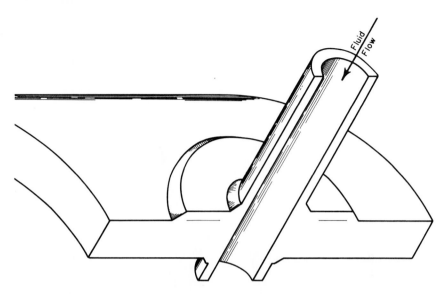

Fig. 1. Example nozzle geometry.

nozzle into the large-diameter thick vessel for a short time. The vessel and nozzle are cooled and then the hot fluid flows into the vessel again. The temperature distribution and thermal stress distribution is nearly axisymmetric about the centerline of the nozzle. Figure 2 shows the approximate temperature distribution in the nozzle at the time of maximum temperature gradient. The objective in this example is to compute the stress and strain in the nozzle in order to estimate thermal cyclic life.

The axisymmetric finite-element model of the nozzle is shown in Fig. 3. The four-noded isoparametric element used was written by a program user and incorporated into the program. A similar element is now in the program library. The material is assumed to be elastic–plastic with linear strain hardening. For this example, the properties are considered to be independent of temperature.

The elastic stresses in the hoop and radial direction, relative to the nozzle centerline, are shown in Figs. 4 and 5. These values are scaled so one element is just at yield. As the fluid temperature is increased, the highly stressed areas yield plastically. The location and extent of the plastic zones for several fluid temperatures are shown in Fig. 6. The strains are computed for each fluid temperature, and the cyclic life can easily be estimated from these strains and a proper fatigue curve.

Operating experience with the nozzle verified the computed zones of high stress and the fatigue life estimations. This agreement leads to a high level of confidence in the MARC program.

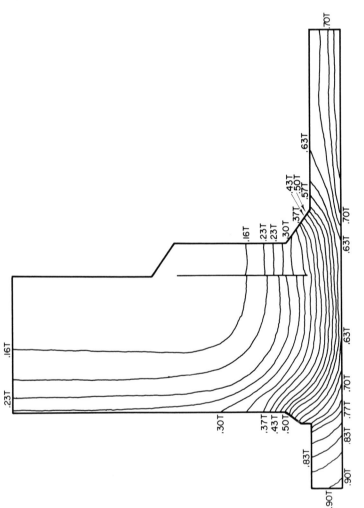

Fig. 2. Temperature distribution at time of maximum temperature gradient. T is the fluid temperature.

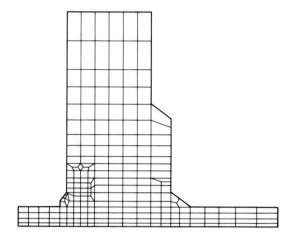

Fig. 3. Finite element model for nozzle.

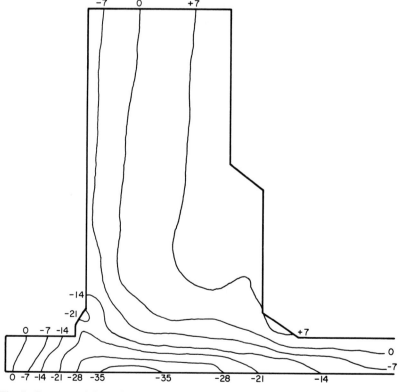

Fig. 4. Contours of thermal stress in hoop direction in nozzle. Note: stresses are in 1000 lb in.$^{-2}$.

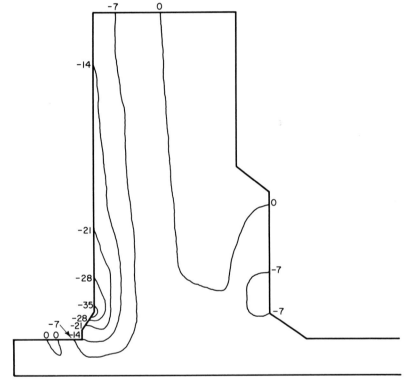

Fig. 5. Contours of thermal stress in radial direction in nozzle. Note : stresses are in 1000 lb in. $^{-2}$.

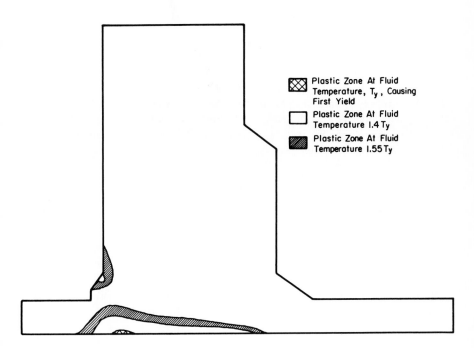

Fig. 6. Plastic zones in nozzle.

The accuracy of the cyclic life predictions is completely dependent on the similarity between the plastic behavior of the nozzle material and the constitutive equation assumed for the analysis, and on the appropriateness of the fatigue curve chosen. This dependency magnifies the material information requirements for an adequate nonlinear analysis to many times that required for an elastic evaluation of the same component.

3. Example 2—Primary Creep Analysis

Many components of boilers, reactors, and associated equipment operate in the temperature range where substantial time-dependent deformation occurs. An example of such a component is a boiler stop valve. The valve was designed for operation at 1200°F with an internal pressure of 5200 lb in.$^{-2}$ A three dimensional photoelastic analysis of the pressure stress in the valve was performed. A section of the photoelastic model is shown in Fig. 7. The redistribution of the stress and the time-dependent deformation is required in order to determine the operating life of the valve.

The valve is a complex three-dimensional body. Because a three-dimensional analysis would have been very expensive, an approximate axisymmetric analysis was attempted. The axisymmetric finite-element model is shown in Fig. 8. The four-noded isoparametric element used is presently in the MARC element library.

The computed hoop and axial stresses are compared to the photoelastic results along the inner surface of the valve in Figs. 9 and 10, respectively. The trends of the stresses are quite similar. This general agreement of the elastic stresses indicates that the axisymmetric model can lead to an understanding of the three-dimensional creep deformation and stress redistribution.

The stress redistribution is illustrated at three sections through the valve. These sections are indicated in Fig. 11. The hoop stresses across sections 1 and 2 for several times after the pressure is applied are shown in Figs. 12 and 13. Creep causes a substantial reduction in these peak stress values. The peak axial stresses on Section 3, however, are shown in Fig. 14 to increase slightly with creep. These figures demonstrate the rapid redistribution of stresses from the elastically determined values which can be expected to occur in the real three-dimensional valve.

If a different primary creep law were assumed, Figs. 12–14 would be different. A large effort must be expended to provide multiaxial creep behavior data in order that the conclusions drawn from a creep analysis will apply to a high-temperature component.

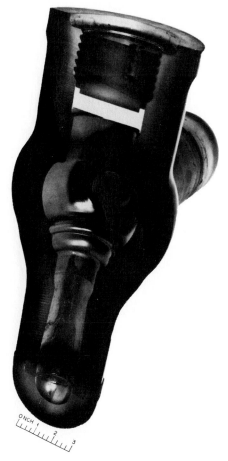

Fig. 7. Boiler stop valve. The internal contour of the valve model with the rubber seal and threaded plug in place.

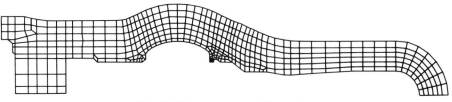

Fig. 8. Finite element model for valve.

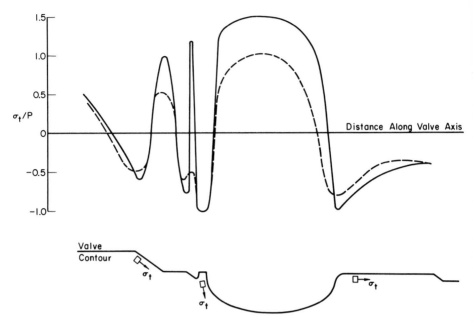

Fig. 9. Ratio of elastic stress tangent to interior surface to pressure in valve. (—): Three-dimensional photoelastic results; (– – –): finite element results.

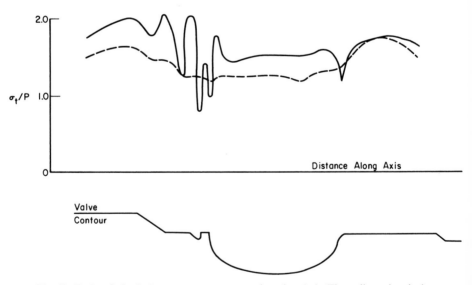

Fig. 10. Ratio of elastic hoop stress to pressure in valve. (—): Three-dimensional photoelastic results; (– – –): finite element results.

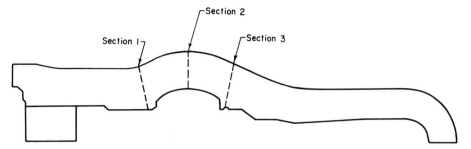

Fig. 11. Sections for examination of stress redistribution.

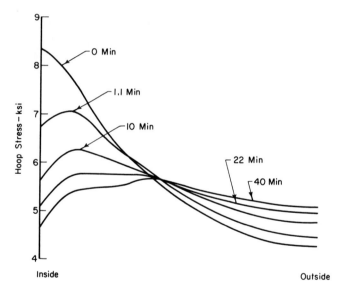

Fig. 12. Redistribution of hoop stress at section 1.

4. Other Examples

Another class of problems involves determining the maximum load that a component can sustain. Limit analysis of simple geometries of elastic–perfectly plastic material can be made fairly easily. More complex geometries or material behavior, however, require a more sophisticated solution. Such a problem is the determination of the maximum moment that a thin tube of elastic–plastic strain-hardening material can sustain.

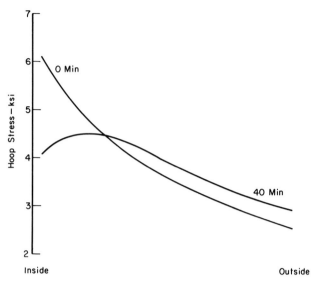

Fig. 13. Redistribution of hoop stress at section 2.

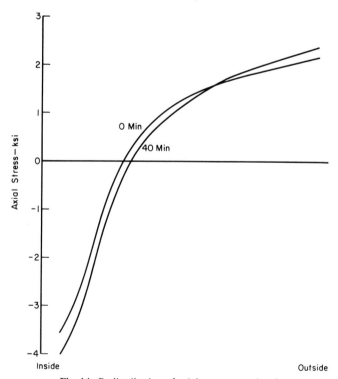

Fig. 14. Redistribution of axial stress at section 3.

The finite-element model of the tube is shown in Fig. 15a. The element used is a three-noded general shell with 9 degrees of freedom at each node [2]. One end of the model is fixed in the axial direction. At the other end, a moment is applied by loading two opposite nodal points. The other nodes in the endplane are tied to the loaded nodes to maintain a linear distribution of axial displacement. Figure 15b illustrates an idealization of the stress–strain curve for the tube material.

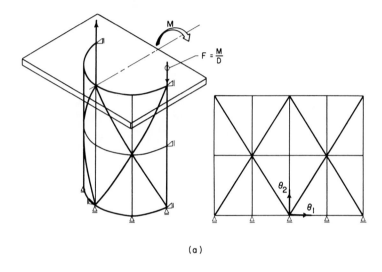

(a)

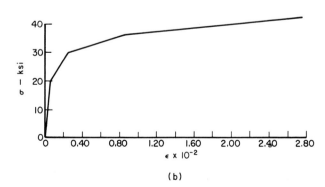

(b)

Fig. 15. (a) *Left:* Finite element model of thin tube showing element application. *Right:* Half tube model in transformed Gaussian coordinates. (b) Linear stress–strain curve for Inconel-600 at 650°F.

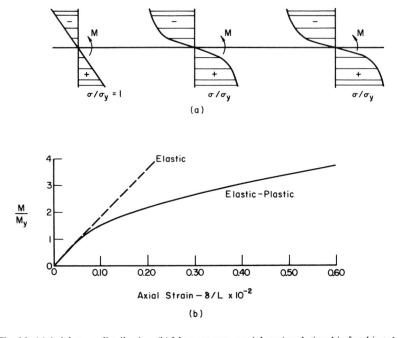

Fig. 16. (a) Axial stress distribution. (b) Moment versus axial strain relationship for thin tube.

The results of the analysis are shown in Fig. 16. The applied moment must always increase for increasing axial strain. The determination of a precise limit moment, the moment which would cause indefinite displacement, therefore, cannot be made. A different criterion, perhaps based on allowable strains, must be employed. This new criterion can only be developed by a combined analytical and experimental effort. The three-dimensional tube model can be used in this effort since loadings in all directions as well as internal or external pressure can be applied to simulate a very general test specimen.

A final example to illustrate the flexibility of the program is the solution of a collapse or contact problem. A thin, rectangular container with rounded corners which is subject to external pressure is shown in Fig. 17. The inward deflection is restricted by the rigid mass which is being contained. The analysis was performed to evaluate several different corner geometries.

The container cross-section was sufficiently unconfined so that plane stress could be assumed. The plane stress finite-element model is shown in Fig. 18. The four-noded isoparametric element is the companion element to the one used in the first example. The material was assumed to be elastic–plastic with strain hardening.

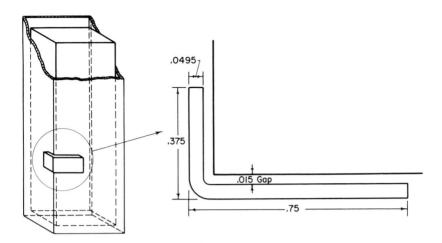

Fig. 17. Geometry of thin container for collapse analysis.

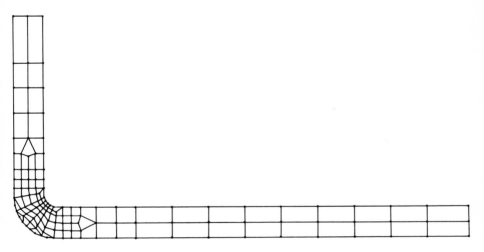

Fig. 18. Fine-element model of container.

The pressure was applied incrementally above the initial yield of the container. If a point reached a displacement so that it was within a prescribed small distance from the rigid mass at the end of a particular increment, that small distance was applied as the displacement during the next loading increment. During all following increments, the displacement normal to the rigid mass was zero.

The effective plastic strain at the design pressure for one of the corner configurations is shown in Fig. 19. The analysis assumed that the strains

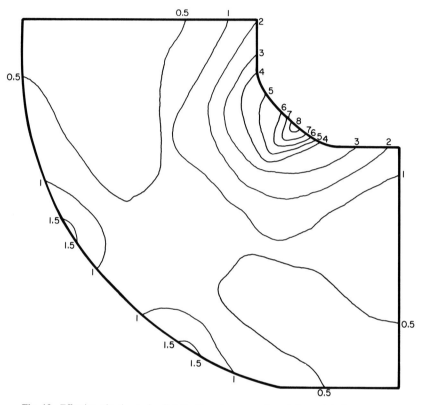

Fig. 19. Effective plastic strain distribution at corner of container at design pressure.

would be small. Since the maximum effective plastic strain is around 8 %, the small strain assumption is suspect. The conclusion can safely be drawn, however, that in this geometry, very high strains exist. Hopefully, another geometry can be found that would produce lower strains.

The MARC program had to be modified to solve the contact poblem. This was done easily and quickly in a fairly general manner. Other contact problems have been solved using variations of this modification. The flexibility of the program and ease of modification allows the user to solve types of problems which the program author did not envision.

5. The Merits of MARC

The MARC program has been shown to have the capability to solve elastic–plastic and creep problems of a general nature. One disadvantage of all finite-element programs, and especially nonlinear ones, is that the output is simply no better than the processed input. Meaningful results require a substantial knowledge of the material property behavior—often a greater knowledge than is presently available. One merit of MARC is that any material description can be input and, by being sufficiently well organized, the implications of a description can be followed through the program to assure that the user's objectives are fulfilled.

A final advantage of MARC is the availability of the author of the program. Although we claim to have a deep understanding of the program, there is no experience comparable to that of the author. Prof. Marcal has been extremely helpful in the resolution of real and imaginary program and system bugs, and in assisting us with our efforts to understand the program.

6. Conclusions

The MARC general-purpose, nonlinear finite-element computer program is a powerful tool for nonlinear stress analysis. The program is oriented toward flexibility of application and user modification. A wide variety of elastic–plastic and creep problems have been solved successfully, thereby increasing confidence in the program.

The MARC program has removed a major obstacle from good plasticity and creep calculations. The stumbling block is clearly the adequate description of material behavior. A substantial effort is now underway to obtain more precise material constitutive equations, and thereby make more realistic our elastic–plastic and creep analysis via the MARC finite element computer program.

References

1. Marcal, P. V., "Finite Element Analysis with Material Nonlinearities . . .". Theory and Practice," U.S.—Japan Seminar on Matrix Methods of Structural Analysis and Design; *JSSC Symp. Matrix Methods Struct. Anal., Tokyo* (August 1969).
2. Dupuis, G., and Goel, J. J., "A Curved Finite Element for Thin Elastic Shells," Brown Univ., Eng. Rep. N00014-0007/4 (December 1969).

PART III

Finite Difference/Finite Elements—
A Merging of Forces

A Survey of Finite-Difference Methods for
Partial Differential Equations

J. P. Wright and M. L. Baron

PAUL WEIDLINGER, CONSULTING ENGINEER
NEW YORK, NEW YORK

1. Introduction

The variety and number of problems and methods associated with finite differences make it virtually impossible to present a comprehensive survey of finite-difference methods in an article of this type. Consequently, it is necessary to limit this paper to some specific areas which may be of general interest in engineering mechanics. Engineering mechanics has moved increasingly toward the solution of nonlinear multidimensional problems, particularly as the computational means for solving such problems have become available. In the present paper, the authors have selected several topics connected with the solution of systems of partial differential equations in multidimensional space or space–time coordinates, which are not yet in common usage in structural mechanics. The topics and references are not intended to be complete, rather, they represent the particular interests of the authors. The majority of the methods that are considered have been developed and have become increasingly popular in the last 10 to 15 years; this, of course, has coincided with the development of high-speed electronic computers which has made the use of many of these approaches practical. It should be emphasized that, although many of these methods were developed in connection with finite-difference applications, they may often be adapted to finite-element methods without changing the basic ideas (although the details may change considerably).

The remainder of the paper is divided into eight sections. Section 2 contains a general discussion of finite-difference methods for solving partial differential equations. Sections 3–5 present a summary of some finite-difference methods for each of the three basic types of equations: hyperbolic, parabolic, and elliptic. Each is dealt with in a separate section since, in many cases, it is necessary to consider the different characteristics of each type of equation and because, historically many of the methods have been developed in this way.

In Section 3, two basic methods for solving hyperbolic equations are discussed: (a) characteristic methods and (b) methods involving viscosity, either numerical or physical. Section 4 presents some of the standard methods for solving parabolic equations including "alternating-direction, implicit" (ADI) methods.

The solution of elliptic equations, discussed in Section 5, typically becomes a problem of solving systems of simultaneous algebraic equations which, in general, may be nonlinear. For linear algebraic systems, direct solution methods by means of elimination are well known and hence, are not discussed in detail. Iterative methods for solving both linear and nonlinear algebraic systems are quite important for elliptic systems, and a survey of some of the standard iterative techniques is presented. The close relationship between iterative methods and initial-value problems is also discussed.

Sections 6 through 9 are directed toward some recent developments concerning the solution of more complicated mixed systems, e.g., hyperbolic–elliptic or hyperbolic–parabolic systems. In Section 6, the "tensor product–fast Fourier transform" method for solving second-order, separable, partial differential equations is discussed in some detail. Section 7 discusses the numerical analysis concept of the "flexibility" of a difference scheme as related to the conclusion that implicit methods are especially important for nonlinear problems, even for hyperbolic systems. Section 8 presents the "method of fractional steps," which is a generalization of the alternating-direction implicit methods. It will be seen that the fractional-step methods provide a unified approach to a class of finite-difference methods for all three types of partial differential equations, and therefore, for more complicated mixed systems. In particular, implicit schemes, which are stable and computationally efficient, can be constructed by means of this method. Finally, since the stability analysis of a time-dependent calculation is an essential part of solving systems of partial differential equations, the so-called "energy" method is discussed in Section 9. This method is particularly useful for analyzing difference schemes for nonlinear problems.

2. General Discussion

Finite-difference methods for partial differential equations, covering a variety of applications, can be found in standard references such as those by Richtmyer and Morton [1], Forsythe and Wasow [2], and Ames [3]. In Richtmyer and Morton [1], initial-value problems are treated quite thoroughly, including applications of finite-difference methods to parabolic and hyperbolic systems, vibrations of a thin elastic beam, and fluid dynamics problems in one and several space variables. In addition, a great many theoretical topics including accuracy, stability, and convergence are considered. Forsythe and Wasow [2] and Ames [3] deal with elliptic as well as hyperbolic and parabolic systems. Other current work in a variety of associated fields can be found in the series, *Methods of Computational Physics* [4, 5], which includes some articles on finite-element methods, as well as finite-difference methods in continuum mechanics.

Numerical finite-difference solutions of boundary-value problems involving partial differential equations in multidimensional space–time coordinates are generally obtained by means of a space grid, using a difference operator of a predetermined order of error. For those cases in which time appears as an independent variable, a numerical step integration in time is used (again of a certain order of accuracy). Such approaches have been widely applied to problems in which the bounding surfaces of the domains coincide with the surfaces of the coordinate system used in the analysis. This has resulted in the development of a series of difference operators in various coordinate systems, such as Cartesian, skew, triangular (both equilateral and nonequilateral), polar, that correspond to regular shapes of interest in structural and more general continuum mechanics applications. Examples of the various types of operators together with their orders of error are given by Salvadori and Baron [6, Chap. V]. The application of such operators to a variety of structural problems involving the buckling of beams, plates, and shells is given by Salvadori [7]. Another interesting structural application involving the use of first-order nonequilateral triangular coordinates in the analysis of skew slabs is given by Jenson [8].

For cases in which the boundaries of the domain of interest are irregular, i.e., the boundaries do not coincide with the coordinate surfaces, the finite-difference formulation of the particular problem can be expressed in terms of unequally spaced difference operators [6], which again were developed to give specific orders of error. It should be pointed out that such operators, when inserted into a finite-difference numerical code, enable one to do the specific problem or problems for which the operator was developed, but that the introduction of other types of irregular boundaries requires a modification of the code. Generally, a series of problems involving a variety of different types of irregular geometries and the same partial differential equation can be more conveniently solved by finite-element methods. It is probably this fact more than any other that is responsible for the current wide interest in finite-element methods in structural analysis and design.

For problems involving derivatives with respect to time, a finite-difference approach leading to step integration in time can be of an explicit or an implicit type. Implicit schemes require the solution of systems of algebraic equations at each time step but are generally stable. Thus the time step in such calculations is limited by accuracy requirements only. In explicit schemes, the unknowns are obtained directly, but the time steps are limited by stability as well as accuracy requirements. The question of the use of explicit versus implicit schemes is very much dependent on the problem to be solved. A more detailed discussion of implicit and explicit schemes is given in Section 7.

In many problems involving localized areas of large gradients, such as stress concentrations or boundary layers, it is often both economical and necessary to concentrate the grid points of the finite-difference mesh in these areas, and to use a sparser grid elsewhere in the domain. The principal difficulty in doing this involves the treatment of points at the interface of the different meshes. This problem is dealt with by means of "transition bands" in "graded nets," as discussed by Forsythe and Wasow [2, pp. 365ff].

Recent interesting work has been done on the simultaneous use of two or more large, finite-difference codes for the purpose of solving complicated multimaterial problems. An example of this is given by Crowley and Barr [9], in which the time-dependent problem of fluid flow through a deformable solid tube is discussed.

3. Hyperbolic Systems

In the solution of hyperbolic systems in two independent variables (one space and one time variable), the characteristic directions and curves of the system play an important role. A finite-difference approach that utilizes characteristics consists of a step integration along a computational mesh composed of arcs that are approximations to the characteristic curves of the system. Since the characteristic curves are not generally known in advance but must be obtained as part of the solution, an approach can be used in which the computational network is built up as the computation proceeds. The general approach of using a characteristic network in problems with two independent variables is discussed by Crandall [10], Lister [11], Massau [12], and Hoskin [13].

The solution of hyperbolic systems in three or more independent variables by characteristic methods is quite complicated in specific cases and, in general, a complete theory, analogous to the case of two independent variables, does not exist. Nevertheless, a large class of equations, including those of fluid dynamics, can be treated by bicharacteristic methods. A number of attempts to solve problems using this approach have been made, e.g., see Butler [14], and Richardson [15]. The main difficulty of using characteristic methods in two or more space dimensions is the geometrical one of locating a priori unknown surfaces (rather than lines as in the case of two independent variables) on which the solution may be discontinuous. The advantage to such an approach, when successful, is that the solution can be made quite accurate with relatively few grid points since discontinuities are treated properly.

If the discontinuity surfaces (characteristics) of a hyperbolic system are not handled properly in a finite-difference calculation, numerical errors can propagate in such a way as to make the calculation worthless, or in fact, to cause numerical instability. An alternative approach to the characteristic methods is to recognize that the surfaces of discontinuity of the hyperbolic system represent, in fact, physical phenomena (e.g., shocks) that can be viewed as regions of rapid but continuous change in the solution, i.e., regions of large gradients. The equations of the problem are then modified so that the new system of equations has such continuous solutions. A version of this approach, which is widely used in the solution of fluid dynamics problems, is the "pseudoviscosity" method of von Neumann and Richtmyer [16], Richtmyer and Morton [1]. The main advantage of this approach is that the regions of rapid change are automatically treated. However, the accuracy of the solutions might not be expected to be as good as those obtained from characteristic methods.

Two computer codes which employ the pseudoviscosity method to solve one-dimensional space–time problems are known as PUFF [17], and WONDY [18]. Tests by Hicks [19], comparing numerical results obtained using the PUFF code with known analytical results for seven problems, show the method to be surprisingly accurate. Hicks also presents a detailed discussion of the usual one-dimensional hydrodynamics model in which it is assumed that the fluid is homogeneous, compressible, inviscid, non-heat conducting, and has a perfect gas equation of state. He then demonstrates that internal friction forces are present across shock fronts and the contradictory assumption that the fluid is inviscid is then modified to state that the viscosity is zero except at the shock fronts. Thus, it is sometimes possible to find a reason, based on physical principles, which guides the selection of the viscosity that is used in the problem solution.

The pseudoviscosity method has been used in multidimensional fluid dynamics calculations by Noh [20, 21], and the method has been used by

Wilkins [22] for multidimensional wave propagation studies in elastic–plastic (von Mises yield condition) solids. Work on generalizing the form of the pseudoviscosity in multidimensional problems has been done by Schultz [23]. The pseudoviscosity method is most commonly employed in problems for which a material (Lagrangian) description is used and a viscosity of some type is needed for numerical stability.

A somewhat similar, but less physical approach to the problem of instability is to write the difference scheme so that a "numerical" viscosity is introduced. This is the source of the stability of the "Lax" method [24], the "donor-cell" method [25], the "Lax–Wendroff" method [26], the "Lelevier" method [1], the "Particle-In-Cell" (PIC) method [27, 28], and many others. A method of investigating the stability of such methods is to use a Taylor series expansion technique as discussed by Hirt in [29], in which he examines the stability of the donor-cell scheme, which is particularly good. One of the main features of Hirt's technique is that it can be used to analyze the stability of nonlinear problems; this is not the case with the "linearized" Fourier stability analysis proposed by von Neumann [30].

In cases where large deformations are relatively unimportant, it is sometimes possible to neglect the nonlinear effects due to material transport and to study stress wave propagation in materials with nonlinear physical properties, such as hysteresis, work hardening, and plasticity. Problems involving the response of soil and rock media to air-blast loadings at appropriate pressure ranges have been studied using the two-dimensional space–time axisymmetric LAYER code [31]. This code has also been used to study effects in linearly elastic materials by calculating (a) Rayleigh wave phenomena near the surface of a half-space and (b) wave propagation in a long, thin, stress-free rod (the Pochhammer problem).

A finite-element approach to this class of problems is presented by Farhoomand and Wilson [32]. In their solution, a banded system of linear equations is solved at each time step and a time integration scheme is used which is stable for any size time step. It is of interest to compare this approach to the explicit finite-difference method that is used in the LAYER code, in which no system of linear simultaneous equations is solved and the time step of the integration scheme is restricted by the Courant–Friedrichs–Lewy (CFL) stability criterion [33]. For problems of this type, it does not appear, for reasons of accuracy, that the time step in the implicit scheme can be increased significantly enough to compensate for the time required to solve a system of algebraic equations at each time step. However, the authors feel that the relative merits of these approaches should be studied further, particularly in light of the importance of implicit methods (see Section 7). It is possible that some of the fractional-step techniques discussed in Section 8 can be used to increase the efficiency of this approach.

4. Parabolic Systems

Most finite-difference methods for solving parabolic systems [1–3, 34, 35] employ "implicit" rather than "explicit" schemes. The main reason for this is that for explicit methods, there is a stability condition which imposes a severe restriction on the time step, whereas implicit methods can be devised which are unconditionally stable. This is perhaps best illustrated by an example. Consider the heat flow equation in one-dimensional space–time

$$\partial u/\partial t = \partial^2 u/\partial x^2 \tag{1}$$

Two possible finite-difference approximations to Eq. (1) are given by

$$\frac{u_j^{n+1} - u_j^n}{\Delta t} = \frac{u_{j+1}^n - 2u_j^n + u_{j-1}^n}{\Delta x^2} \tag{2a}$$

and

$$\frac{u_j^{n+1} - u_j^n}{\Delta t} = \frac{u_{j+1}^{n+1} - 2u_j^{n+1} + u_{j-1}^{n+1}}{\Delta x^2} \tag{2b}$$

The difference operator on the right-hand side of Eq. (2a) is evaluated at time step n, whereas in Eq. (2b), it is evaluated at time step $n + 1$. Equation (2a) has the property that the set of quantities $\{u_j^{n+1}\}$ is given explicitly in terms of the quantities $\{u_j^n\}$, while Eq. (2b) defines an implicit scheme that requires the solution of a linear system of algebraic equations at each time step. The use of Eq. (2a) is subject to the stability condition

$$\Delta t/\Delta x^2 \leqslant \tfrac{1}{2} \tag{3}$$

which means that an extremely large number of time steps may be needed for problems in which Δx must be made small in the interest of accuracy, since the time step Δt is proportional to the square of the space increment, Δx. On the other hand, Eq. (2b) has no associated stability criterion since the scheme can be shown to be stable for any time step, by using the maximum principle [1, 34]. The choice of time step for the implicit scheme, Eq. (2b), is limited only by accuracy requirements. It should be noted that the number of computer operations needed for the implicit scheme is only about twice the number needed for the explicit scheme since Eq. (2b) defines a diagonally dominant, tridiagonal system which can be solved efficiently by means of Gaussian elimination without pivoting [1].

For multidimensional problems, the stability condition of an explicit method can be even more restrictive. For example, the stability requirement of the explicit scheme which is the three-dimensional analog of Eq. (2a) is given by

$$\Delta t[(\Delta x^2)^{-1} + (\Delta y^2)^{-1} + (\Delta z^2)^{-1}] \leqslant \tfrac{1}{2} \tag{4}$$

Such restrictions can be avoided and the efficiency of solving tightly banded systems retained, by employing ADI methods [1, 35], as introduced in 1955 by Peaceman and Rachford [36], and Douglas [37]. A generalization of the ADI methods, called the method of "fractional steps," can also be used for this purpose. This approach is discussed in some detail in Section 8.

The equation for the small transverse (flexural) vibrations of a thin elastic beam,

$$\partial^2 u/\partial t^2 = -\partial^4 u/\partial x^4 \tag{5}$$

is a parabolic equation which is of particular interest in structural mechanics (see Crandall [10]). This equation is treated in detail by Richtmyer and Morton [1, Chap. 11]. The advantages of an implicit scheme are made quite clear. For multidimensional problems such as those involving the flexural vibrations of a thin plate or shell, the obvious conclusion is that ADI methods can be used effectively. For special cases, such as plates with rectangular or circular domains in which the problem can be solved by Fourier analysis methods, fast Fourier transform techniques can be used (see Section 6).

5. Elliptic Systems

The solution of elliptic systems by numerical methods, whether of the Rayleigh–Ritz, finite-difference, or finite-element type, ultimately depends on an ability to solve systems of simultaneous algebraic equations, either linear or nonlinear. The solution of nonlinear algebraic equations usually involves an iterative method that is based on a sequence of approximations to the system which (hopefully) converges to the desired solution. A simplified system of equations, usually linear, is often used at each step in the iteration process. Many of the important results concerning the basic problem of solving linear systems by direct (i.e., noniterative) methods have been summarized by Forsythe and Moler [38]. In this reference, a variety of topics is discussed, including: (*a*) conditioning of systems; (*b*) the need for pivoting during elimination; (*c*) the fact that pivoting is unnecessary when solving positive-definite systems; (*d*) accumulation of inner products in double precision during iterative improvement; (*e*) matrix factorization; and (*f*) banded matrices. They also present program listings in FORTRAN, ALGOL 60 and PL/I, for an algorithm for solving linear systems using matrix factorization with pivoting; the factorization process is equivalent to the well-known Gaussian elimination scheme. Treatment of many of these subjects as well as tridiagonal and block-tridiagonal systems, including operation counts, can be found in Isaacson and Keller [39] and Fox [40].

The use of iterative methods, often involving relaxation techniques, has found wide application in the solution of both linear and nonlinear algebraic equations. Various relaxation methods, including point, line, and block relaxation, are discussed by Varga [35]. As Varga notes, the power of these methods is clearly demonstrated in view of the fact that linear systems with more than 20,000 unknowns were successfully solved by 1962. The rapid development of high-speed computers since that time would certainly permit an order of magnitude increase in this number. The ADI methods (discussed in connection with parabolic systems, Section 4) constitute a special class of block relaxation techniques and are widely used for elliptic systems. More recently, Dupont *et al.* [41], have examined other relaxation methods based on special factorization techniques.

The Newton–Raphson [39, 42], or Newton method, outlined briefly below, is one of the most commonly used methods for solving nonlinear algebraic systems. Consider a system of n equations in n unknowns

$$f_i(x_1, \ldots, x_n) = 0, \qquad i = 1, \ldots, n \tag{6}$$

or, in vector notation, simply

$$f(x) = 0$$

The basic Newton iteration process is defined by the equation

$$x^{(j+1)} = x^{(j)} - J^{-1}(x^{(j)})f(x^{(j)}) \tag{7}$$

where J represents the $n \times n$ Jacobian of f with respect to x, defined by

$$J_{kl} = \partial f_k / \partial x_l, \qquad k, l = 1, \ldots, n \tag{8}$$

and J^{-1} denotes its inverse. Although the inverse matrix is indicated in Eq. (7), the inverse should not be computed as such, but rather the system of equations

$$[J]\,\delta x = -f \tag{9}$$

should be solved for $\delta x = x^{(j+1)} - x^{(j)}$. This point is discussed by Isaacson and Keller [39, p. 34]. It should also be noted that the partial derivatives of the Jacobian can be computed by finite differences in cases where the Jacobian is not known analytically.

A variation of Newton's method, sometimes referred to as the modified Newton's method, is based on the observation that, once matrix factorization is completed, relatively few operations are needed to obtain the solution to a system of equations (see for example, Forsythe and Moler [38, p. 29]). The modified Newton's method thus involves keeping the Jacobian fixed during a certain number of iteration steps so that the equation

$$x^{(j+1)} = x^{(j)} - J^{-1}(x^{(0)})f(x^{(j)}) \tag{10}$$

is used in place of Eq. (7). This type of iteration procedure has been used to solve many problems including the three-dimensional analysis of cable net structures (McCormick and Wright [43]). The cable net problem can be viewed as a finite difference (or finite element) analog of a membrane problem which involves the solution of a nonlinear elliptic equation.

A second variation of Newton's method involves the generalization of Eq. (7) or Eq. (10) to include a relaxation parameter (see Isaacson and Keller [39, p. 120]).

Another important iterative method, referred to as nonlinear over-relaxation (NLOR), is based on the iteration equation

$$x_i^{(j+1)} = x_i^{(j)} - \omega_j \frac{f_i(x_1^{(j+1)}, x_2^{(j+1)}, \ldots, x_i^{(j)}, x_{i+1}^{(j)}, \ldots)}{(\partial f_i/\partial x_i)(x_1^{(j+1)}, x_2^{(j+1)}, \ldots, x_i^{(j)}, x_{i+1}^{(j)}, \ldots)},$$

$$i = 1, \ldots, n \qquad (11)$$

This method, as described by Ames [3], is a successive point relaxation process for Newton's method with the Jacobian approximated by its diagonal elements. It has been used successfully by Perrone and Kao [44], for solving nonlinear problems involving large deflections of a circular, thin-walled tube, a spherical cap membrane, and a uniformly loaded circular membrane. For many structural problems such as these, in which the approximate system of equations is derived from an elliptic operator which yields positive-definite, diagonally dominant matrices, the NLOR method appears to be particularly useful.

It is interesting to note that many iterative methods are equivalent to solving a corresponding time-dependent problem. An example of this equivalence is provided by considering the time-independent equation (not necessarily linear)

$$Ax = b \qquad (12)$$

where A represents a finite difference or finite element approximation to an integro-differential operator. A typical relaxation scheme for this problem involves the iteration equation

$$x^{(j+1)} = x^{(j)} - \omega_j[Ax^{(j)} - b] \qquad (13)$$

where ω_j is a relaxation parameter. The convergence of this procedure depends critically on the choice of the parameters ω_j and therefore, on the eigenvalues of the operator A. The iterative equation, Eq. (13), may be rewritten as

$$(x^{(j+1)} - x^{(j)})/\omega_j + Ax^{(j)} = b \qquad (14)$$

which is clearly a discrete analog of the differential equation system

$$dx/dt + Ax = b \qquad (15)$$

where t corresponds to ω. Thus ADI methods, block relaxation, the Gauss–Seidel iteration method as well as other iterative methods are equivalent to an initial-value problem of the type defined by Eq. (15). This is, of course, not surprising, since the solution of most physical problems not involving time can be obtained as the solution of some equivalent time-dependent problem.

A method which is closely related to the preceding idea is referred to as the "method of continuity," Morrey [45]. This method and the so-called "incremental load method" were used by Thornton and Birnstiel [46] to analyze three-dimensional suspension structures. The incremental load method can be described by considering a system of equilibrium equations for which an equilibrium solution is desired for a given load system. The load system is parametrized so that, by making a sequence of changes in the system, a sequence of equilibrium solutions is obtained, starting from some known equilibrium (obtained, for example, from symmetry) and proceeding to the desired equilibrium configuration. Oden [47], points out that the incremental load method is basically an Euler integration scheme.

6. Tensor Product—Fast Fourier Transform Methods

A direct method, which is useful for solving second-order, separable, partial differential equations, is the "tensor product" method of Lynch et al. [48]. This approach is both conservative of computer storage and fast, particularly when it is combined with the fast Fourier transform (FFT) method of Cooley and Tukey [49]. The main applications thus far have been to problems with regular domains, such as rectangular regions. However, attempts are currently being made to extend the approach to more complicated geometries by combining the method with block relaxation (iterative) techniques. The review article by Dorr [50] contains a rather complete bibliography pertaining to the tensor product–FFT methods. It includes a tabulation of the number of operations required for different methods which shows that ADI methods are faster than the tensor product method, unless FFT or a related method (e.g., Hockney [51]) is employed.

The tensor product method is well suited for solving elliptic and parabolic equations in rectilinear, cylindrical, or spherical coordinates in two or three dimensions. The examples discussed by Cooley and Tukey [49] include Laplace's equation, Helmholtz' equation in spherical coordinates, the heat flow equation, and a special case of the biharmonic equation.

As an example of the tensor product–FFT method, consider Poisson's equation

$$\nabla^2 \psi = -\omega \tag{16}$$

in a rectangular region with the condition $\psi = 0$ on the boundary. Let ψ_{ij} and ω_{ij} represent the values of $\psi(x, y)$ and $\omega(x, y)$ at the grid points $(i\,\Delta x, j\,\Delta y)$ where $i = 0, \ldots, m + 1$; $j = 0, \ldots, n + 1$. The usual five-point second-order finite-difference approximation to this equation is written (using summation convention) as

$$(\Delta x^2)^{-1} A_{ik}\psi_{kj} + (\Delta y^2)^{-1}\psi_{il}B_{lj} = \omega_{i,j} \tag{17}$$

or in matrix notation

$$(\Delta x^2)^{-1} A\psi + (\Delta y^2)^{-1}\psi B = \omega \tag{18}$$

where A and B are symmetric, tridiagonal matrices of order m and n, respectively, with $\{2, \ldots, 2\}$ on the diagonal, $\{-1, \ldots, -1\}$ on the first codiagonals, and zero elsewhere. The eigenvalues and eigenvectors of this matrix are well known and for a matrix of order m, the eigenvalues and normalized eigenvectors are given by

$$\Lambda_l = 2\left(1 - \cos\frac{l\pi}{m + 1}\right), \qquad l = 1, \ldots, m \tag{19}$$

and

$$E_{kl} = \left(\frac{2}{m + 1}\right)^{1/2}\sin\left(\frac{kl\pi}{m + 1}\right), \qquad k = 1, \ldots, m \tag{20}$$

respectively. Thus, one may write

$$A = E\Lambda\tilde{E} \tag{21}$$

where $\tilde{E}$ denotes the transpose of E and

$$E\tilde{E} = \tilde{E}E = I \tag{22}$$

Substituting Eq. (21) for A in Eq. (18) and multiplying by $\tilde{E}$ yields the relation

$$(\Delta x^2)^{-1}\Lambda\bar{\psi} + (\Delta y^2)^{-1}\bar{\psi}B = \bar{\omega} \tag{23}$$

where

$$\bar{\psi} = \tilde{E}\psi \tag{24}$$

and

$$\bar{\omega} = \tilde{E}\omega \tag{25}$$

It should be noted that $\bar{\psi}$ and $\bar{\omega}$ are the Fourier transforms of ψ and ω, respectively. The system of equations (23) is tridiagonal, since Λ is a diagonal matrix, and thus can be solved very efficiently by direct elimination. Thus, the sequence of steps in the tensor product–FFT method is: (i) calculate $\bar{\omega}$, using Eq. (25); (ii) solve the tridiagonal system, Eq. (23) for $\bar{\psi}$; and (iii) calculate $\psi = E\bar{\psi}$. Items (*i*) and (*iii*) are the Fourier analysis and synthesis steps in the method.

It is not necessary that the boundary conditions be as simple as those in the preceding example. Other types of boundary conditions are treated by Williams [52]. This reference deals with a cylindrical three-dimensional problem involving incompressible fluid flow. The time-dependent Navier–Stokes equations are integrated forward in time and the pressure is determined from Poisson's equation at each time step, using the tensor product–FFT method.

It is not even necessary to know the eigenvalues and eigenvectors analytically although the speed advantages of the FFT technique would be lost in such cases. Rather, since the matrix A is tridiagonal, the eigenvalues and eigenvectors can be calculated numerically and, for time-dependent applications such as the one by Williams, this would only have to be done once. On many of today's larger computers, it is not difficult to implement this method since the required procedures (matrix multiplication, tridiagonal systems solver, and FFT routines) can usually be found in user program libraries.

7. Implicit versus Explicit Methods—Flexibility Concept

One of the most difficult decisions to make when solving time-dependent boundary-value problems is whether to use an explicit or an implicit method. Although implicit schemes have been used for parabolic equations for many years, difference schemes for hyperbolic equations have usually been of the explicit type. However, implicit schemes can be used for hyperbolic as well as parabolic equations. In this section, the application of implicit methods is emphasized and the concept of the "flexibility" of a difference scheme is discussed. The main conclusion in connection with this concept is that implicit schemes which are centered in time and space are to be preferred for partial differential equations with variable coefficients.

Implicit schemes for time-dependent partial differential equations of the form

$$\partial u/\partial t = Au \qquad (26)$$

can generally be represented by difference schemes of the form

$$(u^{n+1} - u^n)/\Delta t = A^*[\theta u^{n+1} + (1 - \theta)u^n] \qquad (27)$$

where A^* is a difference approximation to the operator A. In the case $\theta = 0$, u^{n+1} is given explicitly in terms of u^n whereas, for $\theta \neq 0$, a system of algebraic equations must be solved in order to obtain u^{n+1} in terms of u^n.

Consider, as an example, the heat flow equation [Eq. (1)] corresponding to the case $A = \partial^2/\partial x^2$. Equations (2a) and (2b) represent difference schemes which correspond to the cases $\theta = 0$ and $\theta = 1$, respectively. As noted previously, Eq. (2a) has an associated stability criterion given by Eq. (3) whereas, Eq. (2b) has no stability restriction. The more general case of $0 \leqslant \theta \leqslant 1$ for the heat flow equation is discussed in Richtmyer and Morton [1, p. 186], in which it is shown that the stability condition is given by

$$\Delta t/\Delta x^2 \leqslant [2(1 - 2\theta)]^{-1}, \qquad 0 \leqslant \theta \leqslant \tfrac{1}{2} \qquad (28a)$$

$$\text{No restriction,} \qquad \tfrac{1}{2} \leqslant \theta \leqslant 1 \qquad (28b)$$

Thus, for $\theta \geqslant \tfrac{1}{2}$, the implicit scheme is unconditionally stable.

The result that Eq. (27) is unconditionally stable for $\theta \geqslant \tfrac{1}{2}$ holds quite generally whether the operator A is parabolic or of some other type. For example, Varga [40, p. 265], proves this result for the cases $\theta = \tfrac{1}{2}$ and $\theta = 1$. A proof of convergence and stability of an implicit scheme of the Crank–Nicholson type ($\theta = \tfrac{1}{2}$) for hyperbolic systems was given by Ladyzhenskaya [53]. Implicit schemes for hyperbolic equations are also discussed in Richtmyer and Morton [1, p. 263] and Ames [3, p. 443].

The main difficulty with using implicit schemes for multidimensional problems is that a considerable amount of time may be consumed in solving a system of equations at each time step. However, the use of ADI and fractional-step methods provides an effective means for dealing with this difficulty. Lees [54], discusses the use of ADI methods for hyperbolic equations. Chorin [55], applies the fractional-step method to the problem of solving the multidimensional Navier–Stokes equations.

The concept of the "flexibility" of a difference scheme (introduced by Gelfand) is used by Godunov and Ryabenkii [56], in order to classify certain properties of difference approximations. Frequently in the study of difference schemes a Taylor series expansion technique is used to examine the consistency and order of error of the approximation with respect to a particular differential equation (see Richtmyer and Morton [1, page 19]). In time-dependent problems, an assumption is made concerning the relationship between the time increment Δt and the space increment Δx. If, depending on the form of this relationship, the same difference scheme approximates various differential equations, then the scheme is classified as "inflexible" [56, p. 78]. If the difference scheme approximates the same differential equation, the scheme is called "flexible." In general, flexible schemes are desirable, particularly for partial differential equations with variable coefficients [56, p. 219].

Alternatively, one can say that a difference scheme is inflexible if the order of error of the approximation is altered when the relationship between Δt and

Δx is changed. In Godunov and Ryabenkii [56], the choices considered for this relationship were $\Delta t = r \Delta x$ and $\Delta t = r \Delta x^2$. Consider then a difference scheme for which the error terms are proportional to Δt and Δx^2. If the relationship is assumed to be $\Delta t = r \Delta x$, the overall error of the difference approximation is proportional to Δx, whereas, if $\Delta t = r \Delta x^2$ is assumed, the error is proportional to Δx^2. The difference schemes represented by Eqs. (2a) and (2b) are both of this type and thus are classified as "inflexible." The Du Fort–Frankiel method for parabolic equations [1, p. 177; 3, p. 328] is also an inflexible scheme as is Lax's method for the hyperbolic equation $\partial u/\partial t = \partial u/\partial x$ [56, pp. 78, 219]. On the other hand, the Crank–Nicholson scheme which corresponds to $\theta = \frac{1}{2}$ [1] is classified as "flexible." This is so because the error terms of this scheme are proportional to Δt^2 and Δx^2. Thus, whether $\Delta t = r \Delta x$ or $\Delta t = r \Delta x^2$ is chosen, the lowest-order error term is proportional to Δx^2. Thus, it appears that schemes which are centered in time and space ($\theta = \frac{1}{2}$) are flexible if constructed properly. Apparently the concept of the flexibility of a scheme has been extended to include certain properties of stability of a difference scheme. Unfortunately, this point is discussed only briefly in the "Conclusions" section of Godunov and Ryabenskii [56, pp. 215–222]. Nevertheless, this concept seems to merit further study, since Godunov and Ryabenkii have found it to be useful for examining many difference schemes, especially when dealing with problems involving variable coefficients.

8. The Method of Fractional Steps

Some of the most important developments in the numerical solution of partial differential equations have been in the direction of generalizing the ADI methods [57]. A great deal of work along these lines has been done in the Soviet Union and this has led to the development of what is called the method of "fractional steps" [1, pp. 216ff, 58–61]. This method is extremely versatile and can be used to solve nonlinear, multidimensional systems involving different types of operators (hyperbolic, parabolic, and elliptic). The basic approach is to separate a problem into a sequence of manageable sub-problems by means of a technique referred to as "operator splitting." One of the advantages of this approach is that the stability of a complicated system of equations is reduced to the study of the stability of a set of simpler equations. The efficiency of the method usually derives from the fact that implicit schemes involving tightly banded matrices are used, as in the ADI methods. It is felt by the authors that approaches of this type are among the most promising areas of study for solving multidimensional problems.

The essence of the fractional step (FS) method can be understood by considering a system of partial differential equations of the form

$$u_t = Au \tag{29}$$

where the operator A can be expressed as a sum of operators,

$$A = A_1 + A_2 + \cdots + A_n \tag{30}$$

(The subscript of u_t denotes partial differentiation, i.e., $u_t \equiv \partial u/\partial t$.) During each time step, a sequence of equations

$$u_t = A_1 u, u_t = A_2 u, \ldots, u_t = A_n u \tag{31}$$

is solved by choosing approximations (finite difference or finite element) for the operators A_i.

As an example, consider the application of the FS method to the three-dimensional heat flow equation

$$u_t = u_{xx} + u_{yy} + u_{zz} \tag{32}$$

by means of the sequence

$$u_t = u_{xx}, \qquad u_t = u_{yy}, \qquad u_t = u_{zz} \tag{33}$$

Each of these one-dimensional equations can be solved by choosing a discrete approximation. For example, the use of the implicit difference scheme of Eq. (2b) yields the sequence

$$u^{n+1/3} - u^n = (\Delta t/\Delta x^2)\delta_x^2 u^{n+1/3} \tag{34a}$$

$$u^{n+2/3} - u^{n+1/3} = (\Delta t/\Delta y^2)\delta_y^2 u^{n+2/3} \tag{34b}$$

$$u^{n+1} - u^{n+2/3} = (\Delta t/\Delta z^2)\delta_z^2 u^{n+1} \tag{34c}$$

where

$$\delta_x^2 u(x, y, z) \equiv u(x + \Delta x, y, z) - 2u(x, y, z) + u(x - \Delta x, y, z) \tag{35}$$

and similarly for $\delta_y^2 u$ and $\delta_z^2 u$. The fractional superscripts are used to emphasize the fractional-step nature of the scheme. At each step in this sequence, a tridiagonal system of equations must be solved. This scheme is not advocated as being optimal in any sense, and is presented only as an example of the FS method. One of the disadvantages of the above scheme is the fact that the error term is of the order of Δt, instead of order Δt^2 as in the standard ADI method for this problem [1, p. 211ff]. (The spatial order of error is Δx^2 in both cases.) This disadvantage can be eliminated by using a scheme which is centered in time as well as space (i.e., the Crank–Nicholson scheme), as demonstrated by Richtmyer and Morton [1, p. 217].

The particle-in-cell (PIC) method [27, 28] for the solution of problems in hydrodynamics is another example of the FS method (although it was not originally recognized as such). Consider the equations of hydrodynamics

$$\rho_t + \nabla \cdot (\rho u) = 0 \tag{36a}$$

$$(\rho u)_t + \nabla \cdot (u \rho u) = -\nabla p \tag{36b}$$

$$(\rho e)_t + \nabla \cdot (u \rho e) = -p \nabla \cdot u \tag{36c}$$

$$p = p(\rho, e) \tag{36d}$$

where ρ is the density, u is the particle velocity, e is the specific internal energy, and p is the pressure. Equations (36a–c) represent conservation of mass, momentum, and energy, respectively, and $p(\rho, e)$ is the equation of state for the fluid.

The basic idea of the PIC method is to consider a fixed grid of cells with a system of point-mass particles which are distributed throughout the cells. The basic equations are integrated in time by means of the following two-step procedure which is applied during each time step:

Step (i): Calculate the change in momentum and energy due to the terms appearing on the right-hand side of Eqs. (36b) and (36c), respectively; in these calculations, the changes due to the divergence terms are ignored.

Step (ii): Each particle is moved from its current location to a new location in accordance with an interpolated particle velocity which is based on the calculations of Step (i). Mass is exactly conserved by this procedure and a new cell density is calculated. If a particle moves from one cell to another, the momentum and energy quantities are modified in the two affected cells, so as to conserve these quantities.

The PIC method has generally been referred to in the literature as a combined Eulerian–Lagrangian scheme for solving multidimensional space–time problems in hydrodynamics. The Eulerian computational grid is formed by breaking the domain into a finite number of volume elements (cells) which are fixed in space. The Lagrangian part of the procedure consists of the numbering of the individual material particles and the computation of their positions as they move through the fixed Eulerian mesh. The individual mass and position coordinates of each particle are, in effect, the Lagrangian variables.

It is clear that the PIC technique can also be interpreted as a fractional-step method which is applied to a system of equations of the form

$$f_t + \nabla \cdot (uf) = R(f) \tag{37}$$

where f is the vector

$$f = \begin{pmatrix} \rho \\ \rho u \\ \rho e \end{pmatrix} \tag{38}$$

and R denotes the right-hand side of Eqs. (36a–c). (The pressure can be eliminated by means of the equation of state.) The two step PIC procedure thus corresponds to solving, in sequence, the equations

$$f_t = R(f) \tag{39a}$$

$$f_t = -\nabla \cdot (uf) \tag{39b}$$

For many problems, the PIC method is sufficiently accurate if a conservative difference scheme, such as that discussed by Noh [20, 21] is used for Eq. (39b) instead of moving particles in accordance with conservation principles, as in Step (ii) of the PIC method formulation.

It is of interest to consider the nature of the calculations in Step (i) of the PIC method for cases in which the fluid velocities are high or low. The calculational scheme for Eq. (39a) is stable so long as the internal energy remains positive everywhere; this, of course, depends upon the equation of state of the material. The PIC scheme for Eq. (39a) can be shown to be unstable for very low fluid velocities, i.e., low Mach numbers (see Harlow [27, p. 236]). This problem was investigated by Daly [62], who concluded that, although the difference scheme for Step (i) is unconditionally unstable for low Mach number flows, the procedure of Step (ii) is so stable (due to numerical viscosity) that the combined calculation is nevertheless stabilized.

An alternate PIC approach for the low Mach number calculations in which both steps are stable can be obtained by replacing the equation of state with its total time derivative

$$\frac{dp}{dt} = p_\rho \frac{d\rho}{dt} + p_e \frac{de}{dt} \tag{40}$$

By means of Eqs. (36a and c), Eq. (40) can be written as

$$dp/dt = p_t + u \cdot \nabla p = -\rho c^2 \nabla \cdot u \tag{41}$$

where $c = c(\rho, e)$ is the isentropic wave speed which is given by

$$c^2 = p_\rho + (p/\rho^2)p_e \tag{42}$$

A two-step procedure similar to Eqs. (39a, b) can be used. It should be noted that the vector f [Eq. (38)] has the additional (fourth) component p. A stable scheme is obtained for Step (i) by using the standard PIC procedure for the other components, and using a difference approximation for the pressure component which is backward in time (see, for example, Eqs. (8) of Fox [63]). The PIC method has been extensively applied to a number of problems in hydrodynamics including, for example, the analysis of the close-in regions of nuclear explosions, the initiation of armor piercing phenomena and the design of meteorite shields for space vehicles.

Problems involving the formulation and application of boundary conditions in using the fractional step methods are discussed by Chorin [55], and Yanenko [58]. As will be pointed out in Section 9, the problem of applying boundary conditions requires serious study for all computational schemes.

Lions [64] applies the fractional-step method to several different problems. He uses the finite-difference equations to prove the existence of solutions of certain problems.

9. Stability and the Energy Method

Various techniques are available for investigating the stability of finite-difference schemes for time-dependent partial differential equations. Perhaps the best known is the linearized Fourier analysis method, attributed to von Neumann by O'Brien et al. [30]. This method is discussed in Richtmyer and Morton [1], Forsythe and Wasow [2], Ames [3], and Godunov and Ryabenkii [56]. An extension of this method, referred to as the "spectral theory of difference operators" [56, Chap. VI] can be used to study the effects of boundary conditions and the stability of certain equations with variable coefficients. The maximum principle is another method which has been used to show the stability of finite-difference schemes for both parabolic and hyperbolic equations [1, 2, 36, 56, p. 109ff]. An interesting heuristic approach to the stability analysis of certain finite-difference methods was suggested by Hirt [29].

The stability of many finite-difference methods for hyperbolic equations depends on the use of an artificial viscosity of some sort (see Section 3). Numerous examples can be found in the literature of schemes (both explicit and implicit) which are so overly dissipative, due to "numerical" viscosity, that the solution is significantly affected, especially when integrated over many time steps. The study of weather prediction by finite-difference methods leads to problems in which this effect is important, and a great deal of work has gone into the construction of difference schemes which avoid this difficulty (see Arakawa [65] and Martchouk [60]). An overly dissipative, implicit scheme for the wave equation is discussed by Zajac [66].

A great deal of confusion can arise in the study of numerical stability of finite-difference schemes, since alternative definitions of stability are possible as discussed in Richtmyer and Morton [1, p. 95] and Godunov and Ryabenskii [56, p. 227]. One reason for this is that the nature of the solution of an equation may change as the equation is altered. Consider the example [1, p. 71] of the modified heat flow equation

$$\partial u/\partial t = \sigma(\partial^2 u/\partial x^2) + bu, \qquad \sigma > 0, \quad b > 0 \qquad (43)$$

In the case $b \neq 0$, the solution grows exponentially for all time beyond a certain point. A reasonable finite difference approximation to Eq. (43) should reflect this property, and a definition of stability should be used which permits the existence of exponentially increasing solutions. In the case $b = 0$, the solution of Eq. (43) decays as time increases, and a definition of stability consistent with this property should be used.

Another reason for the existence of differing definitions of stability is that different criteria for stability are discovered as the result of the interaction between theory and numerical experiments. For example, an implicit difference scheme for the hyperbolic equation

$$(\partial u/\partial t) + a(x, t)(\partial u/\partial x) = 0 \tag{44}$$

is investigated in Richtmyer and Morton [1, p. 137], where it is shown that the scheme is unconditionally stable. However, computational experiments with this scheme resulted in a sawtooth curve of quite large amplitude, superimposed on the graph of the solution in regions of small values of $a(x, t)$ [56, p. 218]. Thus, this scheme is rejected for computational purposes because, in Godunov and Ryabenkii's terminology, it is "inflexible" with respect to stability. This should be compared with the concept of the "flexibility" of a finite-difference approximation (see Section 7).

The "energy method" is the name given to a group of techniques, based on the use of certain energy-like quantities, for proving existence and uniqueness of solutions in the theory of partial differential equations (see Courant and Hilbert [67, p. 449]). In some cases, these energy-like quantities do, in fact, correspond to the physical energy of the system. In current mathematical terminology, these quantities are called "norms." The key property which norms have in common with energy is that both are positive definite. In the theory of difference schemes, the energy method can be used to prove the stability of a given difference scheme [1, p. 132ff]. The effects of variable coefficients and boundaries on stability can be included in a natural way. The energy method is used by Richtmyer and Morton [1], to study the stability of difference schemes for the equations of coupled sound and heat flow [1, p. 143] and the vibrations of a thin elastic beam with tension [1, p. 285]. One of the main points made in Richtmyer and Morton [1] is that the energy method and the Godunov–Ryabenkii spectral theory are complementary in the sense that the energy method can be used to obtain sufficient conditions for stability, whereas the spectral theory can be used to obtain necessary conditions. The energy method can also be used as a guide in the construction of difference schemes with certain properties. In Morton [68], Morton discusses the use of the method for the design of difference schemes for numerical weather prediction.

The use of the energy method depends mainly on one's ability to find a norm for a given difference scheme. Sometimes the norm can be found by considering the norm of the corresponding continuum problem, although this norm may have to be modified in order to arrive at one which is appropriate to a particular difference scheme. For example, a norm for the one-dimensional wave equation

$$\partial v / \partial t = c(\partial w / \partial x) \tag{45a}$$

$$\partial w / \partial t = c(\partial v / \partial x) \tag{45b}$$

is given by

$$E(t) = \int [v^2 + w^2]\, dx \tag{46}$$

where the range of integration extends over the domain of definition of the problem [1, p. 261ff]. An implicit finite-difference scheme for Eqs. (45a) and (45b) is given by

$$\frac{v_j^{n+1} - v_j^n}{\Delta t} = c \frac{w_{j+1/2}^{n+1} - w_{j-1/2}^{n+1} + w_{j+1/2}^n - w_{j-1/2}^n}{2\Delta x} \tag{47a}$$

$$\frac{w_{j+1/2}^{n+1} - w_{j+1/2}^n}{\Delta t} = c \frac{v_{j+1}^{n+1} - v_j^{n+1} + v_{j+1}^n - v_j^n}{2\Delta x} \tag{47b}$$

An appropriate norm for this implicit scheme can be written as

$$E_n = \sum_j [(v_j^n)^2 + (w_{j+1/2}^n)^2]\, \Delta x \tag{48}$$

where the summation extends over all points of the finite difference mesh. The quantity E_n is clearly a finite difference analog of the quantity $E(t)$, defined by Eq. (46). If the effects of boundary terms are ignored, this scheme can be shown to be unconditionally stable since E_n is positive definite for any value of Δt. On the other hand, for the explicit scheme

$$\frac{v_j^{n+1} - v_j^n}{\Delta t} = c \frac{w_{j+1/2}^n - w_{j-1/2}^n}{\Delta x} \tag{49a}$$

$$\frac{w_{j+1/2}^{n+1} - w_{j+1/2}^n}{\Delta t} = c \frac{v_{j+1}^{n+1} - v_j^{n+1}}{\Delta x} \tag{49b}$$

an appropriate norm is given by

$$E_n' = \sum_j [(v_j^n)^2 + (w_{j+1/2}^n)^2]\, \Delta x + c\,\Delta t \sum_j v_j^n (w_{j+1/2}^n - w_{j-1/2}^n) \tag{50}$$

It can be found that E_n' is positive definite if

$$c\,\Delta t / \Delta x \leqslant 1 \tag{51}$$

by using the inequality

$$\left| \sum_j v_j^n (w_{j+1/2}^n - w_{j-1/2}^n) \right| \leqslant \beta \sum_j (v_j^n)^2 + \beta^{-1} \sum_j (w_{j+1/2}^n)^2 \qquad (52)$$

where β is an arbitrary positive number. Hence, if the effects of boundary conditions are ignored, the usual stability condition [Eq. (51)] is obtained for the explicit scheme.

An interesting case of numerical instability, caused by boundary conditions, is presented in Richtmyer and Morton [1, p. 154ff]. The explicit difference scheme, Eqs. (49a and b), for the wave equation is studied in conjunction with the mixed boundary condition

$$w + \alpha v = 0 \qquad (53)$$

which is written in difference form as

$$w_{-1/2}^{n+1} + \alpha v_0^{n+1} = 0 \qquad (54)$$

where α is a constant. By means of the Godunov–Ryabenkii spectral theory of stability, it is shown that this scheme is unstable for certain, physically meaningful ranges of the parameter α, even though the stability condition [Eq. (51)] is satisfied. Thus, it is clear that the effect of boundary conditions on stability should be investigated whenever possible. However, this is quite difficult to do in complicated, multidimensional problems. In an attempt to simulate "viscous" or "transmitting" boundaries in two dimensional wave propagation problems using the LAYER code, mixed boundary conditions, analogous to Eq. (53), have been used [69]. In certain cases, numerical instabilities were observed which were traced directly to the effect of such boundary conditions.

References

1. Richtmyer, R. D., and Morton, K. W., *Difference Methods for Initial Value Problems*, 2nd ed. Wiley (Interscience), New York, 1967.
2. Forsythe, G. E., and Wasow, W. R., *Finite Difference Methods for Partial Differential Equations*, Wiley, New York, 1960.
3. Ames, W. F., *Nonlinear Partial Differential Equations in Engineering*. Academic Press, New York, 1965.
4. *Methods of Computational Physics*, Vols. 1–9. Academic Press, New York, 1963–1971.
5. "Numerical Solution of Field Problems in Continuum Physics," Vol. II (SIAM–AMS Proc.). Amer. Math. Soc., Providence, Rhode Island, 1970.
6. Salvadori, M. G., and Baron, M. L., *Numerical Methods in Engineering*, 2nd ed. Prentice-Hall, Englewood Cliffs, New Jersey, 1961.
7. Salvadori, M. G., "Numerical Computation of Buckling Loads by Finite Differences," Paper 2441, *Trans. ASCE* **116** (1951).
8. Jenson, V. P., "Analysis of Skew Slabs," Bull. Ser. No. 332, Univ. of Illinois, Eng. Experiment Station, Urbana, Illinois (1941).

9. Crowley, B. K., and Barr, L. K., "TENSOR-PUFL: Boundary Condition Linked Codes," *J. Comput. Phys.* **7**, 167–171 (1971).
10. Crandall, S. H., *Engineering Analysis—A Survey of Numerical Procedures.* McGraw-Hill, New York, 1956.
11. Lister, M., "The Numerical Solution of Hyperbolic Partial Differential Equations by the Method of Characteristics," *Mathematical Methods for Digital Computers,* Vol. 1. Wiley, New York, 1960.
12. Massau, J., *Mémoire sur l'Intégration Graphique des Equations aux Dérivées Partielles,* p. 144. Meyer-van Loo, Ghent, 1899.
13. Hoskin, N. E., "Solution by Characteristics of the Equations of One Dimensional Unsteady Flow," *Methods in Computational Physics,* Vol. 3, p. 265. Academic, Press, New York, 1964.
14. Butler, D. S., "The Numerical Solution of Hyperbolic Systems of Partial Differential Equations in Three Independent Variables," *Proc. Roy. Soc.* **A255**, 232 (1960).
15. Richardson, D. J., "The Solution of Two Dimensional Hydrodynamic Equations by the Method of Characteristics," *Methods in Computational Physics,* Vol. 3, p. 295. Academic Press, New York, 1964.
16. von Neumann, J., and Richtmyer, R. D., "A Method for the Numerical Calculation of Hydrodynamical Shocks," *J. Appl. Phys.* **21**, 232 (1950).
17. Brodie, R. N., and Hormuth, J. E., "The PUFF 66 and P PUFF 66 Computer Programs," Air Force Weapons Lab., Kirtland AFB, New Mexico, Tech. Rep. No. AFWL-TR-66-48 (1966).
18. Holzhauser, P. H., and Lawrence, R. J., "WONDY III—An Improved Program for Calculating Problems of Motion in One Dimension," Sandia Lab., SC-DR-68-217 (June 1968).
19. Hicks, "Hydrocode Test Problems," Air Force Weapons Lab. Kirtland Air Force Base, New Mexico, Tech. Rep. No. AFWL-TR-67-127 (February 1968).
20. Noh, W. F., "CEL: A Time-Dependent Two Space Dimensional, Coupled Eulerian–Lagrangian Code," *Methods in Computational Physics,* Vol. 3, p. 117. Academic Press, New York, 1964.
21. Noh, W. F., "A General Theory for the Numerical Solution of the Equations of Hydrodynamics," in *Numerical Solution of Nonlinear Differential Equations,* pp. 181–211. Wiley, New York, 1966.
22. Wilkins, M. L., "Calculation of Elastic-Plastic Flow," *Methods in Computational Physics,* Vol. 3, p. 211. Academic Press, New York, 1964.
23. Schulz, W. D., "Tensor Artificial Viscosity for Numerical Hydrodynamics," *J. Math. Phys.* **5**, 133 (1964).
24. Lax, P., "Weak Solutions of Nonlinear Hyperbolic Equations and their Numerical Computation," *Commun. Pure Appl. Math.* **7**, 159 (1954).
25. Gentry, R. A., Martin, R. E., and Daly, B. J., "An Eulerian Differencing Method for Unsteady Compressible Flow Problems," *J. Comp. Phys.* **1**, 87 (1966).
26. Lax, P. D., and Wendroff, B., "Difference Schemes for Hyperbolic Equations with High Order of Accuracy," *Commun. Pure Appl. Math.* **17**, 381 (1961).
27. Harlow, F. H., "The Particle-In-Cell Computing Method for Fluid Dynamics," *Methods Comp. Phys.* **3**, 319. Academic Press, New York, 1964.
28. Baron, M. L., Christian, C. E., and Skidan, O., "Particle-In-Cell Method in Shock Propagation Problems," *J. Eng. Mech. Div.,* **EM 6** (December 1966); *Transactions ASCE* **134** (1969).
29. Hirt, C. W., "Heuristic Stability Theory for Finite-Difference Equations," *J. Comp. Phys.* **2**, 339–355 (1968).

30. O'Brien, G. G., Hyman, M. A. and Kaplan, S., "A Study of the Numerical Solution of Partial Differential Equations," *J. Math. Phys.* **29**, 223 (1950).
31. Baron, M. L., McCormick, J. M., and Nelson, I., "Investigation of Ground Shock Effects in Nonlinear Hysteretic Media," in *Computational Approaches in Applied Mechanics.* Amer. Soc. Mech. Eng., 1969.
32. Farhoomand, I., and Wilson, E., "A Nonlinear Finite Element Code for Analyzing the Blast Response of Underground Structures," Rep. to U.S. Army Eng. Waterways Experiment Station, Rep. No. 70-1, Struct. Eng. Lab., Univ. of California, Berkeley, California (January 1970).
33. Courant, R., Friedrichs, K., and Lewy, H., "Über die partieller Differenzen-gleichugen der mathematischen Physik," *Math. Ann.* **100**, 32–74 (1928).
34. Keller, H. B., "The Numerical Solution of Parabolic Partial Differential Equations," *Math. Methods Digital Comput.* **1**, 135 (1960).
35. Varga, R. S., *Matrix Iterative Analysis.* Prentice-Hall, Englewood Cliffs, New Jersey, 1962.
36. Peaceman, D. W., and Rachford, H. H., Jr., "The Numerical Solution of Parabolic and Elliptic Differential Equations," *J. Soc. Ind. Appl. Math.* **3**, 28 (1955).
37. Douglas, J., "On the Numerical Integration of $u_{xx} + u_{yy} = u_t$ by Implicit Methods," *J. Soc. Ind. Appl. Math.* **3**, 42 (1955).
38. Forsythe, G. E., and Moler, C. B., *Computer Solution of Linear Algebraic Systems.* Prentice-Hall, Englewood Cliffs, New Jersey, 1967.
39. Isaacson, H. B., and Keller, E., *Analysis of Numerical Methods.* Wiley, New York, 1966.
40. Fox, L., *An Introduction to Numerical Linear Algebra.* Oxford Univ. Press (Clarendon), London and New York, 1964.
41. Dupont, T., Stone, H. L., and Rachford, H. H., Jr., "Factorization Techniques for Elliptic Difference Equations," Published in Numerical Solution of Field Problems, in *Continuum Physics* (SIAM-AMS Proc.), Vol. 2. Amer. Math. Soc., Providence, Rhode Island, 1970.
42. Traub, J. F., "The Solution of Transcendental Equations," *Math. Methods Digital Comput.* **2**, 171 (1967).
43. McCormick, J. M., and Wright, J. P., Discussion of "Nonlinear Field Analysis of Structural Nets," *J. Struct. Div. Proc. ASCE* **96**, No. ST 3 (March 1970).
44. Perrone, N., and Kao, R., "A General Nonlinear Relaxation Iteration Technique for Solving Nonlinear Problems in Mechanics," *J. Appl. Mech.* **38**, Ser. E, No. 2 (June 1971).
45. Morrey, C. B., Jr., "Nonlinear Methods," in *Modern Mathematics for the Engineer.* McGraw-Hill, New York, 1956.
46. Thornton, C. H., and Birnstiel, C., "Three Dimensional Suspension Structures," *J. Struct. Div., Proc. ASCE* **93**, No. ST 2, Proc. Paper 5196, 247 (April 1967).
47. Oden, J. T., "Finite Element Applications in Nonlinear Structural Analysis," in *Application of Finite Element Methods in Civil Engineering* (*Proc. ASCE Symp.*) (November 1969).
48. Lynch, R. E., Rice, J. R., and Thomas, D. H., "Direct Solution of Partial Difference Equations by Tensor Product Methods," *Numer. Math.* **6**, 185–199 (1964).
49. Cooley, J. W., and Tukey, J. W., "An Algorithm for the Machine Calculation of Complex Fourier Series," *Math. Comput.* **19**, 297–301 (1965).
50. Dorr, F. W., "The Direct Solution of the Discrete Poisson Equation on a Rectangle," *SIAM Rev.* **12**, No. 2, 248–263 (1970).
51. Hockney, R. W., "A Fast Direct Solution of Poisson's Equation using Fourier Analysis," *J. Ass. Comput. Mach.* **12**, 95–113 (1965).
52. Williams, G. P., "Numerical Integration of the Three Dimensional Navier–Stokes Equations for Incompressible Flow," *J. Fluid Mech.* **37**, Part 4, 727–750 (1969).
53. Ladyzhenskaya, O. A., "Solution of Cauchy's Problem for Hyperbolic Systems by the Method of Finite Differences," *Leningrad Gos. Univ. Uch. Zap.* **144** (Ser. Mat.), 192 (1952).

54. Lees, M., "Alternating Direction Methods for Hyperbolic Differential Equations," *J. Soc. Ind. Appl. Math.* **10**, 610–616 (1962).
55. Chorin, A. J., "Numerical Solution of the Navier–Stokes Equations," *Math. Comput.* **22** (1968). [See also *ibid.* **23** (1969).]
56. Godunov, S. K., and Ryabenkii, V. S., *Theory of Difference Schemes—An Introduction.* North-Holland Publ., Amsterdam, and Wiley, New York, 1964.
57. Douglas, J., and Gunn, J., "A General Formulation of Alternating Direction Methods," *Numer. Math.* **6**, 428 (1964).
58. Yanenko, N. N., "The Method of Fractional Steps for the Numerical Solution of the Problems of Mechanics of Continuous Media," in *Fluid Dynamics Transactions*, Vol. 4, pp. 135–147. Inst. of Fundamental Tech. Res., Polish Acad. of Sci., Warsaw (1969).
59. Yanenko, N. N., *Méthode à Pas Fractionnaires.* Librairie Armand Colin, 103, bd. St-Michel, Paris 5ᵉ, 1969.
60. Martchouk, G. I., *Méthodes Numériques pour la Prévision du Temps*, Librairie Armand Colin, 103, bd. St-Michel, Paris 5ᵉ, 1970.
61. Samarskii, A. A., "Economical Difference Schemes for Systems of Equations of Parabolic Type," *Zh. Vychisl. Mat. Mat. Fiz.* **4**, 927 (1964).
62. Daly, B. J., "The Stability Properties of a Coupled Pair of Nonlinear Partial Difference Equations," *Math. Comput.* **84**, 346 (1963).
63. Fox, P., "The Solution of Hyperbolic Partial Differential Equations by Difference Methods," in *Mathematical Methods for Digital Computers*, Vol. 1, p. 180. Wiley, New York, 1960.
64. Lions, J. L., "On the Numerical Approximation of Some Equations Arising in Hydro-dynamics," in *Numerical Solution of Field Problems in Continuum Physics* (SIAM-AMS Proc.), Vol. 2, p. 11. Amer. Math. Soc., Providence, Rhode Island, 1970.
65. Arakawa, A., "Numerical Simulation of Large Scale Atmospheric Motions," in *Numerical Solution of Field Problems in Continuum Physics* (SIAM-AMS Proc.), Vol. 2, pp. 24–40. Amer. Math. Soc., Providence, Rhode Island, 1970.
66. Zajac, E. E., "Note on Overly-Stable Difference Approximations," *J. Math. Phys.* **43**, 51–54 (1964).
67. Courant, R., and Hilbert, D., *Methods of Mathematical Physics*, Vol. 2. Wiley (Interscience), New York, 1962.
68. Morton, K. W., "The Design of Difference Schemes for Evolutionary Problems," in *Numerical Solution of Field Problems in Continum Physics* (SIAM-AMS Proc.), Vol. 2, pp. 1–10. Amer. Math. Soc., Providence, Rhode Island, 1970.
69. McCormick, J. M., Nelson, I., and Ranlet, D., "Investigation of Outrunning Ground Motions from Nuclear Explosions" (In process of publication). Presented at the D.A.S.A. Strategic Structures Vulnerability/Hardening Long Range Planning Meeting, Stanford Res. Inst. Menlo Park, California (February 1971).

Finite-Difference Energy Models versus Finite-Element Models: Two Variational Approaches in One Computer Program

David Bushnell

LOCKHEED MISSILES AND SPACE CO.
PALO ALTO, CALIFORNIA

The finite-difference energy method and finite-element method are compared for stress, buckling, and vibration analyses of shells of revolution. It is shown that certain finite-difference models are equivalent to constant strain elements with normal displacement and slope discontinuities at their edges. The algebraic equations derived by use of finite-difference models with constant grid spacing are shown to be the Euler equations of the variational problem in finite form except at the boundaries of the domain. Two types of curved elements are introduced into the BOSOR3 computer code, a general shell-of-revolution analyzer originally based on the finite-difference energy model. Elastic hemispheres with clamped and free edges are analyzed for stress, buckling, and vibration with both finite-difference and finite-element methods. The various discretization models are compared with respect to convergence properties; computer times required to form global stiffness, "load-geometric," and mass matrices; computer times required to factor and solve the systems of algebraic equations; and number of Gaussian integration points required for adequate determination of local matrices. The rates of convergence are case dependent, the finite-difference method tending to show superior performance with regard to normal displacement and vibration frequencies and the finite-element method tending to lead to faster convergence in cases involving edge disturbances. Computer times required to form global matrices are two to six times greater with the finite-element models than with the finite difference models, primarily because of interpolations, coordinate transformations, evaluation of element energy density at more than one Gaussian integration point, and static reduction, none of which is required in the finite-difference analysis. It is found that with the finite-element model, the most rapidly convergent calculations correspond to use of the lowest number of Gaussian integration points consistent with convergence to the correct solution. Two integration points are required in a finite-element model with linear displacement functions for in-plane displacements and a cubic for normal displacements. Three integration points are required in an isoparametric model with cubic polynomials for all displacement components. None of the finite element models converges to the proper solution if only one "Gaussian" point is used, as is done with the finite-difference energy method. A brief discussion is included concerning extrapolation of the results of the present one-dimensional analysis to two-dimensional problems.

Nomenclature

S	stress resultant vector, Eq. (2)
$\bar{N}_0$	known prestress matrix linearly dependent on load eigenvalue parameter, λ, Eq. (5)
N_{0f}	known prestress matrix of fixed resultants, Eq. (5)
d	vector of displacement components u, v, and w
$\bar{P}$	normal pressure matrix linearly dependent on load eigenvalue parameter, λ, Eq. (7)
P_f	normal pressure matrix of fixed values, Eq. (7)
m	diagonal matrix of mass/area
p_f	distributed load vector consisting of components p_1, p_2, p_3 in u, v, and w, directions, respectively
r	local radius of parallel circle of shell of revolution
s	arc length along meridian
N	stress resultants or number of unknowns/node point, depending on context
M	moment resultants
u	displacement tangent to meridian in direction of increasing s
v	displacement tangent to parallel circle in direction of increasing θ
w	normal displacement, positive to right of increasing s
p	pressure or surface tractions, positive in direction of positive displacement
R	radius of curvature (if subscripted)
n	number of complete circumferential waves
C	matrix of coefficients relating stress and moment resultants to strains and changes in curvature, Eq. (13)
$\bar{u}_1, \bar{u}_2$	additional internal degrees of freedom for meridional displacement within element ② in Fig. 1
$\bar{v}_1, \bar{v}_2$	additional internal degrees of freedom for circumferential displacement within element ② in Fig. 1
q	generalized nodal point displacements
B	matrix relating portion of strains and changes in curvature independent of λ to displacement expansion coefficients α, Eq. (17)

T matrix relating displacement expansion coefficients α to nodal point variables q, Eq. (17)

X matrix relating portion of strains and changes in curvature dependent on λ to polynomial expansion coefficients α, Eq. (17)

D matrix relating displacement components u, v, w to displacement expansion coefficients, α

R matrix relating rotation components χ, ψ, γ to displacement expansion coefficients, α

K_1 global stiffness matrix

K_2 global "load-geometric" matrix, composed of terms multiplied by eigenvalue, λ in Eq. (18)

K_3 global λ^2 matrix, composed of terms multiplied by λ^2 in Eq. (18)

M mass matrix, derived by a consistent method

W energy matrix before transformation to nodal point displacement variables

a_{ij} coefficients used in Eq. (24) and defined in the Appendix

l length of element

i element number

x vector of dependent variables (same as q)

G number of Gaussian integration points used in the finite element models

E Young's modulus

a radius of hemisphere

t shell wall thickness

h element length

Greek letters

ε strain vector, Eq. (3), if unsubscripted. If subscripted reference surface strains given by Eq. (11)

ω rotation vector, Eq. (4)

$\varkappa$ changes in curvature, given by Eq. (11)

χ meridional rotation, Eq. (12)

ψ circumferential rotation, Eq. (12)

γ rotation about normal, Eq. (12)

θ circumferential coordinate

λ load eigenvalue parameter

Ω frequency eigenvalue parameter

α displacement polynomial expansion coefficients, Fig. 1

ϕ polar coordinate

ρ mass density

Subscripts

0   prestress quantity

f   fixed quantity

1   meridional direction

2   circumferential direction

3   direction of outward normal

12   shear, twist

,t   differentiation with respect to time

L   part of strain independent of variable prebuckling rotation

n   part of strain dependent on variable prebuckling rotation

i   element number

b   bending

m   membrane

Superscripts

T   transpose

( )˙   differentiation with respect to circumferential coordinate θ, or with respect to time, depending on context

( )′   differentiation with respect to meridional coordinate, s

(¯)   quantity proportional to load

1. Introduction

Recent applications of finite-difference energy methods to shell stress, buckling, and vibration problems has motivated an interest in comparison of this technique with the finite-element method. It is appropriate to compare formulations, methods of implementation in a computer program, computer times, and rates of convergence with use of various finite-element and finite-difference models, all based on energy and all based on the displacement method, or the principle of minimum potential energy.

Previous comparisons of this nature have been clouded by the use of different computer programs run on various computers with the use of various equation solving techniques. The purpose of this paper is to provide such comparisons in a limited number of cases by use of one computer program [1] with certain of its subroutines altered in a minimum way to accommodate the various methods. In this way the same degree of generality is maintained for the entire analysis: A general-purpose finite-element program is not being compared with a special-purpose finite-difference program.

The subroutines used to solve the systems of equations generated by the various models did not have to be altered. All calculations were performed in double-precision on the Univac 1108 computer, Exec 8 system.

The BOSOR3 computer program was used for the comparisons [1]. This program performs stress, stability, and vibration analyses of shells of revolution. It is based on the finite-difference energy technique.

To generate the desired comparisons it was necessary to "put" a finite element or elements into BOSOR3. This was done as described in the analysis section. Changes in only a few of the subroutines were required to make BOSOR3 a finite-element program instead of a finite-difference program. Several runs were made against known solutions to check for bugs. The program was then used to calculate stresses, buckling loads, and vibration frequencies of a hemisphere with a clamped or free edge, loaded by various pressure distributions.

The intent of this paper is not to present the finite-difference energy technique as new, but to provide a clear physical interpretation of the method and to compare it in a simple one-dimensional context with various well-known finite-element models. The finite-difference energy method was apparently introduced by Courant et al. in 1928 [2] and is discussed in a textbook by Forsythe and Wasow [3]. Several research efforts have been based on the method, including those of Stein [4], Havner and Stanton [5], Budiansky [6], Johnson [7], Almroth et al. [8], Noor [9], and Schnobrich and Pecknold [10]. A comparison of the use of finite-difference and finite-element methods in structural analysis is given by Pian [11].

While the development presented here applies to one-dimensional discretization, the reader is urged to extrapolate the discussions and conclusions to problems involving two and three dimensions.

2. Analysis

The energy functional for a shell of revolution the meridians of which run from A to B is given by

$$H = \tfrac{1}{2} \int_A^B (S^T \varepsilon + \omega^T(\bar{N}_0 + N_{0f})\omega + d^T(\bar{P} + P_f)d - \dot{d}^T m \dot{d} - 2 p_f^T d) r \, ds \quad (1)$$

in which

$$S^T \equiv [N_1, N_2, N_{12}, M_1, M_2, M_{12}] \tag{2}$$

$$[\varepsilon] \equiv \{\varepsilon\}^T \equiv [\varepsilon_1, \varepsilon_2, \varepsilon_{12}, \varkappa_1, \varkappa_2, 2\varkappa_{12}] \tag{3}$$

$$[\omega] \equiv [\chi, \psi, \gamma] \tag{4}$$

$$\bar{N}_0 \equiv \begin{bmatrix} N_{10} & 0 & 0 \\ 0 & N_{20} & 0 \\ 0 & 0 & (N_{10} + N_{20}) \end{bmatrix},$$

$$N_{0f} \equiv \begin{bmatrix} N_{10f} & 0 & 0 \\ 0 & N_{20f} & 0 \\ 0 & 0 & (N_{10f} + N_{20f}) \end{bmatrix} \tag{5}$$

$$d^T \equiv [u, v, w] \tag{6}$$

$$\bar{P} \equiv \begin{bmatrix} -p/R_1 & 0 & -p' \\ 0 & -p/R_2 & 0 \\ -p' & 0 & p(1/R_1 + 1/R_2) \end{bmatrix}, \qquad P_f \equiv [\text{similar to } \bar{P}] \tag{7}$$

$$\dot{d}^T \equiv [u_{,t}, v_{,t}, w_{,t}] \tag{8}$$

$$m \equiv \begin{bmatrix} m & 0 & 0 \\ 0 & m & 0 \\ 0 & 0 & m \end{bmatrix} \tag{9}$$

$$p_f^T \equiv [p_1, p_2, p_3] \tag{10}$$

Quantities with subscript "f" are considered to be fixed, known parameters. Those without "f," such as $\bar{N}_0$, are "variable" eigenvalue parameters. The strain-displacement relations and rotation components corresponding to n

circumferential waves are given by

$$\varepsilon_1 = u' + w/R_1 + \bar{\chi}_0\chi + \chi_{0f}\chi, \qquad \varkappa_1 = \chi'$$

$$\varepsilon_2 = -nv/r + r'u/r + w/R_2, \qquad \varkappa_2 = -n\psi/r + r'\chi/r \qquad (11)$$

$$\varepsilon_{12} = v' - r'v/r + nu/r + \bar{\chi}_0\psi + \bar{\chi}_{0f}\psi, \qquad \varkappa_{12} = -n\chi/r + r'\psi/r + v'/R_2$$

$$\chi = w' - u/R_1, \qquad \psi = nw/r - v/R_2, \qquad \gamma = \tfrac{1}{2}(nu/r - v' - r'v/r) \quad (12)$$

Primes denote differentiation with respect to meridional arc length s; multiplication by n implies differentiation with respect to the circumferential coordinate, θ.

It is assumed that the shell wall material is linear elastic. Therefore,

$$S = C\varepsilon \qquad (13)$$

in which C is a 6×6 symmetric matrix of coefficients of the form given in [12].

The energy functional (1) is the fundamental relationship for determination of stresses, buckling loads and vibration frequencies of shells of revolution. For the sake of simplicity, line loads, thermal loads, and edge loads have omitted. Expressions (1)–(13) form the basis for solution of the stress and eigenvalue problems by displacement methods.

The "variable" prestress quantities $\bar{\chi}_0$ and $\bar{N}_0$ and the "variable" surface loads $\bar{P}$ are considered to be known functions $\chi_0(s)$, $N_0(s)$ and $P(s)$ multiplied by a load factor, λ. The motion of the shell is assumed to be harmonic with frequency Ω. Hence, the energy functional H can be written in the form

$$H = \tfrac{1}{2} \int_A^B (\varepsilon_L{}^T C \varepsilon_L + \omega^T N_{0f}\omega + d^T P_f d + \lambda[\varepsilon_L{}^T C \varepsilon_n + \varepsilon_n{}^T C \varepsilon_L + \omega^T N_0 \omega$$

$$+ d^T P d] + \lambda^2 \varepsilon_n{}^T C \varepsilon_n - \Omega^2 d^T m d - 2p_f{}^T d)r \, ds \qquad (14)$$

In Eq. (14) ε_L is the part of the strain vector independent of $\bar{\chi}_0$ and ε_n is the part dependent on $\bar{\chi}_0$. Stress problems involve only the first and last terms in the integrand of Eq. (14), bifurcation buckling problems involve all terms except the last two, and free vibration problems with prestress included involve all terms except those multiplied by λ and λ^2 and the last one. Expression (14) is basic to the displacement method, independent of what discretization technique is used.

Equation (14) is an integrodifferential form. The objective of the stress or buckling or free vibration analysis is to find the displacement distributions $d(s)$ which render the energy minimum or stationary. In the linear stress analysis there is only one such distribution. It corresponds to equilibrium of the system. In the bifurcation buckling and free vibration analyses such distributions occur only for certain eigenvalues λ or Ω^2. The $d(s)$ for which H is stationary can be determined by means of some search technique, series expansion, the finite-element method, or the finite-difference method. The remainder of this paper is concerned with the last two approaches.

2.1. VARIOUS DISCRETIZATION TECHNIQUES

In both the finite-element and finite-difference energy methods the unknowns of the problem are certain generalized displacement components located at discrete nodes in the domain. Between these nodes the variations of the generalized displacements are expressed as power series in s. Integration can then be performed analytically or numerically. The differing choice of generalized displacement components and locations of the nodes are the only characteristics of the two solution techniques which justify giving them different names. Once the nodes and the appropriate generalized displacement components have been selected, the solution procedure is identical for both methods.

Figure 1 shows seven types of discretization. The nodes are denoted by large dots or crosses. The "element" is defined as the solid line bounded by dots or crosses. Nodal point variables u_i, v_i, w_i, χ_i, etc. are shown next to the nodes with which they are associated. The first three models fall into the category "finite element method," the last four into the category "finite difference method."

The three models ①, ②, and ③ represent standard finite elements such as described in Kotanchik et al. [13], Mebane and Stricklin [14], Adelman et al. [15]. The computer program SABOR5-DRASTIC is based on curved elements of the type ①. A curved element ② with extra internal degrees of freedom (d.o.f.) $\bar{u}_1, \bar{v}_1, \bar{u}_2$, and $\bar{v}_2$ permits rigid body motion without excessive storage of energy. The internal degrees of freedom represent corrections to the linear function. Elements of this type are described in [14]. An alternate way of obtaining higher-order displacement functions is to define more degrees of freedom at the nodes [15]. Element ③ is of this type. The displacements within each of these elements are given by the polynomials shown in the figure. Integration of the energy functional can be performed analytically or numerically. Gaussian quadrature seems to be the most accurate and economical method of integration.

Figure 2 shows schematically a structure consisting of five elements. The displacement function w and its first derivative are continuous throughout the domain. The displacement function for u and v corresponds to model ①. Of course the elements need not be flat. However, it is shown in Fig. 3 that if the element is curved, higher-order displacement functions than linear are required for representation of rigid body motions.

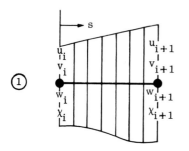

Standard lowest order finite element

$$u = \alpha_1 + \alpha_2 s$$

$$v = \alpha_3 + \alpha_4 s$$

$$w = \alpha_5 + \alpha_6 s + \alpha_7 s^2 + \alpha_8 s^3$$

Energy evaluated at Gaussian integration points.

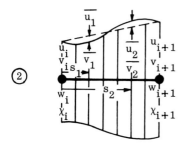

Finite element with extra internal d.o.f.

$$u = \alpha_1 + \alpha_2 s + \alpha_3 s^2 + \alpha_4 s^3$$

$$v = \alpha_5 + \alpha_6 s + \alpha_7 s^2 + \alpha_8 s^3$$

$$w = \alpha_9 + \alpha_{10} s + \alpha_{11} s^2 + \alpha_{12} s^3$$

Energy evaluated at Gaussian integration points.
Static reduction used to get local [K].

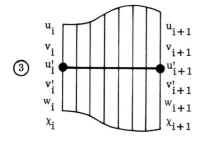

Finite element with extra nodal d.o.f.

$$u = \alpha_1 + \alpha_2 s + \alpha_3 s^2 + \alpha_4 s^3$$

$$v = \alpha_5 + \alpha_6 + \alpha_7 s^2 + \alpha_8 s^3$$

$$w = \alpha_9 + \alpha_{10} s + \alpha_{11} s^2 + \alpha_{12} s^3$$

Energy evaluated at Gaussian integration points

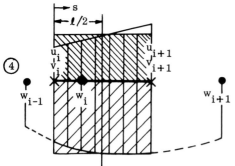

Finite difference with (u_i, v_i) on half-stations

$$u = \alpha_1 + \alpha_2 s$$

$$v = \alpha_3 + \alpha_4 s$$

$$w = \alpha_5 + \alpha_6 s + \alpha_7 s^2$$

Energy evaluated at $s = \ell/2$.

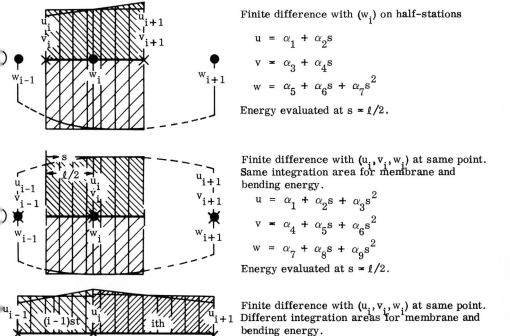

Finite difference with (w_i) on half-stations

$$u = \alpha_1 + \alpha_2 s$$

$$v = \alpha_3 + \alpha_4 s$$

$$w = \alpha_5 + \alpha_6 s + \alpha_7 s^2$$

Energy evaluated at $s = \ell/2$.

Finite difference with (u_i, v_i, w_i) at same point. Same integration area for membrane and bending energy.

$$u = \alpha_1 + \alpha_2 s + \alpha_3 s^2$$

$$v = \alpha_4 + \alpha_5 s + \alpha_6 s^2$$

$$w = \alpha_7 + \alpha_8 s + \alpha_9 s^2$$

Energy evaluated at $s = \ell/2$.

Finite difference with (u_i, v_i, w_i) at same point. Different integration areas for membrane and bending energy.

$$u = \alpha_1 + \alpha_2 s$$

$$v = \alpha_3 + \alpha_4 s$$

$$w = \alpha_5 + \alpha_6 s \quad \text{(membrane energy)}$$

$$w = \alpha_5 + \alpha_6 s + \alpha_7 s^2 \quad \text{(bending energy)}$$

Fig. 1. Various discrete models for energy methods.

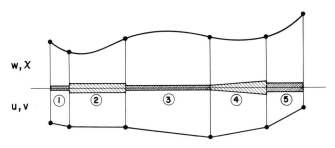

Fig. 2. Displacement functions for finite element model ①.

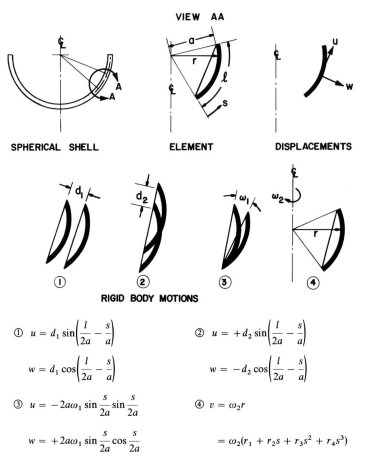

Fig. 3. Rigid body displacements of curved element.

Three characteristics of the finite-difference models ④–⑦ distinguish them from the finite-element models:

1. The rotation χ_i is not a nodal point unknown.
2. The displacement components u_i, v_i, w_i do not necessarily occur at a given node.
3. The displacement polynomial is not necessarily restricted to the domain defined as the "element," shown as a solid line or lines in each of the models.

Models ④–⑥ might be called "finite-difference elements." They can consist of various properties which need not be continuous at element boundaries.

Model ⑦ involves different areas of integration for membrane and bending energy, so that there is no clear physical element boundary.

In the finite-difference energy method, the integrand of Eq. (14) is evaluated at only one point within each element, and the total energy is obtained by multiplication of this value by the element length l. The finite-difference formulas for variable mesh spacing are obtained by Taylor series expansions of the displacements about the centroid of each element. Since first and second derivatives of w and only first derivatives of u and v occur in the integrand of Eq. (14), the appropriate polynomials for the lowest-order difference formulas in each case are shown. As in the case of the finite-element method, the α_i (Fig. 1) can easily be expressed in terms of the nodal point variables.

Most of the finite-difference energy discretization models ④–⑦ have been used in computer programs. Stein [4], Bushnell [12], and Brogan and Almroth [8] used model ④. Attempts were made without success to use model ⑥ [12]. Model ⑦ has been used by Almroth [16] with considerable success. The author is not aware of any use of model ⑤, although for constant mesh spacing, it is of course identical to model ④.

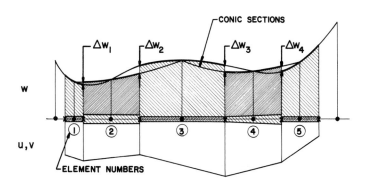

Fig. 4. Displacement functions for finite difference model ⑤.

Figures 4, 6, and 8 show the finite-difference discretizations of the same five-element structure depicted in Fig. 2. In Fig. 4 the elements are assembled at u and v nodal points (crosses). The quadratic w expansions pass through three adjacent w nodes. However, the integration, or "lumping" corresponds only to the areas between adjacent crosses. The in-plane displacements u and v are continuous everywhere. Notice that at element boundaries the normal displacements and derivatives are discontinuous. It can be shown that the displacement discontinuity Δw_i is of maximum order

$$|\Delta w_i| = \frac{h_i h_{i+1}}{8}(h_{i-1} + 2h_i + 2h_{i+1} + h_{i+2}) \qquad (15)$$

and the slope discontinuity is of maximum order

$$|\Delta\chi_i| = \left|\frac{h_i - h_{i+1}}{2}\right|(h_{i-1} + 2h_i + 2h_{i+1} + h_{i+2}) \qquad (16)$$

The finite-difference discretizations ④ and ⑤ are similar to replacement of the actual structure by a structure consisting of elements linked as shown in Fig. 5. The normal displacement w is continuous at the pinned joints and u and v are continuous at the stations where the projections stick into the rounded holes.

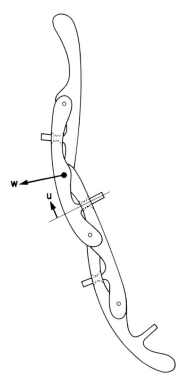

Fig. 5. Structural equivalent of finite difference model ④ or ⑤ (courtesy Carlos Felippa).

At first glance it would seem that this structure is far too flexible to represent the behavior of a continuous shell. However, notice in Fig. 4 that u and v must be continuous at the stations where w is discontinuous. Since the circumferential strain, for example, involves at least both w and v, the Δw at element boundaries must remain small enough to keep this membrane strain component, and hence the energy, at a reasonable level. In other words, the minimum energy state will involve *small* discontinuities in w at the element boundaries.

No such protection is afforded in the case of finite-difference model ⑥, in which u, v, and w are taken at the centers of the elements, as shown in Fig. 6. In this model all variables may be discontinuous at element boundaries. This situation is probably responsible for the poor numerical behavior apparent in Fig. 7 (taken from [12]), which shows the fundamental vibration mode of a simply supported ring-stiffened cylinder, as predicted with the finite-difference model ⑥.

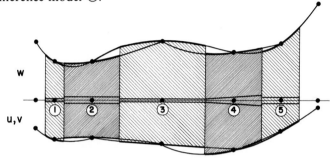

Fig. 6. Displacement functions for finite difference model ⑥.

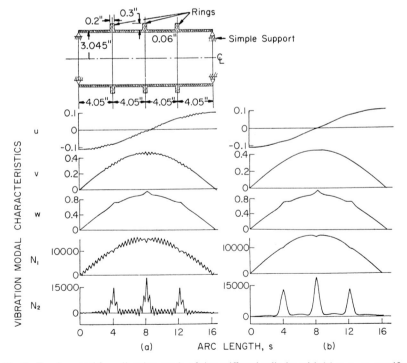

Fig. 7. Fundamental free vibration mode of ring-stiffened cylinder with (a) u_i, v_i, w_i specified at the same point (model ⑥), and (b) u_i, v_i specified midway between w_i and w_{i+1} (model ④).

Figure 8 shows the assembled five-element structure corresponding to discretization model ⑦. All displacement components are taken at the same nodes, as in model ⑥, but different areas of integration are used for the membrane and bending energy. Although this model (termed "whole station scheme" in [16]) has been used with great success in a limited number of

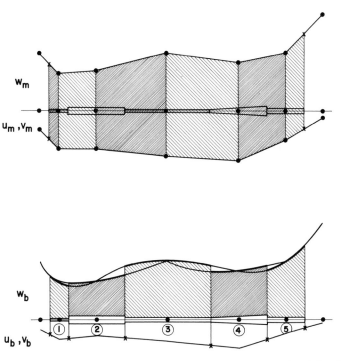

Fig. 8. Displacement functions for membrane and bending energy for finite difference model ⑦.

cases, it suffers from two drawbacks: It is not clear how the method will work in cases for which membrane and bending energies are coupled (e.g., eccentrically stiffened shells in which the stiffeners are "smeared" out), and since different areas of integration are used for the two types of energy, it is not clear how the model could be applied to systems in which sudden discontinuities occur. Also, the bending model, while similar to model ⑤, offers less protection against the "spurious mode" behavior seen in Fig. 7, since enforced continuity of u_b and v_b, the bending portions of u and v, is less demanding of energy than is enforced continuity of total displacements u and v.

2.2. IMPLEMENTATION OF VARIOUS DISCRETE ENERGY MODELS IN A COMPUTER PROGRAM

The procedure for finding stationary values of H is, of course, similar for all of the discretizaton models shown in Fig. 1. In the present investigation this procedure has been carried out for models ①, ②, ④, and ⑤ and implemented in the BOSOR3 computer program [1]. The purpose of the investigation is to compare computer times and rates of convergence with increasing mesh point density for the various discretization models.

The functional H given by Eq. (14) must be expressed in terms of the nodal point variables. The following matrix relations exist:

$$\varepsilon_L = B\alpha = BTq, \qquad \varepsilon_n = X\alpha = XTq,$$
$$\omega = R\alpha = RTq, \qquad d = D\alpha = DTq \tag{17}$$

in which α are the polynomial coefficients shown in Fig. 1 and q are the nodal point generalized displacements. In the finite-difference models ④–⑦ the ε_L, ε_n, ω, and d are expressed directly in terms of q without any separate calculation of the transformation matrix T. In the finite-element method the decision whether to express these variables in terms of the α's and transform after integration or to express them in terms of the q's directly by means of appropriate interpolating polynomials depends on the complexity of the various matrices (particularly B) and on the number of integration points required within each element.

Use of Eq. (17) in Eq. (14) leads to the following relationship:

$$H = \tfrac{1}{2}q^T T^T \left[\int_A^B (B^T CB + R^T N_{0f} R + D^T P_f D \right.$$
$$+ \lambda[B^T CX + X^T CB + R^T N_0 R + D^T PD] + \lambda^2 X^T CX \tag{18}$$
$$\left. - \Omega^2 D^T mD)r\,ds \right] Tq - \int_A^B p_f{}^T Dr\,ds\,Tq$$

Table 1 gives the dimensions of the matrices T, B, C, R, N_{0f}, D, P_f, X, N_0, P, and m for the various discrete models shown in Fig. 1. The dimensions of the local stiffness matrix H corresponding to the ith element are also given. The sizes of these local matrices give some idea of the computer time required to form the local stiffness, mass, "loadgeometric" and λ^2 matrices for each integration point within each element. The stiffness matrix $[K_1]$ is composed of the first three terms of Eq. (18); the "load-geometric" matrix $[K_2]$ is represented by those terms multiplied by λ; the λ^2 matrix $[K_3]$ involves the term multiplied by λ^2; and the mass matrix $[M]$ is the term multiplied by Ω^2. Of course the final transformation from α's to q's is performed outside the integration loop.

TABLE 1

Dimensions of the Local Matrices
Corresponding to Various Discretization Models

Matrix function	Matrix name	Matrix dimensions						
		Model 1	Model 2	Model 3	Model 4	Model 5	Model 6	Model 7[a]
strain-displace-ment	B	6×8	6×12	6×12	6×7	6×7	6×9	6×6 6×7
constitutive law	C	6×6	6×6	6×6	6×6	6×6	6×6	3×3 3×3
rotation-displace-ment	R	3×8	3×12	3×12	3×7	3×7	3×9	3×6 3×7
fixed prestress resultants	N_{0f}	3×3	3×3	3×3	3×3	3×3	3×3	3×3
displacment-generalized q	D	3×8	3×12	3×12	3×7	3×7	3×9	3×6 3×7
pressure-rotation (fixed)	P_f	3×3	3×3	3×3	3×3	3×3	3×3	3×3
strain-displace-ment	X	6×8	6×12	6×12	6×7	6×7	6×9	6×6 6×7
variable prestress resultants	N_0	3×3	3×3	3×3	3×3	3×3	3×3	3×3
variable pressure-rotation	P	3×3	3×3	3×3	3×3	3×3	3×3	3×3
mass/area	m	3×3	3×3	3×3	3×3	3×3	3×3	3×3
transforma-tion alpha-to-q	T	8×8	12×12	12×12	7×7	7×7	9×9	6×6 7×7
local stiffness, mass, load-geometric, or λ^2 matrices for ith element or point	H	8×8	8×8^b	12×12	7×7	7×7	9×9	6×6 7×7

[a] Top dimensions correspond to membrane energy, bottom to bending energy.
[b] After static reduction from 12×12.

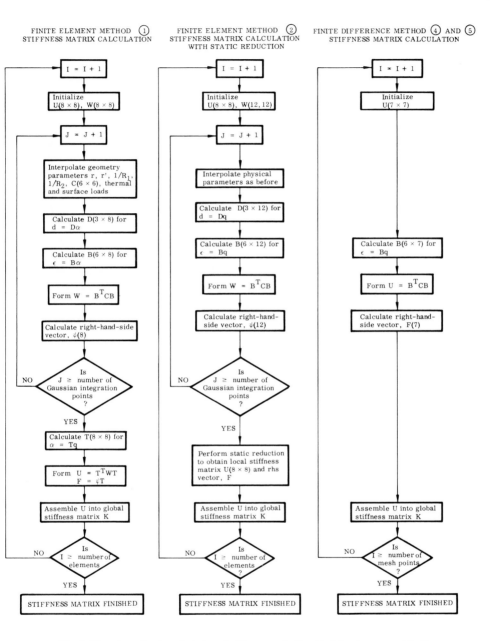

Fig. 9. Flow charts for the calculation of stiffness matrix K_1 with finite-element models ①
and ② and finite-difference models ④ or ⑤.

Figure 9 shows computation flow charts corresponding to stress analysis involving discrete models ①, ②, and ④ or ⑤. The computer time for local stiffness matrix computation is greater for the finite-element method primarily because of the interpolation required for element properties at the Gaussian integration points, the loop over more than one Gaussian point, and the transformation $U = T^T W T$ and/or static reduction outside the integration loop.

In the finite-element calculations element properties required for computation of the various matrices in Eq. (18) [such as geometry parameters r, r', $1/R_1$, $(1/R_1)'$, and $1/R_2$ required for calculation of B] were provided at nodes between elements and appropriate interpolation formulas used to determine these properties at the Gaussian integration points. A cubic polynomial was used for the radius r, as shown in Fig. 3. Thus the finite-element model ② is an isoparametric element. Linear interpolation was used for all other variables except, of course, r', which varies quadratically.

In the finite-difference calculations the "element" properties corresponding to the centroid of each element were provided as input. No interpolation is required because the energy is evaluated at only one point within each element.

The figures and equations in the Appendix give the B matrices for models ①, ②, and ④ (Fig. 1). As can be seen, a great deal more computational effort is required for model ② than for model ④.

2.3. RELATIONSHIP TO EULER EQUATIONS

Differentiation of H with respect to the generalized displacements q yields in the stress analysis

$$[K_1]q = [Q] \tag{19}$$

in which

$$[Q] \equiv \left(\int_A^B p_f^T Dr \, ds \right) T \tag{20}$$

It is instructive to investigate the relationship of Eq. (19) to the Euler equations of the variational problem

$$\delta U = \int_A^B S^T \delta \varepsilon r \, ds - \int_A^B p_f^T \delta d \, r \, ds = 0 \tag{21}$$

which corresponds to the portion of H [Eq. (1)] related to the stress analysis. Insertion of the pertinent equations among Eqs. (17) into Eq. (21) yields

$$\left[\int_A^B (S^T BT - p_f^T DT) r \, ds \right] \delta q = 0 \tag{22}$$

The Euler equation of the variational problem is enclosed in brackets. In finite form Eq. (22) becomes

$$\sum_j [(S_j\{BT\}_{ji} - p_{tj}^T\{DT\}_{ji})r_j \,\Delta s_j]\, \delta q_i = 0 \qquad (23)$$

The summation index j includes all those stations where the integrand of (22) is evaluated which contain q_i, and Δs_j are the Gaussian integration weights. For example, in the finite-difference model ⑤ (referring to Fig. 4) w_3 in the middle of element 3 is included in the expression for energy in elements 2–4. In the finite-element model ① (referring to Fig. 2) w between elements 3 and 4 occurs in the expressions for energy evaluated at all of the Gaussian points in elements 3 and 4.

With use of Eqs. (2), (3), (6), (10) and (23) and the Chart A1 in the appendix, one can show that for finite-difference model ④ the "finite" Euler equation corresponding to w_i is

$$\left(\frac{rN_1}{R_1}a_{23}l\right)_{i-1} + \left(\frac{rN_1}{R_1}a_{22}l\right)_i + \left(\frac{rN_1}{R_1}a_{21}l\right)_{i+1}$$

$$+ \left(\frac{rN_2}{R_2}a_{23}l\right)_{i-1} + \left(\frac{rN_2}{R_2}a_{22}l\right)_i + \left(\frac{rN_2}{R_2}a_{21}l\right)_{i+1}$$

$$+ (rM_1a_{53}l)_{i-1} + (rM_1a_{52}l)_i + (rM_1a_{51}l)_{i+1}$$

$$+ \left[rM_2\left(\frac{-n^2}{r^2}a_{23} + \frac{r'}{r}a_{43}\right)l\right]_{i-1} + \left[rM_2\left(\frac{-n^2}{r^2}a_{22} + \frac{r'}{r}a_{42}\right)l\right]_i \quad (24)$$

$$+ \left[rM_2\left(\frac{-n^2}{r^2}a_{21} + \frac{r'}{r}a_{41}\right)l\right]_{i+1} + \left[2rM_{12}\left(\frac{-n}{r}a_{43} + \frac{nr'}{r^2}a_{23}\right)l\right]_{i-1}$$

$$+ \left[2rM_{12}\left(\frac{-n}{r}a_{42} + \frac{nr'}{r^2}a_{22}\right)l\right]_i + \left[2rM_{12}\left(\frac{-n}{r}a_{41} + \frac{nr'}{r^2}a_{21}\right)l\right]_{i+1}$$

$$= (p_3rl)_i$$

Similar types of equations are obtained corresponding to u_i and v_i. Eq. (24) represents normal force equilibrium at the point labeled E in Chart A1. For constant mesh spacing ($h = k = l$ on Chart A1) and with multiplication by n denoting differentiation with respect to θ (e.g., $nM_{12} = -M^{\cdot}_{12}$), it becomes clear after some algebra that Eq. (24) is equivalent to the analytical form:

$$rN_1/R_1 + rN_2/R_2 + (rM_1)'' + (rM_2)^{\cdot\cdot}/r^2$$
$$- (r'M_2)' - 2(rM_{12})^{\cdot\prime}/r = p_3r \qquad (25)$$

in which the prime ( )′ denotes differentiation with respect to arc length s and ( )$^{\cdot}$ denotes differentiation with respect to circumferential coordinate θ.

This is the differential equation for equilibrium of forces normal to the shell wall. In a similar way, the "finite" equations corresponding to differentiation of H with respect to u_i and v_i can be shown to correspond to the equations of equilibrium of forces in the meridional and circumferential directions, respectively. In the case of these "in-plane" equilibrium equations the correspondence of the "finite" equations (23) to the differential equations of equilibrium is not restricted to the case of uniform mesh spacing. If the mesh spacing varies the only terms which lose correspondence with the Euler equations are those involving a_{21}, a_{22}, and a_{23}. Terms involving first and second s-derivatives (a_{4k} and a_{5k}, $k = 1, 2, 3$) maintain correspondence even if the mesh spacing varies.

The preceding derivation and comments apply only to the finite-difference model ④ and only to the points inside the domain. It is important to consider the form of Eq. (24) at the edges. The "finite" equation associated with the "fictitious" points to the left of element 1 and to the right of element 5 in Fig. 4 (assuming $h = k = l$ at the edges) is:

$$(r/l)M_1 \mp 0.5r'M_2 \pm M_{12}n = 0 \tag{26}$$

in which the top signs are associated with the left-hand boundary. It is seen that for mesh spacing $l \to 0$, Eq. (26) represents the natural boundary condition $M_1 = 0$ at the shell edges. Similar considerations yield the boundary conditions $N_1 = -(M_1/R) = 0$ associated with u at the edges and $N_{12} + (2M_{12}/R_2) = 0$ associated with v at the edges. The equations associated with the w-nodes at the middle of the first and fifth elements in Fig. 4 are also not analogous to the Euler equations of the variational problem, because they contain sets of only two terms each instead of three corresponding to $i - 1$, i, and $i + 1$ in Eq. (24). This anomaly and the inexact expressions for the natural boundary conditions associated with finite meshes cause slower convergence with finer grids than would otherwise be achieved with use of finite-difference energy methods.

The finite-element methods do not correspond in the sense just described to the Euler equations of the variational problem.

3. Numerical Results

The discrete models ①, ②, ④, and ⑤ (Fig. 1) were implemented in the BOSOR3 computer program. Convergence studies were performed for stress, bifurcation buckling loads, and free vibration frequencies. All of the analyses involve a hemisphere with radius $a = 100$ and thickness $t = 1.0$. The edge of the hemisphere is either clamped or free. In the stress analyses the hemisphere is loaded by normal pressure uniform along a meridian and varying around the circumference with two or with six circumferential waves. Buckling loads are computed for a uniformly pressurized clamped hemisphere for $n = 6$ circumferential waves. All of the analyses correspond to the linear branch of BOSOR3. All calculations were performed in double precision on the Univac 1108 computer, EXEC 8 system. Figures 10–22 and Tables 2–4 give the results of these studies.

In all of the studies the mesh spacing is uniform, except at the pole and at the edge. Near each of these stations an additional mesh point is inserted a distance $\frac{1}{20}$th of the original mesh spacing in from the boundary of the domain. In finite-difference model ④ the additional point corresponds to a w-node. In the finite-element models and finite-difference model ⑤ the additional point corresponds to an element boundary, or u-node. The extra points were added to avoid difficulty associated with "fictitious" w-nodes required in the finite-difference analysis for difference expressions of w at the boundaries. Bushnell and Almroth [16] noted that these "fictitious" points lead to the existence of spurious eigenvalues.

3.1. COMPUTER TIMES

Figure 10 shows CP computer times for formation of the stability stiffness matrix $[K_1]$ and the "load-geometric" matrix $[K_2]$ for the bifurcation

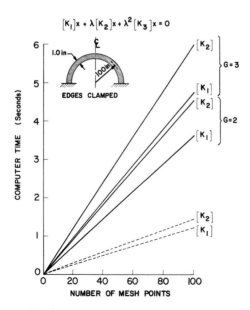

Fig. 10. CP computer times for calculation of the stiffness matrix K_1 and the "load-geometric" matrix K_2 with the finite-element model ① for two and three Gaussian integration points and with the finite difference model ④. (—): Finite element ①; (---): finite difference ④.

buckling problem. Prebuckling rotations χ_0 were assumed to be zero, so that the λ^2 matrix $[K_3]$ is zero. However all calculations involving χ_0 were performed. The prebuckling stress resultants N_{10} and N_{20} were read in as input. The models used were finite-element model ① and finite-difference model ④. The notations $G = 2$ and $G = 3$ refer to the number of Gaussian integration points used in the finite-element model.

The buckling load is the lowest eigenvalue of the system:

$$[K_1]x + \lambda[K_2]x + \lambda^2[K_3]x = 0 \tag{27}$$

The eigenvalues and modes are determined by the inverse power iteration method with spectral shifts. Figure 11 shows the computer time for each factoring of the stability matrix and each inverse power iteration. Each spectral shift requires less than 0.1 sec.

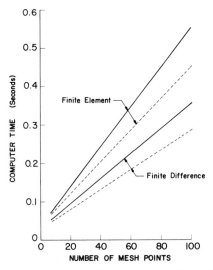

Fig. 11. Computer times for factoring the stiffness matrix and for each inverse power iteration. (——): Factor; (–––): power iteration.

The computer times for factoring and iterating with use of finite-element model ① are somewhat greater than those for the finite difference model because the semibandwidth of the global stiffness matrix is 8 rather than 7, and there are $\frac{4}{3}$ the number of degrees of freedom for the same number of mesh points. The factoring time is insignificant compared to the time required to form the stiffness matrix or the "load-geometric" matrix. This is because the problem is one-dimensional and therefore the bandwidth of the stiffness matrix is very small compared to its rank. The total time for inverse power iterations is the same order as the time to form the stiffness matrix, since 10 to 20 iterations are required for convergence to the eigenvalue and eigenvector.

The computer times plotted in Figs. 10 and 11 apply to the buckling problem. Similar results are obtained for the stress and vibration problems. The time required for each power iteration in the natural vibration analysis is about 30% less, since the eigenvalue problem is of the more standard form

$$[K_1]x - \Omega^2[M]x = 0 \qquad (28)$$

rather than of the form given in Eq. (27).

3.2. STRESS ANALYSIS

Results of two stress analyses of the hemisphere are given in Figs. 12–17 and Table 2. Figure 12 shows inner- and outer-fiber stresses and normal displacement w at $\theta = 0$ as functions of middle surface arc length s for a clamped hemisphere loaded by normal pressure $p = \cos 6\theta$. Table 2 gives the displacements and stresses at $\phi = 45°$ as functions of the number of mesh points. Results from finite-difference model ④ are compared with those from finite-element model ①. Figure 13 shows computer times to form the stiffness matrices and percentage of error in the normal displacement at $\phi = 45°$ as functions of number of mesh points. A comparison is given in Fig. 14 of the rates of convergence of the inner fiber meridional stress at the clamped edge.

As seen from Figs. 13 and 14, the finite-difference model ④ gives a more rapidly convergent solution within the domain, but a more slowly convergent solution at the edge where a rapidly varying discontinuity stress concentration exists. That the finite-difference method yields less accuracy at edges than the finite-element method is one of its disadvantages. The relatively slow convergence at the edge may also arise because of the extreme variation in mesh spacing due to the additional node located $\frac{1}{20}$th of the distance between the edge node and the second node from the edge.

In the finite-element analysis the results with three Gaussian integration points ($G = 3$) are almost identical to those with two ($G = 2$). If one Gaussian point is used, the finite-element analysis converges to the wrong solution. Use of finite-element model ② with $G = 3$ gives virtually identical answers to those listed in Table 2. More will be said about the number of Gaussian integration points later.

Figures 15 to 17 give the results of a stress analysis of a hemisphere with a free edge loaded by a pressure $p = \cos 2\theta$. This case is similar to the "pinched" cylinder problem in that the deformations are almost inextensional. The large deformations and very small membrane strains render the problem ill conditioned. Thus, it provides a good test of curved elements.

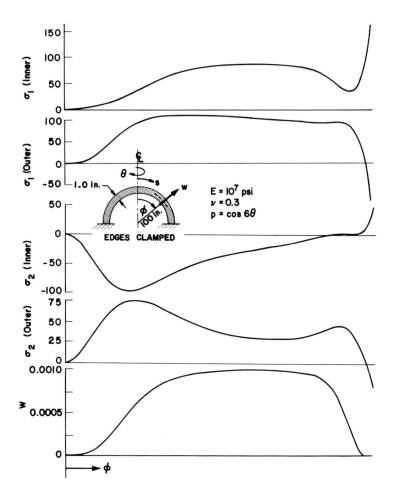

Fig. 12. Stress and normal displacement distributions at $\theta = 0$ for clamped hemisphere with pressure $p = \cos 6\theta$.

Figure 15 shows the meridional and normal displacements at $\theta = 0$. Figure 16 shows the results of convergence studies with finite-element models ① and ② and finite-difference models ④ and ⑤. An additional run was made with a Khojasteh–Bakht-type element [17] improved to handle nonsymmetric displacements. Finite-element model ① is inadequate because the u and v displacement polynomials are of too low an order to reproduce inextensional behavior. Results from the convergence study with the Khojasteh–Bakht type element fall very slightly below the curve for model ①. Finite-element model ② (static reduction) was used with two, three, and five

TABLE 2

CLAMPED HEMISPHERE STRESS ANALYSIS—CONVERGENCE OF SOLUTION AT 45° LATITUDE

Number of elements	Meridional displacement $u \times 10^4$		Normal displacement $w \times 10^4$		Meridional stresses				Circumferential stresses				Effective stresses			
					inner fiber		outer fiber		inner fiber		outer fiber		inner fiber		outer fiber	
	FD[a]	FE[a]	FD	FE	FD	FE	FD	FE	FD	FE	FD	FE	FD	FE	FD	FE
7	0.401	0.355	9.88	9.61	84.1	81.6	109	105	−39.8	−39.1	37.3	35.5	110	107	88.7	92.3
9	0.361	0.361	9.75	9.68	84.7	83.0	108	106	−39.1	−38.1	36.9	37.1	110	107	95.2	93.1
11	0.368	0.364	9.74	9.70	84.0	83.4	107	106	−38.2	−38.0	37.7	37.5	108	108	94.3	93.5
13	0.366	0.365	9.74	9.71	84.0	83.6	107	107	−38.1	−38.0	37.8	37.6	108	108	94.2	93.7
15	0.368	0.366	9.74	9.72	84.0	83.7	107	107	−38.1	−38.0	37.8	37.7	108	108	94.2	93.8
17	0.368	0.366	9.74	9.72	84.0	83.8	107	107	−38.1	−38.0	37.8	37.8	108	108	94.1	93.8
19	0.368	0.366	9.74	9.73	84.0	83.8	107	107	−38.0	−38.0	37.8	37.8	108	108	94.1	93.9
21	0.368	0.366	9.74	9.73	84.0	83.9	107	107	−38.0	−38.0	37.8	37.8	108	108	94.1	93.9
23	0.367	0.366	9.74	9.73	84.0	83.9	107	107	−38.0	−38.0	37.9	37.9	108	108	94.1	93.9
33	0.367	0.366	9.74	9.73	84.0	84.0	107	107	−38.0	−38.0	37.9	37.9	108	108	94.0	94.0
53	0.367	0.366	9.74	9.74	84.0	84.0	107	107	−38.0	−37.9	37.9	37.9	108	108	94.0	94.0
73	0.366	0.366	9.74	9.74	84.0	84.0	107	107	−37.9	−37.9	37.9	37.9	108	108	94.0	94.0
99	0.366	0.366	9.74	9.74	84.0	84.0	107	107	−37.9	−37.9	37.9	37.9	108	108	93.9	94.0

[a] FD = finite difference, FE = finite element.

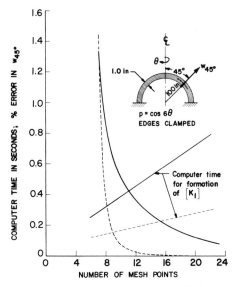

Fig. 13. Computer times and rates of convergence of normal displacement at 45° for clamped hemisphere with pressure $p = \cos 6\theta$. (—): Finite element ①; (– – –): finite difference ④.

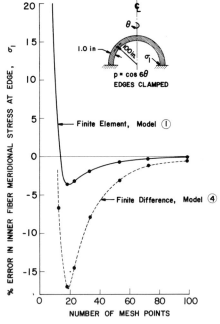

Fig. 14. Rate of convergence of meridional stress at clamped edge of hemisphere with pressure $p = \cos 6\theta$. (—): Finite element ①; (– – –): finite difference ④.

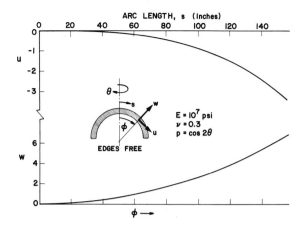

Fig. 15. Meridional and normal displacements at $\theta = 0$ of a hemisphere with a free edge submitted to pressure $p = \cos 2\theta$.

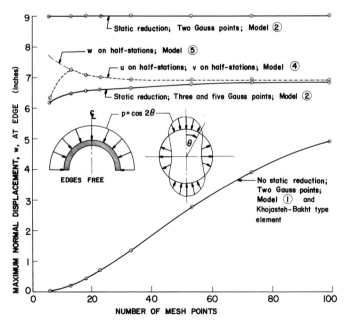

Fig. 16. Normal displacement at free edge of hemisphere with pressure $p = \cos 2\theta$. (——): Finite element; (– – –): finite difference.

Gaussian integration points ($G = 2, 3, 5$). The normal displacement w at the free edge converges extremely rapidly to the wrong answer with $G = 2$. In fact, the solution with $G = 2$ corresponds to very high membrane energy because the tangential displacements u and v are in the incorrect ratios to the normal displacement w.

Model ② results with $G = 3$ and $G = 5$ are virtually identical, with $G = 5$ results corresponding to a very slightly stiffer structure.

This trend of increasing stiffness with increasing number of Gaussian integration points for a given finite-element model was generally the same for all stress, buckling, and vibration problems studied. It was always true that more than one Gaussian point had to be used with the finite element method. Calculations with $G = 1$ always converged to wrong solutions and led to low estimates of the structural stiffness. It always held that the "best" solution, that is the most rapidly convergent solution, was obtained with the minimum number of Gaussian integration points consistent with eventual convergence to the correct answer. This number turned out to be two points with use of model ① and three points with use of model ②. Of course the maximum number of Gaussian points consistent with use of model ① for stress analysis is four [the integrand of Eq. (18) contains linearly varying bending moment squared times linearly varying C times the cubic polynomial r, the product being a sixth-order polynomial]. With model ② the maximum number of Gaussian points is five [the integrand of Eq. (18) contains quadratically varying membrane stress squared times linearly varying C times r]. However, it was determined that even in the sensitive case of the "pinched" hemisphere, it is neither necessary nor advisable to use more than three Gaussian points.

An explanation is necessary as to why one "Gaussian" integration point is adequate for finite-difference models ④ and ⑤, but is not for any of the finite-element models. The reason appears to be that finite-difference models ④ and ⑤ are actually constant membrane and bending strain elements. Therefore, it is inconsistent to use a cubic polynomial for r or to include a linear variation in C over the element length. The energy density within the element is independent of arc length s, and integration therefore amounts to multiplication of the energy density by the element length. This reasoning also tends to explain why two Gaussian points is the minimum number required with use of finite-element model ① and three Gaussian points with model ②. Two points are required to obtain the correct bending energy associated with model ① and three points are required to obtain the correct membrane energy associated with model ②.

Figure 17 shows computer times required to form the stiffness matrix $[K_1]$ and percentage of error in edge displacement for the same problem. The relatively large amount of computer time required for formation of finite-element model ② stiffness matrix is explained by inspection of Table 1,

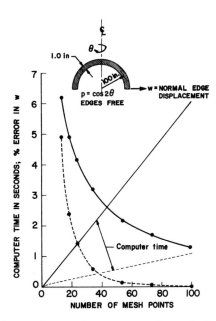

Fig. 17. Computer times and rates of convergence of normal edge displacement for free hemisphere with pressure $p = \cos 2\theta$. (—): Finite element analysis with static reduction model ②; (– – –): finite difference model ④.

Fig. 9, and Chart A3 (p. 334). Also, three Gaussian integration points are required for convergence to the correct solution. One can see from Fig. 17 that for accuracy in w of approximately 1.5% about $\frac{1}{3}$ sec is required for calculation of $[K_1]$ with finite-difference model ④, whereas about 5 sec are required with finite-element model ②. It is expected that this discrepancy would increase in two-dimensional problems because the ratio of numbers of integration points within each element would be 9/1 instead of 3/1.

3.3. BIFURCATION BUCKLING

Table 3 and Figs. 18 and 19 give results for linear bifurcation buckling analysis of a uniformly pressurized clamped hemisphere with $E = 10^7$ lb in.$^{-2}$, $v = 0.3$, and $t/a = 0.01$. The buckling wavenumber n is fixed at 6. All calculations were made with the finite-difference model ④ and the finite-element model ①. From Table 3 it is seen that the finite-elment analysis is more rapidly convergent than the finite-difference analysis, and that the solution with one Gaussian point is entirely incorrect.

Figure 18 shows the normal displacement components of the buckling modes for increasing numbers of nodes. One can see that finite-difference model ④ seems to offer more restraint at the pole, $s = 0$ than does the

TABLE 3
BUCKLING OF EXTERNALLY PRESSURIZED HEMISPHERE

Number of mesh points	Finite-difference analysis	Finite-element analysis		
		One Gauss point	Two Gauss points	Three Gauss points
7	1344	102.5	1411	1532
13	1315	96.47	1282	1295
18	1296	94.71	1264	1265
23	1279	93.74	1251	1251
33	1258	92.96	1237	1236
43	1246	92.68	1230	1229
53	1238	92.54	1226	1226
63	1233	92.47	1223	1223
73	1230	92.43	1222	1222
98	1225	92.36	1220	1220

$$\infty \quad \frac{2Et^2/a^2}{[3(1 - v^2)]^{1/2}} = 1210.45 \text{ (Timoshenko)}$$

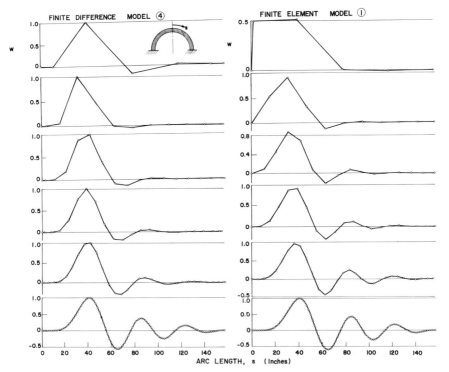

Fig. 18. Normal modal displacements corresponding to $n = 6$ circumferential waves with the finite element ① (*right*) and finite difference ④ (*left*) energy methods applied to buckling of a uniformly pressurized clamped hemisphere.

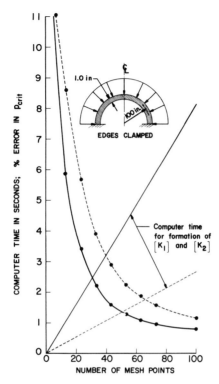

Fig. 19. Computer times and rates of convergence of the buckling load of a clamped, uniformly pressurized hemisphere. (—): Finite element ①; (---): finite difference ④.

finite-element model ①. This increased restraint probably explains why the buckling loads calculated with model ④ are higher than those calculated with model ①. The increased restraint and resultant slow convergence of λ are possibly due to the extra mesh point introduced near the pole as discussed previously in connection with the stress analysis of the clamped hemisphere.

Figure 19 shows for models ① and ④ the computer times to form the stiffness matrix $[K_1]$ and the "load-geometric" matrix $[K_2]$ and percentage of error in the buckling load λ. The finite-element results correspond to two Gaussian points with no static reduction.

3.4. CLAMPED HEMISPHERE VIBRATION

Figures 20–22 and Table 4 gives results from the vibration analysis of a clamped hemisphere with $t = 1.0$ in., $a = 100$ in., $E = 35.925$ lb. in.$^{-2}$, $v = 0.3$, and density $\rho = 0.0001$ lb-sec^2/in.^{4}. The lowest three frequencies were calculated for circumferential waves $n = 0$, 1 and 2. Models ① and ④ were

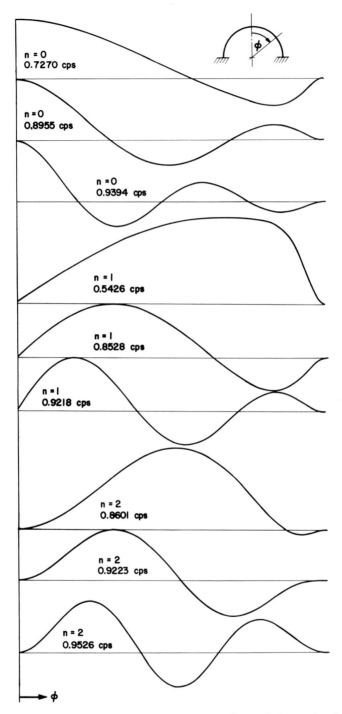

Fig. 20. Vibration normal modal displacement corresponding to the lowest three frequencies with $n = 0, 1,$ and 2 circumferential waves.

used. In the finite-element analysis the number of Gaussian integration points were varied, and from Table 4 it is seen that the optimum value is $G = 2$. The finite difference model ④ leads to more rapidly convergent solutions, probably because the mass is lumped. (One integration point is

TABLE 4

HEMISPHERE VIBRATION FREQUENCIES

Number of mesh points	Circum. waves, n	Finite difference analysis	Finite element analysis		
			One Gauss point	Two Gauss points	Three Gauss points
7	0	0.73615	0.70492	0.75469	0.79027
		0.90115	0.88678	0.93222	0.99693
		0.93342	0.92510	0.96696	1.10154
	1	0.54056	0.52378	0.55308	0.58504
		0.85762	0.83840	0.88511	0.92300
		0.92080	0.91147	0.96384	1.04096
	2	0.86302	0.86038	0.87275	0.89974
		0.92260	0.92356	0.95984	1.01791
		0.93551	0.96889	1.05589	1.17163
13	0	0.72401	0.70187	0.74395	0.74532
		0.89357	0.88219	0.92835	0.92153
		0.93536	0.92204	1.0042	0.99885
	1	0.53988	0.52854	0.54784	0.55392
		0.85000	0.83754	0.87153	0.87376
		0.91900	0.90899	0.96141	0.96452
	2	0.85981	0.85713	0.86605	0.87069
		0.92179	0.91864	0.94229	0.94648
		0.94926	0.94266	0.99680	1.00388
18	0	0.72280	0.70145	0.73445	0.73577
		0.89368	0.88151	0.91138	0.90869
		0.93658	0.92122	0.97156	0.96765
	1	0.53927	0.52905	0.54430	0.54859
		0.85027	0.83744	0.86176	0.86285
		0.91971	0.90852	0.94124	0.94205
	2	0.85950	0.85677	0.86301	0.86536
		0.92168	0.91791	0.93249	0.93421
		0.95052	0.94093	0.97531	0.97789
23	0	0.72354	0.70128	0.73096	0.73220
		0.89404	0.88126	0.90447	0.90333
		0.93726	0.92091	0.95768	0.95564
	1	0.53976	0.52923	0.54315	0.54649
		0.85089	0.83740	0.85784	0.85866
		0.92019	0.90835	0.93289	0.93336
	2	0.85951	0.85664	0.86166	0.86321
		0.92169	0.91764	0.92815	0.92919
		0.95100	0.94034	0.96569	0.96718

TABLE 4 (continued)

Number of mesh points	Circum. waves, n	Finite difference analysis	Finite element analysis		
			One Gauss point	Two Gauss points	Three Gauss points
33	0	0.72477	0.70115	0.72842	0.72943
		0.89444	0.88107	0.89924	0.89898
		0.93787	0.92068	0.94717	0.94647
	1	0.54060	0.52936	0.54235	0.54468
		0.85160	0.83737	0.85485	0.85543
		0.92063	0.90823	0.92648	0.92678
	2	0.85959	0.85654	0.86061	0.86152
		0.92173	0.91744	0.92473	0.92530
		0.95137	0.93993	0.95815	0.95892
53	0	0.72568	0.70106	0.72708	0.72780
		0.89470	0.88096	0.89646	0.89650
		0.93822	0.92055	0.94161	0.94147
	1	0.54125	0.52942	0.54193	0.54339
		0.85202	0.83736	0.85324	0.85362
		0.92089	0.90817	0.92304	0.92323
	2	0.85966	0.85649	0.86004	0.86054
		0.92175	0.91734	0.92286	0.92316
		0.95157	0.93971	0.95406	0.95443
93	0	0.72610	0.70101	0.72654	0.72699
		0.89482	0.88092	0.89536	0.89545
		0.93837	0.92049	0.93942	0.93944
	1	0.54155	0.52945	0.54177	0.54261
		0.85221	0.83735	0.85259	0.85283
		0.92100	0.90814	0.92167	0.92178
	2	0.85969	0.85647	0.85981	0.86007
		0.92176	0.91730	0.92211	0.92226
		0.95165	0.93963	0.95242	0.95260
	0	0.7263	(Zarghamee)		
		0.8948			
∞	1	0.5417			
		0.8523			
	2	0.8598			

used.) The finite-element solution is obtained with a "consistent" mass matrix, as described in the analysis section. All solutions converge to those calculated by Zarghamee [18]. Figure 20 shows the vibration modes corresponding to the converged solution.

Figure 21 shows computer times required to obtain the stiffness matrix $[K_1]$ and the mass matrix $[M]$ and convergence properties of two of the

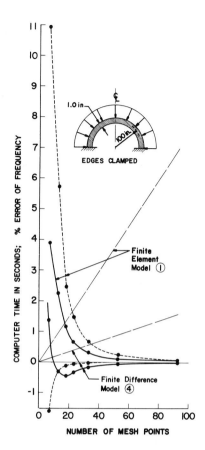

Fig. 21. Computer times and rates of convergence for vibration of a clamped hemisphere. (—): First eigenvalue, $n = 0$; ($-\,-\,-$): third eigenvalue, $n = 2$; ($-\,-$): computer time for formation of $[K_1]$ and $[M]$.

eigenvalues with increasing numbers of mesh points. Figure 22 shows convergence properties of the lowest three eigenvalues corresponding to $n = 0$. It is curious that with finite difference model ④ rapidity of convergence increases as higher harmonics are calculated, whereas the opposite is true with use of finite element model ①. The convergence behavior of the $n = 1$ and $n = 2$ modes is similar.

4. Comments on Application to Two-Dimensional Problems

Four factors of prime importance in numerical analyses of structures are (1) ease of application to practical problems, (2) computer time required

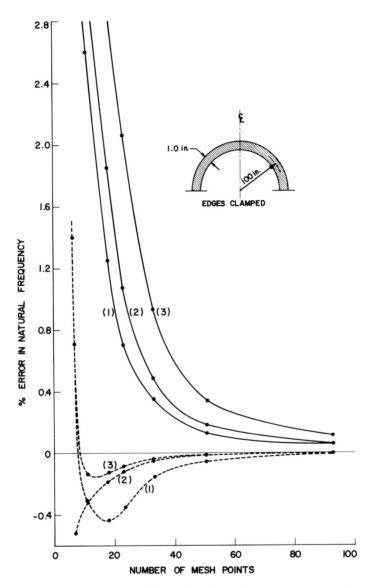

Fig. 22. Rates of convergence of the lowest three frequencies corresponding to axisymmetric vibration of a clamped hemisphere. (——): Finite element ① ; (– – –): finite difference ④ ; curves (1), (2), and (3) correspond to the first three axisymmetric vibration modes.

to form various matrices, (3) computer time required to decompose (factor) the stiffness matrix and solve for eigenvalues (inverse power iterations), and (4) rate of convergence with increasing number of degrees of freedom. It is of interest to "extrapolate" some of the quantitative data just presented in connection with the one-dimensional analyses to obtain a qualitative idea of the relative merits of the finite-difference energy method and the finite element method applied to two-dimensional problems.

4.1. EASE OF APPLICATION

At this point, it appears that the finite-element method is far easier to apply than the finite-difference energy method in two-dimensional problems which involve irregular grids and arbitrary boundaries. The procedure described in the analysis section applies directly, with an additional step not included in connection with curved elements ① and ② used in this study, but generally required: transformation of the various matrices from local element coordinates to global coordinates at the point of assembly of the local matrices into the global matrix. Such a computational step was required in the one-dimensional stress analysis with the Khojasteh–Bakht-type element.

In the finite-difference energy analysis the introduction of "fictitious" w-points at boundaries and the noncoincidence of u, v-mesh points with w-mesh points complicates matters somewhat. Figure 23 shows some possible finite-difference mesh schemes for application to two-dimensional problems. In Scheme (1), which is analogous to Model ④ in Fig. 1, u and v points are taken at the centroids of the quadrilaterals formed by adjacent w-points. The element is the cross-hatched area A. This scheme was used by Johnson [7] and by Almroth et al. [8]. In Scheme (2), which is analogous to Model ⑦ in Fig. 1, u, v, and w are taken at the same points, and different areas are used for membrane energy (A_{im}) and bending energy (A_b). This is the so-called "whole-station" scheme described by Bushnell and Almroth [16]. A suggested Scheme (3) is shown which involves equilateral triangles. The u_i and v_i are associated with element nodes and the w_i are associated with element centroids. A similar scheme might work for arbitrary triangular meshes, so valuable for application to complex engineering structures. However, basic research work needs to be done before the finite difference energy method in two dimensions can be used with facility to solve complicated structures which can presently be analyzed with relative ease by the finite-element method.

4.2. FORMING STIFFNESS MATRICES

One of the major advantages of the finite-difference energy method over the finite-element method is the computer time saved in forming the stiffness matrix, "load-geometric" matrix, and mass matrix. While this saving is not particularly important in one-dimensional problems, it can be considerable in two-dimensional problems, especially for stress analyses involving nonlinear behavior, such as the "pear-shaped" cylinder problem described by Bushnell and Almroth [16].

The differences in computer time for formation of the various global matrices $[K_1]$, $[K_2]$, $[K_3]$, and $[M]$ with the finite-difference energy method and with the finite-element method can be expected to be greater in the two-dimensional case than in the one-dimensional case. There are two reasons for this expectation: The dimensions of the local matrices are larger in the finite-element method because there are more unknowns per grid point; and the difference in number of Gaussian integration points between the two methods is greater. For example, the simplest quadrilateral element involves at least five unknowns at each corner: three displacement components and two rotation components. Thus, the element stiffness matrix has dimensions 20×20. The equivalent finite-difference "elements," such as represented by Scheme (1) in Fig. 23, involves the nine w-points and four (u, v) points shown.

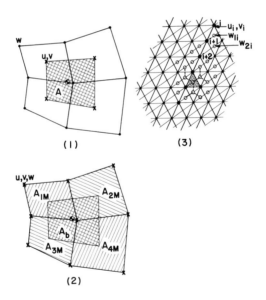

Fig. 23. Three possible finite-difference schemes for two-dimensional analysis.

Hence the local matrix is 17×17. The energy within the quadrilateral is calculated with a minimum of four Gaussian integration points in the finite-element model, whereas the same energy in the finite difference model is obtained by multiplication of the energy density at the element centroid by the elemental area A. The contrast becomes more extreme for more sophisticated two-dimensional finite-elements, such as an element analogous to Model ② in Fig. 1, which requires three Gaussian integration points in each dimension. The stiffness matrices associated with elements which involve more unknowns at each grid point, such as that described by Cowper [19], would require more computer time to generate because of both the increased dimensions of the local matrix and the requirement for more Gaussian integration points consistent with higher-order displacement polynomials.

4.3. FACTORING AND SOLVING

For one-dimensional problems the computer time required to factor the stiffness matrix is insignificant compared to the time required to calculate it. With two-dimensional problems, however, factoring the stiffness matrix represents a more significant portion of the total computer time, since the bandwidth of the matrix is generally much larger in proportion to its rank than is the case with one-dimensional problems. The computer time T_f required to factor a stiffness matrix is proportional to the square of its bandwidth times its rank. For an $m \times n$ array of points the computer time to factor reasonably large stiffness matrices can be expressed as:

$$T_f \alpha B^2 R = (KmN)^2(mnN) = K^2 N^3 m^3 n \qquad (29)$$

in which $m \leqslant n$, K is the difference between extreme column numbers over which coupling exists in the energy functional (these columns each contain m grid points); N the number of unknowns per grid point, m the number of rows of points in the grid, and n the number of columns of points in the grid.

It is of interest to compare the quantity $K^2 N^3$ for various discretization schemes given the $m \times n$ grid. The finite-element formulations lead to coupling over three columns of points, so that $K = 2$. The simplest finite element involves five unknowns at each grid point, $N = 5$. Finite-difference Schemes (1) and (2) in Fig. 23 involve coupling over five columns of points, so that $K = 4$. However, only three unknowns per grid point exist, $N = 3$. The ratios of computer times for factoring the stiffness matrix with these two schemes is therefore $(2 \times 2 \times 5 \times 5 \times 5)/(4 \times 4 \times 3 \times 3 \times 3) = 500/432$, which slightly favors the finite-difference method. Elements with more degrees of freedom per node would require correspondingly more computer time for given grid $m \times n$. However, presumably fewer elements would be required for convergence to a given degree of accuracy. The fact that N is cubed in Eq. (29) would tend to cause one to choose static reduction at the element level as a method of solution in the finite-element method, rather than use of an increased number of degrees of freedom per node, as is done in [15].

With the finite-difference hexagonal Scheme (3) in Fig. 23, there is an average of 4 degrees of freedom u_i, v_i, v_{1i}, and w_{2i} per node; coupling in the energy expression exists over only three columns of grid points (the columns run from upper-left to lower-right, see labels $i, i + 1, i + 2$ in Fig. 23), so that $N = 4$, $K = 2$ in Eq. (29). The ratio of the simplest finite-element scheme to this scheme is $(2 \times 2 \times 5 \times 5 \times 5)/(2 \times 2 \times 4 \times 4 \times 4) = 125/64$, almost a factor of two in favor of the finite-difference method. However, nothing is known about the rate of convergence of this hexagonal finite-difference method. Presumably this ratio would also hold for arbitrary triangular meshes, although it is not clear at this point exactly how the finite-difference model would be set up.

In bifurcation buckling and natural vibration problems rather a large amount of computer time is spent performing inverse power iterations to obtain the lowest few eigenvalues. With each iteration most of this time is spent multiplying various matrices of bandwidth B and rank R. The computer time T_s for such a calculation is proportional to the bandwidth times the rank, or

$$T_s \alpha BR = (KmN)(mnN) = KN^2 m^2 n \tag{30}$$

This formula leads to somewhat shorter estimated computer times for given grid $m \times n$ with the finite difference method than with the finite element method, as the reader can easily verify.

4.4. RATE OF CONVERGENCE

The computer times estimated in the previous section are, of course, all related to the grid size $m \times n$ required for convergence to a given degree of accuracy. The one-dimensional examples given in the section "Numerical Results" sometimes favor the finite-difference method and sometimes favor the finite-element method. At this point one might say that the convergence properties are "about the same" for the two methods. Certainly the results are case-dependent.

The finite-difference method is favored by the fact that for uniform grid spacing, the equations $[K_1]q = Q$ corresponding to finite-difference schemes ④ and ⑤ in Fig. 1 are the Euler equations of the variational problem in "finite" form, as proved in the analysis section. On the other hand, the finite-element method seems to yield superior convergence properties at edges and discontinuities.

5. Conclusions

The similarity of the finite-difference energy method and the finite-element method are demonstrated by application of both methods to problems involving shells of revolution. Curved finite elements are introduced into the BOSOR3 computer program and rates of convergence and computer times are established for stress, buckling, and vibration analyses of an elastic hemisphere.

Several finite-difference discretization models are described. It is shown that these models involving different grid points for displacement components u, v, and w are actually constant strain "elements" with discontinuities in normal displacement w and its derivatives w' and w'' at element boundaries. For constant mesh spacing, the algebraic equations corresponding to the derivative of the energy functional with respect to the nodal point displacements correspond to the Euler equations of the variational problem in finite form.

Two finite elements are introduced into the BOSOR3 computer program: a curved element with linear variation for tangential displacements u and v and cubic variation for normal displacement w, and an isoparametric element with cubic polynomials for u, v, and w. In both cases the energy is expressed in terms of nodal point generalized displacements u_i, v_i, w_i, and χ_i. It is found that with the finite-element method two to six times as much computer time is required for formation of each of the stiffness matrix, "load-geometric" matrix, and mass matrix, as is required for formation of the corresponding matrices with the finite-difference method with the same number of grid points. The difference in time is primarily due to the need in the finite-element method to evaluate the energy density at two or three "Gaussian" points within each element. If too few Gaussian integration points are used in the finite-element method, the calculations converge to the wrong solution as the number of mesh points is increased. It is found that the most rapidly convergent solution is obtained with use of the minimum number of Gaussian integration points consistent with eventual convergence to the correct solution.

Numerical results are obtained with the various discretization models for stress, buckling, and vibration of a hemisphere. Convergence studies indicate that the finite difference method is superior for vibration problems or stress problems not involving rapidly varying edge discontinuity stresses. The finite element method yields more rapidly convergent behavior in the bifurcation buckling problem investigated and in the stress problem involving high meridional stresses at a clamped edge. The convergence properties are clearly case-dependent, and it is felt that no general conclusions about rate of convergence can be drawn at the time of writing.

Appendix

CHART A1

STRAIN–NODAL DISPLACEMENT RELATIONS FOR FINITE-DIFFERENCE MODEL ④

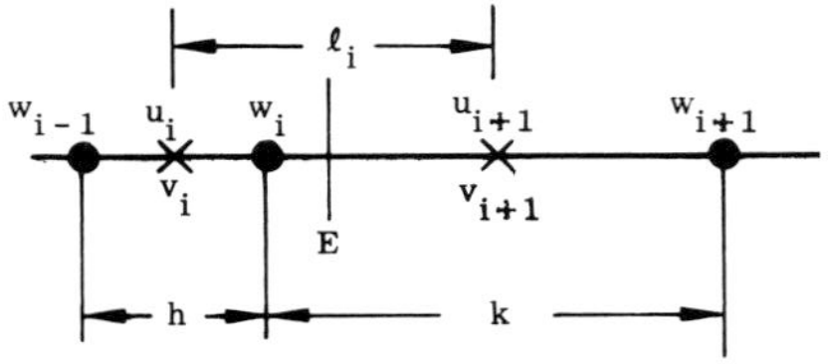

$$w = a_{21} w_{i-1} + a_{22} w_i + a_{23} w_{i+1}$$

$$w' = a_{41} w_{i-1} + a_{42} w_i + a_{43} w_{i+1}$$

$$w'' = a_{51} w_{i-1} + a_{52} w_i + a_{53} w_{i+1}$$

$$a_{21} = (h - k)(k + \ell)/(16\ h\ell)$$

$$a_{22} = (k + \ell)(h + \ell)/(4\ hk)$$

$$a_{23} = (k - h)(h + \ell)/(16\ h\ell)$$

$$a_{41} = -0.5/h$$

$$a_{42} = 0.5\ (1/h - 1/k)$$

$$a_{43} = 0.5/k$$

$$a_{51} = 1/(h\ell)$$

$$a_{52} = -2/(hk)$$

$$a_{53} = 1/(k\ell)$$

	w_{i-1}	u_i	v_i	w_i	u_{i+1}	v_{i+1}	w_{i+1}
ϵ_1	$\dfrac{a_{21}}{R_1} + \chi_{0f}\, a_{41}$	$-\dfrac{1}{\ell} - \dfrac{\chi_{0f}}{2R_1}$	0	$\dfrac{a_{22}}{R_1}$	$\dfrac{1}{\ell} - \dfrac{\chi_{0f}}{2R_1}$	0	$\dfrac{a_{23}}{R_1} + \chi_{0f}\, a_{43}$
ϵ_2	$\dfrac{a_{21}}{R_2}$	$\dfrac{r'}{2r}$	$-\dfrac{n}{2r}$	$\dfrac{a_{22}}{R_2}$	$\dfrac{r'}{2r}$	$-\dfrac{n}{2r}$	$\dfrac{a_{23}}{R_2}$
ϵ_{12}	$\chi_{0f}\, a_{21}\dfrac{n}{r}\,r$	$\dfrac{n}{2r}$	$-\dfrac{1}{\ell} - \dfrac{r'}{2r} - \dfrac{\chi_{0f}}{2R_2}$	$\chi_{0f}\, a_{22}\dfrac{n}{r}\,r$	$\dfrac{n}{2r}$	$\dfrac{1}{\ell} - \dfrac{r'}{2r} - \dfrac{\chi_{0f}}{2R_2}$	$\chi_{0f}\, a_{23}\dfrac{n}{r}\,r$
κ_1	a_{51}	$\dfrac{1}{\ell R_1} - 0.5\left(\dfrac{1}{R_1}\right)'$	0	a_{52}	$-\dfrac{1}{\ell R_1} - 0.5\left(\dfrac{1}{R_1}\right)'$	0	a_{53}
κ_2	$-\dfrac{n^2}{r^2} a_{21} + \dfrac{r'}{r} a_{41}$	$-\dfrac{r'}{2rR_1}$	$\dfrac{n}{2rR_2}$	$-\dfrac{n^2}{r^2} a_{22} + \dfrac{r'}{r} a_{42}$	$-\dfrac{r'}{2rR_1}$	$\dfrac{n}{2rR_2}$	$-\dfrac{n^2}{r^2} a_{23} + \dfrac{r'}{r} a_{43}$
κ_{12}	$-\dfrac{n}{r} a_{41} + \dfrac{nr'}{r^2} a_{21}$	$\dfrac{n}{2rR_1}$	$-\dfrac{r'}{2rR_2} - \dfrac{1}{R_2 \ell}$	$-\dfrac{n}{r} a_{42} + \dfrac{nr'}{r^2} a_{22}$	$\dfrac{n}{2rR_1}$	$-\dfrac{r'}{2rR_2} + \dfrac{1}{R_2 \ell}$	$-\dfrac{n}{r} a_{43} + \dfrac{nr'}{r^2} a_{23}$

CHART A2

STRAIN DISPLACEMENT–POLYNOMIAL COEFFICIENT RELATIONS FOR FINITE-ELEMENT MODEL ①

$$u = \alpha_1 + \alpha_2 s$$
$$v = \alpha_3 + \alpha_4 s$$
$$w = \alpha_5 + \alpha_6 s + \alpha_7 s^2 + \alpha_8 s^3$$

	α_1	α_2	α_3	α_4	α_5	α_6	α_7	α_8
ϵ_1	$-\dfrac{\chi_{0f}}{R_1}$	$\dfrac{1}{\ell} - \dfrac{\chi_{0f}\,s}{R_1}$	0	0	$\dfrac{1}{R_1}$	$\dfrac{s}{R_1} - \dfrac{\chi_{0f}}{\ell}$	$\dfrac{s^2}{R_1} + \dfrac{2\chi_{0f}\,s}{\ell}$	$\dfrac{s^3}{R_1} + \dfrac{3\chi_{0f}\,s^2}{\ell}$
ϵ_2	$\dfrac{r'}{r}$	$\dfrac{r'}{r}\,s$	$-\dfrac{n}{r}$	$-\dfrac{n}{r}\,s$	$\dfrac{1}{R_2}$	$\dfrac{1}{R_2}\,s$	$\dfrac{s^2}{R_2}$	$\dfrac{s^3}{R_2}$
ϵ_{12}	$\dfrac{n}{r}$	$\dfrac{n}{r}\,s$	$-\dfrac{r'}{r} - \dfrac{\chi_{0f}}{R_2}$	$\dfrac{1}{\ell} - \dfrac{r'\,s}{r} - \dfrac{\chi_{0f}\,s}{R_2}$	$\dfrac{n}{r}\chi_{0f}$	$\dfrac{n}{r}\chi_{0f}\,s$	$\dfrac{n}{r}\chi_{0f}\,s^2$	$\dfrac{n}{r}\chi_{0f}\,s^3$
κ_1	$-\left(\dfrac{1}{R_1}\right)'$	$-\left(\dfrac{1}{R_1}\right)'s - \dfrac{1}{R_1\ell}$	0	0	0	0	$\dfrac{2}{\ell^2}$	$\dfrac{6s}{\ell^2}$
κ_2	$-\dfrac{r'}{rR_1}$	$\dfrac{r'}{rR_1}\,s$	$\dfrac{n}{rR_2}$	$\dfrac{ns}{rR_2}$	$-\dfrac{n^2}{r^2}$	$-\dfrac{n^2}{r}\,s$	$-\dfrac{n^2}{r}\dfrac{s^2}{\ell} + 2\dfrac{r'}{r}\dfrac{s}{\ell} - \dfrac{2n}{r\ell}\,s$	$-\dfrac{n^2}{r}\dfrac{s^3}{\ell} + 3\dfrac{r'}{r}\dfrac{s^2}{\ell} - \dfrac{3n}{r\ell}\,s^2$
κ_{12}	$\dfrac{n}{rR_1}$	$\dfrac{n}{rR_1}\,s$	$-\dfrac{r'}{rR_2}$	$-\dfrac{r'}{rR_2}\,s + \dfrac{1}{R_2\ell}$	$\dfrac{r'n}{r^2}$	$-\dfrac{n}{rl}\,s + \dfrac{r'}{rl}\dfrac{n}{r}\,s$ ... $+\dfrac{r'n}{r^2}\,s$	$-\dfrac{2n}{rl}\,s + 2\dfrac{r'}{r}\dfrac{s}{l} + \dfrac{r'n}{r^2}\,s^2$	$-\dfrac{3n}{rl}\,s^2 + 3\dfrac{r'}{r}\dfrac{s^2}{l} + \dfrac{r'n}{r^2}\,s^3$

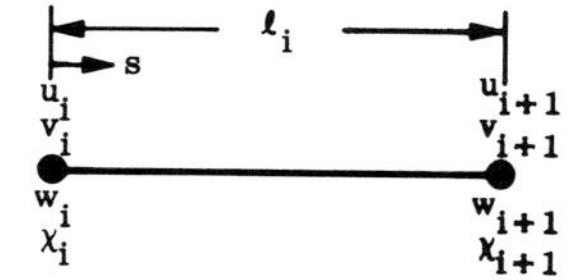

CHART A3

STRAIN–NODAL DISPLACEMENT RELATIONS FOR FINITE-ELEMENT MODEL ②

	u_i	v_i	w_i	χ_i	u_{i+1}	v_{i+1}	w_{i+1}	χ_{i+1}	$\bar{u}_1$	$\bar{v}_1$	$\bar{u}_2$	$\bar{v}_2$
ϵ_1	$a_{31} + \dfrac{a_{21}}{R_1} - \dfrac{\chi_{0f}}{R_1} a_{11} + \chi_{0f} a_{41}$	0	$\dfrac{a_{22}}{R_1} + \chi_{0f} a_{42}$	$\dfrac{a_{23}}{R_1} + \chi_{0f} a_{43}$	$a_{32} + \dfrac{a_{24}}{R_1} - \dfrac{\chi_{0f}}{R_1} a_{12} + \chi_{0f} a_{44}$	0	$\dfrac{a_{25}}{R_1} + \chi_{0f} a_{45}$	$\dfrac{a_{26}}{R_1} + \chi_{0f} a_{46}$	$a_{33} - \dfrac{\chi_{0f}}{R_1} a_{13}$	0	$a_{34} - \dfrac{\chi_{0f}}{R_1} a_{14}$	0
ϵ_2	$\dfrac{r'}{r} a_{11} + \dfrac{a_{21}}{R_2}$	$-\dfrac{n}{r} a_{11}$	$\dfrac{a_{22}}{R_2}$	$\dfrac{a_{23}}{R_2}$	$\dfrac{r'}{r} a_{12} + \dfrac{a_{24}}{R_2}$	$-\dfrac{n}{r} a_{12}$	$\dfrac{a_{25}}{R_2}$	$\dfrac{a_{26}}{R_2}$	$\dfrac{r'}{r} a_{13}$	$-\dfrac{n}{r} a_{13}$	$\dfrac{r'}{r} a_{14}$	$-\dfrac{n}{r} a_{14}$
ϵ_{12}	$\dfrac{n}{r} a_{11} + \chi_{0f}\dfrac{n}{r} a_{21}$	$a_{31} - \dfrac{r'}{r} a_{11} - \dfrac{\chi_{0f}}{R_2} a_{11}$	$\chi_{0f}\dfrac{n}{r} a_{22}$	$\chi_{0f}\dfrac{n}{r} a_{23}$	$\dfrac{n}{r} a_{12} + \chi_{0f}\dfrac{n}{r} a_{24}$	$a_{32} - \dfrac{r'}{r} a_{12} - \dfrac{\chi_{0f}}{R_2} a_{12}$	$\chi_{0f}\dfrac{n}{r} a_{25}$	$\chi_{0f}\dfrac{n}{r} a_{26}$	$+\dfrac{r'}{r} a_{13}$	$a_{33} - \dfrac{r'}{r} a_{13} - \dfrac{\chi_{0f}}{R_2} a_{13}$	$+\dfrac{r'}{r} a_{14}$	$a_{34} - \dfrac{r'}{r} a_{14} - \dfrac{\chi_{0f}}{R_2} a_{14}$
κ_1	$a_{51} - a_{11}\left(\dfrac{1}{R_1}\right)' - \dfrac{a_{31}}{R_1}$	0	a_{52}	a_{53}	$a_{54} - a_{12}\left(\dfrac{1}{R_1}\right)' - \dfrac{a_{32}}{R_1}$	0	a_{55}	a_{56}	$-a_{13}\left(\dfrac{1}{R_1}\right)' - \dfrac{a_{33}}{R_1}$	0	$-a_{14}\left(\dfrac{1}{R_1}\right)' - \dfrac{a_{34}}{R_1}$	0
κ_2	$-\dfrac{n^2}{r^2} a_{21} + \dfrac{r'}{r} a_{41}$	$\dfrac{n}{rR_2} a_{11}$	$-\dfrac{n^2}{r^2} a_{22} + \dfrac{r'}{r} a_{42}$	$-\dfrac{n^2}{r^2} a_{23} + \dfrac{r'}{r} a_{43}$	$-\dfrac{n^2}{r^2} a_{24} + \dfrac{r'}{r} a_{44}$	$\dfrac{n}{rR_2} a_{12}$	$-\dfrac{n^2}{r^2} a_{25} + \dfrac{r'}{r} a_{45}$	$-\dfrac{n^2}{r^2} a_{26} + \dfrac{r'}{r} a_{46}$	$-\dfrac{n^2}{r^2} a_{13}$	$\dfrac{n}{rR_2} a_{13}$	$-\dfrac{n^2}{r^2} a_{14}$	$\dfrac{n}{rR_2} a_{14}$
κ_{12}	$-\dfrac{n}{r} a_{41} + \dfrac{n}{rR_1} a_{11} + \dfrac{r'n}{r^2} a_{21}$	$-\dfrac{r'}{rR_2} a_{11} + \dfrac{a_{31}}{R_2}$	$-\dfrac{n}{r} a_{42} + \dfrac{r'n}{r^2} a_{22}$	$-\dfrac{n}{r} a_{43} + \dfrac{r'n}{r^2} a_{23}$	$-\dfrac{n}{r} a_{44} + \dfrac{n}{rR_1} a_{12} + \dfrac{r'n}{r^2} a_{24}$	$-\dfrac{r'}{rR_2} a_{12} + \dfrac{a_{32}}{R_2}$	$-\dfrac{n}{r} a_{45} + \dfrac{r'n}{r^2} a_{25}$	$-\dfrac{n}{r} a_{46} + \dfrac{r'n}{r^2} a_{26}$	$+\dfrac{n}{rR_1} a_{13}$	$-\dfrac{r'}{rR_2} a_{13} + \dfrac{a_{33}}{R_2}$	$+\dfrac{n}{rR_1} a_{14}$	$-\dfrac{r'}{rR_2} a_{14} + \dfrac{a_{34}}{R_2}$

$u = u_i a_{11} + u_{i+1} a_{12} + \bar{u}_1 a_{13} + \bar{u}_2 a_{14}$

$u' = u_i a_{31} + u_{i+1} a_{32} + \bar{u}_1 a_{33} + \bar{u}_2 a_{34}$

$v = v_i a_{11} + v_{i+1} a_{12} + \bar{v}_1 a_{13} + \bar{v}_2 a_{14}$

$v' = v_i a_{31} + v_{i+1} a_{32} + \ldots$

$w = u_i a_{21} + w_i a_{22} + \chi_i a_{23} + u_{i+1} a_{24} + w_{i+1} a_{25} + \chi_{i+1} a_{26}$

$w' = u_i a_{41} + w_i a_{42} + \chi_i a_{43} + u_{i+1} a_{44} + w_{i+1} a_{45} + \chi_{i+1} a_{46}$

$w'' = u_i a_{51} + w_i a_{52} + \chi_i a_{53} + u_{i+1} a_{54} + w_{i+1} a_{55} + \chi_{i+1} a_{56}$

CHART A4

DEFINITIONS OF THE COEFFICIENTS a_{ij} FOR FINITE ELEMENT MODEL ②

$$u(s) = u_i a_{11} + u_{i+1} a_{12} + \bar{u}_1 a_{13} + \bar{u}_2 a_{14}$$
$$v(s) = v_i a_{11} + v_{i+1} a_{12} + \bar{v}_1 a_{13} + \bar{v}_2 a_{14}$$

$$a_{11} = 1 - s \qquad\qquad F_1 = s_1(s_1 - s_2)(1 - s_1)$$
$$a_{12} = s \qquad\qquad F_2 = s_2(s_1 - s_2)(1 - s_2)$$
$$a_{13} = s(s - s_2)(1 - s)/F_1$$
$$a_{14} = -s(s - s_1)(1 - s)/F_2$$

$$u'(s) = u_i a_{31} + u_{i+1} a_{32} + \bar{u}_1 a_{33} + \bar{u}_2 a_{34}$$
$$v'(s) = v_i a_{31} + v_{i+1} a_{32} + \bar{v}_1 a_{33} + \bar{v}_2 a_{34}$$

$$a_{31} = 1/l, \qquad a_{32} = 1/l$$
$$a_{33} = (2s - s_2 - 3s^2 + 2ss_2)/(F_1 l)$$
$$a_{34} = -(2s - s_1 - 3s^2 + 2ss_1)/(F_2 l)$$

$$w(s) = u_i a_{21} + w_i a_{22} + \chi_i a_{23} + u_{i+1} a_{24} + w_{i+1} a_{25} + \chi_{i+1} a_{26}$$
$$a_{21} = ls(1 - s)^2/R_{1i} \qquad a_{22} = 1 - 3s^2 + 2s^3$$
$$a_{23} = ls(1 - s)^2 \qquad a_{24} = -ls^2(1 - s)/R_{1_{i+1}}$$
$$a_{25} = s^2(3 - 2s) \qquad a_{26} = -ls^2(1 - s)$$

$$w'(s) = u_i a_{41} + w_i a_{42} + \chi_i a_{43} + u_{i+1} a_{44} + w_{i+1} a_{45} + \chi_{i+1} a_{46}$$
$$a_{41} = (1 - 4s + 3s^2)/R_{1i} \qquad a_{42} = (-6s + 6s^2)/l$$
$$a_{43} = 1 - 4s + 3s^2 \qquad a_{44} = (-2s + 3s^2)/R_{1_{i+1}}$$
$$a_{45} = (6s - 6s^2)/l \qquad a_{46} = -2s + 3s^2$$

$$w''(s) = u_i a_{51} + w_i a_{52} + \chi_i a_{53} + u_{i+1} a_{54} + w_{i+1} a_{55} + \chi_{i+1} a_{56}$$
$$a_{51} = (-4 + 6s)/(l R_{1i}) \qquad a_{52} = (-6 + 12s)/l^2$$
$$a_{53} = (-4 + 6s)/l \qquad a_{54} = (-2 + 6s)/(l R_{1_{i+1}})$$
$$a_{55} = (6 - 12s)/l^2 \qquad a_{56} = (-2 + 6s)/l$$

References

1. Bushnell, D., "Stress, Stability, and Vibration of Complex Shells of Revolution: Analysis and User's Manual for BOSOR3," SAMSO TR 69-375, LMSC N-5J-69-1, Lockheed Missiles and Space Co., Sunnyvale, California (September 1969).
2. Courant, R., Friedrichs, K., and Lewy, H., "Über die partiellen differenzengleichungen der mathematischen Physik," *Math. Ann.* **100**, 32–74 (1928).
3. Forsythe, G. E., and Wasow, W. R., *Finite-Difference Methods for Partial Differential Equations.* Wiley, New York, 1960.
4. Stein, M., "The Effect on the Buckling of Perfect Cylinders of Prebuckling Deformations and Stresses Induced by Edge Support," NASA TN D-1510, p. 217. Langley Res. Center, Hampton, Virginia (December 1962).
5. Havner, K. S., and Stanton, E. L., "On Energy-Derived Difference Equations in Thermal Stress Problems," *J. Franklin Inst.* **284**, 127 (1967).
6. Budiansky, B., and Anderson, D. G. M., "Numerical Shell Analysis—Nodes Without Elements," *Int. Congr. Appl. Mech. 12th*, Stanford Univ., August 26–31 (1968).
7. Johnson, D. E., "A Difference-Based Variational Method for Shells," *Int. J. Solids Struct.* **6**, 699 (1970).
8. Almroth, B. O., Brogan, F. A., and Marlowe, M. B., "Collapse Analysis for Elliptic Cones," *AIAA J.* **9**, 32 (1971).

9. Noor, A. K., "Improved Multilocal Finite-Difference Variant for the Bending Analysis of Arbitrary Cylindrical Shells," UNICIV Rep. No. R-63, Univ. of New South Wales (March 1971).

10. Schnobrich, W. C., and Pecknold, D. A., "The Lumped Parameter or Bar-Node Model Approach to Thin Shell Analysis," *Proc. ONR Int. Symp. Numer. Comput. Methods Struct. Mech.* Urbana, Illinois, Sept. 8–10 (1971).

11. Pian, T. H. H., "Variational Formulations of Numerical Methods in Solid Continuum," *Proc. SMD Symp. Comput.–Aided Eng.* May 11–13. Univ. of Waterloo, Waterloo, Ontario, Canada (1971).

12. Bushnell, D., "Analysis of Buckling and Vibration of Ring-Stiffened, Segmented Shells of Revolution, *Int. J. Solids Struct.* **6**, 157 (1970).

13. Kotanchik, J. J., Yeghiayan, R. P., Witmer, E. A., and Berg, B. A., "The Transient Linear Elastic Response Analysis of Complex Thin Shells of Revolution Subjected to Arbitrary External Loadings, by the Finite-Element Program, "SABOR5-DRASTIC," SAMSO TR 70-206, ASRL TR 146-10, Aeroelastic and Struct. Res. Lab., MIT (April 1970).

14. Mebane, P. M., and Stricklin, J. A., "Implicit Rigid Body Motion in Curved Finite Elements," *AIAA J.* **9**, 344 (1971).

15. Adelman, H. M., Catherines, D. S., and Walton, W. C., Jr., "A Method for Computation of Vibration Modes and Frequencies of Orthotropic Thin Shells of Revolution Having General Meridional Curvature," NASA TN D-4972, Langley Res. Center, Hampton, Virginia (January 1969).

16. Bushnell, D., and Almroth, B. O., "Finite Difference Energy Method for Nonlinear Shell Analysis," paper presented at Lockheed Missiles and Space Co. Symp., August 1970, *J. Comput. Struct.* **1**, 361 (1971).

17. Khojasteh-Bakht, M., "Analysis of Elastic-Plastic Shells of Revolution under Axisymmetric Loading by the Finite Element Method," Ph.D. dissertation, Dept. of Civil Eng., Univ. of California, Berkeley, California (1967) (also published as SESM Rep. 67-8).

18. Zarghamee, M. S., and Robinson, A. R., "A Numerical Method for Analysis of Free Vibration of Spherical Shells," *AIAA J.* **5**, 1256 (1967).

19. Cowper, G. R., Kosko, E., Lindberg, G. M., and Olson, M. D., Static and Dynamic Applications of a High Precision Plate Bending Element," *AIAA J.* **7**, 1957 (1969).

Comparison of Finite-Element and Finite-Difference Methods*

Samuel W. Key and Raymond D. Krieg

SANDIA LABORATORIES
ALBUQUERQUE, NEW MEXICO

1. Introduction

The finite-element method and the finite-difference method appear superficially to be quite different but are, in fact, closely related. By examining plane-wave propagation in a half space, the differences in their presentations can be highlighted. The finite-element method starts with a variational statement of the problem and introduces piecewise definitions of the functions defined by a set of meshpoint values. The finite-difference method starts with a differential statement of the problem and proceeds to replace the derivatives with their discrete analogs.

Both methods result in a set of algebraic equations relating a discrete set of variables in place of the relations in the continuous variables. These algebraic equations are remarkably similar and provide the basis for identifying the methods as essentially similar.

A historical review of the literature shows some rather interesting things. Both the structural engineer and the applied mathematician can claim the finite-element method as theirs, and both, with some justification, are correct. How this has come about can be partially explained by examining the literature.

While the two approaches rely on very different descriptions for their discussions, they are, in fact, closely related and more and more frequently

* This work was supported by the United States Atomic Energy Commission.

intertwined in applications. Both approaches are adopting features of the other that prove to be attractive. In dynamics problems, finite-element equations are invariably integrated forward in time with finite-difference techniques rather than using a time dimension on an element and using the integration scheme that results. The finite-difference method has found the variational approach of the finite-element method useful in producing symmetric difference equations, the discrete analog of a system of self-adjoint differential equations, and something which has eluded finite-difference approximations for some time.

2. A Problem in Wave Mechanics

Plane wave propagation in a half space provides a convenient problem for comparing the finite element and finite difference methods. Figure 1 shows a linearly elastic isotropic slab and a one-dimensional coordinate system x.

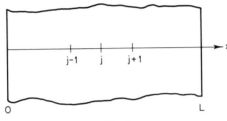

Fig. 1.

For longitudinal wave motion, the global balance of the force equation is

$$\int_A \rho \ddot{U} \, da = \oint_{\partial A} \sigma v \, dl \tag{1}$$

where σ is the stress tensor, acting on the outer unit normal to the boundary v, and ρ and $\ddot{U}$ are the material mass density and acceleration vector, respectively. This balance must hold for all volumes A.

For the one-dimensional linear elastic case, the equation can be written as shown in Eq. (2).

$$\int_{x_i}^{x_j} \rho \ddot{U} \, dx = [E(x)U_x]_{x_i}^{x_j} \tag{2}$$

Here, $E(x)$ is the elastic modulus and the subscript x denotes differentiation in space. In the limit, the differential form in Eq. (3) is obtained.

$$\rho \ddot{U} = (E(x)U_x)_x \tag{3}$$

The initial conditions for this problem are

$$U(x, 0) = u(x) \qquad \text{and} \qquad \dot{U}(x, 0) = v(x) \tag{4}$$

and the boundary conditions at $x = 0$ are either

$$U(0, t) = f(t) \qquad \text{or} \qquad U_x(0, t) = g(t)/E(0) \tag{5}$$

and at $x = L$ are either

$$U(L, t) = h(t) \qquad \text{or} \qquad U_x(L, t) = k(t)(E(L)). \tag{6}$$

This problem may also be stated in terms of a variational principle, Hamilton's principle. The functional in Eq. (7) is to be an extremum on those functions which are continuous in the domain $(0, L) \times (0, t)$.

$$\delta \int_0^t \left\{ \int_0^L [\rho(\dot{U})^2 - E(x)(U_x)^2] \, dx - 2U(0, t)g(t) - 2U(L, t)k(t) \right\} dt = 0 \tag{7}$$

Here it is assumed that the derivative boundary conditions are being enforced.

2.1. A FINITE-ELEMENT SOLUTION

The object in the finite-element method is to construct a family of functions with piecewise definitions over regions, in this case, intervals that are sufficiently general in character that the actual solution may be approximated as closely as desired. The functions must be admissable in the variational principle being used; in this case, they must be continuous in space and time. As an example, consider I uniformly spaced intervals, the general one being (x_i, x_j). Over that interval, the function U is assumed to vary linearly in x.

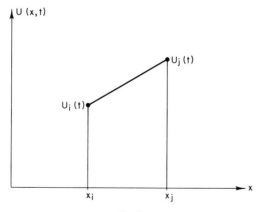

Fig. 2.

The result is pictured in Fig. 2 and using linear interpolation functions, the result is shown in Eq. (8).

$$U(x, t) = U_i(t)\frac{(x_j - x)}{(x_j - x_i)} + U_j(t)\frac{(x - x_i)}{(x_j - x_i)} \qquad x_i \leqslant x \leqslant x_j \qquad (8)$$

With this process repeated in every interval a piecewise linear family of functions defined by the nodal values results. The family is continuous in space and at any time can approximate a function and its first derivative as closely as desired with increasing subdivision of the mesh.

By substituting this family of functions into the variational principle, the spatial description is reduced to a discrete set of values varying in time. (This process may be carried out in time as well, but leads to a prescribed method of time integration which may or may not be desirable.) Consider the kth interval; the quantity to be varied becomes upon the substitution of Eq. (8)

$$\int_{x_i}^{x_j} \frac{1}{(x_j - x_i)^2} \left\{ \rho \begin{bmatrix} \dot{U}_i \\ \dot{U}_j \end{bmatrix}^T \begin{bmatrix} (x_j - x)^2 & (x_j - x)(x - x_i) \\ (x_j - x)(x - x_i) & (x - x_i)^2 \end{bmatrix} \begin{bmatrix} \dot{U}_i \\ \dot{U}_j \end{bmatrix} \right.$$
$$\left. - E(x) \begin{bmatrix} U_i \\ U_j \end{bmatrix}^T \begin{bmatrix} 1 & -1 \\ -1 & 1 \end{bmatrix} \begin{bmatrix} U_i \\ U_j \end{bmatrix} \right\} dx \qquad (9)$$

Here, matrix notation has been introduced as a convenient way to indicate the products involved. As a practical expedient, a constant value for E and ρ are used in each mesh by taking the value at the center.

$$E_k = E\left(\frac{x_i + x_j}{2}\right), \qquad \rho_k = \rho\left(\frac{x_i + x_j}{2}\right) \qquad (10)$$

That this is a permissible approximation that does not destroy the convergence has been discussed by McLay [1]. Thus, the integrals in Eq. (9) can be evaluated to give the results in Eq. (11) for the kth interval.

$$l_k \rho_k \begin{bmatrix} \dot{U}_i \\ \dot{U}_j \end{bmatrix}^T \begin{bmatrix} \frac{1}{3} & \frac{1}{6} \\ \frac{1}{6} & \frac{1}{3} \end{bmatrix} \begin{bmatrix} \dot{U}_i \\ \dot{U}_j \end{bmatrix} - \frac{E_k}{l_k} \begin{bmatrix} U_i \\ U_j \end{bmatrix}^T \begin{bmatrix} 1 & -1 \\ -1 & 1 \end{bmatrix} \begin{bmatrix} U_i \\ U_j \end{bmatrix} \qquad (11)$$

where l_k is the length of the kth interval, $x_j - x_i$.

Hamilton's principle for the entire problem is obtained by summing this result over all intervals as shown in Eq. (12).

$$\delta \int_0^t \left\{ \sum_{k=1}^I \left[l_k \rho_k \begin{bmatrix} \dot{U}_i \\ \dot{U}_j \end{bmatrix}^T \begin{bmatrix} \frac{1}{3} & \frac{1}{6} \\ \frac{1}{6} & \frac{1}{3} \end{bmatrix} \begin{bmatrix} \dot{U}_i \\ \dot{U}_j \end{bmatrix} - \frac{E_k}{l_k} \begin{bmatrix} U_i \\ U_j \end{bmatrix}^T \begin{bmatrix} 1 & -1 \\ -1 & 1 \end{bmatrix} \begin{bmatrix} U_i \\ U_j \end{bmatrix} \right] \right.$$
$$\left. - 2U(0, t)g(t) - 2U(L, t)k(t) \right\} dt = 0 \qquad (12)$$

Here i and j are the nodal values at the left and right ends, respectively, of the kth interval. The sum over all the individual quadratic forms can be re-ordered into a single quadratic form involving all of the mesh point values. The result for a two-element mesh with nodes numbered 1, 2, and 3 is shown in Eq. (13).

$$\delta \int \left\{ \begin{bmatrix} \dot{U}_1 \\ \dot{U}_2 \\ \dot{U}_3 \end{bmatrix}^T \begin{bmatrix} \frac{1}{3}l_1\rho_1 & \frac{1}{6}l_1\rho_1 & 0 \\ \frac{1}{6}l_1\rho_1 & \frac{1}{3}(l_1\rho_1 + l_2\rho_2) & \frac{1}{6}l_2\rho_2 \\ 0 & \frac{1}{6}l_2\rho_2 & \frac{1}{3}l_2\rho_2 \end{bmatrix} \begin{bmatrix} \dot{U}_1 \\ \dot{U}_2 \\ \dot{U}_3 \end{bmatrix} \right.$$

$$- \begin{bmatrix} U_1 \\ U_2 \\ U_3 \end{bmatrix}^T \begin{bmatrix} E_1/l_1 & -E_1/l_1 & 0 \\ -E_1/l_1 & E_1/l_1 + E_2/l_2 & -E_2/l_2 \\ 0 & -E_2/l_2 & E_2/l_2 \end{bmatrix} \begin{bmatrix} U_1 \\ U_2 \\ U_3 \end{bmatrix} \tag{13}$$

$$\left. - 2U_1 g(t) - 2U_3 k(t) \right\} dt = 0$$

The dashed 2×2 submatrices indicate the overlaying and adding of individual element matrices. The process is the same for an I element mesh. If U is a vector of nodal values, M a matrix of mass coefficients, K a matrix of stiffness coefficients, then for I elements Eq. (14) results.

$$\delta \int_0^{t'} \left\{ \dot{U}^T M \dot{U} - U^T K U - 2U(0, t)g(t) \right.$$

$$\left. - 2U(L, t)k(t) \right\} dt = 0 \tag{14}$$

The description is now in terms of a discrete set of variables, and the extremum is found using Lagrange's equations just as in particle dynamics. The result is given in Eq. (15).

$$M \ddot{U} + KU = F(t) \tag{15}$$

Here, F is a vector derived from the linear terms in Eq. (14). The essential ingredients for a dynamic analysis are contained in Eq. (15).

From Eq. (13), the algebraic equation that exists at a general mesh point may be read off. If i represents a general interior nodal point with $i - 1$, the point to the left, and $i + 1$, the point to the right, the result is shown in Eq. (16).

$$\frac{1}{6}(l_{i-\frac{1}{2}}\rho_{i-\frac{1}{2}})\ddot{U}_{i-1} + \frac{1}{3}(l_{i-\frac{1}{2}}\rho_{i-\frac{1}{2}} + l_{i+\frac{1}{2}}\rho_{i+\frac{1}{2}})\ddot{U}_i + \frac{1}{6}(l_{i+\frac{1}{2}}\rho_{i+\frac{1}{2}})\ddot{U}_{i+1}$$

$$- \left(\frac{E_{i-\frac{1}{2}}}{l_{i-\frac{1}{2}}} \right) U_{i-1} + \left(\frac{E_{i-\frac{1}{2}}}{l_{i-\frac{1}{2}}} + \frac{E_{i+\frac{1}{2}}}{l_{i+\frac{1}{2}}} \right) U_i - \left(\frac{E_{i+\frac{1}{2}}}{l_{i+\frac{1}{2}}} \right) U_{i+1} = 0 \tag{16}$$

From a computational standpoint, it is frequently convenient to un-couple the inertia terms. Eq. (17) provides one form of uncoupling, referred to as mass lumping.

$$\tfrac{1}{2}(l_{i-\frac{1}{2}}\rho_{i-\frac{1}{2}} + l_{i+\frac{1}{2}}\rho_{i+\frac{1}{2}})\ddot{U}_i - (E_{i-\frac{1}{2}}/l_{i-\frac{1}{2}})U_{i-1}$$
$$+ \left(\frac{E_{i-\frac{1}{2}}}{l_{i-\frac{1}{2}}} + \frac{E_{i+\frac{1}{2}}}{l_{i+\frac{1}{2}}}\right)U_i - \left(\frac{E_{i+\frac{1}{2}}}{l_{i+\frac{1}{2}}}\right)U_{i+1} = 0 \tag{17}$$

If the meshes are of uniform length l and the modulus E and the density ρ constant, the simplification in Eq. (18) is obtained

$$l\rho\ddot{U}_i + (E/l)(-U_{i-1} + 2U_i - U_{i+1}) = 0 \tag{18}$$

2.2. A Finite Difference Solution

Equation (1), the global balance of force expression, is used as a starting point for the finite-difference method as applied here. This approach is not new in the finite-difference method, and has been used in two-dimensional spatial regions [2, 3]. It is the global equivalent to differential equation (3) and not to be confused with a variational statement.

Again, consider the model where l_1 and l_2 are mesh lengths between the nodes 1 and 2 and nodes 2 and 3, respectively. Equation (2) is applied over the range $(x_L, x_R) = ((x_2 + x_1)/2, (x_3 + x_2)/2)$

$$\int_{x_L}^{x_R} \rho\ddot{U}\,dx = EU_x|_{x_R} - EU_x|_{x_L} \tag{19}$$

If U is quadratic over each element, then the right side can be rewritten directly in the form of Eq. (20).

$$\int_{x_L}^{x_R} \rho\ddot{U}\,dx = E_2\frac{U_3 - U_2}{l_2} - E_1\frac{U_2 - U_1}{l_1} \tag{20}$$

If $\rho\ddot{U}$ is represented as a quadratic function over the range (x_1, x_3), then the integral in the left over the two half elements after some algebra is given in Eq. (21).

$$\int_{x_L}^{x_R} \rho\ddot{U}\,dx = (\rho\ddot{U})_2\left(\frac{l_2}{2} + \frac{l_1}{2}\right) + \frac{a}{2}\left(\frac{l_2^2}{4} - \frac{l_1^2}{4}\right) + \frac{b}{3}\left(\frac{l_2^3}{8} + \frac{l_1^3}{8}\right) \tag{21}$$

where

$$a = \frac{(\rho\ddot{U})_3 l_1^2 + (\rho\ddot{U})_2(l_2^2 - l_1^2) - (\rho\ddot{U})_1 l_2^2}{l_1 l_2(l_1 + l_2)}$$

and

$$b = \frac{(\rho\ddot{U})_3 l_1 - (\rho\ddot{U})_2(l_1 + l_2) + (\rho\ddot{U})_1 l_2}{l_1 l_2(l_1 + l_2)}$$

Equations (20) and (21) can be combined and written in terms of a general nodal point i with the results shown in Eq. (22)

$$
\frac{(-2l^2_{i+\frac{1}{2}} + 2l_{i+\frac{1}{2}}l_{i-\frac{1}{2}} + l^2_{i-\frac{1}{2}})}{24l_{i-\frac{1}{2}}}(\rho_{i-1}\ddot{U}_{i-1})
$$

$$
+ \frac{(2l^2_{i-\frac{1}{2}} + 7l_{i-\frac{1}{2}}l_{i+\frac{1}{2}} + 2l^2_{i+\frac{1}{2}})(l_{i-\frac{1}{2}} + l_{i+\frac{1}{2}})}{24l_{i-\frac{1}{2}}l_{i+\frac{1}{2}}}(\rho_i\ddot{U}_i)
$$

$$
+ \frac{(l^2_{i+\frac{1}{2}} + 2l_{i+\frac{1}{2}}l_{i-\frac{1}{2}} - 2l^2_{i-\frac{1}{2}})}{24l_{i+\frac{1}{2}}}(\rho_{i+1}\ddot{U}_{i+1}) - \left(\frac{E_{i-\frac{1}{2}}}{l_{i-\frac{1}{2}}}\right)U_{i-1}
$$

$$
+ \left(\frac{E_{i-\frac{1}{2}}}{l_{i-\frac{1}{2}}} + \frac{E_{i+\frac{1}{2}}}{l_{i+\frac{1}{2}}}\right)U_i - \left(\frac{E_{i+\frac{1}{2}}}{l_{i+\frac{1}{2}}}\right)U_{i+1} = 0
$$

(22)

If the uncoupling of inertia terms is desirable, it is best to revert back to Eq. (19) and assume $(\rho\ddot{U})$ constant over (x_L, x_R). In this case, the resultant expression is given in Eq. (23).

$$
\frac{1}{2}(l_{i-\frac{1}{2}} + l_{i+\frac{1}{2}})(\rho_i\ddot{U}_i) - \left(\frac{E_{i-\frac{1}{2}}}{l_{i-\frac{1}{2}}}\right)U_{i-1}
$$

$$
+ \left(\frac{E_{i-\frac{1}{2}}}{l_{i-\frac{1}{2}}} + \frac{E_{i+\frac{1}{2}}}{l_{i+\frac{1}{2}}}\right)U_i - \left(\frac{E_{i+\frac{1}{2}}}{l_{i+\frac{1}{2}}}\right)U_{i+1} = 0
$$

(23)

If the meshes are of uniform length l, and the modulus E and density ρ constant, the equation further simplifies to Eq. (18) identically.

2.3. COMPARISON OF THE METHODS

Both methods arrived at discrete analogs to the problem, the finite-element method by minimizing an extremum principle over a family of functions defined piecewise, the finite-difference method by using discrete approximations to integrals and derivatives occurring in balance equations. Even with a variation in mesh spacing and moduli, the same approximations in the space derivatives were obtained although in neither case are these considered unique. Both methods provided a cross-coupling in the inertia terms. However, the averaging involved differed in the two methods. In the finite-difference method, the derivation of the right side of Eq. (20) is straightforward, but the left side requires an assumption of the form of the integrand. An obvious and natural form was taken here, since functions are usually taken to be the highest-order polynomial that can be represented with the available mesh variables. If $(\rho\ddot{U})$ had been taken piecewise linear over the range, or if separate assumptions had been made on the forms of ρ and $\ddot{U}$, then the results change, just as they change when various assumptions are made in the finite-element model. It is a curious result, however, that of the many forms tried for $(\rho \ddot{U})$ and combinations of forms of ρ and $\ddot{U}$ taken separately, that none of the resulting consistant mass matrices had off-diagonal terms as large as the finite-element model unless $\ddot{U}$ was taken to be discontinuous in x. The fact that the two methods produce very similar discrete analogs is not a new observation.* It has become a popular topic and several papers are available on it. Several years ago, Zienkiewicz and Cheung [30] in connection with an application of the finite-element method to Poisson's equation observed the similarities in the algebraic equations that resulted from the two methods. Walz *et al.* [4] undertook a study of convergence in the finite-element method by examining in detail the algebraic equations produced and using approaches common in the finite-difference method showed what forms of convergence could be expected. Goudreau [5] has a rather extensive treatment of this topic including the example used here, although his attention is primarily directed toward the discrete analogs obtained from the finite-element method. Wempner [6] considered a plane stress elasticity problem and showed how the discrete analog of the finite-element method provides a possible finite-difference expression. Croll and Walker [7] have also examined plane stress elasticity and show how several different finite-difference analogs may be obtained from various displacement assumptions, some compatible, some not. Pian [8] in a recent paper has examined Poisson's equation and considers a variety of element patterns and displacement assumptions along with several finite-difference approaches and compares the resulting discrete analogs.

In the case of the inertia coupling that results from the two methods, a great deal can be learned by comparing the solutions to both the differential and discrete equations. The following section carries this out.

2.4. Solution of the Equations

For constant coefficients and fixed ends, the solution for the continuous problem is harmonic in both time and space. This is also true for each of the numerical models that have been developed. In fact, the mode shapes are identical to those of the continuous solution except that only a finite number of modes are represented. The real distinction among the various models shows up in the frequency found for each mode. Unfortunately, the exact temporal solution of the numerical equations is very expensive to obtain for nontrivial problems so that a numerical time integration scheme must also be used in practice. This further distorts the frequency spectrum for each of the methods. It is not the intent of this paper to compare the various

---

* Early applied mathematics papers were concerned almost exclusively with this question [13–17].

time integration methods. For ease of discussion, the ordinary central difference method of time integration is used. The various numerical methods can then be compared to both the exact solution and with each other.

The continuous problem with fixed ends acted upon by an initial velocity has the solution

$$U = \sum_{k=1}^{\infty} A_k \sin \frac{k\pi x}{L} \sin \frac{k\pi Ct}{L} \tag{24}$$

where $C = (E/\rho)^{1/2}$ and the constants A_k are determined from the initial conditions.

The numerical equations from the finite-element and finite-difference methods with consistant mass matrices as well as the lumped mass model can also be solved in the form of Eq. (24) [5, 9]. All of these discrete equations can be written in the form

$$r\ddot{U}_{i+1} + (1 - 2r)\ddot{U}_i + r\ddot{U}_{i-1} = (E/\rho \, \Delta x^2)(U_{i+1} - 2U_i + U_{i-1}) \tag{25}$$

for $i = 1, 2, 3, \ldots, I - 1$, $U_0 = U_I = 0$ and $0 \leqslant r < 1$. As r is varied, a continuous family of mass couplings is obtained, as well as the ones already developed. With central difference expressions in time introduced in Eq. (25), the interior equation for a constant time step size becomes

$$r(U_{i+1}^{n+1} - 2U_{i+1}^n + U_{i+1}^{n-1}) + (1 - 2r)(U_i^{n+1} - 2U_i^n + U_i^{n-1})$$
$$+ r(U_{i-1}^{n+1} - 2U_{i-1}^n + U_{i-1}^{n-1}) \tag{26}$$
$$= \alpha^2(U_{i+1}^n - 2U_i^n + U_{i-1}^n)$$

where

$$\alpha^2 = C^2 \, \Delta t^2 / \Delta x^2 \tag{27}$$

A solution of the form $U_i^n = T^n X_i$ is assumed and substituted into Eq. (26). The variables are separated to obtain Eq. (28)–(30).

$$\frac{T^{n+1} - 2T^n + T^{n-1}}{-rT^{n+1} + (\alpha^2 + 2r)T^n - rT^{n-1}} = \frac{X_{i+1} - 2X_i + X_{i-1}}{X_i} \tag{28}$$

$$X_{i+1} - 2X_i + X_{i-1} = \lambda X_i \tag{29}$$

$$T^{n+1} - 2T^n + T^{n-1} = \lambda[-rT^{n+1} + (\alpha^2 + 2r)T^n - rT^{n-1}] \tag{30}$$

A solution of the form

$$X_i = a \sin \gamma i \, \Delta x \tag{31}$$

is assumed and substituted into Eq. (29). The result, after using sum and difference trigonometric identities, is

$$\sin \gamma i \, \Delta x \cos \gamma \, \Delta x - 2 \sin \gamma i \, \Delta x + \sin \gamma i \, \Delta x \cos \gamma \, \Delta x = \lambda \sin \gamma i \, \Delta x \tag{32}$$

Since this must hold for *all* i, the requirement is

$$\lambda = -2(1 - \cos \gamma \, \Delta x) \qquad (33)$$

Thus, Eq. (31) is indeed a solution and will satisfy the boundary conditions if

$$\gamma = k\pi/I \, \Delta x; \qquad k = 1, 2, 3, \ldots$$

The infinite range on k can be limited by noting that

$$\sin \frac{(2I \pm k)\pi i}{I} = \pm \sin \frac{k\pi i}{I}$$

so that a discrete set of solutions results given by Eq. (34).

$$\gamma = k\pi/I \, \Delta x; \qquad k = 1, 2, 3, \ldots, I \qquad (34)$$

Note that this is the exact spatial solution as given in Eq. (24) without regard for r, except that only the lowest I modes appear. It now remains to find the frequencies associated with these modes, where λ is given by Eq. (33). The temporal equation is taken from Eq. (30) which is rearranged in the form

$$T^{n+1}(1 + \lambda r) - T^n(2 + \lambda \alpha^2 + 2\lambda r) + T^{n-1}(1 + \lambda r) = 0 \qquad (35)$$

The solution is assumed to be harmonic in time,

$$T^n = b \sin \beta n \, \Delta t.$$

This is substituted into Eq. (35), sum and difference trigonometric identities applied, and the result required to hold for all n. This gives the result

$$2(1 + \lambda r) \cos \beta \, \Delta t - (2 + \lambda \alpha^2 + 2\lambda r) = 0$$

Equations (33) and (34) are now used to give the final frequency equation.

$$\cos \beta \, \Delta t = \frac{1 - (\alpha^2 + 2r)[1 - \cos(k\pi/I)]}{1 - 2r[1 - \cos(k\pi/I)]} \qquad (36)$$

This equation holds for the finite-element model with the consistant ($r = \frac{1}{6}$) or lumped ($r = 0$) mass matrices and for the finite-difference model with consistant ($r = \frac{1}{24}$) or lumped ($r = 0$) mass matrices.

The exact frequencies β_e are taken by observation from Eq. (24) to be

$$\beta_e = k\pi C/L = k\pi \alpha \, \Delta x/L \, \Delta t, \qquad \beta_e \, \Delta t = k\pi \alpha/I \qquad (37)$$

For a meaningful comparison of the discrete frequency in Eq. (36) to the exact frequencies in Eq. (37), a time step must be chosen. This choice is restricted by stability considerations in the time integration scheme. The usual Von Neumann definition of stability [37], which requires a bounded solution, can be easily applied to Eq. (36). The magnitude of ($\cos \beta \, \Delta t$)

cannot exceed unity for β to remain real and, hence, the solution bounded. For (k/I) small this is assured. The case where $\cos(k\pi/I) = -1$ is the limiting case where $\cos \beta \, \Delta t$ takes on negative values as α^2 increases for fixed r. The critical $\alpha_c{}^2$ is then given by the solution to the equation:

$$-1 = \frac{1 - 2(\alpha_c{}^2 + 2r)}{1 - 4r} \qquad \text{or} \qquad \alpha_c{}^2 = 1 - 4r \qquad (38)$$

The critical values of α are then

$$\alpha_c = \sqrt{\tfrac{1}{3}} \quad \text{for the consistent finite-element model}$$

$$\alpha_c = \sqrt{\tfrac{5}{6}} \quad \text{for the consistent finite-difference model}$$

$$\alpha_c = 1 \quad \text{for the lumped mass model}$$

The methods are then stable and convergent for $\alpha < \alpha_c$.

If the same element size and time-step size is chosen for the three mass distributions considered above, then α will be the same for all the methods. A value of $\alpha = 0.5$ is stable for all methods and is used for comparison purposes.

Figure 3 is a plot of the ratio of numerical to exact frequencies for various k/I. Several observations can be made. In particular, the consistent mass matrix of the finite-element method gives no better answers than the diagonal mass matrix. Note also that this holds for *all* values of k, even the lower modes. Also, the consistent mass matrix generated by the finite-difference method is the best of the three methods presented. Should it be desirable to reproduce the exact frequencies even more closely, a value of r somewhere between $\tfrac{1}{24}$ and $\tfrac{1}{6}$ could be used to define the mass coupling.

Graphs similar to Fig. 3 could be plotted for numerical time integration methods other than the ordinary central difference integrator used here. It is conjectured that the conclusions would be the same, however. This is based on the striking similarity between Fig. 3 shown here and Fig. 3.1 of Goudreau [5] where he used an exact time integration rather than a numerical time integration.

3. Early Literature

Both the structural engineer and the applied mathematician can claim the finite-element method as theirs, and both with some justification. Early work on the finite-element method was pursued by three separate groups; the applied mathematician, the engineer, and the physicist. The idea appears to have been developed independently in each of these groups and to have served different purposes initially in each case. The applied mathematician

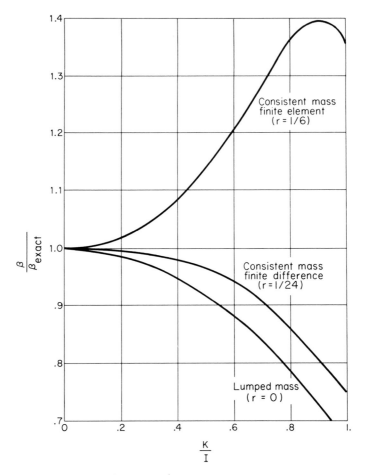

Fig. 3. Ratio of approximate frequency to exact frequency as a function of mode number with *I* the maximum number of modes possible.

was seeking a means of obtaining approximate eigenvalues that bounded the exact eigenvalues above and below. The finite-element method provided one of those bounding values. The engineer was looking for a means of representing the plane stress behavior of shear panels in the stringer, rib, and shear panel constructions common in the airframe industry. The physicists were looking for a means of construction approximate function sequences to get answers in continuum problems. Current interest in the finite-element method is in all three of these applications. The early work of each of these three groups is traced below.

In the applied mathematics literature, one of the early and clear accounts is by Courant [12] in 1943. After a lapse of some 9 years, the notion was picked up again by Polya [13, 16], Hersch [14], and Weinberger [15, 17]. Weinberger [17] claims that the idea is contained implicitly in a paper published in 1938 by Collatz [11]. Courant [21] in a footnote claims to have been motivated by the work of Euler [10]. In any event, the notion was clearly understood and most of the papers at that time were concerned with the algebraic equations that were generated, ignoring the possibility of having arbitrary meshes available for irregular geometries. The idea appears to have lain dormant until Friedrichs [20] in 1962, inspired by the work of White [19], applied the finite-element method to plane elasticity.

The report by White is concerned with what is now called the variational-difference method and is a thorough piece of work from a practical standpoint. It contains a triangular finite element needed to handle the "odd" mesh shapes that irregular geometries generate on regular meshes. Both White and Friedrichs appeared to be unaware of the great strides being made in the engineering community at that time with the finite-element method. Both of them were working with regular meshes making special provisions for the boundaries. With some additional developments, the work of Friedrichs was presented again in 1966 by Friedrichs and Keller [22]. As late as 1967, Birkhoof et al. [23] published a paper on the convergence of the finite-element method without any apparent knowledge of the work going on simultaneously in the engineering community. Now, again, the finite-element method is very much in vogue in the applied mathematics community. A forthcoming book by Fix and Strange [24] contains some 100 references, mostly in applied mathematics.

In an isolated paper in 1959 by Greenstadt [18], the notion of an element as opposed to a mesh point was clearly put forth. He considered breaking the problem into several distinct parts, making independent variable assumptions in each part, and then tying them back together with continuity requirements after evaluating the variational principles in each region. His technique for imposing continuity consisted of adding integrals to the variational principle of Lagrange multipliers times the continuity requirements along the interfaces.

The structural analysis community in the early 1950s applied the ideas of matrix structural analysis developed for discrete beam and frame structures to two-dimensional elasticity problems. What was originally a bookkeeping scheme for large problems became a numerical solution technique at this point. Early accounts of this work are given by Argyris and Kelsey [25] and Turner et al. [26]. In 1960, Clough [27] gave a detailed account of the application of the finite-element method to plane stress elasticity. It became increasingly clear that when used with an energy or variational principle, the finite element method was a form of the Ritz Method. Melosh [28] in 1963 gave a detailed account of the finite element method from this point of view as it is applied to continuum problems. One of the major difficulties in the finite element method is satisfying continuity requirements along element boundaries. Jones [29] in 1964 addressed this question in some depth and provided several new insights into the problem. His work differs only slightly from that of Greenstadt [18], but was carried out without the benefit of this earlier paper. It was at this time that the engineering community began to realize that they had a rather good and rather well developed numerical solution technique on their hands and set about looking for nonstructural applications. In 1965, Zienkiewicz and Cheung [30] treated Poisson's equation in an effort to expand the applications of the finite element method. In the same vein, Herrmann [31] examined Poisson's equation as it applied to torsional problems in elasticity. In the late 1960's, the finite element method achieved a wide acceptance in engineering for its utility and ease of use in problem solving. A review in any detail of its current state in engineering would require a far more extensive treatment than is possible here. An indication of its acceptance can be gained by examining the texts of either Zienkiewicz and Cheung [32] or Przemieniecki [33].

The third collection of papers on the finite element method exists in the physics literature. Synge [34] in 1952 in need of an approximate solution technique to make his bounding theorems applicable to practical problems conceived of the idea of defining functions as piecewise linear over triangular nets. A year later, a student of Synge's, McMahon [35], published a three-dimensional solution using tetrahedral elements and linear displacement assumptions. The work of Synge and his colleagues was summarized in a book in 1957 [36].

The early developments in the finite-element method were carried on more or less independently by these three groups. The engineer made the method immensely practical, leaving its rigorous development to physical insight. The applied mathematician made the method immensely detailed in its theoretical foundations, being satisfied to examine its utility on Laplace's equation on a square domain with equally spaced meshes. Curiously enough, the work of Synge, a physicist, is both practical and rigorous.

References

1. McLay, R. W., On Certain Approximations in the Finite Element Method, Paper No. 70-WA/APM-34, ASME 91st Winter Annual Meeting, New York (1970).
2. Bertholf, L. D., and Benzley, S. E., TOODY II, A Computer Program for Two-Dimensional Wave Propagation, SC-RR-68-41, Sandia Lab., Albuquerque, New Mexico (November 1968).

3. Krieg, R. D., and Monteith, H. C., A Large Deflection Transient Analysis of Arbitrary Shells Using Finite Differences, *Conf. Comput. Oriented Anal. Shell Structures*. Lockheed Palo Alto Res. Lab., Palo Alto, California (August 1970).
4. Walz, J. W., Fulton, R. E., and Cyrus, N. J., Accuracy and Convergence of Finite Element Approximations, *Proc. Conf. Matrix Methods Struct. Mech. 2nd*, AFFDL-TR-68-150, Wright Patterson Air Force Base, Ohio (October 1968).
5. Goudreau, G. L., Evaluation of Discrete Methods for the Linear Dynamic Response of Elastic and Viscoelastic Solids, SESM No. 69-15, Struct. Eng. Lab., Univ. of California, Berkeley, California (1970).
6. Wempner, G. A., Finite Differences via Finite Elements, Panel Discussion of Finite Elements versus Finite Differences, *Conf. Comput. Oriented Anal. Shell Struct*. Lockheed Palo Alto Res. Lab., Palo Alto, California (August 1970).
7. Croll, J. G. A., and Walker, A. C., The Finite Difference and Localized Ritz Methods, *Int. J. Numer. Methods Eng.* **3**, 155–160 (1971).
8. Pian, T. H. H., Variational Formulations of Numerical Methods in Solid Continuum, *SMD Symp. Comput. Aided Eng*. Univ. of Waterloo, Waterloo, Ontario, Canada (May 1971).
9. Krieg, R. D., Phase Velocities of Elastic Waves in Structural Computer Programs, SC-DR-67-816, Sandia Lab., Albuquerque, New Mexico (November 1967).
10. Euler, L., *Methods Inveniendi Lineas Curvas Maximi Minimive Proprietate Gaudentes*. Bousquet, Lausanne and Geneva, 1744.
11. Collatz, L., Konvergenz des Differenzverfahrens bei Eigenwertproblemen Partieller Differentialgleichungen, *Deutsche Math.* **3**, 200–212 (1938).
12. Courant, R., Variational Methods for the Solutions of Problems of Equilibrium and Vibrations, *Bull. Amer. Math. Soc.* **49**, 1–23 (1943).
13. Polya, G., Sur une Interpretation de la Methode des Differences Finies qui peut Fournir des Bornes Superieures ou Inferieures, *C. R. Acad. Sci. Paris* **235**, 995–997 (1952).
14. Hersch, J., Equations Differentielles et Functions de Cellules, *C. R. Acad. Sci. Paris* **240**, 1602–1604 (1955).
15. Weinberger, H. F., Upper and Lower Bounds for Eigenvalues by Finite Difference Methods, *Comm. Pure Appl. Math.* **9**, 613–623 (1956).
16. Polya, G., *Estimates for Eigenvalues*, Studies presented to Richard von Mises," pp. 200–207. Academic Press, New York, 1954.
17. Weinberger, H. F., Lower Bounds for Higher Eigenvalues by Finite Difference Methods, *Pacific J. Math.* **8**, 339–368 (1958).
18. Greenstadt, J., On the Reduction of Continuous Problems to Discrete Form, *IBM J. Res. Develop.* **3**, 355–363 (1959).
19. White, G. N., Difference Equations for Plane Thermal Elasticity, LAMS-2745, Los Alamos Sci. Lab., Los Alamos, New Mexico (1962).
20. Friedrichs, K. O., A Finite Difference Scheme for the Neumann and the Dirichlet Problem, NYO-9760, Courant Inst. of Math. Sci., New York Univ., New York (1962).
21. Courant, R., and Hilbert, D., *Methods of Mathematical Physics*, Vol. 1, 1st English ed., 4th Printing, 1963, p. 177. Wiley (Interscience), New York, 1953.
22. Friedrichs, K. O., and Keller, H. B., A Finite Difference Scheme for Generalized Neumann Problems, *Numerical Solution of Partial Differential Equations* (J. H. Bramble, ed.). Academic Press, New York, 1966.
23. Birkhoff, G., Schultz, M. H., and Varga, R. S., Piecewise Hermite Interpolation in One and Two Variables with Applications to Partial Differential Equations, *Numer. Math.* **11**, 232–256 (1968).
24. Strang, G., and Fix, G., *An Analysis of the Finite Element Method*. Prentice-Hall, in press.

25. Argyris, J. H., and Kelsey, S., Energy Theorems and Structural Analysis, *Aircraft Eng.* **26–27** (October 1954–May 1955).
26. Turner, M. J., Clough, R. W., Martin, H. C., and Topp, L. J., Stiffness and Deflection Analysis of Complex Structures, *J. Aeronaut. Sci.* **23**, 805–823 (1956).
27. Clough, R. W., The Finite Element Method in Plane Stress Analysis, *Proc. ASCE Conf. Electron. Comput. 2nd*, Pittsburgh, Pennsylvania (September 1960).
28. Melosh, R. J., Basis for Derivation of Matrices for the Direct Stiffness Method, *AIAA J.* **1**, 1631–1637 (1963).
29. Jones, R. E., A Generalization of the Direct-Stiffness Method of Structural Analysis, *AIAA J.* **2**, 821–826 (1964).
30. Zienkiewicz, O. C., and Cheung, Y. K., Finite Elements in the Solution of Field Problems, *Engineer* **220**, 507–510 (1965).
31. Herrmann, L. R., Elastic Torsional Analysis of Irregular Shapes, *J. EM Div., Proc. ASCE* **91**, EM6 (December 1965).
32. Zienkiewicz, O. C., and Cheung, Y. K., *The Finite Element Method*. McGraw-Hill, New York, 1967.
33. Przemieniecki, J. S., *Theory of Matrix Structural Analysis*. McGraw-Hill, New York, 1968.
34. Synge, J. L., Triangulation in the Hypercircle Method for Plane Problems, *Proc. Roy. Irish Acad.* **54A21**, 341–367 (1952).
35. Rev. McMahon, J., Lower Bounds for the Electrostatic Capacity of a Cube, *Proc. Roy. Irish Acad.* **55A9**, 133–167 (1953).
36. Synge, J. L., *The Hypercircle in Mathematical Physics*. Cambridge Univ. Press, London and New York, 1957.
37. Richtmyer, R. D., and Morton, K. W., *Difference Methods for Initial Value Problems*, 2nd ed. Wiley (Interscience), New York, 1967.

Incremental Stiffness Method for Finite Element Analysis of the Nonlinear Dynamic Problem

John F. McNamara and Pedro V. Marcal

DIVISION OF ENGINEERING, BROWN UNIVERSITY
PROVIDENCE, RHODE ISLAND

1. Introduction

Many advances have been made in finite element theory which account for both nonlinear geometric and material behavior. These advances have recently been summarized in the literature [1–3]. This nonlinear theory is, in principle, applicable to problems in both the static and dynamic regime. The research effort directed toward investigating the use of the finite-element method in the analysis of structures under dynamic loads compares rather unfavorably with its rapid development in the static regime.

Following the development of a nonlinear, static, finite-element program a companion program was started to study nonlinear dynamic problems. Parallel formulations were made for the latter problem, with some expectations that the dynamic problem would yield to a similar approach. In fact, it turned out that there were sufficient shortcomings in the convergence properties of the element used and in the numerical procedures adopted for the integration of the equations in time and space to cause considerable delays in the progress of the study. This paper presents a synopsis of the results obtained to date which are described more completely in [4]. As in the static problem [2], the writers set out to develop a base on which a general purpose finite-element program could be built for analyzing nonlinear dynamic problems. This procedure seemed the most expedient approach, since much common programming could be utilized from the static program.

2. Review of Literature

It is interesting to note that the original development of the finite-element method was directed to the solution of dynamic problems [5]. The emphasis rapidly switched to static solutions and it is only recently that its potential for solving complex dynamic problems is being realized. The earliest examples in the literature of the solution of a dynamic problem using the finite-element method are those given in Klein and Sylvester [6] and Popov and Chow [7] where the linear solution for a shallow shell cap under a step pressure load was given. A more recent and more general application of the method to linear problems was illustrated in Clough [8] where a nuclear containment vessel, an earth dam, and other complex structures are examined.

Although the geometrically nonlinear static problem was formulated using the finite-element method by Turner *et al.* [9] in 1960, it was not until 10 years later that results were obtained for the corresponding elastic dynamic problem by Stricklin *et al.* [10]. In Turner *et al.* [9], the equations were solved in an incremental form where the behavior was assumed to be linear for each increment and the problem was reduced to a series of linear solutions. In this method a new set of equations was formulated for each increment. The method adopted by Turner *et al.* [10] used the total form of the governing equations and transformed the nonlinear terms into pseudoforces which were then treated as additional loads on the structure. The difficulty arises in that these forces depend on the unknown current displacements and must then be found by extrapolation or iteration techniques. Turner *et al.* [10] concluded, after comparisons with parabolic and cubic extrapolation schemes, that a linear procedure is sufficient for determining these forces. The results given by Turner *et al.* [10] apply only to a shell of revolution with certain asymmetries of the applied load allowed. The approach used by Turner *et al.* has been extended by Klein [11] to include other asymmetric properties of the shell structure and the loading.

The other nonlinearity we wish to discuss is material nonlinearity due to nonlinear constitutive laws as used in modeling the behavior of an elastic–plastic solid. Two general methods for incorporating this behavior into a finite-element analysis have been developed. They are based on the Prandtl–Reuss equations [12] for an elastic–plastic solid, and are known as the initial strain method and the tangent modulus method. The tangent modulus approach corresponds to the incremental method adopted for nonlinear geometry, and therefore is the most convenient and direct manner of accounting for both the geometric and material nonlinearities. The only references known to the writers for dynamic finite-element analysis with nonlinear material behavior are [13, 14] where elastic–plastic wave problems were studied.

The paucity of nonlinear dynamic results contrasts with the large number of results obtained using the finite-difference approach together with numerical integration in time. There is a large body of literature, surveyed in Fulton [15], covering the area of nonlinear elastic deformations, and which is primarily directed to the solution of dynamic elastic shell buckling problems. The works cited by Fulton are concerned with solving numerically the coupled differential equations of motion of the structure. However, a method that is closer to the finite-element approximation was developed initially by Witmer *et al.* [16]. In this approach the structure is discretized initially and equilibrium equations are written for each separate division. This results in an uncoupled system which may be solved directly. Solutions that include large displacements and plastic deformations have been obtained by this method for a variety of problems. It has also been extended to cover general shell structures undergoing three-dimensional deformations [17]. Many experimental programs have been carried out to validate this method, and good agreement has been shown between the theory and experiment [16–18]. The drawback with the preceding methods is that they result in special purpose solutions restricted to certain geometric shapes divided into a regular grid pattern, whereas the finite-element approach allows irregular structural shapes and grids. Approximate techniques for solving dynamic problems, which include large plastic deformations, have been developed. These techniques are based on the assumption that elastic deformations may be neglected, and are described in an extensive review by Symonds [19], who also discusses their relative accuracy and merit. There are also the mode approximation methods [20] based on the bound theorems of Martin [21, 22]. A criterion adopted for the purpose of deciding when these approximate methods may be useful is that the energy applied by the external loading on the structure should be about ten times the amount of energy that could be absorbed elastically by the structure [19].

3. Theoretical Considerations

The basic equations for nonlinear finite-element analysis are well understood and will not be derived here. Instead, we shall quote the equations as our point of departure. By the principle of virtual work we obtain, in terms of initial geometry,

$$\int_{V^0} [N]^T [\rho][N]\{\ddot{u}\}\, dV^0 = -\int_{V^0} [B]\{\sigma\}\, dV^0 + \{P\} \qquad (1)$$

where $[N]$ is an interpolation function that transforms displacements at the nodes to displacement at any point within an element, $[B]$ is the transformation matrix that transforms displacement rates at the nodes to strain rates at any point in an element, $\{\sigma\}$ is the generalized stress vector. $[\rho]$ is the density matrix (the density takes on its matrix form in problems of beams and shells), and $\{P\}$ is the applied load.

We may now linearize the equation by writing it in incremental form:

$$\int_{V^0} [N]^\mathrm{T}[\rho][N]\,dV^0\,\Delta\{\ddot{u}\} = -\int_{V^0} \Delta[B]\{\sigma\}\,dV^0 - \int_{V^0} [B]\,\Delta\{\sigma\}\,dV^0$$
$$+ \Delta\{P\} + O(I) + O(t^m) \tag{2}$$

In Eq. (2) $\Delta\{P\}$ should be understood to include the effects of following loads. The two error terms $O(t^m)$ and $O(I)$ are also included to show that the solution in incremental form contains a discretization error due to the current increment as well as an inherited error due to all previous increments. The error due to discretization in time is shown as a function of time raised to the power m.

We now make use of the linearized incremental stress–strain relations which are written as

$$\Delta\{\sigma\} = [D]\,\Delta\{e\} \tag{3}$$

This equation is appropriate for elastic–plastic behavior and has been outlined for small strain by Marcel and King [24] and for large strains by Hibbitt et al. [25].

Substituting (3) into (2) results in a linearized incremental equation

$$[M]\,\Delta\{\ddot{u}\} = -[K]\,\Delta\{u\} + \Delta\{P\} + O(t^m) + O(I) \tag{4}$$

This equation can be specialized to the static case by neglecting the term on the left. In the static case, convergence to the true solution may be achieved by applying the load in increasingly smaller increments. A parallel procedure was investigated for the dynamic case where the rate of convergence with decrease in time step was examined. The results of this analysis are discussed later.

We now consider the error term $O(I)$ which will be called the residual load correction [26] and consists of writing the residual equation for (1)

$$O(I) = -[M]\{\ddot{u}\} - \int_{V^0} [B]\{\sigma\}\,dV^0 + P \tag{5}$$

It is noted that this error term consists of evaluating the terms at the state before the current increment and that if no numerical errors had been introduced by previous increments the error would be equal to zero. It was shown in [4] that, by including the residual load correction in the dynamic equations, one may obtain convergent solutions using time increments that are relatively large in comparison with the solutions obtained without the correction.

4. Solution Procedure

The selection of an integration scheme for the solution of the incremental equations in the time domain is critical with respect to computational efficiency. A suitable solution scheme which allows a large time step, and yet gives an accurate solution is that developed by Houbolt [27]. The Houbolt scheme is based on the third-order backward difference expression

$$\Delta\{u_n\} = \Delta t^{-2}\{2\,\Delta u_n - 5\,\Delta u_{n-1} + 4\,\Delta u_{n-2} - \Delta u_{n-3}\}$$

Applying this to the condition for incremental equilibrium Eq. (4), we have

$$(2[M] + \Delta t^2[K_n])\,\Delta\{u_{n+1}\}$$
$$= \Delta\{P_{n+1}\}\,\Delta t^2 + [M](5\,\Delta\{u_n\} - 4\,\Delta\{u_{n-1}\} + \Delta\{u_{n-2}\}) + O(I_n)\,\Delta t^2$$

where the subscript denotes the time at which the increment is taken.

This equation is solved for the displacement increment $\Delta\{u_{n+1}\}$ at each step except the first, where a special starting procedure must be employed [27].

5. Note on Solution Convergence

Haisler et al. [23] have reported on studies of numerical integration schemes and their convergence properties in the nonlinear static case. It was shown there that the incremental finite-element formulation gave satisfactory results when the load increments used were small as compared with those adopted in the solutions using the residual load correction term. The nature of the correction procedure was illustrated in [4], where the order of the error for the corrected and uncorrected equations were examined. The result was given for a one-dimensional model, and serves to give an order of magnitude estimate of the error.

In the static case for the solution without the residual load correction and a slowly varying stiffness K, the total discretization error is the sum of the truncation errors for each increment of the incremental approximation.

This error may be expressed as

$$u_N - u_N^* = -\frac{1}{2} \sum_{n=2}^{N} K_{n-1}^{-1} \frac{dK_{n-1}}{du} \Delta u_n^2 + O(\Delta u^3) \tag{6}$$

where u_N is the correct total displacement after N load increments and u_N^* is the displacement obtained by the incremental approach. It is noted that the error is $O(\Delta u^2)$ in the displacement increment. When the residual load correction is included we find that

$$u_N - u_N^c = -\frac{1}{2} K_{N-1}^{-1} \sum_{n=2}^{N/2+1} \frac{dK_l}{du} \Delta u_k^2 + O(\Delta u^3) \tag{7}$$

where for N even $l = m - 1$, $k = m$, and for N odd $l = m$, $k = m + 1$, $m = 2[(2n - 1)/2]$, and fractions are discarded in the computation of the indices. Equation (7) may be described by stating that for even N only the terms involving even displacement increments remain in the series and likewise for odd N and odd displacement increments. In comparing (6) and (7) we see that the inclusion of the residual load correction reduces the number of terms in the series by half. One could state that, approximately speaking, the error is halved in the corrected equations except for the fact that the stiffness quantity is inside the summation sign in (6). The assumption of a slowly varying stiffness K in (6) means the neglect of errors caused by the inherited error terms.

In the dynamic case, the expressions for the discretization errors of the uncorrected and corrected equations at time $N \Delta t$ are, respectively,

$$u_N - u_N^* = -\frac{\Delta t^2}{4} M^{-1} \left\{ \sum_{n-2}^{N} R_{n-1} \frac{dK_{n-1}}{du} \Delta u_n^2 + E^* \right\} + O(\Delta u^3) + O(\Delta t)^4, \tag{8}$$

and

$$u_N - u_N^c = -\frac{\Delta t^2}{4} M^{-1} \left\{ R_{N-1} \sum_{n=2}^{N/2+1} \frac{dK_l}{du} \Delta u_k^2 + E^c \right\} + O(\Delta u^3) + O(\Delta t^4), \tag{9}$$

where $R_n = [1 - (\Delta t^2/2)M^{-1}K_n]$. In Eqs. (8) and (9) E^* and E^c are the truncation errors inherited from the inertia terms and, for the current integration scheme in time, do not appear to be expressible in a general form. However, they are of the same order as the first terms in the brackets in (8) and (9), and it is interesting to speculate that a similar reduction in the error occurs in E^c as compared with E^*.

It has been demonstrated by Haisler et al [23] that the static solutions given by the corrected and uncorrected equations tend to converge as the number of load increments in the uncorrected case are increased. A particular example given by Haisler et al. [23] is a spherical shell cap under a point load at the apex where the uncorrected solution converged using an increment one eighth that required for convergence in the corrected solution. Judging from (8) and (9), one would expect that the convergence rate in the dynamic case would be more rapid both for the corrected and uncorrected solutions, considering the presence of the factor Δt^2 and the fact that the truncation errors for the static and dynamic solutions are approximately of the same order. In the sample problems given later it is shown for the example of a beam under a half sine-wave impulse over the span, that the uncorrected solution converges rapidly with the variation in the time increment. On the other hand, the corrected solution changes very little over a range of time increments. It appears that with the reduced truncation error of the corrected equations, the effect of Δt^2 on the convergence of the solution is diminished.

The advantage in using the corrected dynamic equations is that one can obtain practically convergent results with large time increments. In the numerical examples given later it is shown that convergent solutions to dynamic problems using the corrected incremental equations may be obtained using time increments an order of magnitude greater than those used by other investigators. The other solutions were obtained by using the Houbolt scheme and the total form of the finite-element equations so the comparisons are direct. This fact has important consequences in terms of doing nonlinear problems economically.

6. Computer Program

The system of equations for the dynamic, elastic–plastic analysis with large displacement has been incorporated in a pilot program. A program previously developed at Brown University [2] was used as the basis. Figure 1 gives a flow diagram for the program. The various stiffness and the mass matrices are assembled by the program from stored information, which enables certain element types to be generated. Presently, there are four such elements available in the library, two for beams and two for axisymmetric shells. Controls have been introduced to avoid the generation of the stiffness matrix at every time step and thus uncouple the physical modeling from the small time steps required for numerical accuracy.

7. Case Studies

In this section examples have been selected from [4] in order (a) to make comparisons with results in the literature, (b) to investigate the limits of

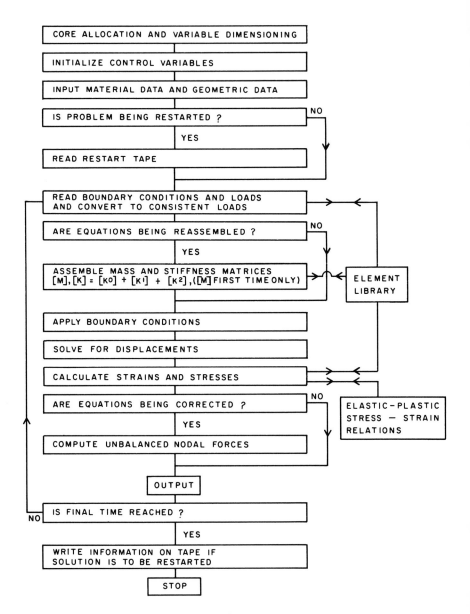

Fig. 1. Flow chart for computer program.

numerical approximations in terms of the frequency of reassembly and residual load correction, and (c) to observe the effect of geometric imperfections in a dynamically loaded sphere. The one-dimensional element used in the following examples is of the isoparametric type and has a rapid rate of convergence even for small numbers of elements. This is due to the fact that it can represent exactly all the rigid body modes of the interpolated surface which is arbitrarily close to the actual structural shape. It has proved to be a very accurate and economic element for use in analyzing dynamic problems. All the current results were obtained on a CDC 6600 using a 60-bit word. The effects of rotary inertia have been included in the equations although the response is not changed noticeably by this term.

7.1. SHALLOW SPHERICAL CAP UNDER A STEP PRESSURE LOAD

A linear example is illustrated in Fig. 2 for the case of a shallow spherical cap under a constant dynamic pressure. The solution using the finite-element method was first given by Klein and Sylvester [6] and later by Stricklin *et al.* [10]. It was noted from Fig. 2 that all three solutions agree, even though different elements and different numbers of elements were used in each case.

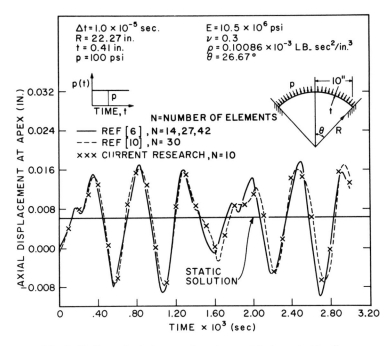

Fig. 2. Shallow spherical cap under axisymmetric dynamical loading.

The current values shown and those from [10] agree more closely in that both used curve elements, whereas Klein and Sylvester approximated the shell by straight segments. A solution [7] based on modal analysis also follows the current results closely.

7.2. Nonlinear Elastic Analysis of a Simply Supported Beam under a Half-Sine Initial Velocity Distribution

The problem illustrated in Fig. 3a was analyzed in order to check the elastic large displacement portion of the method. This problem has been treated theoretically in Eringen [28] and Chu and Herrman [29] where the differential equations of motion for a beam under an initial displacement condition were solved using perturbation techniques. In Eringen [28] more general equations were written which included the effects of large strain, rotary inertia, and transverse shear, whereas Chu and Herrman [29] neglect

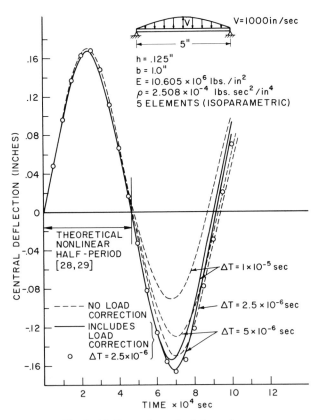

Fig. 3. (a) Comparison of solution schemes.

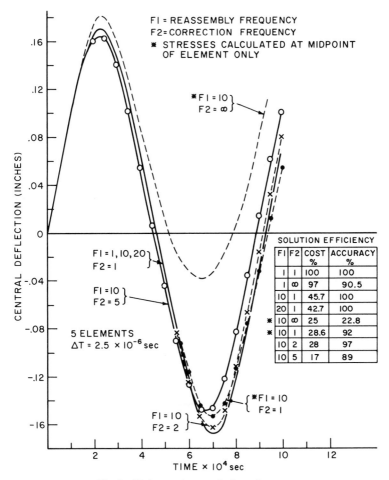

FI = REASSEMBLY FREQUENCY
F2 = CORRECTION FREQUENCY
* STRESSES CALCULATED AT MIDPOINT OF ELEMENT ONLY

*FI = 10⎱
F2 = ∞ ⎰

SOLUTION EFFICIENCY

FI	F2	COST %	ACCURACY %
I	I	100	100
I	∞	97	90.5
10	I	45.7	100
20	I	42.7	100
*10	∞	25	22.8
*10	I	28.6	92
10	2	28	97
10	5	17	89

FI = 1,10,20⎱
F2 = 1 ⎰

FI = 10⎱
F2 = 5⎰

5 ELEMENTS
ΔT = 2.5 × 10⁻⁶ sec

⎱*FI = 10
⎰ F2 = 1

FI = 10⎱
F2 = 2⎰

CENTRAL DEFLECTION (INCHES)

TIME × 10⁴ sec

Fig. 3. (b) Approximate solution schemes.

these effects and solve simpler equations, which consider large rotations but small strains. The final solutions in terms of the variation of the ratio of the nonlinear to the linear period of vibration of the beam with a perturbation parameter are identical for both approaches.

A finite-difference method is used by Krieg and Key [30] to solve the same problem with the initial velocity distribution given in Fig. 3a. The numerical values computed for deflections and generalized forces were found to agree with values found from expressions given in Chu and Herrman [29]. In this case a time step of 1 μsec was used and the beam was divided into 20 meshes.

Since rather exact results are known for this problem, it was chosen as an example for testing the various solution schemes available with the current formulation of the finite-element method. Five isoparametric elements were used, and the most exact finite-element solution is shown by the small circles in Fig. 3a. The equations were reassembled at each time increment and stresses were computed at three integration points per element. The period of vibration agrees with the theoretical result as shown in Fig. 3a and the value for the maximum displacement agrees with that given by Krieg and Key [30].

Shown in Fig. 3a are solutions that were obtained using the corrected and uncorrected equations for a series of time increments. In all cases, the equations were reassembled and corrected at each increment as required. As mentioned previously, the uncorrected equations converge rapidly, whereas the solutions for the corrected cases change very little with the change in time step. A feature of the dynamic behavior is the convergence of all solutions over the first half-period; in static nonlinear solutions the uncorrected solution diverges rapidly from the corrected solution for reasonably sized load increments. This phenomenon bears out the conclusions of the error analysis which indicated the more rapid convergence of the dynamic equations. If we assume that the corrected solution has converged at $\Delta t = 2.5 \times 10^{-6}$ sec we note that the uncorrected solution for the same time increment has converged over practically all the response but still only compares with the corrected solution for $\Delta t = 1 \times 10^{-5}$ sec.

A procedure that results in a great economy of solution time is to reassemble the equations at certain multiples of the basic time step. For the example under consideration, as illustrated in Fig. 3b, the reassembly operation was performed at every increment, every tenth increment, and every twentieth increment with no apparent change in the response. However, the load correction was applied at each step. The effect of varying the latter parameter was also investigated and as shown in Fig. 3b the solution is much more sensitive to changes in correction frequency as compared with changes in reassembly frequency.

An approximation that is used frequently in the static analysis of structures is to consider the stress at the center of the element as being the average stress in the element and to use this value along in computing the stiffness matrix. The results are shown in Fig. 3b, where it is noted that with the load correction, the response compares very favorably with the most exact solution shown. This procedure is very conomical with respect to storage requirements and solution times, and could be usefully employed as an economic first run designed to investigate the probable response of the structure. It would appear unwise, however, to use this approximation without the load correction as evidenced by the response shown in Fig. 3b.

An attempt was made to rate the solution procedures just described with respect to cost in computer time and accuracy as compared with the solution which requires the equations to be corrected and reassembled at each step.

For the sake of comparison the accuracy is determined by the minimum displacement value as a percentage of the absolute minimum. The most efficient solution scheme, judging from the table in Fig. 3b, appears to be the case where the equations are reassembled every tenth increment and corrected every second increment. It will be demonstrated by other examples how effective and accurate this procedure may be.

7.3. Nonlinear Elastic Analysis of a Spherical Shell Cap under a Point Load at the Apex

In order to determine whether the solution scheme chosen as the most efficient for the previous problem is of general use, it was applied to the problem of an axisymmetric spherical shell cap under a concentrated load of infinite duration applied at the apex. The equations were assembled every tenth time increment and the correction term was applied as indicated in Fig. 4. It is noted that the form of the plot does not change but the maximum deflection begins to drop off with the decrease in frequency of application of the correction term. These results give an indication of how often the correction should be applied, but it may be deduced that for an accurate solution should be applied at every step.

The sensitivity of the solution to the correction term may be explained as follows. The correction is computed as an unbalanced force at the end of an increment. Theoretically a correction to the current displacement should be computed by using this force and the equations applying at the beginning of the increment. In the present formulation because of the computational difficulties and expense involved in doing this displacement correction, the unbalanced force is added on as a load over the next increment. In effect, this means that the correction will be computed using the current equations rather than the equations at the beginning of the increment. We may assume that the equations governing the structure will not change appreciably over one time increment and the method just described for applying the correction will be suitable; but it becomes less accurate as the number of intervals between corrections is increased.

Also shown on Fig. 4 is the effect of doubling the time increment where the equations are still reassembled every ten increments and corrected at every increment. Finally a plot is shown of a result obtained with the parameters just described but using the element of Khojasteh-Bakht [31]. It is noted that this element has one less degree of freedom per node but in terms of computational efficiency the isoparametric element takes only approximately 15 % more time for an equal number of elements. The element in Khojasteh-Bakht [31] gave good comparisons with the current results when 30 elements were used.

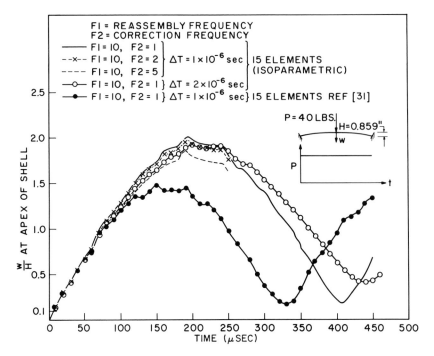

Fig. 4. Effect of various parameters on solution convergence.

The solution arrived at by reassembling every ten steps and correcting every step with a time increment of 1×10^{-6} sec is compared with the results of a finite-element solution given in Strickland *et al.* [10]. This solution was obtained using the total form of the equations which necessitates extrapolating for displacements over an increment and computing the nonlinear portion of the equations as pseudoforces on the right-hand side of the equations. The equations were solved using iteration on Newton–Raphson for the more nonlinear problems and the time integration was performed using the Houbolt scheme.

As shown in Fig. 5 the present solution using 15 elements agrees substantially with the results of Strickland *et al.* [10] for thirty elements. This agreement is interesting in that the approaches used to solve the problem, such as the element type, the time increment (the present increment is eight times that used by Strickland *et al.* [10]), the assembly of the nonlinear terms, and the solution scheme (incremental versus total equations), are quite different in each case.

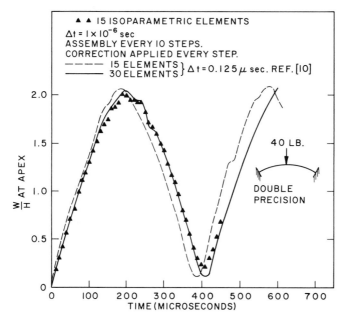

Fig. 5. Solution convergence.

7.4. Elastic–Plastic Beam under a Uniform Initial Velocity over a Portion of the Span

The essentials of the example are given in Fig. 6 where a beam of span 10 in. is given an impulse of 2172 in. $\sec^{-1}$ over the central 2 in. of the span. The material of the beam is elastic–plastic work-hardening and is modeled by a piecewise linear stress–strain curve as shown in Fig. 6. The results for the finite element case were obtained using ten isoparametric elements with three integration points per element. The reassembly of the equations and the correction term were computed every fifth and every time increment respectively, and a time increment of 2.5×10^{-6} sec was used.

In Fig. 6 the results are compared with a finite-difference solution and with experimental data given in Kreig et al [32]. The finite-difference solution was computed using 20 mesh points on the beam and a time increment of 0.33 μsec, which is 80% of the step size necessary for the stability of the solution. The finite-element solution using Houbolt's scheme was unstable only for a time increment greater than 2.5 μsec. A check was made on the solution by running the problem with a time equal to 1 μsec, reassembling every ten steps, and correcting every step, but no significant change occurred.

Since the elastic large-displacement results already presented are in perfect agreement with other finite-difference and finite-element results, the slight

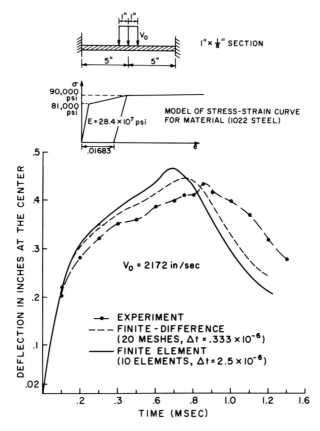

Fig. 6. Comparison of finite-element solution with finite difference solution for beam problem.

discrepancy between the numerical results in Fig. 6 must be attributed to the elastic–plastic portion of the problem. A further check was made on the finite element solution by using five Gaussian integration points per element as opposed to the three used originally but no change was reflected in the result, and so it is assumed that the solution has converged. The differences between the numerical solutions may be encountered for by the fact that in the current solution scheme the elastic–plastic constitutive law is weighted for stress states going from elastic to plastic over an increment. The discrepancy between both numerical results and the experimental values may be accounted for by the fact that the effect of strain rate sensitivity on the yield stress of the steel used in the experiments is neglected in the numerical solutions.

Results obtained by various investigators on the variation of the yield stress of structural materials with increase in strain rate are reviewed in [19] and it is mentioned there that for the steel used in the experiments in question the increase in yield is about 20 % at a strain rate of 100 sec^{-1}. The inclusion in the numerical analysis of this factor on the yield stress would tend to improve the agreement between theory and experiment.

7.5. Elastic–Plastic Buckling of an Imperfect Sphere under a Uniform Constant External Pressure

The geometry of the sphere is shown in Fig. 7 where the imperfection is given as a flat section of the shell of radius R_{imp}, which is taken to be the mean

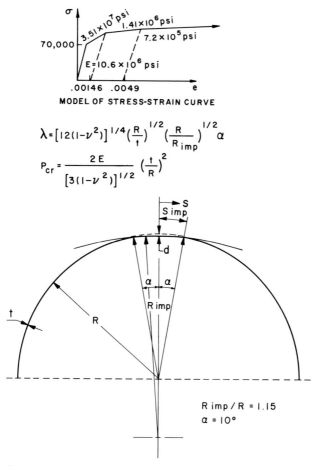

$$\lambda = \left[12(1-\nu^2)\right]^{1/4}\left(\frac{R}{t}\right)^{1/2}\left(\frac{R}{R_{imp}}\right)^{1/2}\alpha$$

$$P_{cr} = \frac{2E}{\left[3(1-\nu^2)\right]^{1/2}}\left(\frac{t}{R}\right)^2$$

Fig. 7. Externally pressurized imperfect hemisphere and material constants.

radius of the oblate portion of the sphere. For the more simple elements, where only displacement continuity is required at the nodes, as in Khajasteh-Bakht [31] and Klein and Sylvester [6], the junction of the imperfect portion with the rest of the shell presents no special problem. But for the higher-order isoparametric element, the displacements and their first derivatives are incompatible. The combination of these higher-order elements across the junction is achieved by applying a constraint relating the displacements at two hypothetical nodes, in the manner of Hibbitt and Marcal [33]. Preliminary analysis of a linear hemispherical shell suggested that a solution with 20 elements and a time increment of 0.1 μsec would be sufficient to cover the response.

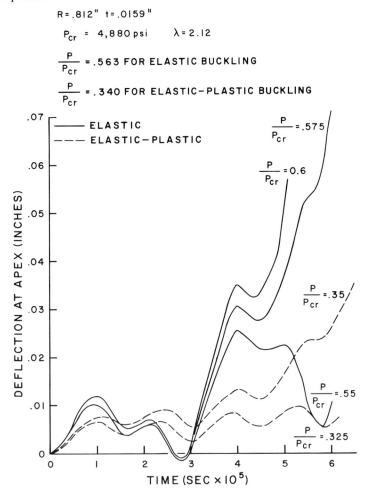

Fig. 8. Deflection profiles for elastic and elastic–plastic buckling of imperfect sphere.

Inserted in Fig. 8 are the expressions for the geometric parameter λ and the critical static buckling pressure P_{crit} for the sphere. Also shown is the model of the stress–strain behavior of the shell material which corresponds to that of an aluminum alloy (7075-T6).

The shell was analyzed both elastically and elastic–plastically and the results for the parameter $\lambda = 2.12$ are shown in Fig. 8, where the deflection at the apex is plotted against time. It is noted that there is a great difference between the elastic and the elastic–plastic values for the ratio of the pressure required to cause dynamic buckling to P_{crit} in the imperfect shell. The form of the response history also is quite different for each material idealization. The structure is deemed to have buckled when the deflection profile increases drastically for a small increment in pressure. As an example of this, consider the divergence between the deflection profiles for the elastic response to the nondimensional pressure values given by $P/P_{\text{crit}} = 0.575$ and $P/P_{\text{crit}} = 0.55$. Nearly all of the curves point to the fact that the response must be traced over a number of oscillations before it is decided whether buckling will occur or not. This is especially true for the elastic example with $P/P_{\text{crit}} = 0.55$ which appears to be just below the buckling pressure threshold.

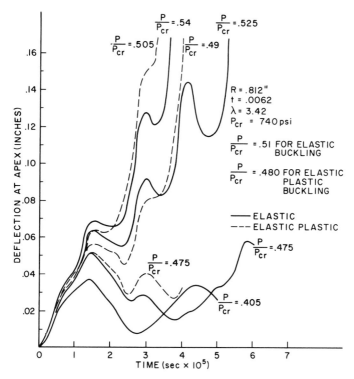

Fig. 9. Deflection profiles for elastic and elastic–plastic buckling of imperfect sphere.

To illustrate the different response characteristics of this shell with the variation of the parameter λ, results are shown in Fig. 9 for $\lambda = 3.42$. In this case the response is mainly elastic and this is reflected in the proximity of the elastic and elastic–plastic buckling values for P/P_{crit} given in Fig. 9. However, the overall response history is very different from that of Fig. 8 in that buckling occurs after the first maximum and not after a number of oscillations.

The problem under discussion was first solved by Bushnell [34] for the static elastic case, and also by Marcal [2] for the elastic–plastic case. These results are now compared with the present elastic and elastic–plastic dynamic results in Fig. 10. Values of the pressure parameters P/P_{crit} which initiate

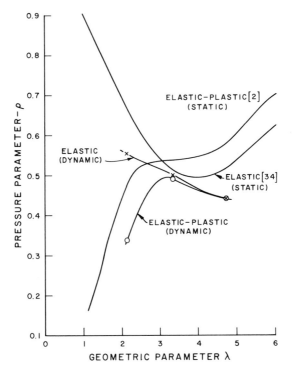

Fig. 10. Buckling pressures for oblate shells; $R_{imp}/R = 1.15$.

buckling are plotted against the geometric parameter λ. The main feature of the results is the manner in which the dynamic elastic–plastic curve resembles the form of the corresponding static result from Marcel [2]. In both the static and dynamic cases, for $\lambda < 3$, the buckling pressure is governed very strongly by the plastic flow of the material. For these values of λ any analysis neglecting nonlinear material behavior would be meaningless. This problem is a good example of a strong interaction between large displacements and nonlinear material behavior.

As observed in Figs. 8 and 9, two different manners of collapse take place for different shell parameters, namely collapse on the first maximum displacement and collapse on subsequent maxima. It is noted from Fig. 10 that, compared to the penalty due to the imperfection, dynamic loading does not impose too large a penalty.

In computing the results just described, the equations were reassembled every ten increments and corrected every second increment. It was felt that these values were sufficiently accurate in that the time increment used of 0.1 μsec was very small, and in that the final buckling value taken was computed as the average of two close values, one of which caused buckling, while the other did not.

8. Discussion and Conclusions

A numerical procedure has been developed whereby the finite-element method may be used to analyze large elastic–plastic deformations of structures under a variety of dynamic loadings. It has been determined that the standard incremental equations are not totally sufficient in the dynamic case and that a correction should be applied to the system in the form of an equilibrium check. The error in the simple incremental method is shown to be of the order of the displacement increment squared coupled with a factor equal to the time increment squared. It is this factor which causes the incremental method to converge rapidly as the time increment is made smaller and smaller. On the other hand, it has been found that the corrected incremental equations give stable and accurate solutions with time increments approximately one order of magnitude greater than corresponding time steps used in finite-difference, and other finite-element work. In this case the reduction of the truncation error of the solution by the correction process favors the convergence of the solution, and makes the incremental approach competitive economically with other methods which require extrapolation, and therefore small time steps, in order to handle the nonlinear terms.

Many approximate procedures, which are often employed in static finite-element analysis, were tested with the purpose of gauging their suitability for use in dynamic work. It was shown for the example in Fig. 3b that computing the stress at one representative point only in the element gives results which agree substantially with the analysis which includes all stress points. This approximation saves time and storage in the computer, and is useful as a first step in an analysis, but it should be used only with the corrected form of the incremental equations. A point to remember about nonlinear analysis, such as demonstrated in the example given for the elastic and elastic–plastic buckling of a sphere, is that it is difficult to estimate the probable response in order to plan a numerical solution. An inexpensive way of investigating the problem, used frequently in this research as a preliminary step, is to reassemble the equations at very wide intervals and likewise with respect to applying the correction terms. Cases in point are the beam example in Fig. 3b and the shell example in Fig. 4, where the reassembly and correction operations were performed every tenth and every fifth increment, respectively, without any great change in the frequency or amplitude. Since most of the solution time is used in the updating procedure, the savings in execution time increase substantially with the frequency of the updating. A word of caution is inserted here to the effect that, while the reassembly of the equations may be performed at will, the correction terms should be computed more frequently as pointed out in the example of the spherical cap under a point load at the apex. This is especially true for the elastic–plastic case where the solution tends to be unstable for too low a frequency of the application of the correction terms. As an example, the solution of the elastic–plastic buckling of the sphere was unstable when the correction was applied every fifth increment, but gave a stable solution when the computation was made every second increment. This is to be expected in a solution where the response varies widely and serves as a built-in indicator of such a response. An added advantage of the incremental method, as opposed to using the total form of the equations and extrapolating for the nonlinear forces, is that the above approximate and inexpensive solutions may be obtained.

A useful development of this research is the general-purpose computer program from which the results described here have been obtained. This forms a versatile research tool due to its modular and flexible construction. Especially important is that any element may be used in this program, and conversely any improvements made in the solution scheme in the body of the program apply to all elements. The capability for node-tying is also available and would be suitable for problems involving combined element analysis where linear constraints must be imposed on the interfaces between the different element types. The solution scheme used, the incremental method, also allows one to add other constitutive relations to the program directly. In using the total form of the equations it is not clear how one would separate the pseudoforces due to material and geometric behavior even for the simplest nonlinear constitutive relation.

In summary, some progress has been made in the solution of nonlinear dynamic problems. Useful approximations have been found to yield economic solution times. The effect of geometric imperfections on the dynamic buckling of an elastic–plastic shell has been observed. Future work will be concerned with extending the analysis to other structure types and to the inclusion of viscoplastic behavior.

Acknowledgments

This project was sponsored by the Structural Mechanics Branch, Office of Naval Research, under Contract No. N00014-67-A-0191-0007. It is a continuation of work initiated with the Department of Structural Mechanics, Naval Ship Research and Development Center under Contract No. N00014-67-A-0191-0008. The authors are grateful for the financial support provided.

References

1. Hibbitt, H. D., Levy, N. J., and Marcal, P. V., "On General Purpose Programs for Nonlinear Finite Element Analysis," *Proc. ASME Sem. Gen. Purpose Finite Element Programs* ASME Winter Ann. Meeting, New York (December 1970).
2. Marcal, P. V., "Finite Element Analysis of Combined Problems of Nonlinear Material and Geometric Behavior," *Proc. ASME Joint Comput. Conf. Comput. Approach Appl. Mech.* Chicago (1969).
3. Oden, J. T., "General Theory of Finite Elements," *Int. J. Numer. Methods Eng.* Pts. I and II **1**, 205, 247 (1969).
4. McNamara, J. F., "Incremental Stiffness Method for Finite Element Analysis of the Nonlinear Dynamic Problem," Ph.D. Thesis, submitted August (1971).
5. Turner, M. J., Clough, R. W., Martin, H. C., and Topp, L. J., "Stiffness and Deflection Analysis of Complex Structures," *J. Aerospace Sci.* **23**, 805–823 (1956).
6. Klein, S., and Sylvester, R. J., "The Linear Elastic Dynamic Analysis of Shells of Revolution by the Matrix Displacement Method," Air Force Flight Dynamics Lab. TR-66-80, pp. 299–329 (1966).
7. Popov, P. V., and Chow, Y. C., Addendum to the above paper.
8. Clough, R. W., "Analysis of Structural Vibrations and Dynamic Response," Japan–U.S. Seminar on Matrix Methods of Structural Analysis and Design, Tokyo, Japan (August 1969).
9. Turner, M. J., Dill, E. H., Martin, H. C., and Melosh, R. J., "Large Deflection Analysis of Complex Structures Subjected to Heating and External Loads," *J. Aerospace Sci.* **27**, 97–106 (1960).
10. Stricklin, J. A., Martinez, J. E., Tillerson, J. R., Hong, J. H., and Haisler, W. E., "Nonlinear Dynamic Analysis of Shells of Revolution by Matrix Displacement Method," Rep. 69-77, Texas Eng. Experiment Station, Texas A & M Univ. (February 1970).
11. Klein, S., "The Nonlinear Dynamic Analysis of Shells of Revolution with Asymmetric Properties by the Finite Element Method," The Aerospace Corp. San Bernardino, California ATR-71(S-9990)-7 (April 1971).
12. Hill, R., *Mathematical Theory of Plasticity.* Oxford Univ. Press, London and New York, 1950.
13. Chang, G. C., and Gilbert, F. S., "Finite Element Approach to Wave Motions in Elastic–Plastic Continua," *Proc. AIAA Aerospace Sci. Meeting, 6th,* paper 68-145, New York (January 1968).
14. Costantino, C. J., "Two Dimensional Wave Propagation Through Nonlinear Media," *J. Comput. Phys.* **4**, No. 2, 147–170 (1969).
15. Fulton, R. E., "Numerical Analysis of Shells of Revolution," paper presented at the *IUTAM Symp. High-Speed Comput. Elastic Struct.* Liege, Belgium (August 29–September 6, 1970).

16. Witmer, E. A., Balmer, H. A., Leech, J. W., and Pian, T. H. H., "Large Dynamic Deformations of Beams, Rings, Plates and Shells," *AIAA J.* **1**, No. 8, 1848–1857 (1963).
17. Leech, J. W., Witmer, E. A., and Pian, T. H. H., "A Numerical Calculation Technique for Large Elastic–Plastic Dynamically-Induced Deformations of General Thin Shells," *AIAA J.* **6**, No. 1, 2352–2359 (1968).
18. Witmer, E. A., Clark, E. N., and Balmer, H. A., "Experimental and Theoretical Studies of Explosive-Induced Large Dynamic and Permanent Deformations of Simple Structures," *Exp. Mech.* 56–66 (February 1967).
19. Symonds, P. S., "Survey of Methods of Analysis for Plastic Deformation of Structures under Dynamic Loading," Brown Univ. Rep., BU/NSRDS/1-67 (June 1967).
20. Martin, J. B., and Symonds, P. S., "Mode Approximations for Impulsively Loaded Rigid-Plastic Structures," *Proc. Amer. Soc. Civil Eng.* **92**, 43–66 (1966).
21. Martin, J. B., "Impulsive Loading Theorems for Rigid-Plastic Continua," *Proc. Amer. Soc. Civil Eng.* **90**, 27–42 (1964).
22. Martin, J. B., "A Displacement Bound Principle for Inelastic Continua Subjected to Certain Classes of Dynamic Loading," *J. Appl. Mech.* **32**, No. 1, 1–6 (1965).
23. Haisler, W. E., Stricklin, J. A., and Stebbins, F. J., "Development and Evaluation of Solution Procedures for Geometrically Nonlinear Structural Analysis by the Direct Stiffness Method," paper presented at the *AIAA/ASME 12th Struct., Struct. Dynamics Mater. Conf.* Anaheim, California, April 19–21 (1971).
24. Marcal, P. V., and King, I. P., "Elastic-Plastic Analysis of Two-Dimensional Stress Systems by the Finite Element Method," *Int. J. Mech. Sci.* **9**, No. 3, 143–155 (1967).
25. Hibbitt, H. D., Marcal, P. V., and Rice, J. R., "A Finite Element Formulation for Problems of Large Strain and Large Displacement," *Int. J. Solids Struct.* **6**, 1069–1086 (1970).
26. Hofmeister, L. D., Greenbaum, G. A., and Evensen, D. A., "Large Strain, Elasto-Plastic Finite Element Analysis," *AIAA J.* **9**, No. 7, 1248–1254 (1971).
27. Houbolt, J. C., "A Recurrence Matrix Solution for the Dynamic Response of Elastic Aircraft," *J. Aerospace Sci.* **17**, 540–550 (1950).
28. Eringen, C. A., "On the Nonlinear Vibration of Elastic Bars," *Quantum Appl. Math.* **10**, No. 4 (1952).
29. Chu, Hu-Nan, and Herrman, G., "Influence of Large Amplitudes on Free Flexural Vibrations of Rectangular Elastic Plates," *J. Appl. Mech.* Paper No. 56-APM-27 (December 1956).
30. Krieg, R. D., and Key, S. W., "Univalve, A Computer Code for Analyzing Dynamic Large Deflection Elastic-Plastic Response of Beams and Rings," SC-RR-66-2682, Sandia Lab., Albuquerque, New Mexico (January 1968).
31. Khojasteh-Bakht, M., "Analysis of Elastic–Plastic Shells of Revolution under Axisymmetric Loading by the Finite Element Method," Ph.D. Dissertation, Univ. of California at Berkeley, SESM 67-68 (April 1969).
32. Kreig, R. D., Duffey, T. A., and Key, S. W., "The Large Deflection Elastic-Plastic Response of Impulsively Loaded Beam: A Comparison Between Computations and Experiment," SC-RR-88-226, Sandia Lab., Albuquerque, New Mexico (July 1968).
33. Hibbitt, H. D., and Marcal, P. V., "Hybrid Finite Element Analysis with Particular Reference to Axisymmetric Structures," *Proc. AIAA 8th Aerospace Sci. Meeting*, New York, January 19–21 (1970).
34. Bushnell, D., "Nonlinear Axisymmetric Behavior of Shells of Revolution," *AIAA J.* **5**, No. 3, 432–439 (1967).

The Lumped-Parameter or Bar–Node Model Approach to Thin-Shell Analysis

W. C. Schnobrich and D. A. Pecknold

DEPARTMENT OF CIVIL ENGINEERING
UNIVERSITY OF ILLINOIS
URBANA, ILLINOIS

1. Introduction

Under current conditions, the stress analyst, when faced with many plate and shell problems, must resort to some form of numerical method, usually a discrete point or element procedure, to obtain a solution appropriate to the conditions of the problem he is trying to solve. As digital computers became available, allowing the practical implementation of such procedures the first method to be applied was the finite-difference concept [19, 27, 28]. The method was viewed in its mathematical sense as a procedure for transforming the governing differential equations over to an algebraic set expressed in terms of the unknowns at selected points on the surface of the structure.

In these early applications, a network of grid lines was selected and placed on the structure. Then all pertinent unknowns were defined at each node point of this grid. The governing differential equations were expressed at each node point in terms of the finite-difference equivalents written using the nodal unknowns. This method of applying finite-difference procedures met with mixed success. A good many shell problems lead to finite-difference solutions of questionable and sometimes even meaningless results.

In the meantime the finite-element procedure started—and from a much different point of view. Using engineering judgment and physical reasoning, a model was developed which related the nodal forces to the nodal displacements. This accomplished the same conversion of the problem as the finite-difference scheme, that is over to a system of algebraic equations which were viewed as the equilibrium equations for the nodal points.

With the development and refinement of the finite-element method, its relationship (kinship) to the variational procedures was recognized. This placed the method on a firmer mathematical foundation and led to a rapid development of the method within the engineering profession [7]. The finite-element method now forms the backbone of many powerful computer programs capable of solving many complex problems.

While the finite-element procedure was being refined and fit into a mathematical cast, the finite-difference procedure took an opposite trend, back to physical analogs and engineering judgment in order to circumvent the accuracy problems that were encountered when the solution process started with some forms of the governing equations. The physical analogs developed in essentially three forms. The first two approaches represented the framework analogy popularized by Hrennikoff and the rigid bar deformable node system invented by Newmark. Their application to shell problems was first dicussed in the 1962 World Conference on Shells in San Fransisco. The Hrennikoff approach which is perhaps closer to finite elements than finite differences has the disadvantage that the computed properties of some of the bars making up the framework have negative stiffnesses. The convergence characteristics of these early Hrennikoff models were very slow.

The bar-node or lumped-parameter model represents an analog to the finite-difference equations. Since the model is developed based on physical concepts it yields a system of equations which are consistent; that is, whether the equilibrium equations or the variational method is used the procedure yields the same equations. In the early application of the model to cylindrical shells it led naturally to the manner of defining displacements and stress resultants at different locations on the shell. The equations developed are the same as those now termed consistent, modified, or half interval.

The development of the governing equations for the lumped-parameter model will be illustrated for the particular case of an elastic isotropic shallow shell. However, the analog model concept is equally applicable to other shell theories, and produces results of superior accuracy for these as well.

The derivation is based on a discretized potential energy formulation. Johnson [20], Havner and Stanton [16], Griffin and Kellogg [14], Noor [25], and many others have utilized this approach in applying finite-difference procedures to various stress analysis problems. In these finite-difference procedures, displacement boundary conditions are satisfied identically, but the stress boundary conditions are satisfied as a part of the minimization procedure and thus do not match the prescribed values identically. This situation also prevails in the conventional finite-element displacement models. In some sense this may be said to represent an advantage since no explicit consideration of the stress boundary conditions is necessary and thus the formulation is simplified. However in some shell problems, the convergence of the stresses to the prescribed values at the boundaries is quite slow. In these cases it seems preferable to satisfy the natural boundary conditions exactly, even at the expense of increased complexity. The analog model procedure is easily adaptable to either option. The formulation is of course slightly more involved when the natural boundary conditions are satisfied.

2. Shallow Shell Equations

The parametric representation of the middle surface of the shell is given by $z(x, y)$ where x, y, z are the Cartesian coordinates of a point on the middle surface. The curvatures are denoted by

$$r \equiv z_{,xx}, \qquad s \equiv z_{,xy}, \qquad t \equiv z_{,yy} \tag{1}$$

Commas followed by a subscript denote partial derivatives.

The shallow shell equations are listed below for later comparison with those of the analogue model.

2.1. STRAIN–DISPLACEMENT RELATIONS

The membrane strain–displacement relations are

$$\varepsilon_x = u_{,x} - rw, \qquad \varepsilon_y = v_{,y} - tw, \qquad 2\varepsilon_{xy} = u_{,y} + v_{,x} - 2sw \tag{2}$$

where u, v, w are the tangential and normal components of displacement. The bending strain–displacement relations are

$$K_x = -w_{,xx}, \qquad K_y = -w_{,yy}, \qquad 2K_{xy} = -2w_{,xy} \tag{3}$$

2.2. STRESS–STRAIN RELATIONS

The stress–strain relations for an isotropic elastic shell are

$$
\begin{Bmatrix} N_x \\ N_y \\ N_{xy} \end{Bmatrix} = N \cdot \begin{bmatrix} 1 & v & 0 \\ v & 1 & 0 \\ 0 & 0 & (1-v)/2 \end{bmatrix} \begin{Bmatrix} \varepsilon_x \\ \varepsilon_y \\ 2\varepsilon_{xy} \end{Bmatrix}
$$
$$
\begin{Bmatrix} M_x \\ M_y \\ M_{xy} \end{Bmatrix} = D \cdot \begin{bmatrix} 1 & v & 0 \\ v & 1 & 0 \\ 0 & 0 & (1-v)/2 \end{bmatrix} \begin{Bmatrix} K_x \\ K_y \\ 2K_{xy} \end{Bmatrix} \tag{4}
$$

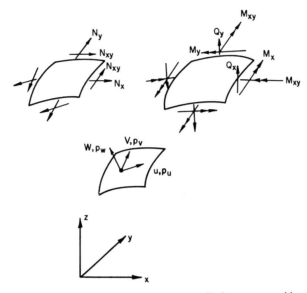

Fig. 1. Sign convention for forces, moments, displacements, and loads.

where

$$N \equiv \frac{Eh}{1 - v^2} \quad \text{and} \quad D \equiv \frac{Eh^3}{12(1 - v^2)}$$

are the membrane and bending stiffnesses, respectively. The positive senses of the stress results are shown in Fig. 1.

2.3. STRAIN ENERGY

The strain energy of the shell is conveniently separated into two parts: U^m, the membrane strain energy and U^b, the bending strain energy. These are given by

$$U^m = \frac{N}{2} \int\int \lfloor \varepsilon_x \varepsilon_y 2\varepsilon_{xy} \rceil \begin{bmatrix} 1 & v & 0 \\ v & 1 & 0 \\ 0 & 0 & (1-v)/2 \end{bmatrix} \begin{Bmatrix} \varepsilon_x \\ \varepsilon_y \\ 2\varepsilon_{xy} \end{Bmatrix} dx \, dy$$

$$\equiv \frac{1}{2} \int\int \varepsilon^{m,T} E^m \varepsilon^m \, dx \, dy \tag{5}$$

$$U^b = \frac{D}{2} \int\int \lfloor K_x K_y 2K_{xy} \rceil \begin{bmatrix} 1 & v & 0 \\ v & 1 & 0 \\ 0 & 0 & (1-v)/2 \end{bmatrix} \begin{Bmatrix} K_x \\ K_y \\ 2K_{xy} \end{Bmatrix} dx \, dy$$

$$\equiv \frac{1}{2} \int\int \varepsilon^{b,T} E^b \varepsilon^b \, dx \, dy$$

The potential energy of the external distributed loading is

$$\Omega^s = - \int \int (p_u u + p_v v + p_w w) \, dx \, dy \tag{6}$$

The potential energy of the boundary tractions is

$$\Omega^b = - \int_{s_t} (N_n u_n + N_t u_t + R_n w - M_n w_{,n}) \, ds \tag{7}$$

where s_t is the portion of the boundary on which stresses are prescribed. N_n and N_t are the normal and tangential components of the membrane force, R_n is the Kirchhoff edge shear, M_n is the normal moment, u_n and u_t are the normal and tangential components of displacement and $w_{,n}$ is the normal slope.

The total potential energy of the shell is

$$V = U^m + U^b + \Omega^s + \Omega^b$$

The equilibrium equations in terms of displacements, obtained from setting $\delta V = 0$ are

$$u_{,xx} + [(1 + v)/2]v_{,xy} + [(1 - v)/2]u_{,yy} - (r + vt)w_{,x}$$
$$\quad - s(1 - v)w_{,y} + P_u/N = 0$$
$$v_{,yy} + [(1 + v)/2]u_{,xy} + [(1 - v)/2]v_{,xx} - (t + vr)w_{,y}$$
$$\quad - s(1 - v)w_{,x} + P_v/N = 0 \tag{8}$$
$$w_{,xxxx} + 2w_{,xxyy} + w_{,yyyy} - (12/h^2)\{(r + vt)u_{,x} + (t + vr)v_{,y}$$
$$\quad + s(1 - v)(u_{,y} + v_{,x}) - [(r + t)^2 - 2(1 - v)(rt - s^2)]w\} - P_w/D = 0$$

3. Lumped-Parameter Model

The discrete model consists of a series of rigid bars and deformable nodes arranged as shown in Fig. 2. The material properties of the continuum are lumped at these deformable nodes. Tangential displacements u and v are defined at the intersections of the rigid bars, which can displace relative to each other in the direction of the normal. The normal displacements w are defined at the deformable nodes (Fig. 2). The external loads are defined at points where the corresponding displacements are defined, i.e., tangential loads are applied at the intersections of the rigid bars, while normal loads are applied at the deformable nodes. A typical "element" consisting of one deformable node and four rigid bars is shown in Fig. 3b. The notation used by Mohraz [23] is used in the development that follows. The deformation of

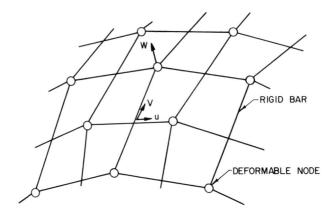

Fig. 2. Lumped parameter or bar-node model for a shell.

the element is described by the u and v displacements and the rotations of "joints" j_1 to j_4, and the normal displacement w of node m_1. The rotations are subsequently eliminated since they are not independent unknowns.

Other rigid, bar-deformable node arrangements suitable for various particular cases [24, 29] can be used, but the present scheme of Mohraz and Schnobrich [22] is the most general. Once the mesh layout has been chosen, finite-difference procedures could be employed directly, leading to so-called "consistent" or "modified" [26] finite-difference approximations.* The analog model concept provides the same equations and in addition aids in the treatment of the boundary conditions.

3.1. DEVELOPMENT OF STIFFNESS MATRICES FOR LUMPED-PARAMETER MODEL

For a typical interior element, the membrane strains at the deformable node are given by

$$\varepsilon_x = \frac{u_{j_4} - u_{j_2}}{L_x} - rw_{m_1}, \qquad \varepsilon_y = \frac{v_{j_3} - v_{j_1}}{L_y} - tw_{m_1}$$

$$2\varepsilon_{xy} = \frac{v_{j_4} - v_{j_2}}{L_x} + \frac{u_{j_3} - u_{j_1}}{L_y} - 2sw_{m_1} \tag{9}$$

or, in matrix form

$$\varepsilon^m = A^m u^m \tag{10}$$

---

* It should be noted that in a consistent, finite-difference formulation, the points of definition of the stress resultants are automatically fixed once the displacement locations are specified, and may not be chosen independently (Croll and Scrivener 8]).

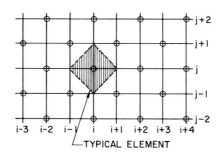

(a) MESH ARRANGEMENT

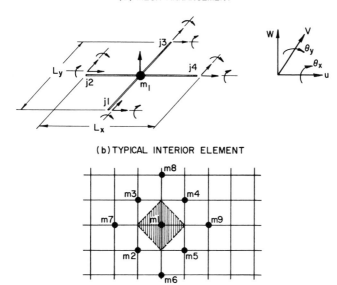

(b) TYPICAL INTERIOR ELEMENT

(c) NUMBERING SCHEME FOR BENDING DISPLACEMENT VECTOR

Fig. 3. Joint numbering system. (a) Mesh arrangement; (b) typical interior element; (c) numbering scheme for bending displacement vector.

where

$$u^m \equiv \{u_{j_1}, v_{j_1}, u_{j_2}, v_{j_2}, u_{j_3}, v_{j_3}, u_{j_4}, v_{j_4}, w_{m_1}\}_{9 \times 1}^T$$

and

$$A^m_{3 \times 9} = \begin{bmatrix} \cdot & \cdot & -L_x^{-1} & \cdot & \cdot & \cdot & L_x^{-1} & \cdot & -r \\ \cdot & -L_y^{-1} & \cdot & \cdot & \cdot & L_y^{-1} & \cdot & \cdot & -t \\ -L_y^{-1} & \cdot & \cdot & -L_x^{-1} & L_y^{-1} & \cdot & \cdot & L_x^{-1} & -2s \end{bmatrix}$$

Similarly, the bending strains at the deformable nodes are given by

$$K_x = \frac{\theta_{yj_4} - \theta_{yj_2}}{L_x}, \qquad K_y = \frac{\theta_{xj_3} - \theta_{xj_1}}{L_y}$$

$$2K_{xy} = \frac{\theta_{xj_4} - \theta_{xj_2}}{L_x} + \frac{\theta_{yj_3} - \theta_{yj_1}}{L_y} \tag{11}$$

or, in matrix form,

$$\varepsilon^b = A^b \theta \tag{12}$$

where

$$\theta \equiv \{\theta_{xj_1}, \theta_{yj_1}, \theta_{xj_2}, \theta_{yj_2}, \theta_{xj_3}, \theta_{yj_3}, \theta_{xj_4}, \theta_{yj_4}\}$$

and

$$A^b_{3 \times 8} = \begin{bmatrix} \cdot & \cdot & \cdot & -L_x^{-1} & \cdot & \cdot & \cdot & L_x^{-1} \\ -L_y^{-1} & \cdot & \cdot & \cdot & L_y^{-1} & \cdot & \cdot & \cdot \\ \cdot & -L_y^{-1} & -L_x^{-1} & \cdot & \cdot & L_y^{-1} & L_x^{-1} & \cdot \end{bmatrix}$$

The rotations are related to the normal displacements of adjacent deformable nodes. With reference to the numbering scheme for the deformable nodes shown in Fig. 3c, this relationship can be written in matrix form as

$$\theta = Cu^b \tag{13}$$

$$u^b \equiv \{w_{m_1}, w_{m_2}, w_{m_3}, w_{m_4}, w_{m_5}, w_{m_6}, w_{m_7}, w_{m_8}, w_{m_9}\}^T$$

and

$$C_{8 \times 9} = - \begin{bmatrix} L_y^{-1} & \cdot & \cdot & \cdot & \cdot & -L_y^{-1} & \cdot & \cdot & \cdot \\ \cdot & -L_x^{-1} & \cdot & \cdot & L_x^{-1} & \cdot & \cdot & \cdot & \cdot \\ \cdot & -L_y^{-1} & L_y^{-1} & \cdot & \cdot & \cdot & \cdot & \cdot & \cdot \\ L_x^{-1} & \cdot & \cdot & \cdot & \cdot & \cdot & -L_x^{-1} & \cdot & \cdot \\ -L_y^{-1} & \cdot & \cdot & \cdot & \cdot & \cdot & \cdot & L_y^{-1} & \cdot \\ \cdot & \cdot & -L_x^{-1} & L_x^{-1} & \cdot & \cdot & \cdot & \cdot & \cdot \\ \cdot & \cdot & \cdot & L_y^{-1} & -L_y^{-1} & \cdot & \cdot & \cdot & \cdot \\ -L_x^{-1} & \cdot & \cdot & \cdot & \cdot & \cdot & \cdot & \cdot & L_x^{-1} \end{bmatrix}$$

At this point, it is appropriate to draw attention to some of the differences between the present method and finite-element techniques. The fact that the bending strains of the element are ultimately related to normal displacements of adjacent elements, as in finite-difference methods, leads to the necessity of special treatment of boundary conditions. This disadvantage is, of course, not present in the finite-element method. However a compensating factor is that much fewer unknowns are involved, three per element in the limit, compared with a normally much larger number in the finite-element method.

The bending strains can now be written in terms of normal displacements only, as

$$\varepsilon^b = A^b C u^b \tag{14}$$

The membrane strain energy of the element is thus

$$U_i^m = (\Delta_i/2) u^{m,\mathrm{T}} A^{m,\mathrm{T}} E^m A^m u^m \tag{15}$$

where Δ_i is the area of the ith element, which is equal to $L_x L_y$ for a typical interior element, and the bending strain energy is

$$U_i^b = (\Delta_i/2) u^{b,\mathrm{T}} C^{\mathrm{T}} A^{b,\mathrm{T}} E^b A^b C u^b \tag{16}$$

or

$$U_i^m = \tfrac{1}{2} u^{m,\mathrm{T}} K_i^m u^m \quad \text{and} \quad U_i^b = \tfrac{1}{2} u^{b,\mathrm{T}} K_i^b u^b$$

where

$$K_i^m \equiv (\Delta_i/2) A^{m,\mathrm{T}} E^m A^m \quad \text{and} \quad K_i^b \equiv (\Delta_i/2) C^{\mathrm{T}} A^{b,\mathrm{T}} E^b A^b C \tag{17}$$

are the membrane and bending stiffness matrices of the ith element, respectively. These are listed in Appendix A. The stiffness matrices K_i^m and K_i^b cannot be added term by term, since they are referred to different displacement components.

For the generation of the structure stiffness matrix, it is convenient to keep the membrane and bending stiffnesses separate. From the topological properties of the structure a membrane "code number" and a bending "code number" can be constructed [35]. The generation of the structure stiffness can then be carried out using procedures very similar to those employed in the direct stiffness method.

3.2. TYPICAL EQUILIBRIUM EQUATIONS FOR AN INTERIOR NODE

Using the membrane stiffness matrix listed in Appendix A, the equilibrium equation at $i, j + 1$ in the y direction for a typical interior node is found to be

$$\frac{v_{ij+3} - 2v_{ij+1} + v_{ij-1}}{L_y^2} + \left(\frac{1-v}{2}\right)\left[\frac{v_{i+2j+1} - 2v_{ij+1} + v_{i-2j+1}}{L_x^2}\right]$$

$$+ \left(\frac{1+v}{2}\right)\left[\frac{u_{i+1j+2} - u_{i-1j+2} - u_{i+1j} + u_{i-1j}}{L_x L_y}\right]$$

$$- (t + vr)\left[\frac{w_{ij+2} - w_{ij}}{L_y}\right]$$

$$- s(1 - v)\left[\frac{w_{i+1j+1} - w_{i-1j+1}}{L_x}\right] + \frac{P_{v_{ij+1}}}{N} = 0 \tag{18}$$

This equation is identical to the consistent finite difference approximation to Eq. (8b).

4. Boundary Conditions

Boundary conditions are of two types: displacement and stress. The displacement boundary conditions are satisfied by deleting appropriate rows and columns from the structure stiffness matrix. In a potential energy formulation, the stress boundary conditions need not be considered explicitly —the specified boundary tractions are merely applied as external loads. For a finite mesh size, the computed boundary forces then do not exactly match the prescribed values. In some shell problems this lack of agreement can seriously affect the accuracy of the results. In these cases, the increased complexity introduced by satisfying the stress boundary conditions exactly is more than offset by the corresponding gain in accuracy. As mentioned previously, the analog model is readily adaptable to either procedure. Two types of modifications must be made to handle boundary conditions: (1) strain-displacement and (2) stress–strain relations must be altered for nodes on or near the boundary.

4.1. MODIFICATION OF STRAIN–DISPLACEMENT RELATIONS

The modification of strain–displacement relations is necessary whether or not explicit consideration is to be given to stress boundary conditions. The membrane strain–displacement relations are modified only for nodes lying on the boundary. However, the bending strain–displacement relations must be changed for nodes one-half interval from the boundary as well as for those on the boundary. As with finite-difference procedures in which no fictitious points are used, averaging procedures are required. For an edge normal to the x axis, the modified membrane strain–displacement relations for a node lying on the boundary are

$$\varepsilon_x = \frac{u_{j_3} + u_{j_1} - 2u_{j_2}}{L_x} - rw_{m_1}, \qquad \varepsilon_y = \frac{v_{j_3} - v_{j_1}}{L_y} - tw_{m_1}$$

$$2\varepsilon_{xy} = \frac{u_{j_3} - u_{j_1}}{L_y} + \frac{v_{j_3} + v_{j_1} - 2v_{j_2}}{L_x} - 2sw_{m_1}$$

(19)

or

$$\varepsilon^m = \bar{A}^m u^m \tag{20}$$

where

$$\bar{A}^m_{3 \times 9} = \begin{bmatrix} L_x^{-1} & \cdot & -2/L_x & \cdot & L_x^{-1} & \cdot & \cdot & \cdot & -r \\ \cdot & -L_y^{-1} & \cdot & \cdot & \cdot & L_y^{-1} & \cdot & \cdot & -t \\ -L_y^{-1} & L_x^{-1} & \cdot & -2/L_x & L_y^{-1} & L_x^{-1} & \cdot & \cdot & -2s \end{bmatrix}$$

The necessary modifications to the bending strain–displacement relations are more complicated when the stress boundary conditions are to be satisfied exactly. In order to account properly for the transfer of the twisting moment to the rigid bar intersection in the case of a free edge, auxiliary rigid bars are introduced into the physical mode [22]. Each auxiliary rigid bar runs parallel to the boundary and connects an edge node to an adjacent rigid bar intersection by means of a frictionless hinge. Thus only a vertical force resulting from the twisting moment can be transferred to the rigid bar intersection. A typical boundary element, Fig. 4, has a modified rotation vector

$$\bar{\theta} = \{\theta_{x1}, \theta_{y1}, \theta_{x2}, \theta_{y2}, \theta_{x3}, \theta_{y3}, \theta^*_{x1}, \theta^*_{x3}\}^T$$

The modified bending strain–rotation relation is then

$$\varepsilon^b = \bar{A}^b \bar{\theta} \tag{21}$$

where

$$\bar{A}^b_{3 \times 8} = \begin{bmatrix} \cdot & L_x^{-1} & \cdot & -2/L_x & \cdot & L_x^{-1} & \cdot & \cdot \\ -L_y^{-1} & \cdot & \cdot & \cdot & \cdot & L_y^{-1} & \cdot & \cdot \\ \cdot & -L_y^{-1} & -2/L_x & \cdot & \cdot & L_y^{-1} & L_x^{-1} & L_x^{-1} \end{bmatrix}$$

Relative normal motion is permitted at the rigid bar intersections. In the interior the relative displacements are determined once the normal displacements of the surrounding nodes are known. However along the boundary

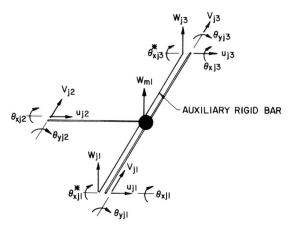

Fig. 4. Element on boundary with auxiliary rigid bars.

these displacements are independent unknowns. Thus the normal displacements of the rigid bars running normal to the boundary are included in the bending displacement vector u^b. For nodes on the boundary

$$\bar{u}^b = \{w_{m_1}, w_{m_2}, w_{m_3}, w_{j_3}, w_{j_1}, w_{m_6}, w_{m_7}, w_{m_8}\}^T$$

These added displacements of the intersections of the rigid bars are only important for the free edge. For other boundary cases they can be taken as the average displacement of the adjoining nodes.

For nodes one-half interval from the boundary

$$\bar{u}^b = \{w_{m_1}, w_{m_2}, w_{m_3}, w_{m_4}, w_{m_5}, w_{m_6}, w_{m_7}, w_{m_8}, w_{j_4}\}^T$$

The relation between the rotation vector and bending displacement vector is

$$\bar{\theta} = \bar{c}\bar{u}^b \tag{22}$$

where, for a node on the edge

$$\bar{C}_{9\times8} = - \begin{bmatrix} L_y^{-1} & \cdot & \cdot & \cdot & \cdot & -L_y^{-1} & \cdot & \cdot & \cdot \\ \cdot & -2/L_x & \cdot & \cdot & 2/L_x & \cdot & \cdot & \cdot & \cdot \\ \cdot & -L_y^{-1} & L_y^{-1} & \cdot & \cdot & \cdot & \cdot & \cdot & \cdot \\ L_x^{-1} & \cdot & \cdot & \cdot & \cdot & \cdot & -L_x^{-1} & \cdot & \cdot \\ -L_y^{-1} & \cdot & \cdot & \cdot & \cdot & \cdot & \cdot & L_y^{-1} & \cdot \\ \cdot & \cdot & -2/L_x & 2/L_x & \cdot & \cdot & \cdot & \cdot & \cdot \\ 2/L_y & \cdot & \cdot & \cdot & -2/L_y & \cdot & \cdot & \cdot & \cdot \\ -2/L_y & \cdot & \cdot & 2/L_y & \cdot & \cdot & \cdot & \cdot & \cdot \end{bmatrix}$$

and for a node one-half interval from the edge,

$$
\bar{C}_{9 \times 8} = - \begin{bmatrix}
L_y^{-1} & \cdot & \cdot & \cdot & \cdot & -L_y^{-1} & \cdot & \cdot & \cdot \\
\cdot & -L_x^{-1} & \cdot & \cdot & L_x^{-1} & \cdot & \cdot & \cdot \\
\cdot & -L_y^{-1} & L_y^{-1} & \cdot & \cdot & \cdot & \cdot & \cdot \\
L_x^{-1} & \cdot & \cdot & \cdot & \cdot & \cdot & -L_x^{-1} & \cdot \\
-L_y^{-1} & \cdot & \cdot & \cdot & \cdot & \cdot & L_y^{-1} & \cdot \\
\cdot & \cdot & -L_x^{-1} & L_x^{-1} & \cdot & \cdot & \cdot & \cdot \\
\cdot & \cdot & \cdot & L_y^{-1} & -L_y^{-1} & \cdot & \cdot & \cdot \\
-2/L_x & \cdot & \cdot & \cdot & \cdot & \cdot & \cdot & 2/L_x
\end{bmatrix}
$$

4.2. MODIFICATION STRESS–STRAIN RELATIONS

The explicit satisfaction of the stress boundary conditions is (with the exception of the transverse edge shear condition) handled by altering the stress–strain relations for nodes on the boundary. As an example, again consider the case of a boundary normal to the x-axis. Suppose the prescribed normal stress on the edge is $\tilde{N}_x$. Then by setting $N_{xij} \equiv \tilde{N}_{xij}$, ε_{xij} is determined in terms of ε_{yij}. Thus the stress–strain relations become

$$
\begin{Bmatrix} N_x \\ N_y \\ N_{xy} \end{Bmatrix}_{ij} = N \begin{bmatrix} 0 & 0 & 0 \\ 0 & 1-v^2 & 0 \\ 0 & 0 & (1-v)/2 \end{bmatrix} \begin{Bmatrix} \varepsilon_x \\ \varepsilon_y \\ 2\varepsilon_{xy} \end{Bmatrix}_{ij} + \tilde{N}_{xij} \begin{Bmatrix} 1 \\ v \\ 0 \end{Bmatrix} \tag{23}
$$

Therefore for a node on the edge

$$
\bar{E}^m \equiv N \begin{bmatrix} 0 & 0 & 0 \\ 0 & 1-v^2 & 0 \\ 0 & 0 & (1-v)/2 \end{bmatrix} \tag{24}
$$

The prescribed boundary tractions are applied as external loads. For example the additional load in the x direction at the intersection of the rigid bar $ij + 1$ is

$$
(\tilde{N}_{xij+2} + \tilde{N}_{xij})(L_y/2) \tag{25}
$$

The membrane stiffness matrix $\bar{K}_i^m$ for a node on the boundary is in this case

$$
\bar{K}_i^m = (L_x L_y/2)\bar{A}^{m,T}\bar{E}^m\bar{A}^m \tag{26}
$$

$\bar{K}_i^m$ is listed in Appendix A. The equilibrium equation in the x direction for node $ij + 1$ becomes, with the use of K_i^m and $\overline{K}_i^m$ as given in Appendix A,

$$
\left(\frac{1-v}{2}\right)\left[\frac{u_{ij+3} - 2u_{y+1} + u_{ij-1}}{L_y^2} + \frac{v_{ij+3} - 2v_{i-1j+2} + 2v_{i-1j} - v_{ij-1}}{L_x L_y}\right]
$$

$$
- s(1-v)\left(\frac{w_{ij+2} - w_{ij}}{L_y}\right) - 2\left(\frac{u_{ij+1} - u_{i-1j+1}}{L_x^2}\right)
$$

$$
- 2v\left(\frac{v_{i-1j+2} - v_{i-1j}}{L_x L_y}\right)
$$

$$
+ 2(r + vt)\frac{w_{i-1j+1}}{L_x} + \frac{\tilde{N}_{xij+2} + \tilde{N}_{xij}}{NL_x} + \frac{P_{uij+1}}{NL_x L_y} = 0
$$

(27)

This equation is identical to that which is obtained by writing the equilibrium equation

$$
\frac{N_{xij+2} + N_{xij} - 2N_{xi-ij+1}}{L_x} + \frac{N_{xyij+2} - N_{xyij}}{L_y} + \frac{P_{uij+1}}{L_x L_y} = 0 \qquad (28)
$$

for the node on the edge and substituting in the prescribed values

$$
N_{xij+2} = \tilde{N}_{xij+2}, \qquad N_{xij} = \tilde{N}_{xij} \qquad (29)
$$

Other specified stress boundary conditions are handled in a similar manner. No modification of the stress–strain relations is used for the transverse edge shear boundary condition. Mohraz has treated the case of beam-supported edges [23].

5. Selection of Proper Model

There are many finite differences based models that have been used to obtain solutions to continuum problems. Some of these have been arrived at in what seems to be a somewhat arbitrary fashion. If the physical model concept is followed, a good bit of this arbitrariness disappears. Furthermore, one can also anticipate behavior characteristics of the system of equations that might not otherwise be apparent.

Using the rigid bar deformable node concept, the equations governing the behavior of a structure, such as a plate or shell, can be readily expressed in discretized form. This was demonstrated for a general shallow shell in the previous section. For shells of zero and of positive curvature, the full model need not be used when the coordinate system is aligned to the principal curvature lines. The method in fact naturally leads to defining the displacement and stress quantities of interest in the manner shown in Fig. 5a. This scheme for defining the displacements and stresses at the points as shown was

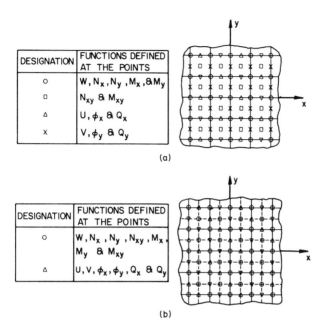

Fig. 5. Points of definition of displacement and stress resultants. (a) Coordinate lines along principal curvature lines; (b) general shell model.

first presented in reference to a cylindrical shell [31]. It was subsequently extended to shallow shells [24], shells of revolution [15, 2], and arbitrary cylindrical shells [26]. Once the manner of defining the quantities at different points has been accepted, the procedure can then be viewed directly from a finite-difference standpoint [2, 26] and the convergence characteristics studied. Furthermore, modifications possible to finite-difference methods (such as the multilocal scheme [25] can easily be included.

Since the positioning was reasoned on physical basis it is not surprising that this same manner of defining quantities was also incorporated in the dynamic relaxation method when it was applied to the cylindrical shell [5] and to shells of revolution [4].

For shells of zero and of positive curvature this arrangement of defining the quantities functions extremely efficiently. No more than the minimum number of equations need be solved. The proper variables are defined at the appropriate places so that no difficulty is encountered in satisfying boundary conditions. Both natural and geometric boundary conditions are readily handled as was shown in the description of the method. For small deflection problems the advantages, in fact at times the necessity, of the bar-node or lumped-parameter form of modified finite-difference procedure has been amply demonstrated [2, 22, 24]. The pin supported cylindrical panel demonstrates the dramatic difference between conventional and model or half-interval finite differences. Figures 6, taken from Chuang and Veletsos [6] show that both the displacement and the stress resultants accuracy are markedly improved by the use of modified equations. A similar although not so dramatic difference is found for a simply supported elliptical parabola [30] Fig. 7.

When applied to nonlinear problems, some difficulties arise with the analog model (modified finite differences). For the geometric nonlinear problem the normal deflection and its second derivative are defined at the

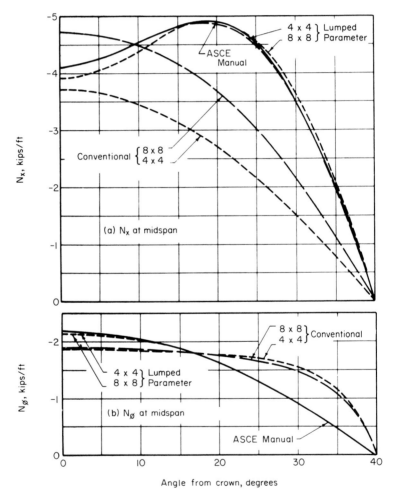

Angle from crown, degrees

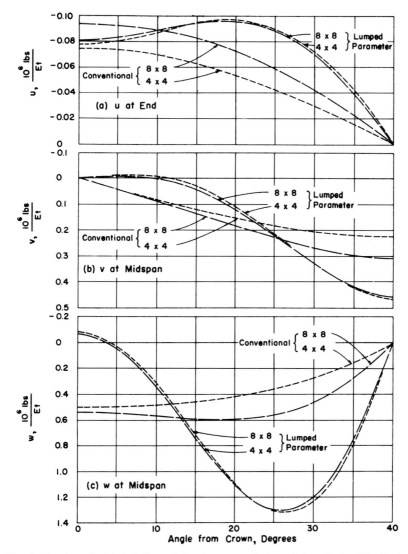

Fig. 6. Circular cylindrical shell panel simply supported along all boundaries. Distributions of stress resultants (opposite page) and displacement components (above) obtained by finite differences.

same points; however, the first derivative terms are not. Therefore it is necessary to establish the needed first derivatives at the required points by an averaging procedure on the first derivative.

Separation of points of definition of the stress resultants, as occurs in the reduced model of Fig. 5a, introduces some difficulty in the application of this

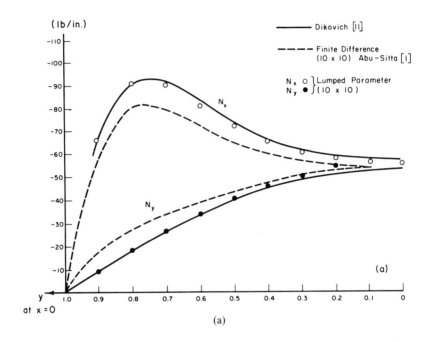

(a)

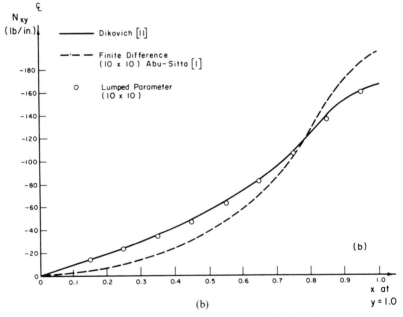

(b)

Fig. 7. (a) N_x, N_y along the center line of a simply supported elliptical paraboloid; (b) N_{xy} along the edge of a simply supported elliptical paraboloid.

minimal or simplified model to problems which include plastic material behavior. One possible solution to the difficulty is to take a second pattern displaced one-half interval in each direction and overlay the original grid. In this fashion the pattern of Fig. 5b results. This pattern is recognized as being the original network derived earlier for general shells when there is twist of the surface along the coordinate lines. Since this pattern results from an overlaying or interlacing of two independent nets, the solution will represent independent solutions of the two nets which have nearly equal but not identical mesh layouts. The shear computed at a stress node will be from one solution net while the direct stress at the same point is from the sister net. When a stress node goes plastic this means one net loses a shear node while the other net loses a direct stress node. This manifests itself as an apparent oscillation in the vicinity of the plastic node with one net out-deflecting the other. This experience has been observed in both plate and shell problems [21, 34].

If the shell surface is of negative Gaussian curvature, the twist of the surface interacts with the normal deflection to produce shear forces. This behavior is not present in the reduced model shown in Fig. 5a but is present in the original model of Fig. 5b. The twist, in effect, couples the two interlacing nets together so that their independent behavior is not possible. As demonstrated in the development, the network for a general shell should contain this model as the basic element. Now all displacement and stress quantities occur along each coordinate line so that if a boundary line falls along a coordinate line points of definition of all stress and displacement quantities occur along that boundary line. This can present some difficulties when joined to a stiffening or a supporting beam in that, the two structural elements will not have all their displacements defined at the same points.

To demonstrate the possible traps one may fall into if one routinely applies a form of the modified finite-difference network without some consideration of the structural type to which it is being applied, a hyperbolic paraboloid shell is considered. The modified finite-difference pattern as printed in the study by Croll and Scrivener [8] in order to ascertain if a modified difference procedure had any value when applied to hyperbolic paraboloid shells is shown in Fig. 8. They found that poor or "at best" no better results were obtained than were achieved using conventional finite-difference equations. An examination of the points of definition of the various quantities in association with the rigid-bar–deformable-node concept demonstrates that the manner of defining the displacements is not consistent with the positions of points where the stress resultants are defined. By deflecting a w point for example it is not possible to alter the normal bending moments. Likewise giving the appropriate nodes u or v displacements does not alter the direct stresses of the nodes. The network used by them only

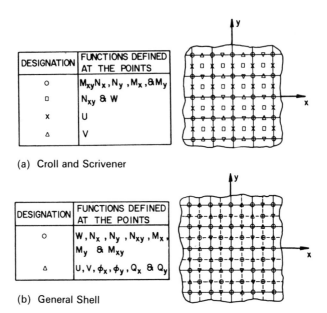

(a) Croll and Scrivener

(b) General Shell

Fig. 8. The modified finite difference model evaluated by Croll and Scrivener [8].

produces the effects of the twist of the surface. This would be adequate for a hyperbolic paraboloid shell if the solution did not also have to satisfy the boundary conditions. It is the boundary condition and the presence of any localized load terms which require the general net presented in this paper.

The use of the bar–node or analog model also eliminates much of the uncertainty or arbitrariness in the selection of the regions over which the integrations are carried out to establish the various energy contributions. This bar–node model suggests points of definition of displacements and regions of integration of the energy which are not among the alternatives considered by Bushnell [3]. Since the reduction of the analog model from the two-dimensional case to the axisymmetric problem leads to equations identical to the half-interval difference equations [2], the solution accuracy can be expected to be effectively the same for the two-dimensional case as for the axisymmetric case. This has been borne out by calculations and the poor results experienced by Bushnell using his half-station system should not occur.

The application of the lumped-parameter model to hyperbolic paraboloids has been reported by Mohraz [22, 23]. The solution accuracy obtained is shown in Fig. 9 where the axial force and bending moments in the beams supporting a hyperbolic shell are compared with the values computed by ordinary finite differences and by finite-element procedures [32]. The agree-

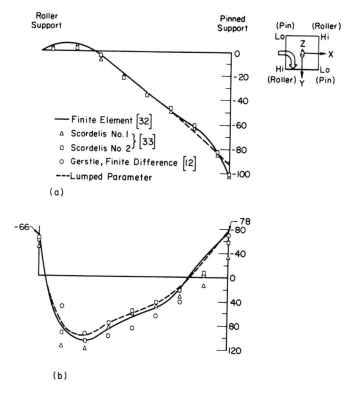

Fig. 9. Square hyperbolic paraboloid panel bounded by characteristic lines and supported by edge beams. (a) Axial force in beam (kips); (b) bending moment in beam (kips-in.).

ment shown between the finite element and the bar–node or lumped-parameter model indicates both methods apply equally well to this problem. Unfortunately a similar comparison cannot be made for the inverted umbrella hyperbolic shell (four hyperbolic paraboloid panels supported from a central column) since a number of the published solutions to this umbrella problem are suspect because of their failure to satisfy natural boundary conditions on the symmetry lines formed by the junction of two panels. Furthermore, for those solutions which provide adequate information, overall equilibrium conditions are not satisfied to any acceptable level. It is anticipated that the lumped-parameter model with the natural boundary conditions included should provide a much more acceptable solution.

One final consideration that should be given to the selection of the model to use in a finite-difference procedure is how well do the points of definition of the quantities of interest match with those needed to establish the proper boundary conditions. The lumped-parameter model reviewed in this paper is developed in a very similar manner to that proposed by Johnson [20]. There is, however, a small difference in the location of points of definition of the in-plane or tangential displacements.. The lumped-parameter model was selected so that all quantities of interest could be placed along a coordinate line and consequently any boundary condition can be satisfied.

A refinement to the bar–node model in the form of the multilocal method demonstrated by Noor [25] to have a beneficial effect on solution accuracy can be readily applied to either the general form or the reduced form of the model. In view of Noor's work the inclusion of this refinement is recommended.

6. Conclusions

The bar–node or lumped-parameter model reviewed in this paper is shown to provide a very efficient solution procedure to a variety of shell problems. The method can be made to satisfy both geometric and natural boundary conditions. When applied to problems involving nonlinear material behavior some uncoupling of the solutions will result unless due care is taken. When developed in a special purpose program the computation efficiency obtainable by this method compares favorably with other published methods.

Appendix A. Lumped-Parameter Element Stiffness Matrices

A.1. Membrane Stiffness Matrix K_i^m for a Typical Interior Element

$$
K_i^m = N L_x L_y
\begin{bmatrix}
\left(\dfrac{1-v}{2}\right)\dfrac{1}{L_y^2} & \cdot & \left(\dfrac{1-v}{2}\right)\dfrac{1}{L_xL_y} & \left(\dfrac{1-v}{2}\right)\dfrac{1}{L_y^2} & \cdot & \cdot & -\left(\dfrac{1-v}{2}\right)\dfrac{1}{L_xL_y} & s(1-v)\dfrac{1}{L_y} \\[6pt]
 & \dfrac{v}{L_xL_y} & \cdot & -\left(\dfrac{1-v}{2}\right)\dfrac{1}{L_xL_y} & -\dfrac{1}{L_y^2} & \dfrac{-v}{L_xL_y} & \cdot & \dfrac{(t+vr)}{L_y} \\[6pt]
 & \dfrac{1}{L_x^2} & \cdot & \cdot & \dfrac{-v}{L_xL_y} & \dfrac{1}{L_x^2} & \cdot & \dfrac{(r+vt)}{L_x} \\[6pt]
 & & \left(\dfrac{1-v}{2}\right)\dfrac{1}{L_x^2} & \cdot & \cdot & \cdot & -\left(\dfrac{1-v}{2}\right)\dfrac{1}{L_x^2} & \dfrac{s(1-v)}{L_x} \\[6pt]
 & & & \dfrac{-(1-v)}{2}\dfrac{1}{L_xL_y} & \cdot & \cdot & \left(\dfrac{1-v}{2}\right)\dfrac{1}{L_xL_y} & \dfrac{-s(1-v)}{L_y} \\[6pt]
 & \text{Symmetrical} & & \left(\dfrac{1-v}{2}\right)\dfrac{1}{L_y^2} & \dfrac{1}{L_y^2} & \dfrac{v}{L_xL_y} & \cdot & \dfrac{-(t+vr)}{L_y} \\[6pt]
 & & & & & \dfrac{1}{L_x^2} & \cdot & \dfrac{-(r+vt)}{L_x} \\[6pt]
 & & & & & & \left(\dfrac{1-v}{2}\right)\dfrac{1}{L_x^2} & \dfrac{-s(1-v)}{L_x} \\[6pt]
 & & & & & & & (r+t)^2 - 2(1-v)(rt - s^2)
\end{bmatrix}
$$

A.2. Bending Stiffness Matrix K_i^b for a Typical Interior Element

$$
K_i^b = DL_xL_y
\begin{bmatrix}
4\left[\dfrac{1}{L_x^4}+\dfrac{2\nu}{L_x^2L_y^2}+\dfrac{1}{L_y^4}\right] & \cdot & \left(\dfrac{1-\nu}{2}\right)\dfrac{1}{L_x^2L_y^2} & \dfrac{-(1-\nu)}{2}\dfrac{1}{L_x^2L_y^2} & \left(\dfrac{1-\nu}{2}\right)\dfrac{1}{L_x^2L_y^2} & -2\left[\dfrac{1}{L_y^4}+\dfrac{\nu}{L_x^2L_y^2}\right] & -2\left[\dfrac{1}{L_x^4}+\dfrac{\nu}{L_x^2L_y^2}\right] & -2\left[\dfrac{1}{L_y^4}+\dfrac{\nu}{L_x^2L_y^2}\right] & -2\left[\dfrac{1}{L_x^4}+\dfrac{\nu}{L_x^2L_y^2}\right] \\[2mm]
& \left(\dfrac{1-\nu}{2}\right)\dfrac{1}{L_x^2L_y^2} & \dfrac{-(1-\nu)}{2}\dfrac{1}{L_x^2L_y^2} & \left(\dfrac{1-\nu}{2}\right)\dfrac{1}{L_x^2L_y^2} & \cdot & \cdot & \cdot & \cdot & \cdot \\[2mm]
& & \left(\dfrac{1-\nu}{2}\right)\dfrac{1}{L_x^2L_y^2} & \dfrac{-(1-\nu)}{2}\dfrac{1}{L_x^2L_y^2} & \cdot & \cdot & \cdot & \cdot & \cdot \\[2mm]
& & & \left(\dfrac{1-\nu}{2}\right)\dfrac{1}{L_x^2L_y^2} & \cdot & \cdot & \cdot & \cdot & \cdot \\[2mm]
& & & & \cdot & \dfrac{1}{L_y^4} & \cdot & \cdot & \cdot \\[2mm]
& & \text{Symmetrical} & & & \cdot & \dfrac{\nu}{L_x^2L_y^2} & \dfrac{1}{L_y^4} & \dfrac{\nu}{L_x^2L_y^2} \\[2mm]
& & & & & & \dfrac{1}{L_x^4} & \dfrac{\nu}{L_x^2L_y^2} & \dfrac{1}{L_x^4} \\[2mm]
& & & & & & & \dfrac{1}{L_y^4} & \dfrac{\nu}{L_x^2L_y^2} \\[2mm]
& & & & & & & & \dfrac{1}{L_x^4}
\end{bmatrix}
$$

A.3. Membrane Stiffness Matrix $\bar{K}_i^m$ for a Typical Element on a Boundary with Prescribed N_x

$$
\bar{K}_i^m = (1 - \nu^2)N\frac{L_x L_y}{2}
\begin{bmatrix}
\cdot & \cdot & \cdot & \cdot & \cdot & \cdot & \cdot & \cdot \\
 & \dfrac{1}{L_y^2} & \cdot & \cdot & \cdot & -\dfrac{1}{L_y^2} & \cdot & \cdot & \dfrac{t}{L_y} \\
 & & \cdot & \cdot & \cdot & \cdot & \cdot & \cdot & \cdot \\
 & & & \cdot & \cdot & \cdot & \cdot & \cdot & \cdot \\
 & & & & \cdot & \cdot & \cdot & \cdot & \cdot \\
 & & & & & \dfrac{1}{L_y^2} & \cdot & \cdot & -\dfrac{t}{L_y} \\
 & & & & & & \cdot & \cdot & \cdot \\
 & & & \text{Symmetrical} & & & & \cdot & \cdot \\
 & & & & & & & & t^2
\end{bmatrix}
$$

References

1. Abu-Sitta, S. H., "Finite Difference Solutions of the Bending Theory of the Elliptic Paraboloid," *IASS Bull.* No. 20 (December 1964).
2. Abu-Sitta, S. H., "A Finite Difference Solution of the General Novozhilov Equations," *Proc. Int. Colloq. Prog. Shell Structures Last Ten Years*, IASS, Madrid (1969).
3. Bushnell, D., and Almroth, B. O., "Finite Difference Energy Method for Nonlinear Shell Analysis," *Conf. Comput. Oriented Anal. Shell Struct.* Lockheed Palo Alto Res. Lab., Palo Alto (August 1970).
4. Cassell, A. C., "Shells of Revolution under Arbitrary Loading and the Use of Fictitious Densities in Dynamic Relaxation," *Proc. Inst. Civil Eng.* (January 1970).
5. Cassell, A. C., Kinsey, P. J., and Sefton, D. J., "Cylindrical Shell Analysis by Dynamic Relaxation," *Proc. Inst. Civil Eng.* (January 1968).
6. Chuang, K. P., and Veletsos, A. S., "A Study of Two Approximate Methods of Analyzing Cylindrical Shell Roofs," SRS 258, Univ. of Illinois (October 1962).
7. Clough, R. W., "The Finite Element Method in Structural Mechanics," in *Stress Analysis* (Zienkiewicz, O. C., ed.). Wiley, New York, 1965.
8. Croll, J. G. A., and Scrivener, J. C., "Convergence of Hypar Finite Difference Solutions," *J. Struct. Div. ASCE* (May 1969).
9. Cyrus, N. J., and Fulton, R. E., "Finite Difference Accuracy in Structural Analysis," *J. Struct. Div., ASCE* (December 1966).
10. Das Gupta, N. C., "Using Finite Difference Equations to Find the Stresses in Hypar Shells," *Civil Eng. Public Works Rev.* **56** (1961).
11. Dikovich, V. V., *Analysis of Shallow Shells of Revolution with Rectangular Planform.* Gosstroiizdat, Moscow, 1960 (in Russian).
12. Gerstle, K. H., private communication.
13. Gilles, D. C., "The Use of Interlacing Nets for the Application of Relaxation Methods to Problems involving two dependent Variables," *Proc. Roy. Soc. London* 407–433 (1948).

14. Griffin, D. S., and Kellogg, R. B., "A Numerical Solution for Axially Symmetrical and Plane Elasticity Problems," *Int. J. Solids Struct.* 3, No. 5 (1967).
15. Hashish, M. G., and Abu-Sitta, S. H., "Free Vibration of Hyperbolic Cooling Towers," *J. Eng. Mech. Div., ASCE* (April 1971).
16. Havner, K. S., and Stanton, E. L., "On Energy-Derived Difference Equations in Thermal Stress Problems," *J. Franklin Inst.* **284**, No. 2 (August 1967).
17. Hrennikoff, A. P., "Solution of Problems in Elasticity by Framework Method," *J. Appl. Mech.* **8**, No. 4 (December 1941).
18. Hrennikoff, A. P., and Tezcan, S. S., "Analysis of Cylindrical Shells by the Finite Element Method," *Large Span Shells* (IASS Congr. Leningrad 1966), Tsinis, Moscow, 1968.
19. Hubka, R. E., "A Generalized Finite-Difference Solution of Axisymmetric Elastic Stress States in Thin Shells of Revolution," STL (now TRW) Rep. EM 11-19 (June 1961).
20. Johnson, D. E., "A Difference-Based Variational Method for Shells," *Int. J. Solids Struct.* **6**, 699–724 (1970).
21. Lopez, L. A., and Ang, A. H.-S., "Flexural Analysis of Elastic–Plastic Rectangular Plates," SRS 305, Univ. of Illinois (May 1966).
22. Mohraz, B., and Schnobrich, W. C., "The Analysis of Shallow Shell Structures by a Discrete Element System," Civil Eng. Studies, SRS 304, Univ. of Illinois (March 1966).
23. Mohraz, B., "A Lumped Parameter Element for the Analysis of Hyperbolic Paraboloid Shells," in *Int. J. Numer. Methods Eng.* (to be published).
24. Noor, A. K., "Analysis of Doubly Curved Shells," Ph.D. Thesis, Univ. of Illinois (August 1963).
25. Noor, A. K., "Improved Multilocal Finite-Difference Variant for the Bending Analysis of Arbitrary Cylindrical Shells, UNICIV Rep. No. R-63, Univ. of New South Wales (March 1971).
26. Noor, A. K., and Khandelwal, V. K., "Improved Finite-Difference Variant for the Bending Analysis of Arbitrary Cylindrical Shells," UNICIV Rep. No. R-58, Univ. of New South Wales (December 1969).
27. Radkowski, P. P., Davis, R. M., and Bolduc, M. R., "Numerical Analysis of Equations of Thin Shells of Revolution," *ARS J.* **32** (1962).
28. Singhal, A. C., Villaveces A., Utku, S., "A Computer Analysis of Thin Shells of Revolution," *Proc. World Conf. Shell Struct.*, National Research Council, Pub. No. 1187, San Francisco (1962).
29. Schnobrich, W. C., "A Physical Analogue for the Analysis of Cylindrical Shells," Ph.D. Thesis, Univ. of Illinois (June 1962).
30. Schnobrich, W. C., and Gustafson, W. C., "Multiple Translational Shells," *Int. Congr. Appl. Thin Shells Architecture*, IASS, Mexico City (September 1967).
31. Schnobrich, W. C., and Melin, J. W., "Model Analogue for Cylindrical Shells," *Proc. World Conf. Shell Struct.*, National Research Council Publ. No. 1187, San Francisco (1962).
32. Schnobrich, W. C., "Analysis of Hyperbolic Paraboloid Shells," *Symp. Concrete Thin Shells*, New York, ACI Special Publ. No. 28 (1970).
33. Scordelis, A. C., private communication.
34. Shoeb, N. A., and Schnobrich, W. C., "Analysis of Elasto–Plastic Shell Structures," SRS 324, Univ. of Illinois (August 1967).
35. Tezcan, S. S., "Computer Analysis of Plane and Space Structures," *J. Struct. Div., ASCE* (April 1966).
36. Utku, S., and Norris, C. H., "Utilization of Digital Computers in the Analysis of Thin Shells," RILEM Bull. No. 19 (June 1963).

PART IV

Large Interactive Data Bases

Design Philosophy of Large Interactive Systems

*Steven J. Fenves**

DEPARTMENT OF CIVIL ENGINEERING
UNIVERSITY OF ILLINOIS
URBANA, ILLINOIS

Until very recently, a researcher or research group could design an independent, standalone program to operate only in a prescribed local computer environment. A number of developments during the past few years have made this approach impractical, if not self-defeating. Among these developments are internal demands for interaction among various program segments and desire for alternate access modes, external demands for more intimate program exchange across different machine configurations, and nationwide developments in new computer access concepts. As a result, a fundamental change in program design approaches is taking place, and the "traditional" concept of the program as the central ingredient is giving way to a new philosophy, where the data base takes on the central role, with the individual programs or program steps looked upon as processors operating within this central data base.

In this paper, a design philosophy applicable to large-scale structural mechanics program systems is outlined and exemplified. The following major ingredients of such systems are discussed:

1. design of program segments of processors as operators on data files, independent of input–output;
2. design of conceptual data structures for interfacing with program segments independent of actual file structure;
3. design of systems or utility programs to provide data management and control facilities for the overall system.

* Present address: Department of Civil Engineering, Carnegie-Mellon University, Pittsburgh, Pennsylvania.

The design considerations affecting portability of the system and orderly growth potential of the entire system are discussed. An attempt is made to define the demarcation line between the areas of responsibility of the application-oriented mechanics programmer or analyst and the cooperating systems programmer.

1. Introduction

Engineering applications software suffers from a major shortcoming. In contrast to hardware, and, more recently, certain types of systems and utility software, applications software is seldom produced in an orderly manufacturing process of design, development, and fabrication [13]. Engineering application programs in structural mechanics appear to arise in two fashions: In universities and smaller organizations, an individual or a small group deal simultaneously with all design and implementation considerations, while in larger organizations, a division of labor is made according to functional lines (engineer versus programmer) rather than manufacturing stages.

In either approach, it is a practical impossibility for an outsider to check a program and attempt to separate well-established mathematical and physical relationships from built-in approximations, assumptions and limitations on one hand, and coding tricks introduced to circumvent hardware or software constraints on the other. Furthermore, programs tend to be such slavish codification of the manner of thinking of the originators that they invariably fail when released to be used by an outside organization. Finally, every change in the hardware and operating system is likely to have profound repercussions in the entire program structure. It is no wonder that computer center personnel consider engineer–programmers the most conservative and hide-bound group with which they must deal.

The preceding shortcomings are especially damaging in the present situation, where engineer–programmers are becoming increasingly subject to three classes of additional pressures:

1. pressures on the profession as a whole, evidenced by the increased complexity of the projects, often encompassing environmental and ecological considerations not previously considered, combined with shorter lead-times, partly caused by the escalation of costs;

2. financial pressures within organizations, manifested to a great extent by demands for greater machine independence, so as to be able to shop around for alternate computing facilities and access mechanisms without exorbitant conversion costs; and

3. pressures from users for greater flexibility, alternate modes of access, and faster and more orderly "internalization" of research results, such as new finite-element capabilities.

It is fair to say that every engineering mechanics computer shop is straining under these pressures. In the area of structural mechanics applications, many of these shops have abandoned portions or all of in-house programming, and have turned to one or more of the large-scale systems available, some of which are discussed in this volume. The available systems have been summarized and compared by Gallagher [5]. It is fair to say, however, that these systems, their impressive capabilities notwithstanding, have not lived up to our early expectations [1, 11], and have not solved all of our problems, for three apparent reasons.

1. Even in the most versatile systems, the addition of new capabilities is such a complex process that, for all practical purposes, users are constrained to operate within the capabilities provided by the developers.

2. In all these systems, versatility is achieved by deciding at execution time which of the available options to exercise. This, of course, necessitates a machine large enough to hold all control decisions and program and data segments, thereby eliminating a large class of potential users having access to smaller equipment only.

3. Notwithstanding the best talent expended on writing users' manuals, the sheer size and complexity of these systems is such that most potential users flounder around in attempting to model and describe their problem to the system.

Therefore, it appears that, in order to satisfy the profession's varied needs in the future, a much more intimate merger of in-house developed and externally supplied capabilities will be needed. The design of such systems requires a new design philosophy, involving a critical examination of each component of the overall system.

2. Components of Data Base

In this paper, the term "data base" is taken in its widest possible sense, to include a collection of files residing in some secondary storage device and accessible in some fashion to the user. The files may be of two types—program files or data files. The former can be further subdivided into: "ready-to-run" complete programs, and program components (subroutines, load modules, etc.) which must be assembled, either by means of user commands or automatically, to produce an executable program. The data files can similarly be subdivided into several categories, such as: (a) reference or read-only files, containing material for common reference by several programs or projects; (b) active or project files, containing global information shared by several programs about the current status of a given project; and (c) scratch or buffer files containing temporary information shared by programs or segments working on a given task.

The collection of files just described can be represented by a two-phase graph, with circles representing data files and rectangles representing processes (programs) [8], as illustrated in Fig. 1. Note that such diagrams are not

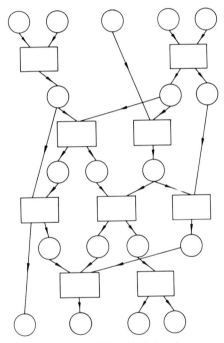

Fig. 1. Graph representation of files (circles) and processes (rectangles).

flow diagrams or even block diagrams in the conventional sense: The branches (arrows) represent ingredience or precedence relations, and not the flow of control. The latter is either implicit, if the graph represents a complete process, or must be indicated explicitly if the graph represents a collection of available subprocesses from among which the user can select a specific path to accomplish a particular objective.

In his pioneering work, Langefors developed a powerful matrix representation and generalized matrix algebra which enables one not only to compute such things as total memory requirements and transport (I/O) times for a collection of subprocesses, but also to design an optimal system by grouping of processes and/or files, subject to memory and transport time constraints. Langefors' work will not be pursued further here, for two reasons: first, most of his system algebra deals with implicit control flow, i.e., systems representing a complete process; and second, the design decisions he addresses himself to pertain to the fabrication stage, i.e., the proper "packaging" of process and data files for optimal execution performance. His graphic notation, however, provides an excellent context for the discussion of the design stages of the various files. Also, his multiple warnings about preventing system incompatibility by proper design of component interfaces should be heeded by all software designers.

Another formal approach to the general problem of structuring data bases is given by Johnson [7]. While his work is also restricted primarily to "business" data-processing applications, he also presents a useful and powerful graphic notation to describe the various components of the data base.

3. Program Design

The term "program" is loosely used today to describe any collection of executable code, ranging from small, self-contained "packages" designed to perform a specific function, to enormously complex systems of great flexibility. In the latter, it has long become commonplace that the programs are segmented both horizontally (by option or user function) and vertically (by internal function) into layers of subprograms, load modules, or tasks.

For proper operation within a data base, it is mandatory that even the smallest program be designed in a modular fashion, and be functionally and operationally segmented into the following three phases:

1. an input routine or editor, which accepts all input data, checks them for consistency and reasonableness, substitutes assumed values if no data are specified, and performs any other input editing required;

2. a processing program proper, which operates only on the edited input data, independent of its source and the degree of editing and preprocessing that has taken place. The processing program produces an internal output file which contains the minimal basic data from which all conceivable result types can be generated;

3. an output routine or report generator, which produces the requested results, including an echo-print of the input, either by directly copying from the internal output file or by performing various postprocessing operations.

The functional relationship between the three phases is shown in Fig. 2. It is important to recognize the basic philosophy of the system, namely that

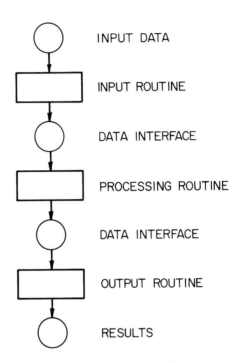

INPUT DATA

INPUT ROUTINE

DATA INTERFACE

PROCESSING ROUTINE

DATA INTERFACE

OUTPUT ROUTINE

RESULTS

Fig. 2. Functional organization of a program.

the processing program can operate independently of the other two routines, provided it is supplied with equivalent input file. Some of the other possible environments for the processing program are shown in Fig. 3.

Larger programs will, of course, consist of many more input, computation, and output segments, and their design must reflect the same general organization. It is particularly important that two philosophies be followed in their design:

1. All machine-dependent functions should be programmed as separate subroutines which can be readily replaced. These functions include all secondary storage access, character and other symbol conversions, and computational procedures likely to be sensitive to word length (e.g., matrix triple products).

2. It is to be recognized that many operations which are automatically performed in a loop in a batch environment may be performed only selectively, at the user's command, in an interactive environment. For example, in a finite-element program, in a batch mode one typically requests that all element stresses be printed, while in an interactive mode one may want to see the stresses for only a few critical elements. Thus, the basic computational

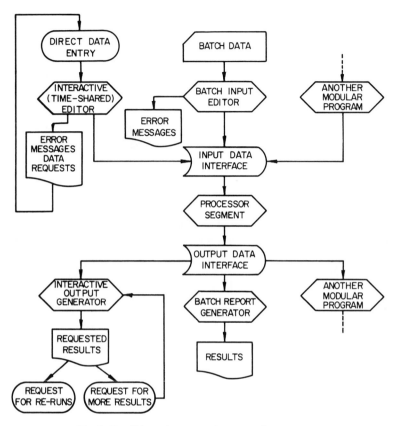

Fig. 3. Possible environments for processing program.

unit is not the stress computation loop, but the stress computation for the *i*th element.

4. Data Structure Design

The basic philosophy that should underlie the design of data files in a data base is the recognition that data structure design involves two distinct considerations:

1. The design of the conceptual data structure, regardless of its manner of storage. In this category are included considerations of content (i.e., how many data items of what type are needed to define a given entity) and organization (i.e., the types of arrays, lists, tables, hierarchies, etc., needed to relate the individual entities to each other);

2. The design of the file structure, that is, the mapping of the conceptual structure into physical secondary storage, so as to make the needed data available at the right time to the right process.

If this philosophy is followed, it is a relatively easy job to find the most appropriate file structure for a given conceptual data structure. Even if the "best" file structure is not available in a given situation, there is usually sufficient flexibility in the conceptual definition that one of the available file structures can be adapted to the task without significant loss in processing efficiency. A good discussion of available implementation strategies is given by Lefkovitz [9].

In a previous paper [6] it was argued that the variety and complexity of conceptual data structures one must deal with in a large-scale, integrated data base are so large that it is impossible to design programs to interface with every possible data structure. Rather, the data structure must carry its

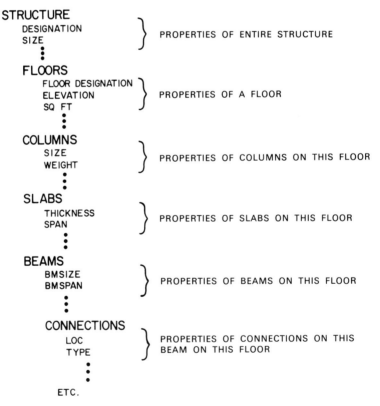

Fig. 4. Hierarchical outline of hypothetical data structure.

own description or definition, and utility routines must be provided to retrieve such data and pass them to the application programs.

A conceptual structure which appears particularly attractive for engineering project-design applications is one of the heirarchy of tables, where a conceptual data item may be either a terminal datum (word, vector or matrix) or a subtable containing additional data, including sub-subtables, pertaining to constituent subelements of the element described in the table. A hierarchical outline of a hypothetical data structure is shown in Fig. 4.

The type of data structure described is well suited for two of the file categories already mentioned, namely, reference files accessed by different programs in one functional category (e.g., a table of element or material properties) and project files accessed by programs operating on the different aspects of the file (e.g., stress, weight and clearance programs). It should be recognized, however, that a very high penalty is being paid for this flexibility : All data accesses must be executed interpretively by consulting the file description in order to locate the data. Thus, this type of data structure is prohibitively expensive for scratch or buffer files used by computational routines, so that these must exist in the data base in such a form that pre-programmed references to the data may be made by the programs [10].

In summary, the data base must allow for two types of data files : conventionally structured data (vectors, matrices, etc.) designed for rapid and efficient communication with a highly restricted set of program segments, and self-documenting, loosely structured data designed to be accessible by an unlimited set of program segments. In both cases, however, the design must proceed in two steps, by first determining the conceptual structure best suited for the application, and only after that selecting the best file storage and access mechanism to support it.

5. Control System Design

Given a collection of data and program files residing in the data base, it is necessary to develop a control mechanism which, on the basis of suitable commands, assembles the necessary components to produce an executable sequence of operations, including the implicit or explicit data access and storage operations.

This topic has been widely treated in both civil engineering and computer science literature (e.g., Roos [14], Flores [4], and Wegner [16]), although the discussion has too often been slanted toward the external control language and implementation strategies rather than the actual functions triggered by the users' commands. It appears, therefore, that as with the previously discussed aspects, it is convenient to discuss the three separate components of the problem:

1. the definition of the control functions needed, without regard to implementation or control language;
2. the implementation of these functions; and
3. the design of the external language.

The actual number of distinct control functions needed is surprisingly small; a dozen or so program and data file management routines will generally suffice. Furthermore, many of these are standard functions in most operating systems. By contrast, the number of implementation alternatives for each function is extremely large, ranging from interpretive execution through preprogrammed calls to precompilation. Finally, the range of levels of possible external control languages is even larger, ranging from natural language-like commands through mnemonics or code numbers to predefined, fixed sequences.

Both implementation and user language are highly environment-dependent, the first being heavily influenced by the organization and operating system of the computer on which the system is to be run, the latter depending on the operational procedures and preferences of the user organization. As it appears unlikely that either of these environments will become standardized in the near future, it appears imperative that any system be so designed that it can be operated in a variety of machine environments using a variety of control languages. This is, in essence, what is meant by the currently fashionable terms of "program mobility" [12], and "program portability" [15].

6. Who Will Do It

In the preceding sections, the three components of the data base, namely programs, data, and control system, were examined in some detail. In each of the three, a clear-cut distinction emerges between those aspects which pertain to the engineering content and those which are independent of it. Thus, the design of a program segment is an engineering problem, but its interconnection with its data file is not; the conceptual layout of the data structure is application-dependent, but its mapping into secondary storage is not; and the design of the control language is influenced by the engineer–user environment, but the implementation of control functions is not.

In a previous paper [2], it was argued that program development is to be segmented to distinguish clearly between the development stage to be performed by engineers and the production stage best entrusted to professionals specializing in that area. In a later paper [3], this concept was carried further, and the design life of a program or system subdivided into the states shown in the center of Fig. 5. It appears clear that the professional activities shown on the left are the province of the application user community. But most of the steps in the technical production process, shown on the right, transcend the application field and are primarily in the domain of software engineering.

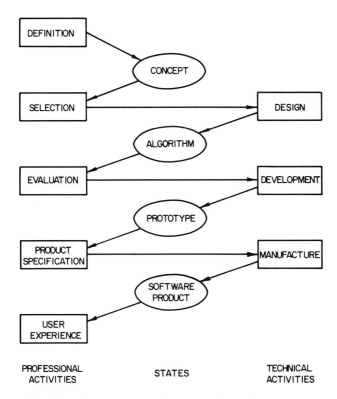

Fig. 5. Development process of an engineering software product.

7. Conclusions

The combination of sophistication and expertise acquired and of demands imposed from the outside has brought us to the point where previous ideas about programs and programming systems are no longer tenable. We are rapidly moving into an era where we will be dealing with large data bases, accessible through a variety of interaction modes by large groups of people, ranging from users through program segment developers to system developers. The data base must support all these levels and media of access.

The design of such data bases requires engineering design input at the highest level. In particular, it requires an analytic approach to the definition and structuring of the problem. A beginning of such a structure was attempted.

The most significant factor arising from this modeling attempt is that each aspect of the design naturally subdivides itself into a part wholly within the domain of the application area, and another in which a different profession has established predominant expertise. It is hoped that approaches such as the one outlined here will eventually bring about a healthier interaction and understanding between these two disciplines.

References

1. Fenves, S. J., "STRESS—A Computer Programming System for Structural Engineering Problems," Dept. of Civil Eng., MIT Rep. T63-2 (June 1963).
2. Fenves, S. J., "Scenario for a Third Computer Revolution in Structural Engineering," *J. Struct. Div. ASCE* **97**, No. ST1, 3–11 (January 1971).
3. Fenves, S. J., "Software Products for Engineering Applications," *APEC J.* **6** (1), 9–12 (May 1971).
4. Flores, I., *Computer Software*. Prentice-Hall, Englewood Cliffs, New Jersey, 1965.
5. Gallagher, R. H., "Large-Scale Computer Programs for Structural Analysis," *Symp. General Purpose Finite Element Comput. Programs* ASME, pp. 1–14 (1970).
6. Hatfield, F. J., and Fenves, S. J., "The Information Organizer: A System for Symbolic Table Manipulation," *Comput. Struct.* **1**, 85–102 (1971).
7. Johnson, L. R., *System Structure in Data, Programs and Computers*. Prentice-Hall, Englewood Cliffs, New Jersey, 1970.
8. Langefors, B., *Theoretical Analysis of Information Systems*, 2 vols. Studentlitteratur, Lund, Sweden, 1966.
9. Lefkovitz, D., *File Structures for On-line Systems*, Macmillan, New York, 1969.
10. Lopez, L. A., "POLO: Problem-Oriented Language Organizer." *Comput. Struct.* **2**, 555–572 (1972).
11. Miller, C. L., "Man–Machine Communication in Civil Engineering," *J. Struct. Div. ASCE* **89**, No. ST4, 5–36 (August 1963).
12. Poole, P. C., and Waite, W. M., "Machine Independent Software," *Proc. ACM Symp. Operating Syst. Principles, 2nd*. Princeton, New Jersey (1969).
13. Ross, D. T., "Fourth Generation Software: A Building-Block Science Replaces Hand-Crafted Art," *Comput. Decisions* **2**, No. 4, 32–38 (1970).
14. Roos, D., *ICES System Design*. MIT Press, Cambridge, Massachusetts, 1966.
15. Waite, W. M., "Building a Mobile Programming System," *Comput. J.* **13**, No. 1, 28–31 (1970).
16. Wegner, P., *Programming Languages, Information Structures, and Machine Organization*. McGraw-Hill, New York, 1968.

Integrated Design of Tanker Structures

Johannes Moe

TECHNICAL UNIVERSITY
TRONDHEIM, NORWAY

This paper presents a brief discussion of some important features concerning the structural behavior of tankers. It is suggested that the design task may, with sufficient accuracy for preliminary design purposes, be subdivided into an iterative sequence of fairly simple jobs. These jobs are discussed in some detail with emphasis on their interrelationships. Some of the jobs involve design work, which may be performed by means of automated design and optimization programs. Four examples of such programs are presented.

In order to use the job programs efficiently it is necessary to work in an environment that offers:

(*i*) easy administration of problem-oriented programs;
(*ii*) easy administration of data through a common data base; and
(*iii*) easy communication between user and computer in an interactive mode of operation.

The BOSS system described in this paper is designed to meet these requirements, and it is shown how the complete design task may be organized in an integrated system.

1. Introduction

The designers of large tanker structures are responsible for tremendous expenditures. The weight of the hull of a 250,000-ton-dead-weight tanker may typically be approximately 30,000 tons. This hull is a steel structure that, regarding size and cost, finds few competitors ashore. The cost of the hull structure is likely to exceed 50 % of that of the completed ship. One should therefore expect that in the highly competitive atmosphere of the shipbuilders of the world, utmost importance is attributed to the development of cost-effective structures. It is not the author's impression that such are always achieved. The designers are all too often forced to complete their preliminary designs within very tight time schedules that offer little time for thorough structural synthesis, and even to place the steel orders on the basis of highly preliminary studies.

Most of the potentials for savings through careful design studies—as well as losses through inefficient designs—are usually encountered during the initial design stages. Therefore, the designers ought to be able to study a number of different alternative solutions at this stage, in a search for potential savings. This is, of course, a rather time consuming job if traditional methods of design are used.

This paper presents an outline of a computer-aided design system that may hopefully meet some of the needs of the designers. The system is based on the application of a number of automated design and optimization programs, and offers additional software to facilitate interactive communication between designer and computer. The system is not completed yet, but the more important modules of the system are operative.

A brief description of the structural problem is followed by a presentation of the synthesis scheme and the software designed to facilitate the interactive mode of operation.

2. Presentation of the Structural Problem

The principal dimensions of a tanker hull are such that it may in a first approximation be treated as a box beam. For a 250,000-ton-dead-weight tanker one may typically find the following overall dimensions: length, 330 m; breadth, 52 m; depth, 26 m; draught, 20 m. Figure 1 shows a cross section through the cargo tank area. Only the longitudinal strength material is shown. If distortions of the cross section are disregarded, it is a fairly elementary task to design this cross section once external longitudinal moments and shearing forces are known and corresponding maximum stresses are decided upon. One has, of course, to super-impose effects of bending about the horizontal and the vertical axes as well as torsion. This is done in a statistical sense, by assuming that the dynamic effects are statistically independent. Local loads also must be considered simultaneously.

Figure 2 shows a more complete picture of the interior of such a hull. The cargo tank area is subdivided into numerous tanks by means of transverse

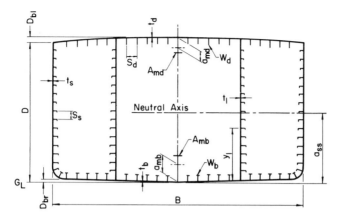

Fig. 1. Cross section of tanker.

as well as longitudinal bulkheads. The distances between the transverse bulkheads may today be as much as 10% of the total length of the ship. Every second bulkhead may be made rather light and with openings, their primary purpose being to reduce the sloshing of the fluid in the tanks. Hence the total distance between the heavier oil-tight bulkheads may be up to 20% of the ship length, which for the biggest ships is more than 60 m. It is likely that the maximum tank sizes, and hence the distances between the oil-tight bulkheads, will have to be reduced significantly in the near future in order to reduce the risk of oil pollution in cases of accidents.

The stiffened panels shown in Fig. 1 are supported at intervals of 4–5 m by heavy transverse frames. These frames must be designed to withstand the effects of the lateral pressures against the plate panels, caused by external or

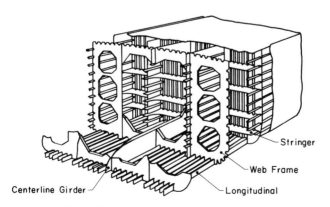

Fig. 2. Typical internal structure of tanker.

internal overpressure. A number of different loading conditions have to be considered with different patterns of full and empty tanks, as well as different draughts, see Fig. 3.

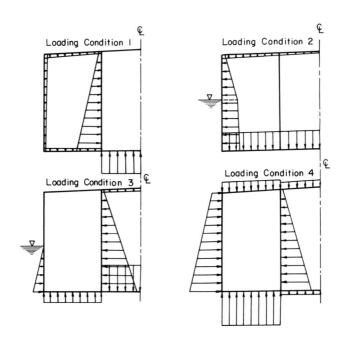

Fig. 3. Typical loading conditions for transverse frames.

The frames may be considered to be supported in the vertical direction at the ship sides as well as at longitudinal bulkheads and eventual longitudinal girders in bottom and deck. These supports deflect differently. It is therefore necessary to assume that the frames are provided with elastic supports, that at the ship sides and the longitudinal bulkheads are distributed along the depth of the hull girder. It is, in other words, not permissible to disregard the distortion of the hull girder cross section, except in the first approximation of the longitudinal strength analysis mentioned earlier.

By now, the reader will probably appreciate that an accurate stress analysis of a tanker hull girder is quite complicated. Finite-element analyses employing a superelement technique to cope with the extremely large system may seem to be the only answer to this problem. Such analyses have already been performed occasionally [1, 2], and will certainly become more common as the development of user-oriented program systems progresses further.

3. Design Procedure

In the initial design phase it is necessary to use a set of reasonably accurate approximate methods of analysis, that are simple enough to allow the designer to investigate many alternatives with moderate computational effort. As has been already pointed out, this initial design stage is extremely important. Most of the potentials for savings through optimization lie here. Rapid and reliable methods of initial design are also of great importance as competitive tools in the sales efforts, allowing the company to prepare well-founded prize bids for alternative designs at short notice.

There is not yet available any well-established preliminary design process for this purpose. A brief description of the approach selected for the integrated tanker design system under development at the Department of Ship Structures, Trondheim, will be presented in the following sections.

Let us first take a look at the different design and analysis jobs that must be done. The interplay between these jobs in the synthesis will be discussed later. We shall assume here that external loads as well as limiting stresses and deflections are known. Similarly we assume that proper buckling and vibration constraints have been selected. Table 1 then presents a list of the most important individual jobs to perform, together with an indication of the more important groups of decision variables that have to be selected prior to the execution of these jobs. The table also lists output from each job. A brief discussion of each job follows.

a. *Job 1*

This job involves the use of simple box beam theory on the hull girder. Stresses due to local hydrodynamic pressures on the plate panels have to be added to those caused by the longitudinal bending of the hull, taking into account statistical independency of the various components of the dynamic effects. Plates and stiffeners must be designed with ample safety with respect to buckling. The resulting dimensions of stiffeners are, of course, strongly influenced by the preselected spacing between transverse frames. The position of the longitudinal bulkhead is also an important decision variable.

b. *Job 2*

This job involves the application of elastic or plastic grillage theory in the design of the transverse bulkheads. Prior to the execution of this job one will have to decide whether to use corrugated or plane bulkheads and whether to run corrugations or stiffeners vertically or horizontally. The parameters describing the general topology of the grillage that supports the bulkhead, such as number of girders in each direction, also belongs to the category of decision variables. The methods of analysis are fairly straightforward, but

TABLE 1
OUTLINE OF DESIGN AND ANALYSIS JOBS

Job no.	Decision variables (input)	Job	Output
1	Principal dimensions of ship. Position of longitudinal bulkhead. Spacing of transverse frames. Positions and dimensions of longitudinal girders.	Design of longitudinal strength members for combination of hull girder stresses and local stresses, assuming undistorted hull girder cross section. (Fig. 1) Program: LANOPT	Plate thicknesses and stiffener arrangement and sizes for deck, bottom, ship sides and longitudinal bulkhead.
2	Types of bulkheads, types of grillages.	Design of transverse bulkheads. (Fig. 10) Program: KOROPT	Plate thicknesses, stiffeners, grillage dimensions.
3		Grillage analysis of large portion of the ship for various loading arrangements. (Fig. 4)	Estimates of relative deflections between supports of transverse frames.
4		Design of transverse frames. (Fig. 7) Programs: RAMOPT GIROPT	Dimensions of frame members including local stiffeners.
5		Frame analyses. (Fig. 8)	Stiffness properties of the frames, to be used in job 3.
6		Finite element study.	

structural interaction with adjacent longitudinal elements has to be taken into account. In the final design it will normally be required that stiffeners and girders on the bulkhead land on corresponding longitudinal elements in deck, bottom, ship sides, and longitudinal bulkheads (output of job 1).

c. *Job 3*

It is now necessary to investigate the effects of distortions of the hull girder cross section, through a study of the relative vertical displacements of ship sides, longitudinal bulkheads and girders. For this purpose a grillage analysis is performed on a model of the type indicated in Fig. 4b. Effects of

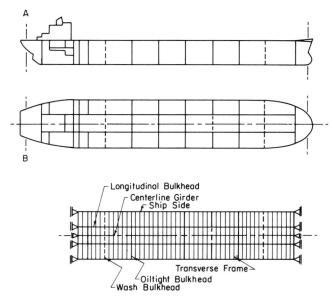

Fig. 4. Model for grillage analysis of the cargo tank portion of the hull.

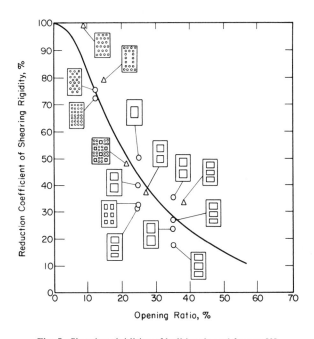

Fig. 5. Shearing rigidities of bulkheads and frames [3].

shearing deformations are for some members of the grillage more important than flexural deformations, and have to be included in this analysis. At this stage of the design process, very little information is available about the stiffness properties of the transverse frames, which have not yet been designed. It is therefore necessary to use available approximate data. Figure 5 shows an example of derived empirical curves for shearing rigidities of wash bulkheads with various amounts of cut-outs as well as typical frames in side tanks of tankers. Similar curves have been presented by Roberts [4] and others. Job 3 may have to be rerun after jobs 4 and 5 have been completed, on the basis of the new data generated. Additional information about stiffness properties of the bulkheads, based on, for instance, finite-element studies, may also be desirable at this stage.

The most important output from this job is the information about relative displacements between the various supports of the transverse frames. But also the obtained information about distribution of shearing forces between longitudinal bulkheads and ship sides, are important. This distribution is not in agreement with the simple box beam theory. Stresses in longitudinal girders caused by local loads, are also derived at this stage.

d. *Job 4*

This job involves the design of the transverse frames. Before starting on the job, the general topology of the frames has to be decided. Plate thicknesses as determined under job 1 together with the selected frame spacing are used to determine areas of the plate flanges of the frame members, see Fig. 6. The job

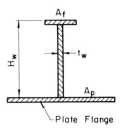

Fig. 6. Typical girder cross section of transverse frames.

involves selection of girder heights, web thicknesses, and flange areas as well as tripping brackets, web stiffeners, etc. Constraints with respect to stresses, deflections, stability and vibration characteristics have to be complied with.

The proportions of the transverse frames (see Fig. 7) are often such that one may question the applicability of the conventional frame theory. It is believed that by means of intelligent modeling schemes, the taking into account of effects of overlapping regions between joining elements as well as

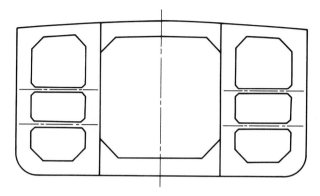

Fig. 7. Typical transverse frame topology.

knee plates, one may obtain the stresses at most of the critical points with fair accuracy. Systematic calibration with results obtained by means of the finite-element technique will have to be carried out in order to supply a safe basis for this approach. Hybrid models consisting of a combination of frame members and regions covered by finite elements may in the future replace the simple frame models.

e. *Job 5*

Having obtained a transverse frame design under job 4, it is now easy to derive more accurate information about the stiffness properties of the frames that may be used in a rerun of job 3. Figure 8 indicates how such stiffness data are obtained by successively imposing unit relative displacements of the various frame supports.

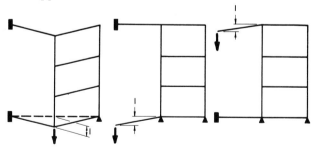

Fig. 8. Determination of stiffness characteristics of frames.

f. *Job 6*

After the completion of the preliminary design jobs it will often be desirable to carry out finite-element stress analyses covering a major portion of the hull. Finite-element analyses of alternative designs of structural details may also be desirable at this stage. It is desirable to have data generators that can provide the required input for the finite-element analysis from the information produced during the previous jobs described here.

4. Synthesis

In the previous section we divided the design task into six individual jobs and discussed some of the interrelationships among them. Due to these relations they cannot be completed entirely in one sequence. The solutions to jobs 1 to 5 must be found in an iterative manner. The designer may also want to vary several of the decision variables in his search for the optimum solution. It is, therefore, desirable to create a design environment that allows the designer to carry out the jobs repeatedly and in an arbitrary sequence, preferably through remote access to the computer. All of the programs used in these jobs should therefore communicate with a common data base, as schematically shown in Fig. 9. The amount of common data for the various jobs 1 to 5 is fortunately moderate. Such a data base may therefore be fairly simple if tailored only to the needs of the task discussed here.

A design system of the type indicated in Fig. 9 is presently being developed. Only jobs 1–5 will be implemented in the first phase of the project. Jobs 1, 2,

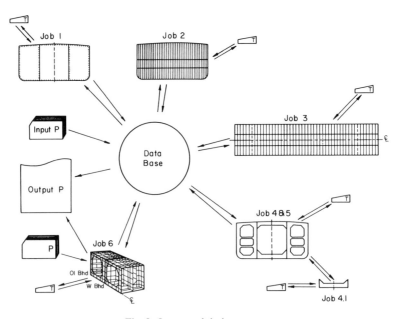

Fig. 9. Integrated design system.

and 4 are of special interest, since they involve design work. Automated design routines have already been developed to perform jobs of these types. Work on the coordination of these programs and the implementation of the data base is in progress. Much work remains to be done in order to develop a complete package of practical design and analysis programs.

The following automated design programs, which are described briefly in Appendix 1, constitute presently the most important parts of the system:

LANOPT: Optimum design of midship sections of tankers (Fig. 1).

KOROPT: Optimum design of vertically corrugated transverse bulkheads (Fig. 10).

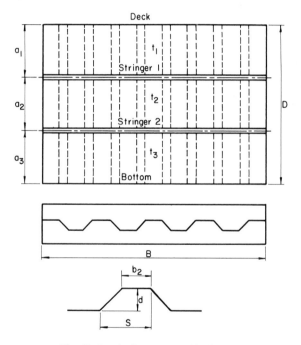

Fig. 10. Vertically corrugated bulkhead.

RAMOPT: Optimum design of statically indeterminate frames (Fig. 7).

GIROPT: Optimum design of steel girders (Fig. 11).

Several of these programs are well-documented elsewhere [6–10]. Emphasis is here placed on features of interest in judging their applicability as parts of an integrated program system. The automated design programs are all employing the sequential unconstrained minimization technique (SUMT) for the solution of the nonlinear constrained optimization problems [5].

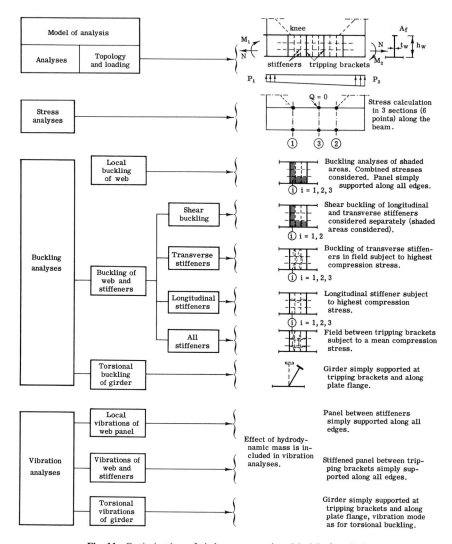

Fig. 11. Optimization of girder cross section. Model of analysis.

All of the three first mentioned programs employ the same general search routine [11], which is based on either Powell's or Rosenbrock's direct search methods at the users choice. The fourth program uses a modified version of this general search program, in which integer requirements for some or all of the variables may be accounted for [12].

In relation to the integrated design process discussed here, it is important to note that any combination of the design variables may be fixed at preselected values during special runs of these automated design programs. As the design process progresses, the designer will make decisions involving fixation of an increasing number of the variables. Consider, for instance, the design of a transverse bulkhead. In the initial runs positions of girders and stiffeners may, for example, be considered to be free variables. In later stages these positions will have to be coordinated with adjoining longitudinal structural elements in the hull girder. Similarly, one may consider all of the member sizes of the transverse frames as free variables in the first run. Later on, one may decide to link some of these variables in order to reduce the number of different elements. One may, alternatively, link member heights but leave the other dimensions free to be designed on an individual basis.

In the frame design program RAMOPT only one free variable is introduced for each member. On the basis of the starting point for the search process, the members are scaled according to a certain law, and the scaling factors are maintained as the variables. Several alternative scaling laws are built into the program. The starting values have to be supplied by the designer or may in the future possibly be fetched from the data base. Assistance in the selection of optimum cross sectional forms of the individual members may be obtained from the program called GIROPT. This program needs information from RAMOPT about member end forces and other loads as well as plate flange areas. It optimizes girder cross sections including stiffener arrangement. The cost figures account for materials and labor, including such items as corrosion protection either by paint or by increased plate thicknesses. This program deals with integer variables such as number of stiffeners in each direction and web plate thicknesses in increments of 0.5 mm. The field of integer and mixed-integer nonlinear programming is still in its infancy. GIROPT, therefore, does not employ well-established methods and is not yet a very practical design tool. It also is rather inefficient concerning computer time in comparison with the other automated design programs mentioned here (see Appendix I).

In the iterative design process described here, it may be desirable to have several programs that perform the same job, on different levels of sophistication. Starting for instance with simple programs based on conventional frame theory and scanty information, one progresses through gradual refinements ending up with rather complete finite-element studies. It may be rather inefficient to apply highly accurate and complex methods of analysis on insufficient data. The system should also allow the designer to copy preliminary data directly from files containing information about earlier designs, so that the designer may work on and modify these instead of generating all of the data from scratch. This possibility is implemented in the data base of the system discussed here.

5. Software System

In order to draw the full benefits from an integrated design system as outlined here, it is necessary to communicate interactively with the computer. In an effort to facilitate the interactive administration of programs and data, a special software system, called BOSS, is being developed [13, 19]. The functions performed by BOSS may be summarized as follows:

(*i*) Administer problem-oriented programs, involving such jobs as search for information about available programs and activating the selected ones.

(*ii*) Administer data base, intermediate storages, and communication with user. This may be done from user programs or directly from the terminal via special commands.

(*iii*) Administer, update, and change the internal BOSS system data. This involves, for instance, the addition of new problem-oriented programs and commands into the system and definition of new data bases.

The BOSS system offers the programmer several routines designed to simplify his task in writing problem-oriented programs. These routines include:

(*i*) routines that establish contact between the problem-oriented programs and BOSS;

(*ii*) routines for easy communication between program and external storages; and

(*iii*) routines for easy and flexible communications between program and user, including facilities for directing large output quantities to a printer.

Most of these routines work differently depending on the mode of execution (batch or demand). Thus the problem-oriented programs may be operated efficiently in batch as well as demand.

In Appendix II an example of a short BOSS-session is presented. The words typed by the user are underlined. The example demonstrates how an inexperienced user might approach the system. Experienced users are offered to specify the desired parameters immediately, and thereby to save considerable time. It should be noted that the programmer of a problem oriented program will activate the sequence between * WANT DATA TABLE DISPLAYED? and the answer NO in the example by means of only one single statement in his program.

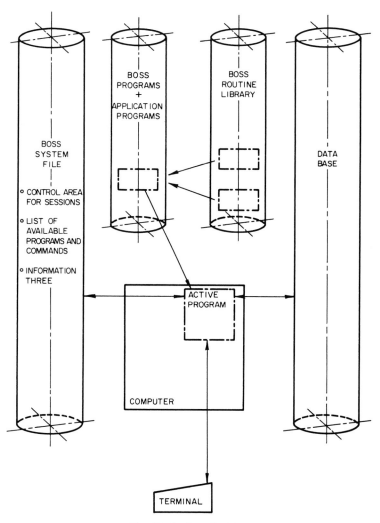

Fig. 12. Outline of BOSS.

Figure 12 indicates how the BOSS-system is organized. The system consists of the following three files:

   (*i*) the library of BOSS-routines;
   (*ii*) the program library, containing the BOSS-program as well as problem-oriented programs; and
   (*iii*) system file with internal system data.

In addition, each specific data base will reside on a file.

6. Concluding Remarks

The design jobs described in this paper are only a part of the total ship design task, which also involves determination of optimum principal dimensions, hull form parameters, propulsion plant, etc. A number of papers have been published recently on the application of computers to various groups of these other jobs [14–16]. An interesting integrated system for computer-aided design of ships has been described by Yuille [17]. In his software package most of the job routines implemented so far are working on problems related to hull surface definition, hydrostatics, capacity calculations, etc. A similar package of programs is Prelikon. This package is delivered as part of the newest version of Autokon, a program system designed to assist in the generation of all of the information about hull form and structure that is required for numerical control of flame cutters in the preparation of the steel plates [18]. Prelikon and Autokon are at present only run in the batch mode of operation. It is believed that program systems of the type described in this paper will in the future have to be incorporated into such larger systems. But the data handling problem increases dramatically if one attempts to obtain complete integration through the use of one common data base.

Acknowledgment

The author should like to acknowledge the many important contributions to the development of the system described in this paper given by a number of present and previous collaborators in the Department of Ship Structures, Trondheim.

Appendix I. Automated Design and Optimization Programs

PROGRAM NAME: LANOPT

(a) *Task*
Optimum design of midship sections of tankers (job 1)

(b) *Input Dimensions*

Length of ship, breadth, draft, and depth
Length of tank section
Block coefficient of ship
Camber, dead rise and bilge radius
Areas of eventual longitudinal center and side girders
Distances from center line to side girders
Distance from center line to longitudinal bulkhead
Frame spacings

(c) *Variables in Automated Design Process (Fig. 1)*

Spacings of longitudinals in deck (s_d), bottom (s_b), sides and longitudinal bulkheads (s_s).

Plate thicknesses in deck (t_d), bottom (t_b) and ship sides (t_s).

Section moduli of deck longitudinals (W_d) and bottom longitudinals (W_b).

Plate thicknesses in longitudinal bulkheads and section moduli of longitudinals on sides and bulkheads are calculated as functions of the free variables.

(d) *Additional Information*

The designs are required to comply with the rules of Det norske Veritas, including considerations to practical maximum thicknesses of plates.

Weight or cost of a tank section, excluding transverse frames and bulkheads, are used as object functions. A fairly old version of this program is described in [6].

Typical CPU-time on UNIVAC 1108: 15–20 sec.

PROGRAM NAME: **KOROPT**

(a) *Task*

Optimum design of vertically corrugated transverse bulkheads (job 2).

(b) *Input Dimensions*

Length of ship, breadth and height of bulkhead.

Distance between bulkhead and wash bulkhead.

Number of horizontal stringers.

(c) *Variables in Automated Design Process (Fig. 10)*

Width of profile flange (b_2) and depth of profile (d).

Length of corrugation (s).

Distances from deck to stringers $(a_1, a_2, \ldots)$.

Thicknesses of plates between stringers $(t_1, t_2, \ldots)$.

(d) *Additional Information*

The bulkheads are designed to comply with the rules of Det norske Veritas.

Weight of bulkheads, excluding stringers are presently used as object function. Further description of program is given in [7].

Typical CPU-time on UNIVAC 1108: 10 sec.

PROGRAM NAME: RAMOPT

(a) *Task*

Optimum design and analysis of statically indeterminate frames (jobs 4 and 5).

(b) *Input Dimensions*

Description of the frame topology such as node coordinates, codes for support conditions and spring constants at the supports.
Description of the elements, such as codes for cross section types, and plate flange areas for elements in sides, deck and bottom.
Loads
Automatic generation of loads and approximate spring constants at supports is presently being implemented.

(c) *Variables (Fig. 6)*

One variable for each member, for instance the cross sectional areas of the webs of the elements (A_w). During the optimization all of the members are scaled individually according to one of several implemented schemes, with starting values of cross sectional shapes as basis.

(d) *Additional Information*

Restrictions are based on equivalent stresses computed by means of von Mises yield theory.
Weight of frames (excluding plate flanges) is used as object function. Further descriptions of the program are presented in Lund [8], and Lund and Moe [9].
Typical CPU-time on UNIVAC 1108: 40 sec.

PROGRAM NAME: GIROPT

(a) *Task*

Optimum design of steel girders (job 4.1 of Fig. 9).

(b) *Input Dimensions*

Length of girder, sizes of brackets.
Distances between tripping brackets.
End loads and distributed loads (maximum four independent loading conditions).
Area of plate flange.

(c) *Variables in Automated Design Process* (*Figs. 6 and 11*)

Height of web (H_w), area of top flange (A_f), and thickness of web (t_w).
Moments of inertia of transverse and longitudinal stiffeners.
Numbers of transverse and longitudinal stiffeners.

(d) *Additional Information*

The stress constraints are based on equivalent stresses computed according to von Mises yield theory.

Buckling constraints as well as contraints with respect to the natural frequencies of the girders are also included.

Weight or cost of the girder is used as object functions.

Further description of program is presented in Tønnessen [10], see also Fig. 11.

The numbers of stiffeners are treated as integer variables.

The program is operational, but problem formulation as well as method of solution need to be developed further.

Typical CPU-time on UNIVAC 1108: 40 sec.

Appendix II. Example of BOSS Session

```
@XQT BOSS*BOSS:PROGRAM,START
** BOSS SESSION STARTED * DATE 070271  * TIME 134527
* DO YOU KNOW HOW TO USE THE SYSTEM BOSS?
* IF NOT, TYPE NO AND PUSH BUTTON 'RETURN'  ELSE TYPE YES
NO
****** DEPARTMENT OF DESIGN TECHNIQUES, SINTEF TRONDHEIM NORWAY ******
.
****** BOSS - A SYSTEM FOR INTERACTIVE PROGRAM AND DATA HANDLING ******
.
THE SYSTEM IS IN GENERAL OPERATED VIA COMMANDS OF THIS FORM:
.
'COMMANDCODE PARAMETER1,PARAMETER2,PARAMETER3,........,PARAMETER20'
    WHERE BOTH COMMANDCODE AND THE PARAMETERS EACH MAY BE 12 CHARACTERS
    OR LESS. THE COMMANDCODE IS  A CHARACTER STRING AND THE PARAMETERS MAY
    BE EITHER CHARACTER STRINGS OR NUMERIC VALUES OF ARBITRARY FORMAT.
IF YOU ARE NOT FAMILIAR WITH THE SYSTEM, YOU MAY GIVE THE COMMANDCODE
AND THE PARAMETERS WILL BE ASKED FOR LATER.
IF YOU DON'T KNOW HOW TO START, WRITE 'SEARCH'.
.
INPUT TO THE SYSTEM IS ONLY TO BE GIVEN WHEN ASKED FOR BY A
    'SPECIFY COMMAND', A 'SPECIFY ANSWER' OR BY A QUESTION.
.
AN INPUT LINE IS TERMINATED BY A PUSH ON THE 'RETURN'-BUTTON.
**************************** GOOD LUCK  ****************************************
** SYSTEM DATA CREATED BY RUN-ID: 61383
    ** PROJECT: ST-BOSS        ON DATE* 070271 AT TIME: 130339
* IS$COMLOG$$I   IS READY
* 31             IS READY
* WANT TO USE A DATA BASE?  WRITE BASE OR NOBASE

* SPECIFY COMMAND
BASE MS-FUTURA
* DATA BASE ON FILE MS-FUTURA    IS ACTIVE
* SPECIFY COMMAND
```

SEARCH
* SEARCH IS ACTIVATED
* A   APPLICATION PROGRAMS
* B   COMMANDS
* C   BOSS ROUTINES
* SPECIFY COMMAND SEARCH WITH THE ONE REFERENCE CHARACTER ABOVE
* IDENTIFYING THE SUBJECT YOU ARE INTERESTED IN,

* SPECIFY COMMAND
SEARCH B
* A   GENERAL SERVICE COMMANDS
* B   EXECUTION CONTROL COMMANDS
* C   GENERAL DATA FILE COMMANDS
* D   DATA BLOCK COMMANDS
* E   DATA BASE COMMANDS
* F   DAREK DATA BASE COMMANDS
* G   SYSTEM UTILITY COMMANDS
* H   APPLICATION COMMANDS

* SPECIFY COMMAND
SEARCH A
* A   SEARCH (S)   SEARCH FOR INFORMATION
* B   CPUT   GIVE ELAPSED CENTRAL PROCESSOR NIT TIME
* C   MSG   SEND MESSAGE TO OPERATOR
* D   ASKOP   SEND QUESTION TO OPERATOR AND PRINT HIS REPLY

* SPECIFY COMMAND
S A
* A   WRITE ONE OR MORE REFERENCE CHARACTERS AS PARAMETERS
* B   TO RESTART SEARCH GIVE PARAMETERS: TOP,REFERENCE-CHARS
* C   OR GIVE PARAMETERS: BACK,NO-OF-LEVELS,REFERENCE-CHARACTERS

* SPECIFY COMMAND
S,BACK
*
** COMMANDCODE NOT SEPERATED FROM PARAMETERS BY BLANK(S)

* RESPECIFY COMMAND
S BACK,2,B
* A   START   START EXECUTION OF APPLICATION PROGRAM
* B   STOP   STOP EXECUTION OF ACTIVE PROGRAM
* C   GOODBYE (BYE)   TERMINATE BOSS-SESSION

* SPECIFY COMMAND
S BACK,1,E
* A   CREATEBASE   - CREATE A BOSS DATA BASE
* B   CHANGEVERS (CHGV) CHANGE ACTIVE VERSION
* C   LISTVERS (LV)   LIST VERSION(S) IN DATA BASE
* D   COPYVERS (CPYV) COPY ALL VALUES FROM ONE VERSION TO ANOTHER
* E   COPYDATA (CPYD) COPY DATA ITEM VALUES FROM ONE VERSION TO ANOTHER
* F   DISPLAYDATA (DD) PRINT VALUES FROM DATA ITEM IN ACTIVE VERSION
* G   WRITEDATA (WD)   INSERT DATA INTO DATA ITEM IN ACTIVE VERSION
* H   DISPALYVERS (DV) PRINT THE VALUES IN THE DATA ITEMS IN VERSION

* SPECIFY COMMAND
S TOP,A
* A   STRUCTURAL DESIGN
* B   CONTROL THEORY

* SPECIFY COMMAND
S A
* A   OPTIMIZATION
* B   FINITE ELEMENT ANALYSIS
* C   OTHER DESIGN PROGRAMS

* SPECIFY COMMAND
S A
* A   CORBULKOPT   - OPTIMIZATION OF CORRUGATED TRANSVERSE BULKHEAD

```
* B   PLANETRUSS    - OPTIMIZATION OF PLANE TRUSS
* C   LONGMEMBOPT   - OPTIMIZATION OF LONG. STRENGTH MEMBERS OF TANKER

  * SPECIFY COMMAND
START
  * SPECIFY PROGRAM NAME
CORBULKOPT
  * EXECUTION OF PROGRAM: CORBULKOPT  IS STARTED

  *

OPTIMIZATION OF VERTICALLY CORRUGATED TRANSVERSE BULKHEAD
FOR OIL TANKERS, SATISFYING 1969 RULES OF DET NORSKE VERITAS

GIVE A TITLE OF YOUR RUN
DEMO
MAIN PARTICULARS CONCERNING BULKHEAD DESIGN
  * WANT DATA TABLE DISPLAYED?

YES
  * A =           4   NUMBER OF STRINGERS
  * B =    312,00     LENGTH OF SHIP      (M)
  * C =    26,700     DEPTH OF BULKHEAD   (M)
  * D =    25,000     DISTANCE BETWEEN (SWASH) BULKHEADS    (M)
  * E =    46,400     WIDTH OF BULKHEAD   (M)
  * F =    12,500     MINIMUM PLATING THICKNESS   (MM)

  * SPECIFY CHANGE OR NOCHANGE

  * SPECIFY COMMAND
CHANGE
  * GIVE NEW VALUES IN THIS WAY:
  * CHANGE B=12,C=,8E-7

  * SPECIFY COMMAND
CHANGE A=1
  * A=           1
  * ANY MORE CHANGES OF ABOVE DATA?
NO

DO YOU WANT TO CHANGE BUILT-IN BULKHEAD DESIGN PARAMETERS?
NO

FREE VARIABLES OF DESIGN PROBLEM
  * WANT DATA TABLE DISPLAYED?
YES
  * A =    ,86000     FLANGE WIDTH        (M)
  * B =    ,86000     CORRUGATION DEPTH   (M)
  * C =    1,2000     SPACING OF CORRUGATION  (M)
  * D =    13,000     THICKNESS OF PLATE  1 FROM DECK  (MM)
  * E =    7,5000     DISTANCE FROM DECK TO STRINGER 1 (M)
  * F =    15,000     THICKNESS OF PLATE 2

  * SPECIFY CHANGE OR NOCHANGE

  * SPECIFY COMMAND
CHANGE C=1,5,E=8,2
  * C=      1,5000
  * E=      8,2000
  * ANY MORE CHANGES OF ABOVE DATA?
NO

DO YOU WANT TO KEEP ANY OF THE VARIABLES AT A FIXED VALUE
DURING THE OPTIMIZATION?
STOP
  ** PROGRAM IS TERMINATED
```

```
 * SPECIFY COMMAND
BYE

** TERMINATION OF BOSS SESSION  * DATE 070271  * TIME 140?07

* DO YOU WANT TO HAVE A PRINTOUT ON CENTRAL PRINTER
* OF POSSIBLE GENERATED PRINTER OUTPUT?
NO
 ** CPU-TIME USED IN THIS SESSION  1,150 SECONDS

** HAVE BEEN PLEASED TO HELP YOU, COME BACK SOON  **
*
END
```

References

1. Dukes, T. P., and Dixon, J. R., "Analysis of Ship Structures by The Finite Element Method," *Proc. ISD/ISSC Symp. Finite Element Techniques, Stuttgart, June 1969* pp. 62–98.
2. Araldsen, P. O., Holtsmark, G., and Røren, E. M. Q., "Analysis of tankers with SESAM-69, Comparison of calculations and measurements," Course on Practical Use of the Finite Element Method, Trondheim (January 1971).
3. The Shipbuilding Res. Ass. of Japan. Rep. No. 68, "Studies on the Transverse Strength of Huge Ships," Tokyo (June 1970).
4. Roberts, W. J., "Strength of Large Tankers," *North East Coast Inst. Eng. Shipbuilders, Trans.* 86 **4**, 93–108 (March 1970).
5. Fiacco, A. V., and McCormick, G. P., *Nonlinear Programming, Sequential Unconstrained Minimization Technique.* Wiley, New York, 1968.
6. Moe, J., and Lund, S., "Cost and Weight Minimization of Structures with Special Emphasis on Longitudinal Strength of Tankers," *Trans. Roy. Inst. Naval Arch.* 110 **1**, 43–70 (1968).
7. Hagen, E., Leegard, F. O., Lund, S., and Moe, J., "Optimization of Hull Structures" (in Norwegian), Dept. of Ship Struct., Trondheim, Meddelelse SKB II/M11 (November 1968).
8. Lund, S., "Tanker Frame Optimization by Means of SUMT-Transformation and Behaviour Models," Dept. of Ship Struct., Trondheim, Meddelelse SKB II/M17 (December 1970).
9. Lund, S., and Moe, J., "Application of Behaviour Models in Structural Optimization" (to be published).
10. Tønnessen, A., "Girder Optimization Program, GIROPT—A Description of the Model of Analysis and Documentation of the Program," Dept. of Ship Struct., Trondheim (April 1971).
11. Lund, S., "Direct Search Program for Optimization of Nonlinear Functions with Nonlinear Constraints," Users Manual SK/P13, Dept. of Ship Struct., Trondheim (1971).
12. Gisvold, K. M., "A Method for Nonlinear Mixed-Integer Programming and its Application to Design Problems," Dept. of Ship Struct., Meddelelse SKB/M18, Trondheim (1970).
13. Beyer, E., "BOSS—System for Interactive Administration of Data and Program" (in Norwegian), Course on Practical Use of the Finite Element Method, Tech. Univ. of Trondheim (January 1971).
14. Gallin, C., "Entwurf wirtschaftlicher Schiffe mittels Elektronenrechner," Jahrbuch der Schiffbautechn. Gesellschaft, H.61, 270–306 (1967). Also BSRA Transl. No. 2945.
15. Nowacki, H., Brusis, F., and Swift, P. M., "Tanker Preliminary Design—An Optimization Problem with Constraints," Soc. of Naval Arch. and Marine Eng. Paper No. 9, Ann. Meeting, New York (November 1970).
16. Getz, J. R., Erichsen, S., and Heirung, E., "Design of a Cargo Liner in Light of the Development of General Cargo Transportation," Soc. Naval Arch. and Marine Eng., Diamond Jubilee Int. Meeting, New York (June 1968).

17. Yuille, I. M., "A System for On-Line Computer Aided Design of Ships—Prototype System and Future Possibilities," *Trans. Roy. Inst. of Naval Arch.* **112**, 443–463 (1970).
18. Mehlum, E., and Sørensen, P. F., "Example of an Existing System in the Shipbuilding Industry: The Autokon System," *Proc. Roy. Soc.* **321**, No. 1545, 219–233 (1971).
19. Beyer, E., Hansen, H. R., and Gisvold, K. M., "Outline of a General User-Oriented Computer Aided Design System, as Applied to Ship Design—BOSS," Dept. of Ship Struct., Meddelelse SKB II/M15, Trondheim (July 1969).

The STORE Project
(The Structures Oriented Exchange)

J. M. McCormick and M. L. Baron

WEIDLINGER ASSOCIATES
NEW YORK, NEW YORK

and N. Perrone

OFFICE OF NAVAL RESEARCH

A discussion and description of the STORE (Structures Oriented Exchange) system developed under the direction of the Office of Naval Research is presented. The project presents a method for the development, exchange and use of computer programs and information in structural mechanics analysis, design, and research. The unique features of the system include the active interaction of the users with the system designers and the use of an information space. The present pilot study has been limited to structural mechanics, but the general approach can certainly be extended to other disciplines.

1. Introduction

This report presents a discussion and description of the STORE (Structures Oriented Exchange) system developed under the direction of the Office of Naval Research. Utilizing today's computer time-sharing technology as a convenient tool, the project has aimed at a method for the cost-effective development, exchange, and use of computer programs and information in structural mechanics analysis, design and research.

STORE is based essentially on the principle that by entering a number of such programs into a central computer with time-sharing capabilities, an entire group of users could have on-line at their ready disposal a library of programs that no single user could afford to develop. Moreover, unlike most program depositories, STORE allows the user both to seek and to use the program immediately, through the use of a time-sharing computer, and to comment continuously on the programs in the system for their possible updating by the STORE monitors. Specifically, it is this latter interactive capacity that makes STORE unique, insofar as the user becomes an active member of the system, employing the terminal not only to utilize the programs already on line, but to input directly into the system both his comments based upon this use (for the guidance of both the system managers and other users) and his needs for new programs (for the guidance of research planners). The system thus grows and is fashioned by the same group of users that it serves, thus providing an effective communications medium between various workers in the same field and an easy and effective method of exchanging scientific information on an up-to-date basis.

The program library for the current prototype system has been limited to a single discipline: structural mechanics. The user can utilize the time-sharing computer system by means of a relatively inexpensive teletypewriter unit. Such units can be remotely located, so that the users of the system can be at distant locations from the large computer itself. One must recognize that STORE is not intended to provide a total capability in computing the solutions to structural mechanics problems. It would not, in its present phase, be used for extremely large scale computations such as multistory space frames. The local use of computers in existing installations would not be infringed upon. Rather, STORE would be used as a convenient tool for increasing the usefulness of key engineers and scientists by providing them with essentially a desk capability of using a continuously updated library of current research results and methods in structural mechanics.

The STORE project was originally conceived by Dr. Nicolas Perrone, Director of the Structural Mechanics Program, Office of Naval Research. After a survey which established the availability of appropriate computer hardware [1], the system was directed by that Program and defined and developed by the firm of Paul Weidlinger, Consulting Engineer,* which served as the STORE system manager. Significant contributions to the work were also made by the T.R.W. group [2]. A team of Office of Naval Research Contractors has contributed programs to the system, and used the system on an experimental basis. The more general "user community" envisioned consists of interested groups in the Navy and other United States Government Agencies. During its formative period, management and computing costs were paid by the Office of Naval Research for the prototype system use and testing.

It should be emphasized that STORE is completely user-oriented and

* Now Weidlinger Associates.

requires no programming of any kind on the part of the user. All instructions for the use of the system are given on both sides of one $8\frac{1}{2} \times 11$ in. sheet, which is supplied to the user. This approach has proved to be very successful during the testing and trial period of the STORE system.

It should be noted that in addition to providing the user with the capability of solving specific problems in structural mechanics, the system allows for a relatively simple and efficient means of exchanging up-to-date technical information, i.e., a continuous updating of the system. Moreover, through its interactive phase, the system allows a concentration of resources in those areas of structural mechanics which the system managers and the entire community of users feel are most urgently needed.

Section 2 discusses in detail the basic concept of STORE. Section 3 provides a description of the STORE system itself in both its information retrieval and program use and execution aspects. Section 4 describes in some detail the use of the system and Section 5 outlines costs. Some remarks on the basic differences between STORE and other time-sharing systems are given in Section 6.

2. Basic Concepts of STORE

The basic concept of STORE can best be understood by comparing it to a library of technical handbooks and textbooks in a particular discipline, which is shared by a community of users. It is therefore useful to list the activities which normally take place during the use of such a technical library:

1. (a) The user first searches in the index and/or the table of contents of one or several handbooks to locate the subject of his interest.
   (b) The user will look at the appropriate sections of the book, read chapter headings and parts of the text to ascertain that the subject found in the book matches the one that he had in mind.
   (c) After the appropriate chapter is located, the user will probably copy the pertinent information and formulas, or he may directly copy information from tables and charts contained in the book.
2. With these activities, the use of the reference book has been completed. The numerical solution of the problem itself now proceeds. This numerical solution may involve trivial or frequently elaborate calculations.
3. It is not unusual that the information obtained by the user contains some typographical errors or that it is not up-to-date. Normally, the user will have access to corrected data and/or updated information only through subsequent editions of the book, which normally occur in intervals of several years.

The STORE concept simulates and replaces all three of the activities described above:

1. It is an information retrieval system.
2. The system consists of a library of working programs stored in a central computer system, accessible from dispersed user terminals, and puts at the disposal of the users a library of programs that no single user could afford to develop.
3. The STORE system is dynamic insofar as it responds to the comments, corrections, and observations of the user. With the minimum of effort, the programs in the system may be continuously corrected and updated, so that the system is equivalent to the continuous publication of new updated technical handbooks and textbooks.

It is especially significant to recognize that STORE adds these other dimensions to the library concept. The system continuously grows as it is fashioned by the community of users that it serves. Such an advance has great potential insofar as it provides an effective communications medium between workers in the same field, it enhances their exchange of information, and it puts in the hands of the practicing engineer and scientist, up-to-date research results that are usually available only in published books or paper form at a considerably later date and in forms requiring integration or conversion for engineering use. STORE may perhaps best be likened to a continuously updated electronic textbook in which the user, through the use of computer time-sharing technology, can economically and conveniently make use of current research and development results in a given field or fields.

It must be noted that the computer time-sharing technology is the mechanism that permits a large number of individuals to use the library inexpensively. With only an inexpensive teletypewriter unit located at the user's convenience, the system can be employed by a large number of users at distant locations. From a practical viewpoint, the program library was generally limited to a single discipline, e.g., structural mechanics. This was done to cut down on the enormous computer system which would be required if several major disciplines were combined. With the present computer technology available, it would seem advisable to set up a separate STORE system for each major discipline, rather than one overall supersystem for several disciplines. The reason for this will become apparent when the information retrieval and program aspects of the STORE system are described in Section 3.

It should be recognized that the STORE concept is not intended to replace specialized computer systems or the simple acquisition of "software" and that, therefore, it does not compete with in-house computer centers of possible user groups, nor does it enter into the areas normally serviced by "software companies." It is also not intended to provide what might be termed a "total capability" for computing in structural mechanics. The STORE system is based on the principle that there exists, in the discipline for which the system is set up, e.g., structural mechanics, a common grouping of programs that the user community is willing to build up and share. The user of STORE is *not required* to have any specialized knowledge in the areas of computer programming or computer usage. The only prerequisite for the user of STORE is knowledge in the technical area of specialty of the user.

3. Description of the STORE System

The STORE system can best be described by considering two basic aspects: (1) the information retrieval aspect and (2) the program use and execution aspect.

3.1. INFORMATION RETRIEVAL ASPECT OF STORE

The information retrieval aspect of STORE consists of an information space and a series of system programs as follows:

(a) *Information Space*

The information space enables the user to identify, to a high degree of specialization, the particular area in which he is interested. It provides the user with the coordinates in the information space that correspond to the problem or problems that are to be solved. The information space, as set up for structural mechanics, consists of the following subject areas. Each subject area is designated by three ordered letters, one from each of the subdescriptor lists given in Table 1. Thus, for example, EAA designates an area concerned

TABLE 1
COORDINATES OF INFORMATION SPACE

TYPE OF STRUCTURE (COLUMN 1)		LOAD/ENVIRONMENT (COLUMN 2)		ANALYTICAL MODEL (COLUMN 3)	
BEAM/ROD/CABLE	A	STATIC/CONCENTRATED	A	ELASTIC	A
SHAFT	B	STATIC/UNIFORM/PRESSURE	B	PERFECTLY PLASTIC	B
COLUMN	C	STATIC/OTHER	C	STRAIN HARDENING	C
GRID	D	IMPULSIVE	D	LIMIT ANALYSIS	D
FRAME/TRUSS	E	PERIODIC	E	VISCOPLASTIC/STRAIN RATE EFFECTS	E
RING/ARCH	F	PRESSURE PULSE	F	VISCOELASTIC	F
MEMBRANE	G	CENTRIFUGAL/INERTIAL	G	INHOMOGENEOUS	G
PLATE-CIRCULAR	H	NATURAL VIBRATIONS/EIGENVALUES	H	ORTHOTROPIC	H
PLATE-RECTANGULAR	I	THERMAL	I	GEOMETRIC NONLINEAR	I
PLATE-OTHER	J	ACOUSTIC	J	MATERIAL NONLINEAR	J
TUBE/PIPE	K	ELECTRO/MAGNETIC	K	INSTABILITY/BUCKLING	K
SHELL-CYLINDRICAL	L	RADIATION	L	STRESS CONCENTRATION	L
SHELL-SPHERICAL	M	RANDOM	M	WAVE PROPAGATION	M
SHELL-ROTATIONALLY SYMMETRIC	N	ARBITRARY FORCING FUNCTION	N	FRACTURE	N
SHELL-OTHER	Y	MULTIPLE	Y	FATIGUE	O
CONTINUUM	Z	OTHER	Z	ACOUSTIC	P
COMBINATION				OPTIMIZATION	Q
OTHER				COMBINATION	Y
				OTHER	Z

with frames or trusses, under static concentrated loads, where an elastic analysis is performed.

(b) *System Programs*

A series of six system programs make up the information retrieval aspect of STORE. A brief description of each of these programs follows:

(1) *Programs* This program provides the STORE user with a listing of all programs in the STORE system which correspond to the portion of the information space requested. The program is entered with the coordinates of the information space.

(2) *Search* This program is also entered with the coordinates of the information space. However, it gives the user all programs in that portion of the information space that are known to the STORE managers. This includes programs already in STORE (as given in Programs), programs under development but not yet in STORE, and programs in this area which exist on other systems.

(3) *Comments* This program allows the user to comment on any program or group of programs in STORE that he has used. The comments can later be incorporated into the program description by the STORE system manager during periodic reviews of the Comment file. This interaction between the program users and the STORE system management allows the programs to be updated continuously in the light of the experience of the users, and thus allows future users to share the benefits of this experience. It is the essential feature of the system which allows the user to take an active and important part in its improvement and updating.

(4) *Need* Based upon the needs of the STORE users and their experience with existing programs in the program file, this program allows the STORE user to make requests for new programs and for modifications of the programs already included in the STORE library.

(5) *Bulletin* This program gives general announcements, suggestions, and notes of interest to the STORE participants. It may include, for example, a set of recommended procedures for solving particular problems of interest.

(6) *Master* This program gives the user a complete list of all the programs in the STORE library.

3.2. PROGRAM USE AND EXECUTION ASPECTS OF STORE

This phase of the system is entered with the knowledge of the program or group of programs that are to be executed. The program is known by its name, since it has been obtained by the user from his search of the information space.

(a) Execute (FORTRAN)

This program calls on the program that is to be used from its position in the STORE library. The user may elect to use any one of four execution options as follows:

Option (1): Execute with Data This option would be used by one who is thoroughly familiar with the program requested and had used it previously. It enables the user to enter his data and execute the program directly. No description of the program is printed out under this option.

Option (2): One Paragraph Description This option gives the user a brief one paragraph description of the program. This description is more detailed than the simple description which appears in the "Programs" file.

Option (3): Complete Documentation Printout This option gives the user a complete printout of all 18 program information suboptions which are obtained in the program documentation format. These suboptions are listed under Option (4). Slower teletypewriter units may require a considerable amount of time to complete this printout (often more than 30 min). This, however, adds to the terminal costs, rather than to the considerably higher computer costs.

Option (4): Selected Documentation Printout One or more of 18 suboptions in the program documentation format may be printed out at the request of the user. The eighteen possible suboptions are listed in Table 2.

TABLE 2
DOCUMENTATION SUB-OPTIONS

PROGRAM DOCUMENTATION FORMAT AND USER SUB-OPTIONS	
SUB-OPTION 1.	DEFINITION OF STRUCTURAL PROBLEM AREA(S) IN DESCRIPTOR LETTERS
SUB-OPTION 2.	TITLE, CONFIDENCE INDEX, AND SIMPLE ONE-LINE DESCRIPTION
SUB-OPTION 3.	NAME, ADDRESS, TELEPHONE NUMBER OF PROBLEM SOLVER
SUB-OPTION 4.	ONE PARAGRAPH DESCRIPTION OF PROGRAM
SUB-OPTION 5.	DETAILED STATEMENT OF PROBLEM SOLVED
SUB-OPTION 6.	SOLUTION METHOD
SUB-OPTION 7.	PROGRAM LIMITATIONS
SUB-OPTION 8.	ASSUMPTIONS AND SIGN CONVENTIONS
SUB-OPTION 9.	DEFINITIONS OF SYMBOLS AND TERMS
SUB-OPTION 10.	DESCRIPTION OF INPUT PROCEDURES AND FORMAT
SUB-OPTION 11.	DESCRIPTION OF OUTPUT PROCEDURES AND FORMAT
SUB-OPTION 12.	DISCUSSION OF NUMERICAL CHECK-OUT AND SAMPLE PROBLEM
SUB-OPTION 13.	CORRELATION WITH EXPERIMENTAL EVIDENCE
SUB-OPTION 14.	IDENTIFICATION OF PROGRAMMING LANGUAGE AND SYSTEM
SUB-OPTION 15.	ELAPSED TIME AT TERMINAL FOR TYPICAL RUNS
SUB-OPTION 16.	COMPUTER CENTRAL PROCESSOR TIME USED FOR TYPICAL RUNS
SUB-OPTION 17.	REFERENCES
SUB-OPTION 18.	NOTES AND REVISION DATES

The programs listed and discussed in this section provide a description of the basic STORE system. The use of the system is described in Section 4. An illustrative example of the suboption documentation is presented in Section 7.

4. Use of the STORE System

The STORE system is completely user-oriented, requiring no programming of any kind on the part of the user. The only prerequisite is that he be knowledgeable in the specialty field in which he wishes to solve the problem, in this case, some aspect of structural mechanics.

A prime ground rule which was maintained in developing STORE is that it minimize paperwork exchange and be simple to use. To implement this, it was required that all instructions for the use of the system be restricted to both sides of an $8\frac{1}{2} \times 11$ in. sheet. This plastic-coated sheet is the only paperwork that the user sees, other than that which is printed at his terminal. Reproduction of both sides of the instruction guide is presented in Figs. 1 and 2.

---

1 GETTING STARTED - $FORT STORE

OBTAIN AN AUTHORIZED TELEPHONE NUMBER, USER CODE NUMBER, AND USER CODE PASSWORD FROM YOUR STORE REPRESENTATIVE. PRESS THE "ORIG" KEY TO THE RIGHT OF YOUR TERMINAL KEYBOARD AND DIAL YOUR ASSIGNED NUMBER. IF YOU HAVE DIFFICULTY MAKING THE CONNECTION, WAIT A FEW MINUTES AND TRY AGAIN. WHEN YOU ARE CONNECTED THE COMPUTER WILL REQUEST YOUR USER CODE NUMBER; TYPE IT IN AND PRESS THE "RETURN" KEY. SIMILARLY, TYPE IN YOUR USER CODE PASSWORD FOLLOWED BY A "RETURN". IF YOUR CODE NUMBER AND PASSWORD ARE VALID, THE SYSTEM WILL PRIN "READY", OTHERWISE YOUR LINE WILL BE PROMPTLY DISCONNECTED FROM THE COMPUTER. WHEN THE SYSTEM PRINTS "READY TYPE $FORT STORE (HOLD DOWN THE "SHIFT" KEY FOR THE DOLLAR SIGN - $), PRESS THE "RETURN" KEY. SPECIAL MESSAGES AN ACCOUNTING DATA FOR YOUR PARTICULAR USER GROUP WILL BE PRINTED.

PLEASE READ BOTH SIDES OF THIS GUIDE CAREFULLY IN ORDER TO ACQUAINT YOURSELF WITH THE MANY FEATURES PROVIDED FOR YOU. REMEMBER TO HOLD DOWN THE "SHIFT" KEY WHEN TYPING IN A DOLLAR SIGN ($) AND ALWAYS PRESS THE "RETURN KEY AFTER COMPLETING AN ENTRY. NOTE: FAILURE TO TYPE $FORT STORE MAY RESULT IN EXCESSIVE CHARGES.

2 BULLETINS FROM THE STORE PROJECT OFFICE - $FORT BULLETIN

THE "BULLETIN" FILE CONTAINS ANNOUNCEMENTS, SUGGESTIONS, AND NOTES OF INTEREST TO STORE PARTICIPANTS. TYPE $FORT BULLETIN. THE COMPUTER WILL PRINT THE VARIOUS BULLETINS BEGINNING WITH THE MOST RECENT. WHEN YOU WISH TO TERMINATE THE PRINTING, PRESS THE "BREAK" KEY AND THEN PUSH THE "BREAK-RELEASE" (BRK-RLS) BUTTON, THE SYSTEM WILL RETURN TO "READY".

3. SEARCHING THE STORE LIBRARY FOR WORKING PROGRAMS - $FORT PROGRAMS

WORKING PROGRAMS IN THE LIBRARY ARE CLASSIFIED BY "SUBJECT AREA". A SUBJECT AREA IS DESIGNATED BY 3 ORDERED LETTERS, ONE FROM EACH OF THE SUB-DESCRIPTOR LISTS BELOW. THUS, EAA DESIGNATES AN AREA CONCERNED WITH FRAMES OR TRUSSES UNDER STATIC/CONCENTRATED LOADS WHERE AN ELASTIC ANALYSIS IS PERFORMED.

	TYPE OF STRUCTURE (COLUMN 1)		LOAD/ENVIRONMENT (COLUMN 2)		ANALYTICAL MODEL (COLUMN 3)
A	BEAM/ROD/CABLE	A	STATIC/CONCENTRATED	A	ELASTIC
B	SHAFT	B	STATIC/UNIFORM/PRESSURE	B	PERFECTLY PLASTIC
C	COLUMN	C	STATIC/OTHER	C	STRAIN HARDENING
D	GRID	D	IMPULSIVE	D	LIMIT ANALYSIS
E	FRAME/TRUSS	E	PERIODIC	E	VISCOPLASTIC/STRAIN RATE EFFECTS
F	RING/ARCH	F	PRESSURE PULSE	F	VISCOELASTIC
G	MEMBRANE	G	CENTRIFUGAL/INERTIAL	G	INHOMOGENEOUS
H	PLATE-CIRCULAR	H	NATURAL VIBRATIONS/EIGENVALUES	H	ORTHOTROPIC
I	PLATE-RECTANGULAR	I	THERMAL	I	GEOMETRIC NONLINEAR
J	PLATE-OTHER	J	ACOUSTIC	J	MATERIAL NONLINEAR
K	TUBE/PIPE	K	ELECTRO/MAGNETIC	K	INSTABILITY/BUCKLING
L	SHELL-CYLINDRICAL	L	RADIATION	L	STRESS CONCENTRATION
M	SHELL-SPHERICAL	M	RANDOM	M	WAVE PROPAGATION
N	SHELL-ROTATIONALLY SYMMETRIC	N	ARBITRARY FORCING FUNCTION	N	FRACTURE
O	SHELL-OTHER	Y	MULTIPLE	O	FATIGUE
P	CONTINUUM	Z	OTHER	P	ACOUSTIC
Y	COMBINATION			Q	OPTIMIZATION
Z	OTHER			Y	COMBINATION
				Z	OTHER

THERE MAY BE SEVERAL PROGRAMS STORED IN A SUBJECT AREA, OR THERE MAY BE NONE. TO FIND OUT, TYPE SPORT PROGRAMS
THE SYSTEM WILL PRINT SUBJECT AREA= . TYPE IN THE 3-LETTER DESIGNATION FOR THE AREA OF INTEREST. IF THERE ARE
PROGRAMS IN THAT AREA, THEIR TITLES AND BRIEF DESCRIPTIONS WILL BE PRINTED; OTHERWISE, THE SYSTEM WILL PRINT, E.G.
"THERE ARE NO PROGRAMS SAVED IN AREA ABC." IF YOU CHOOSE TO DISCONTINUE YOUR SEARCH, PRESS THE "BREAK" KEY,
THEN PUSH THE "BREAK-RELEASE" (BRK-RLS) BUTTON. THE SYSTEM WILL PRINT "READY". TRY IT, IT'S REALLY QUITE SIMPLE.
THE NUMBER IN PARENTHESES RIGHT AFTER A PROGRAM TITLE IS THE "CONFIDENCE INDEX", SEE ITEM 8 ON THIS INSTRUCTION
SHEET FOR A FURTHER DISCUSSION OF THE INDEX.

---

4. OBTAINING INFORMATION ABOUT EXISTING PROGRAMS NOT IN THE STORE SYSTEM - SPORT SEARCH

THERE ARE HUNDREDS OF EXISTING PROGRAMS, NOT IN STORE, FOR WHICH LIMITED INFORMATION IS AVAILABLE. THIS DATA
IS STORED IN THE SEARCH FILE AND MAY BE RETRIEVED BY SPECIFYING THE NAME OF A PROGRAM, OR THE AUTHOR/PROBLEM
SOLVER, OR SUBJECT AREA AS DESCRIBED IN SECTION 3. TYPE SPORT SEARCH AND SUPPLY THE APPROPRIATE DATA. THE INFOR-
MATION IN THE "SEARCH" FILE CAN BE USED IN CONJUNCTION WITH THE "NEED" FILE DESCRIBED BELOW. WHEN YOUR SEARCH
IS FINISHED, PRESS THE "BREAK" AND "BREAK-RELEASE" BUTTONS IN THAT ORDER.

---

5. REGISTERING YOUR NEED FOR A PROGRAM OR OTHER INFORMATION - SPORT NEED

THIS FILE IS OF FUNDAMENTAL IMPORTANCE BECAUSE IT PROVIDES A MECHANISM FOR BUILDING A LIBRARY DIRECTLY REFLECTING
YOUR INTERESTS, THE STORE USER. IF YOU HAVE A NEED FOR A PROGRAM OR STRUCTURAL MECHANICS INFORMATION NOT
PROVIDED IN THE SYSTEM, IT IS VERY SIMPLE TO MAKE YOUR REQUIREMENT KNOWN. TYPE SPORT NEED, THE COMPUTER WILL
ASK YOU FOR YOUR NAME, ADDRESS, TELEPHONE NUMBER, THE SUBJECT AREA, AND WHEN THE EQUAL SYMBOL APPEARS YOU
MAY THEN TYPE IN YOUR REQUEST ON AS MANY LINES AS IS NECESSARY, WAITING FOR THE EQUAL SYMBOL FOR EACH LINE.
WHEN ALL OF YOUR REQUEST IS TYPED IN, PRESS THE "RETURN" KEY AND THEN THE "BREAK" KEY FOLLOWED BY PUSHING THE
"BREAK-RELEASE" BUTTON.

---

6. COMMENTING ABOUT PROGRAMS, SYSTEM OPERATION, OR OTHER REMARKS - SPORT COMMENT

THIS FILE, OPERATING MUCH LIKE THE "NEED" FILE (ITEM 5 ABOVE), GIVES YOU THE OPPORTUNITY TO POINT OUT ERRORS IN
PROGRAMS, MAKE SUGGESTIONS, ASK QUESTIONS, COMPLAIN, AND GENERALLY STATE YOUR OPINIONS CONCERNING ANY
ASPECT OF STORE. TYPE SPORT COMMENT, SUPPLY THE INFORMATION REQUESTED, AND THEN TYPE IN YOUR COMMENT AFTER
EACH EQUAL SYMBOL APPEARS. WHEN THROUGH, PRESS "RETURN", "BREAK", AND "BREAK-RELEASE" IN THAT ORDER.

NOVEMBER
1968

Fig. 1. Instructions for use of the STructures ORiented Exchange (STORE).

The following paragraphs contain a summary description of the actual use of the STORE system.

1. The only equipment that the user requires is a console such as a typewriter terminal or teletypewriter connected to the central computer system. To use the system, each user is assigned a code number and an appropriate password. These are both entered through the console and the security of the password is protected by the fact that it is not reproduced (typed out) on his record of the transaction.

2. After the user is connected with the center, he may, upon request, obtain a bulletin containing up-to-date notes of interest to the STORE system participants (such as information regarding the updating of programs already in the system and the availability of new programs in the system).

3. The working programs in the STORE library are currently identified within the framework of the information space by three sets of descriptors, which are sufficient for the identification of a subject area. The user locates his subject area by entering only three letters, each corresponding to a particular word in each of the three sets of descriptors. The console will respond by typing the titles and a brief description of the available programs corresponding to the subject as defined by the descriptors. The user can then discontinue his search if he is presented with a title which corresponds to his

area of interest, or he can continue to "browse" through related topics in the library. Once the specific program has been requested, the user has the option to obtain increasing details regarding the program, in accordance with his needs and interest.

4. When the user has found the appropriate program, his search of the STORE information space is completed, and he enters into the execution phase of the problem. Since this execution of the program takes part in the computer system itself, all that is theoretically required is that the data be entered through the terminal; the result of the computations will be available on the console as they are computed by the central computer. If the user is thoroughly familiar with the program and has used it previously, this, in general, will be all that is required.

---

7. MASTER DIRECTØRY ØF ALL WØRKING STØRE PRØGRAMS - SFØRT MASTER

THE "MASTER" FILE IS A DIRECTØRY ØF ALL WØRKING PRØGRAMS IN THE STØRE SYSTEM. TYPE SFØRT MASTER, THE CØMPUTER WILL PRINT THE TITLES AND ØNE-PARAGRAPH DESCRIPTIØNS ØF EACH PRØGRAM IN THE SYSTEM, BEGINNING WITH THE ØNE ADDED MØST RECENTLY. THE FILE WILL BE EVER-GRØWING. SEARCHING FØR A PRØGRAM BY REVIEWING THIS DIRECTØRY WILL BECØME LESS AND LESS PRACTICAL AS PRØGRAMS ARE ADDED BECAUSE IT WILL BE SØ LARGE (THE PRØGRAMS FILE SHØULD BE USED). HØWEVER, "MASTER" WILL PRØVIDE A MEANS FØR THE USER TØ FIND ØUT WHAT IS IN THE SYSTEM IN TØTAL. IF YØU WISH TØ TERMINATE PRINTING, PRESS THE "BREAK" KEY FØLLØWED BY THE "BREAK-RELEASE" (BRK-RLS) BUTTØN.

---

8. THE CØNFIDENCE INDEX (HIGH - 1, LØW 5)

THE NUMBER IN PARENTHESES BESIDE THE TITLE ØF A WØRKING PRØGRAM IS THE CØNFIDENCE INDEX ASSIGNED BY THE STØRE PRØJECT ØFFICE. IT IS A NUMBER ARRIVED AT AFTER PRØGRAM REVIEWS BY TECHNICAL SPECIALISTS AND TYPICAL USERS. THE INDEX DISTINGUISHES PRØVEN PRØGRAMS ØF HIGH QUALITY BASED ØN ACCEPTED THEØRY AND NUMERICAL SØUNDNESS FRØM THØSE ØF A PØSSIBLY QUESTIØNABLE NATURE. THE PURPØSE ØF THE INDEX IS TØ ALLØW THE USER TØ ASSESS, AT A GLANCE, THE EXTENT TØ WHICH A GIVEN PRØGRAM HAS BEEN PRØVEN SØUND. THE RATING IS AS FØLLØWS:

$$1 \quad \text{-HIGH CØNFIDENCE}$$
$$\cdot$$
$$\cdot$$
$$5 \quad \text{-LØW CØNFIDENCE}$$

MØST NEW PRØGRAMS ENTERING THE SYSTEM WILL BE ASSIGNED A PRELIMINARY CØNFIDENCE INDEX ØF 3. SUBSEQUENT RAISING ØR LØWERING ØF THE CØNFIDENCE INDEX DEPENDS ØN THE EXPERIENCES ØF USERS.

---

9. RETRIEVAL AND USE ØF A SPECIFIC PRØGRAM - SFØRT NAME

A PRØGRAM TITLE MAY HAVE AS MANY AS 8 CHARACTERS. IF YØU WISH TØ RETRIEVE A PRØGRAM ENTITLED, E.G. BUCKLING, TYPE SFØRT BUCKLING. IF THE PRØGRAM IS, IN FACT, STØRED IN THE LIBRARY THE CØMPUTER WILL PRINT ØPTIØN= . YØU MUST RESPØND BY TYPING EITHER 1, 2, 3, ØR 4. THESE NUMBERS CØRRESPØND TØ THE 4 ØPTIØNS DISCUSSED BELØW. NØTE THAT ØPTIØN 4 REQUIRES YØUR SELECTING CERTAIN SUB-ØPTIØNS. UPØN CØMPLETIØN ØF ANY ØNE ØPTIØN, THE CØM-PUTER WILL AGAIN PRINT ØPTIØN= WHEN FINISHED, PRESS "BREAK" AND THEN "BREAK-RELEASE".

ØPTIØN=1 - DIRECT PRØGRAM EXECUTIØN. ASSUMING YØU KNØW HØW TØ RUN A PRØGRAM, THIS ØPTIØN LETS YØU IMMEDIATELY START SUPPLYING DATA ØR DEFINING AN INPUT FILE.

ØPTIØN=2 - ØNE PARAGRAPH DESCRIPTIØN. THE TELETYPEWRITER PRINTS A ØNE PARAGRAPH DESCRIPTIØN ØF THE PRØGRAM (SAME AS SUB-ØPTIØN 4 BELØW). IT IS MØRE DETAILED THAN THE SIMPLE DESCRIPTIØN IN "PRØGRAMS".

ØPTIØN=3 - CØMPLETE DØCUMENTATIØN PRINTØUT. A PRINTØUT IS PRESENTED ØF ALL 18 SUB-ØPTIØNS IN THE PRØ-GRAM DØCUMENTATIØN FØRMAT LISTED BELØW. SLØWER TELETYPE UNITS MAY REQUIRE MØRE THAN 30 MINUTES TØ CØMPLETE THE PRINTØUT.

ØPTIØN=4 - SELECTED DØCUMENTATIØN PRINTØUT. ØNLY THØSE NUMBERED SUB-ØPTIØNS SPECIFIED WILL BE PRINTED. FØR EXAMPLE, SUPPØSE YØU ØNLY WANTED PRINTØUT ØF INFØRMATIØN IN SUB-ØPTIØNS 2, 11, AND 13 ØF THE DØCUMENTATIØN FØRMAT, THE SYSTEM REQUESTS AND YØUR RESPØNSES WØULD BE:

ØPTIØN = 4
SUB-ØPTIØNS, HØW MANY? = 3
LIST SUB-ØPTIØN NUMBERS = 2, 11, 13

PROGRAM DOCUMENTATION FORMAT AND USER SUB-OPTIONS
SUB-OPTION 1.   DEFINITION OF STRUCTURAL PROBLEM AREA(S) IN DESCRIPTOR LETTERS
SUB-OPTION 2.   TITLE, CONFIDENCE INDEX, AND SIMPLE ONE-LINE DESCRIPTION
SUB-OPTION 3.   NAME, ADDRESS, TELEPHONE NUMBER OF PROBLEM SOLVER
SUB-OPTION 4.   ONE PARAGRAPH DESCRIPTION OF PROGRAM
SUB-OPTION 5.   DETAILED STATEMENT OF PROBLEM SOLVED
SUB-OPTION 6.   SOLUTION METHOD
SUB-OPTION 7.   PROGRAM LIMITATIONS
SUB-OPTION 8.   ASSUMPTIONS AND SIGN CONVENTIONS
SUB-OPTION 9.   DEFINITIONS OF SYMBOLS AND TERMS
SUB-OPTION 10.   DESCRIPTION OF INPUT PROCEDURES AND FORMAT
SUB-OPTION 11.   DESCRIPTION OF OUTPUT PROCEDURES AND FORMAT
SUB-OPTION 12.   DISCUSSION OF NUMERICAL CHECK-OUT AND SAMPLE PROBLEM
SUB-OPTION 13.   CORRELATION WITH EXPERIMENTAL EVIDENCE
SUB-OPTION 14.   IDENTIFICATION OF PROGRAMMING LANGUAGE AND SYSTEM
SUB-OPTION 15.   ELAPSED TIME AT TERMINAL FOR TYPICAL RUNS
SUB-OPTION 16.   COMPUTER CENTRAL PROCESSOR TIME USED FOR TYPICAL RUNS
SUB-OPTION 17.   REFERENCES
SUB-OPTION 18.   NOTES AND REVISION DATES

10. STOPPING THE SYSTEM AT ANY TIME - BREAK

THERE WILL BE INSTANCES IN WHICH YOU WILL WANT TO STOP THE COMPUTER FROM DOING WHATEVER IT IS DOING: PRINTING, EXECUTING, ETC. TO DO THIS, PRESS THE "BREAK" KEY, THEN THE "BREAK-RELEASE" (BRK-RLS) BUTTON. THE SYSTEM WILL RETURN TO A "READY" CONDITION.

11. DELETING TYPING ERRORS (A LINE: "CONTROL" X, A CHARACTER: "SHIFT" L)

THERE ARE TWO METHODS FOR CORRECTING ON-LINE ERRORS:
     A. AN ENTIRE LINE MAY BE DELETED BY HOLDING DOWN THE "CONTROL" KEY (ON THE LEFT) AND THEN PRESSING THE X KEY. THE SYSTEM WILL PRINT "DELETED" AND YOU MAY THEN RE-TYPE THE LINE.
     B. THE "\" CHARACTER (HOLD DOWN THE SHIFT KEY AND PRESS L) WILL DELETE ONE PRECEDING CHARACTER FOR EACH REVERSE SLASH.
THESE ARE TIME-SAVERS THAT YOU SHOULD NOT HESITATE TO USE WHENEVER A TYPING ERROR IS MADE.

12. SIGNING OFF - $FORT STORE, $END

WHEN YOUR WORK IS FINISHED, TYPE $FORT STORE AS YOU DID TO GET STARTED. DO NOT FAIL TO DO SO, OTHERWISE YOUR GROUP MAY BE CHARGED FOR TIME NOT ACTUALLY USED. THEN TYPE $END, THE TERMINAL WILL BE PROMPTLY DISCONNECTED.

13. RESPONSIBILITIES OF STORE PARTICIPANTS

THE PROJECT STORE GOALS CAN BE REACHED ONLY THROUGH PARTICIPATION BY ALL CONCERNED. IF YOU NEED A PROGRAM OR OTHER STRUCTURAL MECHANICS INFORMATION, CALL UP THE "NEED" FILE AND MAKE YOUR REQUEST. IF YOU HAVE A COMPLAINT, A CRITICISM, OR OTHER COMMENT, CALL UP THE "COMMENT" FILE AND LET IT BE KNOWN. THESE FILES ARE EASY TO USE AND YOU ARE ENCOURAGED TO DO SO.

Fig. 2. (Continuation of Fig. 1.)

In many cases, however, the user is not familiar with the program and will be using it for the first time. He will then use one of the four executive options in the Execute (FORTRAN) program. These options, described in Section 3, range from a one paragraph description of the program to a complete documentation printout which includes a meaningful illustrative example. He will then enter the required input data and the program will be executed.

5. The results of the executed problem will be received by the user through his terminal.

The STORE system has been made interactive purposely in order to give the user various options that can be used to register his needs for new programs and requests for the updating of old programs. Through the Comments program, he may enter his need for programs that are not found in the library or he may register observations and suggestions regarding the quality and applicability of the programs used, as well as corrections of the information he has obtained. It should be noted that if the desired program is not available within the STORE library, the user may be provided with information regarding similar programs obtainable from other sources. The STORE search file contains data on over 500 programs in existence throughout the country. Moreover, the system contains a built-in capacity for continuously updating itself by registering observations, future and current needs of the user, and calling these needs to the attention of the system managers. As a result of such requests, the program library can be extended in a manner tailored to the users' needs and appropriate corrections and additions can be made as required.

It should be recognized that the ability to request and comment on specialized information is an important part of the STORE system. If a request is recognized as legitimate, it may be registered in the Bulletin file and hopefully read and responded to by numerous users who are familiar with the subject area. As the STORE community of users grows, this becomes an important means of exchanging and updating technical information. This technical information exchange, in addition to the ability to solve specific problems in structural mechanics, is a prime aim of the STORE concept.

5. Costs

The costs involved in the use of a time-sharing system vary widely with the system used. During the period of the development of the STORE system, the GE 605 computer and its time-sharing system was utilized, together with a group of standard teletypewriter (TTY) consoles as remote terminals. The monthly rental for each of the TTY Consoles is $116.

The basic system cost is determined largely by the random access storage space rental charge. This storage space is required for the STORE library programs and the systems programs. During the time of the development and demonstration of STORE, a maximum of 2400 links of storage were used, at a storage cost of $1 per link per month. Each link consisted of 512 four-character words, hence, 2048 characters per link. The breakdown of the storage link usage follows:

(a) System program   600 links
(b) System files (needs, comments, etc.)   100 links
(c) Reservoir file—Search   300 links
(d) Working area   1400 links (maximum)
                                            ——————
                        Total = 2400 links of storage

Toward the end of the project, the basic storage charge was reduced to 0.04¢ per character per month, or $0.82 per link per month.

Two basic costs were encountered by the STORE system user during the running of a problem. These involved direct running costs of two types:

(1) Central processing unit cost   40¢ per second
(2) Connection cost $9.00 per hour

Input and output time was primarily connection cost time and hence the user of a slow terminal was essentially charged at the rate of $9.00 per hour, rather than at the considerably higher central processing unit cost.

As a typical example, the direct cost of running a two-dimensional structural frame problem with 22 joints was as follows:

Central processing unit cost: 11 sec @ 40¢/sec = $4.40
Terminal connect time cost: 17 min @ $9.00/hr = $2.51
(includes input–output)   ——————
                                          Total   = $6.91

The total cost of $6.91 does not include storage link costs or the user's own remote terminal rental costs, which must be amortized over the total usage in a monthly period.

The development of Project STORE to the prototype demonstration state described herein has been accomplished over a period of about 18 months and at a total cost of approximately $175,000 of funds provided by the Structural Mechanics Program of the Office of Naval Research.

6. Distinguishing Characteristics of the STORE System

The differences between STORE and other time-sharing systems are of two types: differences in concept and differences in degree.

6.1. CONCEPTUAL DIFFERENCES

(a)   The STORE system is probably the only existing system that is organized on the basis of a detailed information space, in which the machine itself searches and gives the user specific information and programs corresponding to the input coordinates of the information space. While all time-sharing systems have the equivalent of the master program which prints out a complete listing of all programs in the system, we know of no other existing system which presents the user with a group of programs from an identifiable area of the information space.

(b)   Through the use of the Search program, the user is also brought up-to-date with what exists his field of interest. This includes programs that (1) are in the STORE library at the present time, (2) in the process of development for future entry into the STORE library, and (3) not available from STORE, but available from other sources which are identified.

(c) Through the use of the Comments and Needs programs, the user becomes an active participant in an interactive system in which the existing programs are continuously being upgraded and new programs being added in accordance with the indicated needs of the system users. This ability for interaction between the users and the system managers is perhaps the most important conceptual difference between STORE and other time-sharing systems. In the usual time-sharing system, the user can only call for a listed program and use it. The STORE concept gives the user an active role in the system itself. If problems arise in the use of the program, or if the user wishes to relate his experience in using the program to solve problems, he may do so through the Comments program. If he feels that important programs are missing, he may ask for these through the Needs program. While all time-sharing systems have the Bulletin and Master program options, it is felt that STORE is unique in its Comments and Needs options which provide a means for the continuous updating of the system and its programs. It must be realized that the users of the system will be scientists and engineers who will require such interaction and active participation in the system.

6.2. DIFFERENCES IN DEGREE

(a) While all time-sharing systems necessarily give a brief program description and instructions for the program use, the documentation options in the STORE system appear to be more complete by an order of magnitude. The user may elect to call for any or all of the 18 documentation suboptions. The system, through this detailed documentation thus becomes suitable for use by both experienced users who do not use the documentation options but directly execute the program, and new users who require various levels of documentation. In most time-sharing systems, a great deal of this is necessarily done with the technical people of the system—a process which is both time consuming and frustrating. In STORE, this is done directly by the user on his console. The convenience and flexibility of the STORE approach is an important asset which can best be appreciated by those who have gone through the time delays and frustrations of the alternate approach. The users of STORE have definitely indicated their support and appreciation for the flexibility of the system.

7. Illustrative Example—Typical STORE Program Documentation

This section gives, as an illustrative example, the complete 18 suboptions documentation of a typical program in Project STORE. The problem considered involves the calculation of buckling loads for a ring stiffened cylindrical shell under hydrostatic pressure. This problem was contributed to Project STORE by the Thompson–Ramo–Wooldridge Corporation, Redondo Beach, California. It was developed by Mr. Sesto J. Voce, Staff Engineer.

In what follows, each suboption is presented as a separate numbered paragraph.

1 Definition of structural problem—descriptor letters
LBH, LBK

2 Character title and simple one-line program description
Buckling, calculated general and local instability—ring stiffened cylinders under hydrostatic pressure.

3 Name, address, telephone number of problem solver
Sesto J. Voce, Engineering Mechanics Laboratory, Staff Engineer, Room 1004, Building R 1, One Space Park, Redondo Beach, California 90178, Telephone Number 213 + 679-8711, Extension 63601. Also can contact J. Hrzina, TRW Systems participant, Project STORE, 1735 Eye Street, N.W., Washington, D.C. 20006, 202 + 265-1616, Extension 336.

4 One paragraph description of program
This program calculates the hydrostatic pressures at which general instability and shell local instability between rings occur. The cylinder is circular with evenly spaced equal strength circular ring frames. The program accounts for orthotropic properties of the cylinder wall in the longitudinal and hoop directions. The theory confirming this program is an extension of the energy method of Kendrick. Input required consists of cylinder and ring geometrical and material properties assuming elastic behavior. The outputs are the pressures at which general and local instabilities occur and corresponding parameters describing the buckled mode shape.

5 Detailed statement of problem solved
This program calculates the external hydrostatic pressure at which general instability occurs and also that at which shell local buckling between rings occurs.

The cylinder wall can have orthotropic properties in the circumferential and longitudinal directions. It is stiffened with an arbitrary number of evenly-spaced, uniform-strength rings.

The original analysis [1] was instigated by the attractiveness of ring-stiffened filament-wound cylinders for deep submersible structures for which the general instability mode is a primary design consideration.

6 Solution method
Voce [1] contains a full discussion. The problem is dealt with by an energy method. Total potential energy is expressed as the sum of
  (1) Extensional strain energy of shell
  (2) Bending strain energy of shell
  (3) Extensional strain energy of ring frames
  (4) In-plane bending strain energy of ring frames
  (5) Potential energy due to external hydrostatic pressure
Nonlinear strain-displ rels used were those of Kendrick [5] and Kempner [7], and genl instab. modes used were those recommended in Kendrick [8]. The total potential energy becomes an expression involving five arbitrary displ constants. The energy expression is minimized with respect to these constants which results in a quintic equation in a pressure parameter PHI, which is a function of shell and ring geometry and material properties as well as the parameters

Q, N. For all acceptable values of Q, the value of N is determined that produces the lowest root of the quintic eqn. The critical pressure is the lowest of all possible alternatives considered in all of the various mode shapes.

Shell local buckling between rings is predicted through use of classical buckling deformations for simply-supported end conditions.

7 Program limitations

Definition of parameter ranges over which the program gives good results is extremely difficult and must be dealt with in the "good judgment" of the user. Limited test results (see item 13) indicate very good correlation with experiments in the low radius/thickness range (less than 100) which is unusual. It is expected that good results would be obtained in the higher ranges too, but test results are not available.

The program should give consistently conservative buckling pressures because it considers many different buckling modes. This is obviously a very good feature for design purposes.

8 Assumptions and sign conventions

Voce [1] contains a full discussion. Out-of-plane bending, torsional and warping strain energies of ring frames excluded because they were found [5, 6] to have negligible effects for isotropic ring-stiff'd cyls and are not expected to be significant for the orthotropic case.

Prebuckling ring and cylinder stresses were obtained assuming uniform axial and radial contraction. The general instability mode assumed is that recommended in Kendrick [8]. Such a mode is accompanied by interbay buckling and the effects of the end bulkheads or rigid supports are confined to Q frame spaces at each end of the cyl. Investigators [8, 9] have found this mode, while containing slope discontinuities at the rings, gives lower buckling values than the classical buckled deformations usually assumed. For $Q = (NR + 1)/2$ the mode reduces to that used in Kendrick's [8] first solution for isotropic ring-stiff'd cyls which, to date, has been in good agreement with the scarce experimental data. Shell local buckling between rings is predicted through use of classical buckling deformations for simply supported end conditions.

All input and output quantities have positive values. The pressure is external hydrostatic.

9 Definitions of symbols and terms

The following quantities appear in the solution of the problem. Input quantities comprise the first thirteen items. The last four quantities (N, M, Q, P CRITICAL) are outputs.

1. H = Cylinder wall thickness, inches
2. A = Shell mean radius, inches
3. E = Ring cross section eccentricity (distance from C.G. of ring cross section to centerline of cyl. wall.), inches
4. $I - XO$ = Ring cross section area moment of inertia, inches **4
5. $E - X$ = Modulus of elasticity of cyl wall in X direction, psi
6. $E - THETA$ = Modulus of elasticity of cyl wall in theta direction, psi
7. $MU - THETA X$ = Poisson ratio theta $- X$
8. $MU - X THETA$ = Poisson ratio $X -$ theta
9. $G - X THETA$ = Shear modulus of elasticity of cyl wall, psi
10. $E - F$ = Modulus of elasticity of ring frame matl, psi
11. $A - F$ = Ring cross sectional area, inches **2
12. $L - F$ = Ring spacing, inches
13. NR = Number of rings
14. N = Number of buckle half-waves in circumferential direction
15. M = Number of buckle half-waves in longitudinal direction
16. Q = Buckling modal parameter describing longitudinal variations in general instability mode (integer ranging from unity to $(NR + 1)/2$ for NR odd and unity to $NR/2$ for NR even).

17. *P* CRITICAL   = Hydrostatic pressure at which general instability and local instability
                      occur, psi

10   Description of input procedures and format
There are 13 input quantities for which numerical values must be prescribed. See item 9 for their
definitions. Order your input values on 4 lines as below:

$A, H, L-F$
$E-$THETA, $E-X, G-X$ THETA
MU$-X$ THETA, MU$-$THETA X, NR
$A-F, I-XO, E, E-F$

When you specify user option 1, the computer will ask you to type in the appropriate values in
the above order.

11   Description of output procedures and format
General instability is dealt with first. Several different cases will be printed out, each with a
different value of *Q* (the modal parameter). The correct buckling pressure is the smallest one in
the *P* CRITICAL column, but not necessarily the last one.

General instability

Q	*N*	*P* CRITICAL
1	6	.22771924 $E+04$
2	2	.15300623 $E+04$

Shell local buckling between rings has a similar format with *M* replacing *Q* above.

    When the calculations are finished, the computer will print "OPTION?." You can select
option 1 and directly run another case if you so desire.

12   Discussion of numerical check-out and sample problem
Numerical results have been checked-out thoroughly. In the isotropic case, numerical results
are very nearly the same as those calculated in Kendrick [8]. The original program ran on the
IBM 7094. Results from it and those from the G. E. 605 differ by less than one tenth of one
percent.

    In the following, the sample output shown is what you should get if you input the values below.
Sample problem input:

$$H = .065$$
$$A = 1.7825$$
$$E = .125$$
$$I-XO = .527\,E-04$$
$$E-X = .55\,E+07$$
$$E-\text{THETA} = .6\,E+07$$
$$\text{MU}-\text{THETA}\,X = .14$$
$$\text{MU}-X\,\text{THETA} = .13$$
$$G-X\,\text{THETA} = .1\,E+07$$
$$E-F = .65\,E+07$$
$$A-F = .0185$$
$$L-F = 1.025$$
$$NR = 9$$

Sample problem output:

General instability

Q	*N*	*P* CRITICAL
5	2	.10507545 $E+04$

Shell local instability

M	*N*	*P* CRITICAL
1	6	.19349133 $E+04$

13 Correlation with experimental evidence

Test results reported in Hom and Couch [4] and discussed in Voce [1] for a filament-wound cylinder gave an 18300 psi genl instability pressure. This program gives a corresponding prediction of 18600 psi.

For two isotropic aluminum cylinders which failed in general stability at 2950 psi and 2930 psi, the program predicts buckling at 2500 psi.

So far, correlation is very good.

Please report any applicable test results of your own.

14 Identification of programming language and system

FORTRAN on the G.E. desk side time-sharing service (G.E.605).

15 Elapsed time at terminal for typical runs

The terminal time goes up if the system has very many users. You can expect to be at the terminal for about 10 min per run.

16 Computer central processor time used for typical runs

Typical execution runs use 10–15 sec of desk-side time sharing service CPU time.

17 References

1. Voce, S. J., "Buckling under external hydrostatic pressure of orthotropic cylindrical shells with evenly spaced equal strength circular ring frames," TRW Syst. Rep. No. EM 17-21 (October 1967).
2. Myers, N. C., Lee, G. D., Wright, F. C., and Daines, J. V., "Investigation of structural problems with filament wound deep submersibles," AD-440273 (January 1964).
3. Pellini, W. S., "Status and projections of developments in hull structure materials for deep ocean vehicles and fixed bottom installation," NRL Rep. 6167 (November 4, 1964).
4. Hom, K., and Couch, W. P., "Investigation of filament reinforced plastics for deep-submergence application," DTMB Rep. 1824 (November 1966).
5. Kendrick, S., "The buckling under external pressure of circular shells with evenly spaced equal strength circular ring frames—Part 1," NCRE (England) Rep. R211, Undex 300 (February 1953).
6. Voce, S. J., "On the general instability of ring-stiffened cylindrical shells under the action of radial external pressure and axial compression," UCLA thesis (1965).
7. Kempner, J., "Unified thin-shell theory," Pibal Rep. No. 566 (March 1960).
8. Kendrick, S., "The buckling under external pressure of circular cylindrical shells with evenly spaced equal strength circular ring frames—Part III," NCRE (England) Rep. R244, Undex 312 (September 1953).
9. Cox, H. L., "Buckling of thin plates in compression," Aeronaut. Res. Committee, R and M No. 1554.
10. Hom, K., Couch, W. P., and Willner, A. R., "Elastic material constants of filament—wound cylinders fabricated from E-HTS/E787 and S-HTS/E787 prepreg rovings, DTMB Rep. 1823 (February 1966).

18 Notes and revision dates

Revision dates for suboptions are as follows:

Suboption	Revision date	Suboption	Revision date
1	Nov 15, 1968	5	Nov 15, 1968
2	Nov 15, 1968	6	Nov 15, 1968
3	Nov 15, 1968	7	Nov 15, 1968
4	Nov 15, 1968	8	Nov 15, 1968

Suboption	Revision date	Suboption	Revision date
9	Nov 15, 1968	14	Nov 15, 1968
10	Nov 15, 1968	15	Nov 15, 1968
11	Nov 15, 1968	16	Nov 15, 1968
12	Nov 15, 1968	17	Nov 15, 1968
13	Nov 15, 1968	18	Nov 15, 1968

Ready

8. Present Status, Conclusions, and Recommendations

The feasibility, flexibility, and potential value of the STORE system to organizations such as the United States Navy which are extensively active in structural design has been successfully demonstrated.

The STORE system is ready for exploitation by interested Navy and other government and professional society activities concerned with structural analysis and design. Operation of such a system by some nonprofit group would have the possible advantage of being able to attract many industrial participants and existing programs, some of which would of course require conversion to a suggested format. A useful byproduct of the STORE effort was the compilation of a list of about 500 programs for possible inclusion in the STORE system [3].

It is strongly recommended that serious consideration be given by the Department of Defense and other U.S. Government activities involved in structural analysis and design, to the implementation of a STORE-type system to upgrade substantially their engineering research and design capability and efficiency. Moreover, it is quite apparent that the basic STORE-type system can be used advantageously in other disciplines. It is further recommended that such uses be given serious consideration.

Appendix A. List of Programs in Project STORE

The following is a list of the completed programs in Project STORE at this time.

1. BUCKLING
2. FRAME2D
3. FRAME3D
4. OVAL
5. STRUT
6. DISC
7. RING
8. BEAMBEND
9. COMPOSIT
10. BRANCH
11. TRANS
12. BRANS

References

1. Wiener, J., "Project STORE System Evaluation for the Office of Naval Research," Mandate Systems Inc., New York (June 28, 1968).
2. Hrzina, J., "Office of Naval Research Computer Programs Project (STORE) Final Report," T.R.W. Corp. Syst. Group, Redondo Beach, California (May 5, 1969).
3. "Compendium of Structural Mechanics Computer Programs," Office of Naval Research (April, 1971).

PART V

New Capabilities for Computer-Based Analysis

Symbolic Computing

Andrew Ka-Ching Wong

CARNEGIE-MELLON UNIVERSITY
PITTSBURGH, PENNSYLVANIA

1. Introduction

Symbolic computing, in a general sense, refers to software processing techniques for manipulating symbolic expressions that relate symbolic entities which may not necessarily have numeric values. Its problem solving domain extends from algebraic manipulations of mathematical formulas to processing of abstract symbolic representation of a wide class of problems. Its problem solving potential spans a wide spectrum from the formulation of physical problems to the reduction and manipulation of their mathematical expressions to yield closed form or numerical solutions. This paper will present some of the symbolic mathematical computation concepts developed by the author and others, as well as discuss the role and potential of symbolic computation in the general area of engineering mechanics and applied mathematics.

In the area of mathematics, considerable success has been achieved in the use of computers to prove theorems [1, 2], perform symbolic integration and complex algebraic manipulation [3–6], etc. The same may be said about physics [7–9] and engineering [10–13]. Of the programs developed for special symbol manipulation tasks, a more general approach has been presented by the author and others [10, 11], according to which the programming aspects for problem solving are "unified" through utilizing symbolic computing facilities at various levels of algebraic manipulation. In such a system, each fundamental unit built in is not just a symbolic entity, but also an independent processing unit which can serve as a building block for the construction of compound expressions or procedures, or as an argument of another function.

By introducing some simple control mechanism, algebraic manipulation activities on mathematical expressions can be carried out uninterrupted at various levels, for example, from general tensors to special tensors to scalar representation of tensors, to their numerically solvable forms. Furthermore, with the capability of recognizing formula patterns and performing deductive inference or formula transformations accordingly, special problem-solving schemes can be built into the system for solving certain particular classes of problems.

The capability and potential of symbolic computing approach in solving engineering mechanics problems are largely determined by the degree of generality and flexibility of the organization of data and the data processors. In the case of general list processing or high-level formula manipulation languages, the data structure and the processors are such that symbolic expressions or list expressions* can be stored as a single unit, recognized and classified according to their patterns, retrieved, combined, and transformed in a manner similar to that performed by human problem solvers. Such programming environments provide facilities for altering, combining formulas, replacing variables by designated formulas, substituting numerical or logical values or other formulas for variables in an expression, and for evaluating formulas or formula components. Each function or formula thus constructed may be used as subexpression in other formulas or as arguments in other functions, and used recursively within itself. The pattern recognition facilities allow s-exp* to be transformed by prescribed rules, according to the pattern of the intermediate expression recognized in the due course of their successive transformation. Heuristics may also be introduced to aid such transformation. Extensible languages [14–16] with even more flexible data structures and processing capabilities have also been developed. With the introduction of these languages, more intricate data structures can be built for more sophisticated algebraic systems.

2. Organization of Data and Data Processors

2.1. An Overall View

In order to illustrate the activities of symbolic mathematical computation, a model adapted from [17] is shown in Fig. 1. Here the space labeled A represents the mathematical entities for which symbolic mathematical computation attempts to provide a concrete embodiment. This space can be subdivided into "expression space" and "transformation space"; the former is the totality of all mathematical expressions, whereas the latter is the totality of all transformation rules. Blocks C and D represent the pro-

---

* Hitherto s-exp is used to stand for symbolic expressions, which are equivalent to list expressions or formulas.

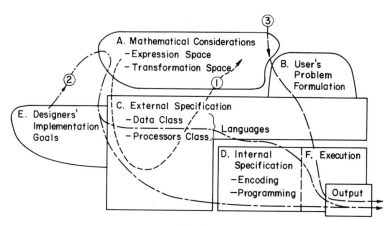

Fig. 1.

gramming environment in which the form of data and processors are specified both externally in terms of programming languages, and internally in machine codes. In many aspects, the "data class" and the "processor class" in the programming environment correspond to the "expression space" and "transformation space" in A, and both together characterize the symbol manipulation languages. The "flow arrows" describe various activities involved in the design, implementation, development of the representational and processing facilities, as well as the use of the symbolic computing system. Arrow 1 indicates how the class of data and processors are abstracted from the "expression space" and the "transformation space" and how they can be represented formally back in the mathematic space. Arrow 2 indicates the design process of symbolic computing systems, showing that mathematical formalism for formalizing external and internal specifications. Arrow 3 indicates that when the data structure and basic processors are provided explicitly in the form of programming languages, they can be used for performing certain specific tasks desired by the user.

Since different designers have different implementation goals, both the external and internal specifications vary from language to language, and there are presently about 20 symbol manipulation languages. In this paper, the most general data structures and processing methodology employed in symbolic computing will be discussed as the author attempts to show how formula expressions can be constructed, recognized, and transformed in accordance with specific set of rules.

In general, languages for symbolic computing can be categorized into (1) general list processing languages (e.g., LISP 1.5 [18], IPL V [19]), (2) more specific formula manipulation languages (e.g., FORMULA ALGOL [20],

FORMAC [21]), and (3) extendable languages [14–16] (still in the experimental stage) according to the emphasis of their implementation. There are, however, several basic processing features which characterize all effective symbolic computing systems. The most basic feature is that they all deal with special kinds of data which requires special data structures. Another common feature is that their memory space for data structures need not be preassigned since it is allocated only when needed. Furthermore, these symbolic computing systems provide a set of basic processors for accessing, replacing, manipulating, and transforming s-exp or sub-s-exp. A great number of them also provide recursive operations for constructing or processing s-exp in a more sophisticated manner.

In this paper LISP 1.5 and FORMULA ALGOL are chosen to illustrate the symbolic programming concepts. While the former represents a very general list processing language, the latter represents a high-level formula manipulation and limited processing language which provides facilities for the user to construct formula and formula transformation processors in a more direct manner. Though FORMULA ALGOL is now no longer available, the important formula manipulation concepts which it possesses have been adopted by and further developed in some of the more general flexible extendable languages.

2.2. DATA STRUCTURE

In most of the list processing or high-level symbol manipulation languages (hitherto denoted by LP) the data contain symbolic, numeric, and Boolean logic information, which is the relational structure of its symbolic, numeric, or Boolean logic content. The basic forms of programming formalism in most LP are infix notations, Polish notations, and dotted pairs. A formula is most directly represented by infix notation. Thus $T = L/GJ$ can be expressed in the form of

$$T = L \div (G * J) \tag{1}$$

for a general list processing language such as LISP 1.5, the Polish notation is the basic processing form where data can be conveniently accessed and constructed by a few basic processors [18]. With this notation, the equation (1) then becomes

$$\text{(EQUALS } (T\text{(DIVIDES } (L\text{(TIMES } (GJ)))))) $$
$$\quad 1 \qquad\qquad 2\ 3 \qquad\qquad 4\ 5 \qquad\quad 6\quad 6\,5\,4\,3\,2\,1$$

where the numbers mark the depths of the list enclosed by the brackets. Internally, a list structure can be represented by a binary tree in accordance with the heirarchical organization of the structure. Externally, these relational structures can be represented by dotted pairs. Such programming formalism enhances the construction and parsing of s-exp. However, all the notations mentioned are interchangeable and translators can be constructed to map data of one form into another (e.g., [11]).

In some of the formula-manipulation languages, instead of having the symbolic expressions treated externally in the form of list structure, a set of arithmetic and Boolean operators is provided for organizing the formula in the form of a binary tree. For instance, in binary tree structure, formula (1) becomes

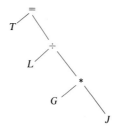

With such a hierarchical organization, each sub-tree branching from an operator node represents a subexpression that can be treated as a unit in the process of manipulation.

Some LP provide a more direct way of representing different types or forms of data structure than others. In FORMULA ALGOL and all extendable languages, data of various forms can be constructed, and processors or functions of various types can be declared, thus enabling the system to perform various specific tasks in a direct and flexible manner. For example, data can be constructed not only as a single formula but as arrays of formulas. Its generality and flexibility make it possible for the users of these languages to represent tensors in both their indexed and scalar form. For instance, $A[I, J]$ can represent a matrix with specific elements assigned as follows

$$A[1, 1] \leftarrow 1; \qquad A[1, 2] \leftarrow 3*\text{SIN } X$$
$$A[2, 1] \leftarrow 0; \qquad A[2, 2] \leftarrow C*\text{EXP } X \tag{2}$$

$A[IJ]$ then becomes the index notation of the matrix

$$\begin{bmatrix} 1 & 3 * \text{SIN } X \\ 0 & C * \text{EXP } X \end{bmatrix} \tag{3}$$

$A[IJ]$ then can be used as an index notation in another formula which can be expanded into full scalar form if desired by the user. The capability of these data structures, and the concepts of both processing and control will be discussed in greater detail later.

In some high-level LP, pattern structure construction facilities are provided for specifying formula patterns. One could, for example, recognize that the complex number $a + i * f(x)$ is of the pattern:

$$(\text{Constant}) + i * (\text{function of } x) \tag{4}$$

In a language like FORMULA ALGOL, this pattern or class of formulas can be represented by

$$A: \ (\text{OF CONST}) + I * F: \ (\text{OF } FX) \tag{5}$$

which means that the symbol A is of the type "constant" and F being a function of X. Thus this pattern will match formulas such as

(a) $3 + I * \text{SIN } X$
(b) $4 + I * (10 + X)$
(c) $C + I * (B + \text{COS } X)$

For LP where these pattern structure facilities are not available, the user has to construct his own parsing algorithm to determine class membership of s-exp.

2.3. PROCESSING OF SYMBOLIC DATA

At the programming level, the most fundamental processes for manipulating s-expressions include the building of s-exp from smaller s-exp, and the extraction of subexpressions from given expressions. Languages like LISP 1.5 provide some elementary functions for performing these tasks. In languages that are more formula-manipulation-oriented, subexpressions can be extracted from their formula pattern structures. Since in most of the LP, any s-exp can be constructed (1) as a processing unit like functions SIN, EXP in FORTRAN, or (2) as a distinct unit, the value of which can be treated as augments of a function, of (3) as variables in a s-exp, more complex s-exp—which may be formulas or processors—can be constructed from smaller units. In general, most LP's not only enable the users to construct complex expressions from simpler ones but also provide them with the means to create new functions. Besides, recursive operations are provided to enable a function to call itself during execution.

In LISP 1.5, a function can be defined by using a formal expression called lambda expression, λ-exp. The formal syntax of the expression is given as:

(LAMBDA (⟨variable list⟩) (⟨procedures operate on the variable list⟩))

The definition of a function then takes the form of

DEFINE ((
            (⟨function name⟩ ⟨λ-exp⟩))))

For example, if INT(F X) is given as a function which performs symbolic integration on the integrand F with respect to the integrating variable X, then the volumetric integral [suppose it is desired to be represented in the form of INTV(F X Y Z)] can be defined as:

DEFINE ((
(INTV(LAMBDA (F X Y Z)
(INT((INT((INT(F X))Y))Z))))))

Note that the variables (F, X, Y, Z) are bounded to those in the procedure. In a language like FORMULA ALGOL, INTV can be defined as

FORM PROCEDURE INTV(F,X,Y,Z)
VALUE F, X, Y, Z; FORM F, X, Y, Z;
BEGIN
  INT$V \leftarrow$ INT(INT(INT(F, X), Y), Z)
END

Once the function is defined (in both LISP 1.5 and FORMULA ALGOL) it can be applied to sets of argument like ($W *$ SIN U, U, V, W) which will be bounded to (F, X, Y, Z), respectively, inside the procedure to yield closed-form value of the integral

$$\int \int \int w \sin u \, du \, dv \, dw \tag{6}$$

Another important feature of a LP is its capability for searching, matching, and recognizing the relational structure of s-exp and its content. In many situations, the recognition of s-exp and subexpression is associated with subsequent operations the programmer would like to perform on that s-exp. In FORMULA ALGOL, a general transformation formalism called Markov algorithm is provided for transforming s-exp in accordance with a specific set of rules such as symbolic integration. A Markov algorithm is a list of transformation rules which transform a formula in a specific way according to the pattern of the formula as recognized by the algorithm. A transformation rule is a pair of formulas connected by a transformation arrow $\rightarrow$ such as:

$$P_i \rightarrow T_i \tag{7}$$

in which P_i is a specific formula pattern structure and T_i is the transformand expression containing some variables bounded to those which occur in P_i. Hence

$(F: \text{FORM} \div (B: \text{OF NONZERO}) = C: \text{OF CONST}) \rightarrow .F = .C \times .B$

is a transformation rule which specifies that if F is a formula, B is a nonzero number, and C is a constant, then any formula that matches the left-hand side formula pattern can be transformed into the right-hand side expression, $.F = .C \times .B$. The dots are used to hold up the evaluation of the s-exp F, C, B during the transformation phase. Here F, C, B are symbols representing the s-exp assigned to them. If dots are absent, evaluation of those expressions will be executed in the process of transformation. Thus a Markov algorithm can be expressed as

$$M \leftarrow [[P_i \rightarrow T_1 \cdots P_n \rightarrow T_n]] \tag{8}$$

and if a formula $F1$ is going through this set of possible transformations according to its pattern (like integration simplification), then the result of the transformation T can be obtained by executing the assignment

$$T \leftarrow F1 \downarrow .M \tag{9}$$

which causes $F1$ to be transformed under the set of rules M.

In a language like LISP 1.5 where this formal processing scheme is not provided; this type of transformation can be performed by utilizing conditional expressions of the form:

$$(COND(P_1\ e_1) \cdots (P_n\ e_n)) \tag{10}$$

in which P_i is a predicate function yielding Boolean logic value and e_i is an execution statement. Hence pattern-drive-type formula transformation can still be carried out, but in a rather indirect or less efficient manner.

The dotted expression in Markov transformation illustrates an important aspect of the evaluation and processing control in a LP. In many instances, a s-exp has to remain as it is at a certain level before it is evaluated, and most LP's provide means to hold and release evaluation at various operational levels. Other basic operations such as replacement, insertion, and deletion of subexpressions in a s-exp and composition of s-exp are also provided in most LP, thus rendering the language a convenient tool for formula manipulation.

3. The Role of Symbolic Computation in Mechanics

To identify the role and capacity of symbolic computing in solving engineering mechanics problems, it is essential to consider the problem-solving activities involved. In general, these activities may be divided into the following steps:

(i) formulation of physical problems in mathematic expressions
(ii) reduction of the general formulation to a certain known solvable form
(iii) organization and execution of a sequence of necessary algebraic manipulations and evaluations to yield the final solution.

The following sections will discuss the role and capacity of symbolic computing in these three phases of problem-solving activities. Illustration will be drawn from the work developed by the author and others [10, 11] especially the symbolic computing concepts in mechanics introduced in the system CONFORM, a system developed by Bugliarello and the author. Also included in these sections are a few general problem-solving paradigms that combine all the three phases and that apply to a large class of problems such as boundary-value problems. The capability of these systems may demonstrate the potential of symbolic computing in the area of mechanics.

3.1. Problem Formulation

The formulation of an engineering mechanics problem requires both physical insight and mathematical modeling technique. At present, symbolic computing does not have the capability for generating new physical insight (though occasionally users may find that their interaction with a symbolic computing system does lead to some new programming techniques—representational as well as computational). Nevertheless, once certain physical entities are modeled by mathematical formalism, one could employ a symbolic computing language to define and construct a backlog of mathematical expressions, functions, and operators which may serve as building blocks for formulating various types of mechanics problems.

A good example may be found in the construction of a mathematical expression called alternator, ε_{ijk}.* In CONFORM ε_{ijk} is defined and constructed as an operational unit of the form ALT(I, J, K). It can then be used for formulating the determinant of a 3×3 matrix, a Jacobian and a curl of a vector, etc. as shown in Fig. 2. In a like manner, various forms of governing equations of a general problem class can be conveniently formulated by using the backlog of various independent units (or knowledge cells) that the user has defined within his own system. These units or cells can be called by symbols in a very flexible manner. Within the system, each of these cells defined is more than merely a convenient symbolic representation for it is also an operational unit since its full functional expressions can be retrieved, expanded to various degrees of detail (e.g., from general tensor to full Cartesian representation), and evaluated to yield closed-form expressions or in numerical value. Thus, in affording both an algebraic representation of physical concepts and the capability of manipulating of these formulas or symbols by the computer, a high-level symbolic computing system can function as a mechanized mathematical language for formulating and solving

* ALTERNATOR ALT (I, J, K) is defined as : 1 IF I, J, K IS AN EVEN PERMUTATION OF 1, 2, 3 ; -1 IF I, J, K IS AN ODD PERMUTATION OF 1, 2, 3 ; and 0 IF I, J, K IS NOT A PERMUTATION OF 1, 2, 3.

Conventional mathematical expressions	Conform expressions
$\text{Det } A = \sum_{i=1}^{3} \sum_{j=1}^{3} \sum_{k=1}^{3} \varepsilon_{ijk} a_{1i} a_{2j} a_{3k}$	$\text{DET}(A) \leftarrow$ $\text{SUM3 }(I,J,K, \text{ALT}(I,J,K)*AA[1,I]*$ $AA[2,J]*AA[3,K])$
$\text{Curl } \mathbf{v} = \sum_{i=1}^{3} \sum_{j=1}^{3} \sum_{k=1}^{3} \varepsilon_{ijk} \mathbf{e}_i \dfrac{\partial y_k}{\partial x_j}$	$\text{CURL}(V) \leftarrow$ $\text{SUM3 }(I,J,K, \text{ALT}(I,J,K)*E[I]*$ $D(VV[K], XX[J]))$
$J(y, x) = \sum_{i=1}^{3} \sum_{j=1}^{3} \sum_{k=1}^{3} \varepsilon_{ijk} \dfrac{\partial y_1}{\partial x_i} \dfrac{\partial y_2}{\partial x_j} \dfrac{\partial y_3}{\partial x_k}$	$J(Y, X) \leftarrow$ $\text{SUM3 }(I,J,K, \text{ALT}(I,J,K)*$ $D(YY[1], XX[I]*$ $D(YY[2], XX[J])*$ $D(YY[3], XX[K]))$

Fig. 2.

mechanics problems, shared and manipulatable by man and computers alike.

3.2. REDUCTION OF GENERAL FORMULATION TO SOLVABLE FORM

Once a problem class is identified and once the problem is formulated, it can be transformed or reduced to solvable forms by an algorithm or through heuristic procedures. Such transformation or reduction has been successfully tackled by many early symbolic computing systems [5, 22, 23]. At this stage of solution the main objective of symbolic computation is to eliminate monotonous manual algebraic manipulation activities such as coordinate transformations, algebraic simplification, and polynomial expansion and reduction. A simple example of this transformation can be found in CONFORM. If an equation is given in general tensor or Cartesian tensor form, it can be transformed into another form by GENERATE, a formal problem-solving function in CONFORM, by specifying which curvilinear form one desires. Figure 3 shows how the function GENERATE can be used to generate both the CARTESIAN and cylindrical form of the equation of equilibrium, EQEQ. The Cartesian form is obtained by simple expansion of the Cartesian tensor, whereas the cylindrical coordinate form is obtained by first mapping it into general curvilinear coordinate form and then reducing the general curvilinear form to specific coordinate form. In CONFORM, this type of transformation is carried out by using Markov algorithm. Other

INPUT:
BEGIN
PRINT(EQEQ);
$E1 \leftarrow$ GENERATE (EQEQ, CARTESIAN);
$E2 \leftarrow$ GENERATE (EQEQ, CYLINDRICAL);
PRINT($E1,E2$);
END

OUTPUT:
(1) CARTESIAN TENSOR
$$D(SS[I,J], XX[J]) + FF[J] = 0$$
(2) CARTESIAN
$$D.(SXX,X) + D.(SYX,Y) + D.(SZX,Z) + FX = 0$$
$$D.(SXY) + D.(SYY,Y) + D.(SZY,Z) + FY = 0$$
$$D.(SXZ,X) + D.(SYZ,Y) + D.(SZZ,Z) + FZ = 0$$
(3) CYLINDRICAL
$$D.(SRR.R) + 1/R * D.(SRDA,DA) + D.(SRZ,Z)$$
$$+ SRR/R - SDADA/R + FR = 0$$
$$D.(SRDA,R) + 1/R * D.(SDADA,DA) + D.(SDAZ,Z)$$
$$+2/R * SRDA + FDA = 0$$
$$D.(SRZ,R) + 1/R * D.(SDAZ,DA) + D.(SZZ,Z)$$
$$+1/R * SRZ + FZ = 0$$

Fig. 3.

applications of Markov algorithm in CONFORM include (1) symbolic differentiation, (2) symbolic integration, (3) transformation of a given partial differential equation into various finite difference form, (4) solution of ordinary and matrix equations, and (5) other algebraic simplification procedure.

3.3. FORMULA EVALUATION AND CONTROL

In order to manipulate s-exp in a flexible manner, a symbolic computing system should provide the users with not only a set of basic processors for performing evaluation, substitution, replacement, and transformations of s-exp, but also the programming facilities for controlling evaluation or manipulation at various operation levels. The functions provided in LISP 1.5 for such controls are QUOTE, EVAL, and EVALQUOTE. QUOTE is a special function of the argument which is unevaluated when the evaluation function EVAL is applied to the s-exp. When EVALQUOTE is executed instead, the s-exp within the function QUOTE will be evaluated. In FORMULA ALGOL, a dotted expression is used in the place of QUOTE.

Here evaluation implies the substitution into the s-exp the values (s-exp, symbols, Boolean or numeric) assigned to the bound variables and then the performance of the necessary arithmetic, Boolean, or algebraic manipulation so as to reduce the s-exp to its simplest form. An example of formula evaluation can be illustrated from the derivation of the values of the radial and tangential bending moments of a fixed-edge circular plate at $r = 0$ and $r = a$, respectively, from the deflection w by using the following formulas:

$$M_r = D\frac{d^2w}{dr^2} + \frac{0.3}{r}\frac{dw}{dr}, \qquad M_t = D\frac{1}{r}\frac{dw}{dr} + 0.3\frac{d^2w}{dr^2} \qquad (11)$$

Here $w = Q(a^2 - r^2)/64D$ is obtained from the program TRIALSOLU-TIONS which will be described in the next section. Hence in the program TRIALSOLUTIONS, if the input is submitted in LISP infix notation as

> DEFLECTION
> $((W = Q[A \uparrow 2 - R \uparrow 2] \uparrow 2/[64\,D]))$
> EVALUATE
> $(((MR = D[D2WR + 0.3\,DWR/R)A\ \text{NIL})$
> $((MT = D[DWR/R + 0.3\,D2WR])0\ \text{NIL}))$

where A and 0 denote the replacement of r by a and 0, respectively, after differentiation, the final evaluated values of MR and MT are then given as

$$MR = 1.250000E - 01\ Q\ A \uparrow 2$$
$$MT = -8.124999E - 02\ A \uparrow 2Q$$

If A and Q are assigned numerical values, then the evaluated values of MR and MT will be numerical values only.

The control of evaluation over several operational levels can be well illustrated by an example taken from finite-element method. In the finite-element method, one may perform formula manipulation at various operational levels—direct notation, index notation and full scalar manipulation. During the process of matrix algebraic manipulation the integrating process over the matrix elements has to be held up until the execution of integration is desirable. For instance, if the closed form value of the beam stiffness, k, is to be derived from the following relations [24]:

$$\sigma = DBA^{-1}\Delta, \quad \bar{\varepsilon} = BA^{-1}\bar{\Delta}, \quad s = \int_0^l \bar{\varepsilon}^T\sigma\,dx, \quad k = [\bar{\Delta}^T]^{-1}S\,\Delta^{-1} \qquad (12)$$

the manipulation can be carried out at direct notation level by substitution and reduction as follows:

$$k = [\overline{A}^{\mathrm{T}}]^{-1} S \, \Delta^{-1}$$

$$= [\overline{A}^{\mathrm{T}}]^{-1} \left(\int_0^l \bar{\varepsilon}^{\mathrm{T}} \sigma \, dx \right) \Delta^{-1}$$

$$= [\overline{A}^{\mathrm{T}}]^{-1} \left(\int_0^l [BA^{-1} \overline{\Delta}]^{\mathrm{T}} [DBA^{-1} \Delta] \, dx \right) \Delta^{-1}$$

$$= [\overline{A}^{\mathrm{T}}]^{-1} \left(\int_0^l \overline{\Delta}^{\mathrm{T}} [A^{-1}]^{\mathrm{T}} B^{\mathrm{T}} DBA^{-1} \Delta \, dx \right) \Delta^{-1}$$

$$= [\overline{A}^{\mathrm{T}}]^{-1} \overline{\Delta}^{\mathrm{T}} [A^{-1}]^{\mathrm{T}} \left(\int_0^l B^{\mathrm{T}} DB \, dx \right) A^{-1} \Delta \, \Delta^{-1}$$

$$= [A^{-1}]^{\mathrm{T}} \left(\int_0^l B^{\mathrm{T}} DB \, dx \right) A^{-1}$$

In carrying out this deductive process, both the integration and expansion of the matrix into full scalar form have to be held up by declaring the Boolean SCALAR (scalar expansion) as false and POSTINT (postpone integration) as TRUE. In addition, the type of each symbol has also to be known in the reduction process. Both the rank and the element assignments to each of the matrices are necessary if expansion into scalar form is desired. Algebraic manipulations at the scalar form level (such as transposition, inversion and multiplication of matrices) as well as integration process can then be carried out. Note that the integration over the elements of $B^{\mathrm{T}}DB$ can only be executed after the matrix algebraic operations have been performed, and that the matrix operations on

$$[A^{-1}]^{\mathrm{T}} \left\{ \int_0^l B^{\mathrm{T}} DB \, dx \right\} A^{-1}$$

will be carried out at the next level (i.e., after integration).

Though CONFORM provides all necessary operations for the performance of the above-described operations, yet because of the complexity of the control and transformation process in the transference of data from one form into the other the full implementation of such processes requires a system that is more flexible in handling of data structure of various forms as proposed in some of the extendable languages.

3.4. FORMAL PROBLEM SOLVING PROGRAMS

The convenience of symbolic system can be further enhanced if its capabilities are incorporated in general problem-solving paradigms, for then the solutions of a general problem class can often be cast. In CONFORM, there are two of such paradigms.

1. GENERATE (DOMAIN, GOAL), i.e., given a set of equations or expressions (a DOMAIN), generate a new set to suit some specific purpose (a GOAL).

2. INFER (PREMISES, PROPOSITION), i.e., given a set of PREMISES (theorems, identities, etc.), infer from them a new PROPOSITION.

Examples of problems that can be solved by these two procedures are:

GENERATE (EQ, CYLINDRICAL), i.e., generate from the Cartesian tensor form of a given equation (EQ) its cylindrical coordinates form.

GENERATE (EQ,FINITE < CARTESIAN), i.e., generate the Cartesian form of a partial differential equation (EQ) and transform the result into finite difference form.

INFER ($V \ll$ VEC(SPACE), $VV[I] =$ INN($V,E.[I]$)), i.e., deduce from the premises that V is in a vector space, the proposition that the Cartesian components $VV[I]$ of V are equal to the inner product of V and the corresponding bases, $E[I]$.

The problems represented by these general paradigms can be called *formal problems*. A formal problem is a general problem-solving procedure in which a particular type of problem statement is accepted as a variable. This concept of formal problem-solving paradigms would increase the variability to program developed for solving a wider class of problems.

Two other formal problem-solving paradigms called LORDE and TRIALSOLUTIONS written in LISP 1.5 have been developed by Ossenbruggen *et al.* [11]. They are more specific attempts to solve linear and partial differential equations with constant linear boundary conditions that may result from one of a great variety of equilibrium-type boundary value problems.

The program LORDE is constructed to yield a closed-form solution of linear differential equations with constant coefficients through symbolic computing. It can also be extended to the solution of a set of simultaneous linear differential equations, or the perturbated solution of slightly nonlinear differential equations.

The LORDE program is based on the theory of differential equations and it follows the usual procedure for solving ordinary differential equations. The activities are divided into three separate parts: (1) the generation of the homogeneous solution; (2) the generation of a particular solution; and (3) the determination of a unique solution for the given set of boundary conditions.

A typical example of using LORDE to solve an ordinary differential equation is given as follows:

INPUT: LORDE

$(D2WX - 9W = E \uparrow X \, [\text{SIN } X])$

OUTPUT: (ANSWER IS: $W = WH + WP$)

$WH = C1 \, E \uparrow 2.999999E\text{-}00 \, X + C2 \, E \uparrow$
$\quad\quad - 2.999999E\text{-}00 \, X$

$WP = 02.342940E\text{-}02 \, E \uparrow X \, \text{COS } X +$
$\quad\quad - 1.058823E\text{-}01 \, E \uparrow X \, \text{SIN } X$

(UNKNOWN CONSTANTS ARE DETERMINED FROM BOUNDARY CONDITIONS)

INPUT: 0 ARGUMENTS FOR EVALQUOTE...

BOUNDARYCON

$(((X = 0 \quad W = 0) \quad (X = 0 \quad DWX = 0)$
$(X = 0 \quad D2WX = 0) \quad (X = 0 \quad D3WX = 0)))$

OUTPUT: $C1 = 0$

$C2 = 2.343772E\text{-}02$

$C3 = 04.687533E\text{-}02$

$C4 = 0$

The program TRIALSOLUTION is constructed to solve the class of differential equation expressed in the form

$$\sum_{i=0}^{n} A_{n-1}(x, y) w_{x,y}^{(i)} = R(x, y) \tag{13}$$

in which $A_{n-1}(x, y)$ and $R(x, y)$ are, in general, functions of x and y, and $w_{x,y}^{(i)}$ is the ith partial derivative of $w = w(x, y)$ with respect to x and y where $w^{(0)}$ refers to w itself. The boundary conditions are given by a set of n equations

$$\sum_{i=0}^{p} B_{iq}(x, y) w_{x,y}^{(i)} = f_q(x, y) \qquad q = 1, 2, \ldots, n \tag{14}$$

in which $B_{iq}(x, y)$ and $f_q(x, y)$ are, in general, functions of x and y for each value of q corresponding to the qth boundary condition, and $w_{x,y}^{(i)}$ is the ith partial derivative of $w = w(x, y)$ with respect to x and y, where $i \leqslant p < n$.

The preliminary attempt to solve this problem symbolically is based on variational technique (Galerkin's method, in particular) of determining the unknown parameters of each feasible trial solution by minimizing the residual of the governing differential equation. Galerkin's method requires that a trial solution satisfy all boundary conditions and have the form

$$w(x, y) = w_0(x, y) + \sum_{j=1}^{m} C_j w_j(x, y) \tag{15}$$

where C_j are undetermined parameters $w_0(x, y)$ and $w_j(x, y)$ are linearly independent functions. In selecting the feasible trial solution, the program automatically checks whether the equation satisfies all the boundary conditions, and then determines the residual of the governing differential equation by

$$r(x, y) = R(x, y) - \sum_{i=1}^{n} A_{n-i}(x, y)w_{x,y}^{(i)} \tag{16}$$

If a trial solution is the exact solution, the residual will equal zero; if not, the residual $r(x, y)$ is expressed in terms of the undetermined parameters by substituting the trial solution into Eq. (16), and the residual in this form is denoted by $\bar{r}(x, y)$. Then the residual is minimized by solving for the undetermined parameters C_j from a set of m simultaneous definite integral equations of the following form:

$$\int\int_{D} w_j(x, y)\bar{r}(x, y)\,dx\,dy = 0, \qquad j = 1, 2, \dots, m \tag{17}$$

where $w_j(x, y)$ is the jth term of the trial solution in Eq. (15) and the definite integral is integrated over the domain D.

Following is an example of solving the governing differential equation for a uniformly loaded circular plate with fixed edge by TRIALSOLUTION. The governing differential equation is given as

$$x^2\frac{d^3w}{dx^3} + x\frac{d^2w}{dx^2} - \frac{dw}{dx} = \frac{qx^3}{2D} \tag{18}$$

with boundary conditions

$$\frac{dw}{dx} = 0 \quad \text{at} \quad x = 0; \qquad w = 0; \quad \frac{dw}{dx} = 0 \quad \text{at} \quad x = a \tag{19}$$

where the variable x represents the radius of the plate and D the thickness of the plate. The trial solution submitted as a test is the exact solution

$$w = A_1(a^2 - x^2)^2 \tag{20}$$

The input format and the output of TRIALSOLUTION of this problem is shown as follows:

INPUT:

```
GOVERNINGEQS
((X ↑ 2 D3WX + X D2WX − DWX = 0.5 Q X ↑ 3 / D)
NIL
((X = 0 DWX = 0) (X = A W = 0 DWX = 0)))
TRIALSOLUTIONS
(((W = A1[A ↑ 2 − X ↑ 2] ↑ 2)))
```

OUTPUT:

(TRIAL SOLUTION IS:)
$(W = A1[A \uparrow 2 - X \uparrow 2] \uparrow 2)$
(VALUE OF UNKNOWN PARAMETER IS:)
$A1 = 1.562499E\text{-}02\ Q\ D \uparrow -1$
(THE BEST TRIAL SOLUTION IS THE ONE WITH THE SMALLEST PHI VALUE)

The significant contribution of these programs lies not only in their current computability but also in their flexibility and extensibility. Furthermore, on the set of LISP functions they embody, new symbol manipulating programs can be built for implementing more complex problems.

4. Discussion and Conclusion

In the various programming experiences mentioned, the capacity of symbolic computing has been demonstrated, but more direct and uniform programming language systems as well as more extensive operational programs are still to be developed. We need a language which is (1) general enough for defining various data structures; (2) effective enough for numerical computation; and (3) direct and simple enough for the average users. Such a language would unify the problem formulation as well as formula reduction and manipulation with their numerical counterparts in solving complex engineering problems.

The introduction of extendable symbolic languages and the recent emphasis on simplicity and uniformity in the design of programming languages are, in fact, steps toward this goal. In an extendable language such as EXTENDABILITY [25], the emphasis of the language design shifts from the system features toward the user's features. Such a system provides flexible programming facilities at various programming and algebraic operational levels—including syntax, semantics, data structures, the lexicon, lexical structures, input/output facilities, character sets (for both variable and operators), and system configurations. In addition to this extendable and flexible programming setting, the system also provides a general parser for parsing data structure with greater variability and generality. The realization of such a system would free users from many programming language constraints which confine the user's programming capability in carrying out a complex task, consequently forcing the user to choose long detour or weird programming tricks to get the problem across. Once the programming flexibility at user's level is acquired, the programming languages at the user level can then be conformed more directly to the mathematical representation desired by the user without losing processing generality.

So far as engineers are concerned there are many classes of problems, the solutions to which can be totally or partially computerized. For example, finite-element methods which involve extensive differentiation, integration, and matrix manipulation in their formulation (as illustrated in one of the previous sections) are ideal processes to be tackled by symbolic computing. In the area of continuum mechanics, generally laborious work is required to reduce formulation to its computational level. But many of these difficulties would vanish with the use of symbolic computing, which provides the automatic translation from the conceptual level of continuum mechanics to any desirable operational levels. In solving slightly nonlinear differential equations by the method of perturbation or other methods, again laborious repetitive performances of differentiation, integration, solving ordinary differential equations, as well as other algebraic manipulations are normally required. This is another area where symbolic computation may help to provide a unified computational system. In the method of trial solutions, a symbolic computing system may take over the laborious work of testing the solutions and deriving the underdetermined parameters from a few possible forms of solution. Furthermore, the capability of building in the system a backlog of functions and formulas which can be readily called by the user in a most flexible way would make symbolic computing system an ideal teaching and knowledge-building tool particularly when the language is implemented on an on-line interactive conversational system.

In conclusion, by using symbolic computing approach, a system can be incorporated to afford both an algebraic representation of engineering mechanics concepts (in the form of their mathematical expressions), and their manipulation by computer. The system thus developed not only provides an extendable backlog of mathematical expressions, but also enables some of the expressions to function as formula manipulator. Thus, it is obvious that the potential of application of symbol manipulating programs is significant and merits further investigation for future professional application.

References

1. Norton, L. M., ADEPT—*A Heuistic Program for Proving Theorems of Group Theory*, Mass. Inst. Technol. (Sept. 1966).
2. Cohen, J., "An Interactive System for Proving Theorems in the Predicale Calculus," *Proc. Symp. Symbolic Algebraic Manipulation, 2nd* 268–280 (1971).
3. Moses, J., *Symbolic Integration*, Massachusetts Inst. of Technol. (December 1967).
4. Moses, J., "The Integration of a Class of Special Functions with Risch Algorithm," *SIGSAM Bull.* No. 13, 14–27 (December 1969).
5. Tobey, R. G., Robrow, R. J., and Zilles, S. N., "Automatic Simplification in FORMAC," *Proc.—Fall Joint Comput. Conf.* (1965).

6. Barton, D., Bourne, S. R., and Fitch, J. P., "An Algebra System," *Comput. J.* **13** (1) (February 1970).

7. Hearn, A. C., "Applications of Symbolic Manipulation in Theoretical Physics," *Proc. Symp. Symbolic Algebraic Manipulation, 2nd* 17–21 (1971).

8. Flodmark, S., and Blokker, E., "Use of Computers in Treating Non-numerical Mathematics and Group Theoretical Problems in Physics," *SIGSAM Bull.* No. 15, 28–63 (July 1970).

9. Howard, C. J., "Relativistic Applications of Symbolic Mathematical Computation," *SIGSAM Bull.* No. 15, 64–83 (July 1970).

10. Wong, A. K. C., and Bugliarello, G., "Artificial Intelligence in Continuum Mechanics," *Proc. ASCE* **96**, No. EM6 (December 1970).

11. Ossenbruggen, P. J., Wong, A. K. C., and Au, T., "Symbolic and Algebraic Manipulation for Solving Differential Equations," *Proc. 1971 Nat. Conf. Ass. Comput. Machinery* 717–735 (1971).

12. Howard, J., "Computer Formulation of the Equations of Motion Using Tensor Notation," *Commun. Ass. Comput. Machinery* **10**, No. 9, 543–548 (1967).

13. Tobey, R. G., "Eliminating Monotonous Mathematics with FORMAC," *Commun. Ass. Comput. Machinery* **9**, 742–751 (1966).

14. Garwick, J. V., "GPL, a Truly General Purpose Language," *CACM* **11**, No. 9, 634–638 (September 1968).

15. Irons, E. T., "Experience with an Extensible Language," *CACM* **13**, No. 1, 31–40 (January 1970).

16. Newell, A., McCracken, D., Robertson, G., DeBenedetti, L., "L*(F) Final Version," Carnegie-Mellon Univ., Users' Manual (January 1971).

17. Tobey, R. G., "Symbolic Mathematical Computation—Introduction and Overview," *Proc. Symp. Symbolic Algebraic Manipulation, 2nd* 1–16 (1971).

18. Weissman, C., *LISP 1.5 Primer*. Dickenson Publ., Belmont, California, 1967.

19. Newell, A. *et al.*, *Information Processing Language-V Manual*, The Rand Corp. Prentice-Hall, Englewood Cliffs, New Jersey (April 1965).

20. Perlis, A. J., Iturriago, R., and Standish, T. A., "A Definition of Formula Algol," *Commun. Ass. Comput. Machinery* **9**, 549 (1966).

21. Tobey, R. G. *et al.*, *PL/I-FORMAC Interpreter User's Reference Manual*. 1967.

22. Slagle, J. A., "A Heuristic Program that Solves Symbolic Integration Problems in Freshman Calculus," *Computers and Thought* (Feifenbaum, E. A., and Feldman, J. eds.), pp. 191–203. McGraw-Hill, New York, 1963.

23. Wang, H., "Toward Mechanical Mathematics," *IBM J. Res. Develop.* **4** (January 1960).

24. Jenkins, W. M., *Matrix and Digital Computer Methods in Structural Analysis*. McGraw-Hill, New York, 1969.

25. Krutar, R., "Extentability," Ph. Dissertation, Comput. Sci., Carnegie-Mellon Univ. (1971) (manuscript in preparation).

A Review of the Capabilities and Limitations of Parallel and Pipeline Computers*

William R. Graham

R & D ASSOCIATES

SANTA MONICA, CALIFORNIA

1. Introduction

Several different approaches have been taken to improving the speed with which computers solve problems. The most obvious approach is to increase the speed at which each component can operate. This approach has brought about the large speed increases seen to date. There is at least another factor of 10 to 100 in speed improvement achievable through this approach, but beyond that problems of size, cooling, and transit time appear to increase rapidly [7]. Another approach is to improve the speed with which the arithmetic operations are constructed from elementary logical operations. However, recent studies [9] have established that most modern computers operate within a factor of two of the theoretical limiting speed for addition, and within a factor of three of the limit for multiplication; large speed increases are not to be found here. More complicated operations, such as generating trigonometric functions, may have the potential for a greater speed increase.

Another approach is to search for faster algorithms that will solve the problems of interest. Even when successful, this approach can prove to be very expensive in terms of capital and human resources. It is an unfortunate

---

* Any views expressed in this paper are those of the author. They should not be interpreted as reflecting the views of R & D Associates or the official opinion or policy of any of its government or private research sponsors.

and seldom-appreciated fact that one is forced into the tasks of finding new approaches to problems as a logical consequence of taking a seemingly unrelated approach to improving computer speed : modification of the basic organization of the machine.

Although a wide variety of modifications are possible and several are under consideration, two machines having radically new organizational schemes are in the final stages of construction. Since they manifest many of the advantages and disadvantages of the organization approach to greater computational speed, they will be referred to throughout the following discussion. The two new computers are the University of Illinois' parallel-organized ILLIAC IV and the Control Data Corporation's pipeline-organized STAR.

The price paid for the substantial speed improvement possible with these machines is extracted from the user in a subtle but definite way [1]. To appreciate the advantages and disadvantages, one must first understand the new organizational schemes [3, 4]. The following descriptions do not do justice to the subtleties in the designs of either machine, but review only the most basic features of each.

2. The Parallel Computer

The parallel computer, such as the ILLIAC IV, has for its conceptual basis the notion that two conventional computing machines can work at twice the rate of one machine. The major deficiency in this approach is that two machines also *cost* twice as much as one machine. To overcome this fundamental drawback, the parallel design economizes by having many identical copies of the conventional computer's arithmetic unit driven with only one control unit. The control unit is responsible for obtaining, decoding, and issuing instructions, and generally assuring that the machine executes the program. Since the control unit is rather sophisticated and expensive, a considerable amount of money is saved by having only a single control unit in charge of W identical arithmetic units. (Here W is called the "width" of the parallel processor.) The price paid for getting by with only a single control unit is that each of the arithmetic units must do the same thing at the same time or else be completely inhibited and do nothing. It will be shown that this condition places definite constraints on computational procedures when efficient use of the resources available is a consideration.

The parallel processor must be organized so that it is impossible for two or more arithmetic units to attempt to change the same location in memory at the same time. It is achieved most simply by allocating to each arithmetic unit an exclusive block of memory, not directly accessible to any other arithmetic unit. The result is that if arithmetic unit J needs a number stored in the memory of arithmetic unit K, the control unit must have K recall the number and then transmit it, over specially provided channels, to J. The number of unidirectional channels that would be required to interconnect directly all of the arithmetic units is $W^2 - W$. If the designer wishes to sacrifice transmittal time, a smaller number of channels may be used. For example, in the ILLIAC IV, if one imagines the arithmetic units strung on a circle, then only the two nearest units and the two seven units away are in direct communication. Another way to visualize the interconnections is to arrange the arithmetic units on a toroid as shown in Fig. 1. Then only the four nearest neighbors (two on the circles shown and two along the toroid) can communicate directly. In the ILLIAC arrangement, eight transmissions are required for communication between the most distant arithmetic units.

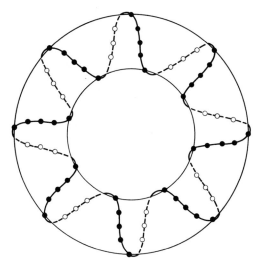

Fig. 1. ILLIAC IV memory interconnection.

3. The Pipeline Processor

A quite different approach underlies the design of the pipeline processor, such as the STAR, which is being developed by the Control Data Corporation. In the conventional computer, the time required to retrieve operands from the memory, execute the operation, and return the result to memory must be at least as great as a time equal to the distance traveled by the information divided by the speed of light, and several other less inevitable factors. The pipeline processor, shown schematically in Fig. 2, gains its advantage by starting the retrieval of a second set of operands, each located in memory adjacent to the first, before the first result has been returned to the memory. Thus, a pipeline begins to fill, and when the pipeline is full, the round-trip distance divided by the speed of light no longer limits the apparent cycle time.

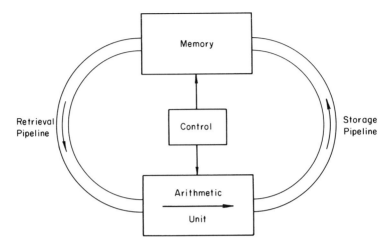

Fig. 2. The pipeline processor.

So that the arithmetic unit will not be an obstruction in the flow, it too is built as a pipeline. It can receive and start processing a second set of operands before finishing the calculation with the first set. To complete the memory-to-memory pipeline, the arithmetic unit must deliver the results back into memory at the same time and rate that it is receiving new pairs of operands from the memory.

It is sometimes desirable to skip an operation on certain operands in the pipeline. This can be done through the use of a control vector, which contains one bit that is associated with each operand set in the pipeline. Depending on whether the associated control vector bit is one or zero, the operation is either performed or skipped. If no operation is performed, the memory location for the result does not have its previous content changed.

The logical control processes that must take place for the number stream in the pipeline to flow smoothly are sufficiently complex that the foreseeable pipeline machines will permit only one type of arithmetic operation per stream (e.g., add or multiply or divide) and only one stream at a time. (An impressive exception is the vector inner-product operation, which may be implemented on the STAR.) Furthermore, the pipeline memory retrieval and storage locations for each operand string are constrained to lie on consecutive linear sweeps through the memory. As with the parallel processor, both the maximum computing rate and the opportunities for inefficient operation are greatly increased by this design. The hardware required to make the arithmetic and memory units operate as an integrated pipeline is not only complex but also expensive, so that if the pipeline processor is to be economical, the pipe must be kept full a substantial part of the time. To achieve this condition, careful consideration must be given to the form of the solutions.

4. Parallel and Pipeline

To compare the two types and determine which problems are solved most efficiently using the parallel computer, and which using the pipeline computer, it is first necessary to have a conceptual framework for considering computational procedures in general.

Consider the hierarchy of computer activities as starting with the basic *step*; the retrieval of a set of operands from memory, the operation, and the return of the results to memory. At the next level is the parallel or pipeline *stage*. The stage is the collection of the steps which could be done in parallel or in the same pipeline stream, without violating the program logic or the machine constraints for execution. A stage is a property of the logic of the method of solution and the computer logical design, but not of the computer width or pipeline capacity. Finally, all of the stages for a solution, connected in the proper order, constitute a computer *program*.

The following conditions result from the constraints of logical simplicity being imposed in the present designs. Some number N of computing steps, $S_1, S_2, \ldots, S_N$, form a parallel stage if (1) no step depends upon the result of any other in the stage, (2) all steps require either the same operation or no operation to be performed, and (3) the operands are properly distributed among the arithmetic unit memories.

The N computing steps form a pipeline stage if (1) no step depends upon the result of any other in the stage, (2) all steps require either the same operation or no operation to be performed, and (3) the elements of each operand string are packed in successive memory locations. Comparing these two sets of conditions, one sees that (1) and (2) are the same; only (3), memory location assignment, is different for the parallel and pipeline machines.

Considering the complete dissimilarity in the parallel and pipeline organization plans, it might appear remarkable that the two sets of conditions for a stage are so similar. The second condition is required only for reasons of hardware simplicity. The first condition is only a manifestation of a single deeper principle which is responsible for the great speed of the two computers. Stated most applicably in its negative form, the principle is: When information does not have to flow from the result of one operation to the input of the next operation, then the speed of the hardware for performing sequential operations does not limit the rate at which operations can be performed.

Substantial constraints result from these three conditions. The first condition means that many implicit differencing schemes, including, for example, the usual two-pass algorithm for solving the Crank–Nicholson equation, will reduce pipeline and parallel computers to conventional one-step-per-stage operation, with the associated long execution time and low machine efficiency. The third condition, proper storage allocation, makes performing such a simple operation as multiplying a matrix by itself an interesting though solvable problem in parallel or pipeline operation. The parallel design requires that the operands for a stage be distributed throughout its arithmetic units' memories in a two-dimensional arrangement (the arithmetic unit index representing one dimension, the local memory address the other), while the pipeline computer requires that the same operand strings each be packed together tightly in the one-dimensional memory. This difference in memory allocation, the special routing needed for the parallel arithmetic units to communicate with each other, and the ease with which operand strings are offset in the pipeline computer are the dominant differences that the user sees between the two designs.

Assuming that the three conditions have been met, one may then proceed to compare the machines on the basis of execution time and the efficiency with which resources are used. One overriding fact to keep in mind through the following discussion is that one-step-per-stage (sequential) operations will greatly diminish the computers' performance. If half of the operations to be executed are so well matched to the machine that they take essentially no time to execute, but the other half must be done one step per stage, then the total program execution time is at best an unimpressive factor of two less, than that required by a conventional machine. As a first approximation, the ratio of the parallel or pipeline program execution time to the conventional computer's time is the ratio of the number of steps that must be performed sequentially, one at a time, to the total number of steps. To make efficient use of the parallel or pipeline computer's resources, it is not sufficient that a few or even half of the steps of the program be suited to the design ; nearly all of the program steps must be part of large stages—ones that contain many steps.

5. Parallel and Pipeline Execution Times

In a parallel processor, the three parts of a computing step—memory retrieval, operation, and memory storage—may be diagrammed as shown in Fig. 3. The step time $T_{\parallel}$ depends upon the specific operation being performed.

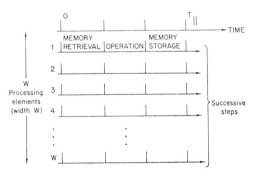

Fig. 3. Parallel computer cycle.

For a stage of N steps to be executed on a computer of width W (i.e., having W arithmetic units), the time required to execute steps 1 through W is $T_{\parallel}$, steps $W + 1$ through $2W$ is $2 \cdot T_{\parallel}$, etc. The execution time is shown as a function of the number of steps in the stage in Fig. 4. Note that twice as much time is required to execute $W + 1$ steps as is required for W steps; adding one more step nearly halves the mean rate of execution.

The operation of the pipeline machine can be diagrammed as shown in Fig. 5. The time required for the pipeline computer to execute a stage of N

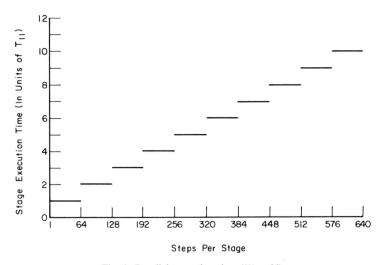

Fig. 4. Parallel execution time ($W = 64$).

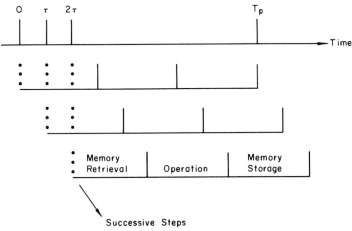

Fig. 5. Pipeline computer cycle.

steps is the time to fill the pipeline T_p plus one pipeline sequencing time τ ($\tau \ll T_p$) for each subsequent step, as is shown in Fig. 6. If the parameters $T_{\|}$, W, T_p, and τ of the two machines are known, the execution times may be calculated as a function of the number of steps in the stage N. For example, when $N = 1$, the parallel machine time is just $T_{\|}$ and the pipeline machine time is T_p. At the other extreme, as N tends to infinity, the parallel machine time per step tends to $T_{\|}/W$, and the pipeline machine time per step tends to τ.

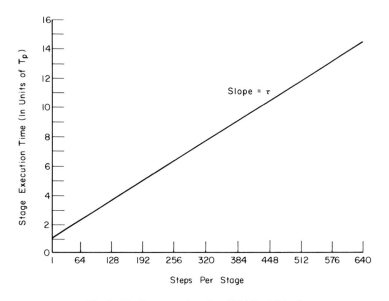

Fig. 6. Pipeline execution time (STAR addition).

A comparison of the parallel and pipeline execution times is graphed in Fig. 7. Depending upon the values of the computer parameters $T_{\|}$, W, T_p, and τ, the pipeline machine can take more time to execute a stage of any size (corresponding to the upper curve), can take less time to execute a stage of any size (corresponding to the lower curve), or the advantage can shift back and forth between the two machines as the number of steps per stage increases (corresponding to the middle curves).

As the program size increases, it will ultimately fill the main memory, and secondary memory will have to be accessed, at the cost of increased execution time and programming complexity.

Using the preliminary information in Table 1, one may deduce that the STAR and ILLIAC IV quadrant have comparable speeds for performing addition, with the ILLIAC having the advantage when only a few sums are to be calculated. In multiplication, the ILLIAC IV is always faster than the

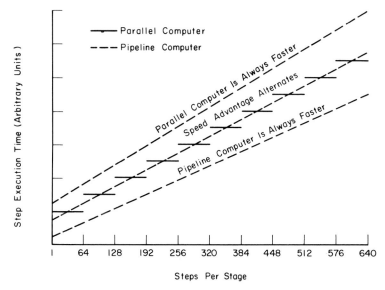

Fig. 7. Parallel/pipeline comparative execution times.

TABLE 1

64-Bit Precision Characteristic Parameters[a] for the CDC STAR
and the ILLIAC IV

Operation	CDC STAR		ILLIAC IV	
	T_p (μsec)	τ	$T_{\parallel}$ (μsec)	W (One quadrant)
Addition	1.76	20	1.28	64
Multiplication	1.76	40	1.45	64
Division	1.80	80	3.76	64

[a] The figures are preliminary and for illustrative purposes only.

STAR, while in division, the STAR is about twice as fast as the ILLIAC in finding a single quotient, but the ILLIAC is about 1.4 times as fast as the STAR in calculating a long sequence of quotients. Both designs will also perform 32-bit floating point operations at correspondingly higher speeds. Of course, accurate determinations of the hardware speeds must await the final stages of machine development.

TABLE 2

64-Bit Precision Computation Speeds[a] (Memory to Memory) in Millions of Operations
per Second

Operation	Steps per stage	IBM 360/75	IBM 360/195	CDC 7600	CDC STAR	ILLIAC IV (One quadrant)
Addition	$N = \infty$	0.24	4.6	5.2	50	50
	$N = 1$	0.24	0.55	1.6	0.57	0.78
Multiplication	$N = \infty$	0.14	4.6	5.2	25	44
	$N = 1$	0.14	0.53	1.5	0.57	0.69
Division	$N = \infty$	0.096	1.7	2.0	12.5	17
	$N = 1$	0.096	0.43	0.93	0.56	0.27

[a] For the ILLIAC IV and the STAR, the figures are preliminary and for illustrative purposes only. For the 360/195 and 7600, the numbers are sensitive to the way the smaller, faster memories are used.

The maximum computing rates for large problems are shown in Table 2 for the IBM 360/75, the CDC 7600, and the IBM 360/195 as well as for the ILLIAC and the STAR. Although the 7600 and 195 are much faster than the familiar 360/75, they obviously do not approach the STAR and the ILLIAC IV's performance on long sequences of identical calculations. However, except for problems well suited to parallel or pipeline manipulation, the 7600 and 195 are as fast or even faster than the STAR and ILLIAC IV.*

6. Parallel versus Pipeline: Efficiency

The cost of executing a stage is more closely related to the efficiency with which each computer's resources are used than it is to the execution time. For the parallel computer, the maximum computing rate is reached when all arithmetic units are used. At the maximum, $W/T_{\parallel}$ steps per second are executed. For the pipeline computer, the maximum computing rate is reached when the pipeline is filled; then the rate is $1/\tau$ steps per second.

Efficiency for a stage may be defined as the ratio of the actual computing rate in steps per unit time to the maximum computing rate; this efficiency always lies between zero and one. For a stage of one step, the efficiency of the parallel machine is $1/W$ and that of the pipeline machine is τ/T_{p}. The efficiency of the parallel computer reaches unity whenever the number of steps per stage N is an integer multiple of W, but the efficiency is not a monotonically increasing function of N, whereas the efficiency of the pipeline

* The use of carefully designed 7600 stack loops will, in some cases, increase the speed of the 7600 by a factor of two over those shown in the table [6].

processor increases monotonically with N, but approaches unity only asymptotically as N tends to infinity. The relationships between efficiency and stage size are shown in Fig. 8.

Fig. 8. Parallel and pipeline 64-bit multiplication efficiency.

The ILLIAC IV has a quadrant width $W = 64$, so it can execute a stage of 64 steps with unit efficiency. The STAR can execute a stage of this size with an efficiency lying between 0.42 (for addition) and 0.75 (for division). For a stage of 65 steps, the STAR efficiency would increase slightly, while the ILLIAC IV efficiency would drop to slightly more than 0.5.

In the worst case, the parameters given in Table 1 indicate that the one-step efficiency for the STAR is between 1 and 4.5 %, depending on the operation, and for the ILLIAC IV is always about 1.5 %. The possibilities for inefficient use of computer resources which the parallel and pipeline designs give to the analyst who specifies the algorithms to be applied, the programmer, and to the compiler have the potential to exceed by far any habitual excess yet seen in the computing world.

The efficiency with which a complete program is executed is the average of the efficiency of each stage weighted by the total time required to complete that stage. Stated simply, the program efficiency is the ratio of the time required to execute a program if the computer is operating at unit efficiency, as already defined, to the actual execution time. For a program which has half of its operations executed at unit parallel or pipeline efficiency, and the other half executed one step per stage, memory to memory, the overall program efficiency would lie between 2 and 9 % for the STAR and be 3 % for the ILLIAC IV.

7. Languages for the Parallel and Pipeline Computers

The ease and expense with which the very large computers are brought to bear on problems depend strongly upon the language through which the user dictates his program. Those familiar with the history of computers will not find it remarkable that a definitive high-level language does not yet exist for either the ILLIAC or the STAR. The lack of definitive language designs stems from the fact that high-level languages are not the products of a science but rather the result of a craft that always lags behind hardware construction.

The increased demands which the parallel and pipeline designs have put on software and the paucity of applicable software research have caused this lag to increase. Ideally, a language structure would encourage the user to do those operations which the computer does best and avoid those which it does not so well. Consideration of the adaptability of most users has instead dictated language compromises for both machines which blend array operations into the structure of FORTRAN. Particularly when a less than ideal high-level language is to be used, an important asset of a language compiler would be the ability to discover automatically and take advantage of the inherent parallelism in any given program [2]. If a program contains no conditional statements, then the extent of parallelism possible can be known before the program is executed. This is also true for programs containing some types of conditional statements, but there are other programs in which the degree of parallelism allowed by the logic can only be known after the execution—information of more limited value. In practice, however, the high-level language advantage of being concise is instead offset by the excessively circumspect program which the compiler produces. Both the advantages and disadvantages of various levels of language are shown in Table 3, derived from programming a back substitution matrix operation for the ILLIAC IV [5].

TABLE 3
PROGRAM CHARACTERISTICS FOR THREE LEVELS OF LANGUAGE[a]
(Relative to the machine language level)

Language level	Relative time required to write program	Relative number of source statements	Relative number of resulting machine language statements	Relative inner loop execution time
High	0.13	0.28	17.6	8.9
Intermediate	0.67	0.39	2.7	3.0
Machine	1	1	1	1

[a] From [5].

Of course, the basic machine languages exist for both computers. In fact, the ILLIAC IV has two machine languages: one for the control unit and the other for the arithmetic unit. Both of these languages operate on a conventional elementary level. The compiler and operating system used by the ILLIAC IV reside on a separate but attached computer of a more conventional design. This arrangement has led to a proposal that would overcome the inefficiency resulting from one-step-per-stage operation of the parallel machine. The attached computer would be used to perform the sequential computations while the ILLIAC would perform the parallel parts. With sufficiently skillful multiple programming and memory protection, the proposal could be effective, but the sequential machine could easily become a severe bottleneck and pace the overall speed of computation.

The wired-in machine level language used by the STAR is, except for memory allocation and addressing, a much higher level language than FORTRAN. Although high level, the language is very well matched to the machine—efficiencies and inefficiencies inherent in the design of the machine and of the language seem to coincide. Limited studies in which the same problems were programmed in machine language for both the ILLIAC and the STAR have indicated that the STAR requires about one-sixth as many instructions as the ILLIAC and that on the average, each STAR instruction is executed six times more slowly.

8. Application Programming

A considerable amount of application programming has been completed for the parallel and pipeline machines without benefit of a guiding theory. One of the first problems encountered was summing a list of N numbers. It might appear that this operation is necessarily sequential and consists of $N - 1$ stages of one step each. However, a clever algorithm has been devised for both the parallel and pipeline machines [11], in which the terms are summed in pairs, and then the sums summed in pairs, and so on until the complete sum is formed. The number of stages required is $\log_2(M)$, where M is the smallest integer greater than or equal to N and also a power of two. When N is a power of two, the ratio of stages required by the two methods is $(N - 1)/\log_2(N)$, which increases with increasing N.

Both the parallel and the pipeline designs are well suited to doing matrix operations. The time required to do vector addition has already been given. The vector inner product may be done by performing one vector multiplication stage and then $\log_2(N)$ additional vector stages (when N is assumed to be a power of 2), the first of which requires $N/2$ additions and each succeeding one requiring a factor of two fewer additions.

The matrix × vector product and the matrix × matrix product are elaborations of the vector inner product, but with more interesting storage allocation problems. A discussion of the parallel matrix multiply and various storage allocation schemes has been given elsewhere [4]. The linear storage of the pipeline design requires in the case of the matrix product $A \times B$ that the rows of A and the columns of B be stored sequentially. If B is initially stored by rows, then its transpose must be created and stored by rows before the multiplication can begin. For this reason, matrix transposition is a wired-in intruction in the STAR design and runs at pipeline speed.

9. Performance on Large Problems

To understand the performance of the parallel and the pipeline machines when dealing with large deterministic programs, the Air Force had a series of programs written for solving the same problems on the one quadrant ILLIAC [11] and the STAR [10]. Parts of two large programs, HEMP and SC, were coded in ILLIAC IV machine language and STAR machine language. Only the central 5% of the complete HEMP and SC programs were coded for the parallel and pipeline machines, but that 5% accounted for 95% of the running time on the present generation machine.

HEMP was an electromagnetic source and field calculation involving the time dimension and one spatial dimension. Its five sections are as follows:

Section 1: Combination of input and intermediate values which result in an input to Section 2.

Section 2: Search for a match and subsequent interpolation of the values resulting from Section 1. This is a logistics problem not directly related to the mathematical algorithm.

Section 3: Convolution integrals.

Section 4: Interpolation or restructuring of data resulting from convolution. This is also a logistics problem not directly related to the algorithm.

Section 5: Calculation of the electromagnetic fields.

After the same parts of these sections were coded for the ILLIAC IV and STAR, the execution times were determined to be the values shown in Table 4. The efficiency was calculated for the ILLIAC IV, and was derived approximately for the STAR from the data reported. The efficiencies are also shown in Table 4.

For the HEMP problem, the STAR is about 65 times as fast as the present generation machine, while the ILLIAC IV is about 43 times as fast. The high efficiency of the STAR results in part from the very powerful operation code of the machine, and in part from "brute force" techniques that are natural to use on the STAR for some operations, such as linear search table lookups. The last point is substantiated by the similarity in the total execution times, in light of the similar hardware speeds of the two machines. The variation in the ILLIAC IV efficiency among the various sections of HEMP gives some

TABLE 4
HEMP TIMING AND EFFICIENCY

HEMP section	ILLIAC IV quadrant		STAR (50-nsec cycle)		Present generation machine (≈ 6600) Time (sec)
	Time (sec)	Efficiency (%)	Time (sec)	Efficiency (%)	
1	0.11	90	0.301	93	
2	8.80	22	1.539	99.8	
3	6.40	51	8.067	98	
4	0.22	2	0.009	67	
5	0.35	29	0.672	74	
Totals	15.88	32 overall	10.59	98	688

notion of the control which the programmer and the algorithm have on the computer performance.

SC was an electrodynamics program which solved Maxwell's partial differential equations in the time dimension and two spatial dimensions. In it, there are variables defined by a relation of the form

$$T_{IJ} = T_{I(J-1)} * C_J$$

Straightforward evaluation of this formula for a fixed I would seriously degrade the performance of both the pipeline and the parallel computers. Fortunately, it was found that the equations were uncoupled for fixed J over an appropriate choice of I's, and that by performing the operation for a range of I's and a fixed J, efficient use could be made of the machines.

Had J been the only subscript, the T_J's could still have been computed efficiently by first noting that

$$T_1 = C_1 T_0, \qquad T_2 = C_2 T_1 = C_2 C_1 T_0, \qquad T_3 = C_3 T_2 = C_3 C_2 C_1 T_0$$

and then forming the following products by the operations noted [8].

Given the array:

$c_1 \ c_2 \quad c_3 \quad c_4 \qquad c_5 \qquad\qquad c_6 \qquad\qquad c_7 \qquad\qquad c_8$

Multiply by self shifted right 1:

$c_1 \ c_1 c_2 \ c_2 c_3 \quad c_3 c_4 \qquad c_4 c_5 \qquad c_5 c_6 \qquad c_6 c_7 \qquad c_7 c_8$

Multiply by self shifted right 2:

$c_1 \ c_1 c_2 \ c_1 c_2 c_3 \ c_1 c_2 c_3 c_4 \ c_2 c_3 c_4 c_5 \quad c_3 c_4 c_5 c_6 \qquad c_4 c_5 c_6 c_7 \qquad c_5 c_6 c_7 c_8$

Multiply by shelf shifted right 4:

$c_1 \ c_1 c_2 \ c_1 c_2 c_3 \ c_1 c_2 c_3 c_4 \ c_1 c_2 c_3 c_4 c_5 \ c_1 c_2 c_3 c_4 c_5 c_6 \ c_1 c_2 c_3 c_4 c_5 c_6 c_7 \ c_1 c_2 c_3 c_4 c_5 c_6 c_7 c_8$

Each element of the final array of products is then multiplied by T_0 to form the T_J array. As in the cumulative sum problem, the number of stages required is $\log_2[M]$ rather than $M - 1$ as would be required in the straightforward approach. The set of equations

$$T_J = C_J * T_{J-1}$$

should prove an interesting test of algorithms designed to detect parallelism in problems.

TABLE 5
SC TIMING AND EFFICIENCY

ILLIAC IV quadrant		STAR (50-nsec cycle)		Present Generation machine (≈ 6600)
Time (sec)	Efficiency (%)	Time (sec)	Efficiency (%)	(sec)
6.55	85	15.2	90	590

The results of the SC coding are shown in Table 5. For the SC problem, the STAR is about 39 times as fast as the present generation machine, while the ILLIAC IV is about 90 times as fast. The overall efficiency of the ILLIAC IV is considerably improved over its value for the HEMP code; the efficiency of the STAR is slightly decreased. It was pointed out in the original work that a slight change in the ILLIAC IV memory allocation scheme would double the total SC running time and halve the efficiency, and it is probable that similar small errors in allocation would have greatly reduced the STAR program efficiency. This again shows the sensitivity of these machines to the user's skill.

The results reviewed here should be interpreted cautiously. Only the central 5% of the HEMP and SC programs were coded for the ILLIAC and the STAR. Although these portions of the programs required 95% of the running time on the present generation machine, the division of time may be considerably different on the parallel and pipeline machines, as was noted earlier. It was not inconceivable that if the remaining 95% of the programs were to be coded, it might result that the additional code would dominate the running time, and reduce the overall efficiency to the range of 10 to 40%.

10. Conclusion

The parallel and pipeline designs produce computers that when used to their maximum, will be impressively faster than more conventional computers. However, formulating problems and writing programs which will use the new machines to anything approaching their maximum capabilities will prove a severe and perhaps, on occasion, an overwhelming challenge to the creativity of all concerned.

Acknowledgments

The author gratefully acknowledges the many interesting and stimulating observations made to him by Dr. R. N. Byrne, Lawrence Radiation Laboratory; Dr. T. C. Chen, International Business Machines Corporation; Mr. D. E. McIntyre, University of Illinois; Mr. L. J. Sloan, University of California; Mr. M. Warshaw, The RAND Corporation; and Mr. J. E. Wirsching, Symbolic Control, Inc.

References

1. Chen, T. C., The fact that the parallel and pipeline computer designs have inherent limitations which render their performance very sensitive to problem and programming formulation was first pointed out to the author by Dr. Tien Chi Chen, a result which Dr. Chen derived from his extensive unpublished research on the subject, Private Communication, IBM San Jose Res. Lab., San Jose, California (1969).
2. Gozalez, M. J., and Ramamoorthy, C. V., "Program Suitability for Parallel Processing," *IEEE Trans. Comput.* **C-20** (6), 647–654 (1971).
3. Graham, W. R., "The Parallel and the Pipeline Computers," *Datamation* 68–71 (April 1970).
4. McIntyre, D. E., "An Introduction to the ILLIAC IV Computers," *Datamation* 60–67 (April 1970).
5. McIntyre, D. E., "ILLIAC IV Language Evaluation—A Preliminary Report," Univ. of Illinois, ILLIAC IV Document No. 213 (May 1970).
6. McMahon, F. H., and Sloan, L. J., Private Communication, Lawrence Radiation Lab., Livermore, California (1971).
7. Ware, W. H., "On Limits in Computing Power," RAND Corp., P-4208 (October 1969).
8. Watson, V., Private Communication, Ames Res. Lab., Sunneyvale, California (1971).
9. Winograd, S., "On the Time Required to Perform Addition," *J. Ass. Comput. Machinery* **12**, No. 2, 277–285 (1965); idem, "On the Time Required to Perform Multiplication," *J. Ass. Comput. Machinery* **14**, No. 4, 793–802 (1967).
10. Wirsching, J. E., and Alberts, A. A., "Application of the STAR Computer to Problems in the Numerical Calculation of Electromagnetic Fields," A. F. Weapons Laboratory, Tech. Rep. No. AFWL-TR-69-165 (April 1970).
11. Wirsching, J. E., Alberts, A. A., McIntyre, D. E., and Carroll, A. B., "Application of the ILLIAC IV Computer to Problems in the Numerical Calculation of Electromagnetic Fields," Air Force Weapons Lab. Tech. Rep. No. AFWL-TR-69-91 (March 1970).

Equation-Solving Algorithms for the Finite-Element Method

Bruce M. Irons and David K. Y. Kan

UNIVERSITY OF WALES
SWANSEA, WALES, U.K.

1. Introduction

This paper discusses finite-element solution algorithms with particular reference to programming techniques. It is too widely assumed that an algorithm must succeed or fail regardless of the skill, care, and time given to implementing it. It is also too widely assumed that an algorithm of proved efficiency is worth implementing, regardless of ease of programming, ease of debugging, size of program, convenience of running in a given installation, etc. We recall that the "isoparametric" elements owe their popularity chiefly to their ease of programming, modularity, and decisiveness of debugging.

In truth, much of the algebra that we habitually write is mere philosophizing; for so many decisions lie between the algebra and the final coding. It is hoped that this paper will guide the programmer in the sort of choices he can make.

2. Types of Record Encountered

The most important choice is the system of housekeeping to be adopted for the various arrays. After the choice of algorithm, it is the one that must be considered most carefully, because a mistake is expensive to rectify.

To introduce the systems, we now list the items of data that must be accommodated in certain cases:

497

1. Element stiffness matrices. There are also equivalent direct methods of computing residual forces which, if numerical integration is used, require the following:
2. Shape functions at the integrating points, and their derivatives, and
3. the Jacobean matrices relating x, y, z and ξ, η, ζ coordinate systems, the integrating factors, etc.
4. Element type, element degrees of freedom, node numbers, destination vectors, indicators of first or last appearances of nodes, indicators for the timing for output by elements, etc.
5. Material properties, temperatures, pressures, etc.
6. Stresses at integrating points within an element, locked-up stresses, plastic strains, etc.
7. Stresses at nodes, including some means for averaging them.
8. Stresses at interelement boundaries which are not node points, e.g., at midface, again including some means for averaging them.
9. Extrapolated and/or averaged stresses: the maximum variation of stresses between elements at a point.
10. Coordinates, base vectors, normal directions, etc.
11. Vectors of nodal deflections, external forces as nodal equivalents, residual forces, error forces due to round-off, etc.
12. Vectors of displacements, etc. at a previous iteration.
13. Assembled stiffness matrix.
14. Equations for back-substitution.
15. Prescribed deflections, reactions to earth; linear constraints in general.
16. Error diagnostics with deferred action.

Such a list cannot be exhaustive, because different workers have different requirements. It is thought-provoking. For example, it is natural to store element contributions to external loads; because external loads are logically similar to diplacements? Why not store deflections according to element?

3. Principal Types of Organization

We now discuss some standard housekeeping systems, that is, methods of spreading an array between high-speed core storage and low-speed backing storage, so as to economize in high-speed storage, amount of data transferred, the number of transfers (for certain large installations are beginning to charge according to the number of transfers alone*), and perhaps in the future—as backing storage comes to mean low-speed core—in the total low-speed storage requirement. The systems now outlined can be applied to arrays other than the assembled stiffness matrix. This fact is not always appreciated.

* If adequate buffers are provided these amount to the same thing.

1. The whole record can remain on core. We encourage students to learn how element stiffnesses assemble by treating the simplest case with the whole stiffness matrix on core. But cases arise where it would be uneconomical not to have the whole array on core. For example, in the prefrontal routine, which is entered before a frontal elimination, we must discover the last appearance of each node, which necessitates stringing the element node numbers together into a long vector—it would be difficult anyway to subdivide this vector. However, should we keep the whole deflection vector on core? Yes, perhaps, if there is only one. But consider Cornell's method of optimization [6], here we have a deflection vector for each applied load case, and the same again for each allowed design change, so that the "deflections" can be converted into design sensitivities.

2. In the "creeping band" technique only a semibandwidth N remains on core. To illustrate this and other techniques we consider the simple problem of accumulating element external loads X^e into a vector X. Suppose $N = 4$. At some time we could have the subvector of X comprising, say $(X_4 X_5 X_6 X_7)$. Into this we accumulate certain element contributions, taken in an acceptable order. Eventually one of two things happens: (a) We meet an element which uses X_4 for the last time. (b) We meet an element which requires X_8. These represent alternative programming strategies. In either case, we shift all the values one place to the left, we write X_4 to backing storage (or perhaps we put it into a buffer area) and hence we make a space for X_8. The subvector becomes $(X_5 X_6 X_7 X_8)$.

The name "creeping band" implies, correctly, that this method is slow. The values are shifted too frequently.

3. The "cycling band" approach avoids the shifts. Here $(X_3 X_4 X_5 X_6)$ would become $(X_7 X_4 X_5 X_6)$ on losing X_3. Then it would become $(X_7 X_8 X_5 X_6)$ on losing X_4, and so on.

This is still not a good idea for the assembled stiffness matrix, because the bandwidth usually varies. If the subarray is sufficiently large for the maximum bandwidth, there will be wasteful zero operations when the bandwidth is less.

4. The "band plus buffer" approach is attractive. We now underline the active variables: $(\underline{X_2 X_3 X_4 X_5} X_6 X_7)$ becomes $(X_2 \underline{X_3 X_4 X_5 X_6} X_7)$ becomes $(X_2 X_3 \underline{X_4 X_5 X_6 X_7})$. Now X_8 is required, assuming strategy (b). We discard $(X_2 X_3)$, the prebuffer, to backing storage. We then shift everything two places: $(\underline{X_4 X_5 X_6 X_7} X_8 X_9)$ becomes $(X_4 X_5 \underline{X_6 X_7 X_8 X_9})$, etc. However, before operating on X_8 and X_9 it may sometimes be necessary to retrieve some previous record concerning $(X_8 X_9)$, i.e., a buffer from some previous iteration. In this case it will be necessary to make the buffer length some submultiple of N, or equal to N.

It is possible, by equivalencing etc., to give the prebuffer and the postbuffer their own variable names, so as to transfer them without implied DO-loops. This is quicker.

5. Frontal housekeeping [4, 5] is equally straightforward to program, but different in philosophy. For clarity of explanation, we use the idea of "destination vectors," as we use them in the program itself. These are abbreviated Boolean matrices. To illustrate, we consider element data $(X_2 X_5 X_3)$ which is to be assembled into a longer subvector according to the destination vector (4, 9, 6). This means that X_2 assembles into position 4 of the subvector, X_5 into position 9, and X_3 into position 6.

We now consider a real example, involving three triangular elements with 1 degree of freedom per node.

Element I with nodal vector (1, 2, 6), i.e., contributions $(X_1 X_2 X_6)^\text{I}$.

The destination vector is found, by a process described for element II, to be (1, 2, 3). Thus the assembled subvector X is $(X_1 X_2 X_6 \cdots)$ due to element I.

We now observe that node 1 appears here for the last time. Thus X_1 already takes its final value. We remove it to backing storage and leave a space in the assembled subvector: $(0 X_2 X_6 \cdots)$.

Element II with nodal vector (2, 3, 5), i.e., contributions $(X_2 X_3 X_5)^\text{II}$.

The destination vector is now derived. X_2 has a place, the second. X_3, however, is new, and we give it the first vacant place, the first. X_5 is also new, so we give it the next vacant place, the fourth. Thus the destination vector is (2, 1, 4) and the assembled subvector is $(X_3 X_2 X_6 X_5 \cdots)$.

We now observe that node 3 appears for the last time, so the first position again becomes vacant: $(0 X_2 X_6 X_5 \cdots)$.

Element III with nodal vector (2, 5, 6), i.e., contribution $(X_2 X_5 X_6)^\text{III}$.

The variables are active already and the destination vector is (2, 4, 3). After assembly the forces are all completely summed, because this is the last element.

This brief example adequately describes the housekeeping. New and improved housekeeping systems are now appearing, as described in Appendices I and II.

It is both convenient and efficient to compute the destination vectors and other related quantities before they are needed. At the same time PREFRONT can collect together all the available data for each element and also perform *data checks*, to print *diagnostics* and perhaps to allocate dynamic storage.

Frontal housekeeping is better than band housekeeping for several reasons [6]:

(a) It allows the engineer to use any consistent node numbering system.
(b) It uses a smaller subvector if the elements possess midside or midface nodes.
(c) It allows certain variables to be retained at the end. Elimination to reduce the size of an eigenvalue problem demands some housekeeping akin to the frontal system.

6.   We should have a file which stores data as one complex record per element. This is trivial: Element stiffnesses, external forces, material properties, etc. are called together. If in this list we include deflections, we can afford to keep 10 or 20 different deflection vectors with negligible core storage penalty.

Yet there is a penalty, which must be appreciated. If a node appears in four elements, an item of information is recorded in backing storage four times instead of once. In general, the total backing storage requirements and the data to be transferred are multiplied by the "mean nodal valency." Evidently, for nodal data, we can retrieve frontally, at some cost in core storage and in program complexity. However, nonnodal intraelement data such as stresses at integrating points will certainly go in the element file.

There is an interesting housekeeping opportunity if the residual forces are calculated by direct integration [1]. The virtual work per unit volume is:

$$\sigma_{ij} \, d\varepsilon_{ij} = \sigma_{ij} \, d(\partial u_i/\partial x_j)$$

$$= \sigma_{ij} \, d(\partial u_i/\partial \xi_k)(\partial \xi_k/\partial x_j)$$

$$= [(\partial \xi_k/\partial x_j)\sigma_{ji}] \, d(\partial u_i/\partial \xi_k) \tag{1}$$

where x_j denotes x, y, z and ξ_k denotes ξ, η, ζ, the local coordinates for, say, an "isoparametric" element. We interpret the item in square brackets as a 3×3 unsymmetric stress matrix, $[\sigma^*_{ik}]$. The first row of σ^* multiplied into the ξ, η, ζ derivatives of the shape functions gives the integrand for the force in the x direction, and so on. It becomes possible to keep in core the shape functions with their ξ, η, ζ derivatives at the integrating points, and thus reduce greatly the amount of data transferred.

4. Gaussian Reduction

Almost all finite-element expenditure is concerned with symmetric, positive-definite equations, and most of this time is in Gaussian reduction. When x_s is eliminated, K_{ij} becomes

$$K^*_{ij} = K_{ij} - (K_{is}/K_{ss})K_{sj} \tag{2}$$

It is usual to store a symmetric matrix, e.g., K_{ij}, as columns of an upper triangle strung into a vector. Thus

$$K_{ij} = \text{STIF}(NFUNC(I,J))$$

where $I \leqslant J$ and $NFUNC$ is defined as the function

$$NFUNC(K, L) = L * (L - 1)/2 + K \qquad (3)$$

The innermost loop (2) does not, of course, use $NFUNC$. Instead it is programmed as the single statement:

$$\text{STIF}(N) = \text{STIF}(N) - \text{FACTOR} * EQN(N + NDELTA) \qquad (4)$$

where N is a simple DO-loop variable, and FACTOR is the bracketed term in (2) and, like $NDELTA$, is calculated before entering the loop. Although (4) is fast, a very small machine language subroutine may be twice as fast, in the authors' experience. There is no risk and no loss of clarity if the original innermost loop is retained, but modified to become two comment statements. In case of failure, the CALL statement becomes a comment instead. It is foolish not to take advantage of an increase of efficiency that is so easily won.

Frontal housekeeping appears best in Gaussian reduction, but it is not easy to compare its efficiency with certain partitioning techniques.

5. Error Diagnostics

The experienced programmer knows the importance of adequate diagnostics. This section is intended for those who appreciate the need, but have not discussed the matter with a programmer experienced in finite elements. Among the diagnostics that will not be obvious are those relating to round-off error, and to recognizing the "flying structure" that the engineer has forgotten to earth, or which contains some mechanism. The following tests are very easily introduced and are usually adequate. They cost virtually nothing in computing time.

1. The diagonal terms decrease monotonically as the reduction proceeds, but remain positive. If the final, pivotal, value is small compared with the preceding values, then we could have a conditioning problem. If we assemble as we reduce, we should choose between two measures of the effective size of the diagonal: (a) the assembled value K_{ij}, (b) the root-sum-square value

$$t_i = [\Sigma \, (\text{value of } K_{ii})^2]^{1/2} \qquad (5)$$

where every time the diagonal is modified, either by assembling an element or by reduction, a contribution is made to t_i. The criterion, $K_{ii}/(\text{pivot})_i$ or $t_i/(\text{pivot})_i$ is calculated. If it exceeds, say, 10^5, then the solution has a chance and should proceed, but a warning message should be printed. If it exceeds, say, 10^{10}, the run should be aborted.

2.   The strain energy is $U = \frac{1}{2}\delta^T K \delta$ and an approximate measure of its variability is [7]:

$$\frac{\sigma(U)}{U} = 2^{-n+1} N^{1/2} \frac{\Sigma K_{ii} \delta_i^2}{\Sigma K_{ij} \delta_i \delta_j} \tag{6}$$

or

$$= 2^{-n+1} N^{1/2} \frac{\Sigma t_i \delta_i^2}{\Sigma K_{ij} \delta_i \delta_j} \tag{7}$$

where N is the semibandwidth and σ denotes standard deviation. The arithmetic uses n binary places. Equation (6) or (7) gives a good overall measure of the error, certainly better than Eq. (1).

Evidently the numerators are easily calculated. An artifice, now described, is needed to calculate the denominator economically. In Gaussian reduction, K is factored into $L_1 P U_1$ where $L_1 = U_1{}^T$ is a lower triangular matrix with unit diagonals: P is a diagonal matrix containing the pivots. After reduction the equations $K\delta = X$ become

$$PU_1 \delta = L_1^{-1} X = X^*, \quad \text{say} \tag{8}$$

The reduced forces X^* are not usually deemed worthy of discussion. But consider the sum of the reduced forces squared and divided by the diagonals:

$$(X^*)^T P^{-1} X^* = X^T K^{-1} X = \delta^T K \delta \tag{9}$$

It is equally possible to compute, say, $\delta_1{}^T K \delta_2$ where δ_1 is the deflection vector for the previous iteration, again at negligible cost. This could have applications in acceleration techniques when the problem is nonlinear [8].

3.   Diagnostics are most useful when they refer to particular variables. It is better, therefore, if individually large contributions to (6) or (7) are reported. This is possible, because Eq. (9) calculates the denominator during the elimination that precedes the calculation of δ.

It is also useful to have values of diagonal stiffnesses in the printout.

We have discussed only the simplest tests, which are cheap enough to be applied to every problem. A costlier test is to calculate in double precision the error forces $R = X - K\delta$. The only really valid test is costlier still, and involves another solution, $K^{-1}R$, to be compared with δ.

6. The Conjugate Gradient Algorithm

The conjugate gradient algorithm is iterative, so that the computing time cannot be accurately forecast. The tests carried out by the authors suggest that it does not compete with the frontal Gaussian solution. Yet it has undoubted attractions, foremost of which is its remarkably small storage requirements. This property will appeal to the company with a small and as yet under-utilized computer, and with problems much too large for an orthodox solution.

The algorithm [3] minimizes the potential energy along a certain series of directions, P, so that an iteration takes the form:

$$\delta_{i+1} = \delta_i + \alpha_i P_i \tag{10}$$

where α_i is a scalar. Each P_i is orthogonal to its predecessor, in the sense that $P_{i+1} K P_i = 0$—indeed it is orthogonal to all its predecessors. The P are updated as follows:

$$P_{i+1} = Z R_i + \beta_i P_i \tag{11}$$

where $R_i = X - K\delta_i$ is the residuals, Z is normally 1, and β_i is a scalar chosen to give $P_{i+1}^T K P_i = 0$. To start the process, $\delta_0 = P_0 = 0$. If there are M equations the process should converge exactly in M iterations. In practice it requires 15–20 % of M in finite element calculations. The authors hoped that this would decrease markedly for large problems, but were disappointed.

The authors' program is somewhat different in that it uses a trial displacement:

$$t_i = \delta_i + \gamma_i P_i \tag{12}$$

where γ_i is an arbitrary constant that must, however, be chosen with some care. The δ_i are never calculated, and the residuals $R_i = X - K\delta_i$ are calculated as weighted means between R_{i-1} and $T_{i-1} = X - Kt_{i-1}$. If a bad choice of γ is made, then round-off error results, and it becomes necessary to compute another trial residual along the direction $P_i (Z = 0$ here). The process is described in detail in Kan [1].

Before iteration starts, it is necessary to normalize the equations so that the diagonals are unity. This requires the calculation of diagonal stiffnesses. However, it is possible to avoid all other stiffness calculations, and merely calculate residuals—for the solution routine automatically scales these. In principle, only small changes would be needed to process geometrically nonlinear problems.

The program uses "band-plus-buffer" housekeeping for the current and for the previous test residuals. A problem with 132 parabolic brick elements, each with 60 degrees of freedom, and 2646 assembled degrees of freedom, semibandwidth 339, would have used only 18,000 words of 24 bits on the ICL 1905 E computer. However, convergence was not quite complete after 460 iterations, or 90 min on the IBM 360/75, whereas a Gaussian frontal solution took only 17 min. An indication of the conditioning is that $\delta^T \delta / \delta^T K \delta = 1600$, which is not bad.

The results of a conjugate gradient solution reveal some characteristics of interest to engineers rather than to mathematicians; they may well be peculiar to finite element problems. Given a deflection vector δ, the ratio $\delta^T\delta/\delta^T K\delta$ is recognized from the conditioning criterion in (6), bearing in mind that K has been normalized. Again, the smallest and the largest possible values of this ratio are the extreme eigenvalues of K^{-1}. The small values of the ratio correspond to "noisy" deflection patterns that would distort our many elements into a zigzag pattern, whereas a large value corresponds to a smooth deformation. Tests have not been done, but it appears probable that real structures loaded in the usual way give nearly the smoothest deflected shape.

It is an easy matter to calculate $\delta^T\delta/\delta^T K\delta$ as the iteration proceeds, and also $P^T P/P^T KP$, a measure of the noisiness of the direction. The early shapes of P are noisy. For example, a point load will give a particularly noisy P_1. Then, as iteration proceeds, $P^T P/P^T KP$ increases, although not monotonically, to quite large values, before dropping again when convergence is almost complete. This provides an interesting background to the observation

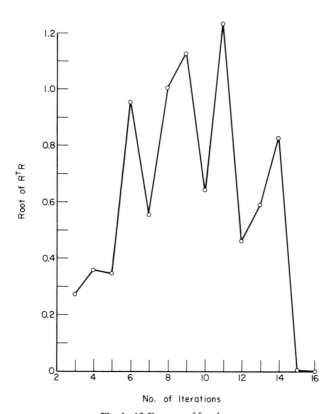

Fig. 1. 12 Degrees of freedom.

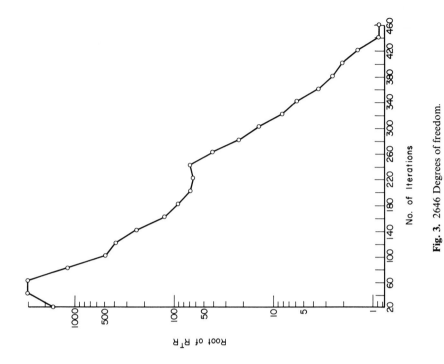

Fig. 3. 2646 Degrees of freedom.

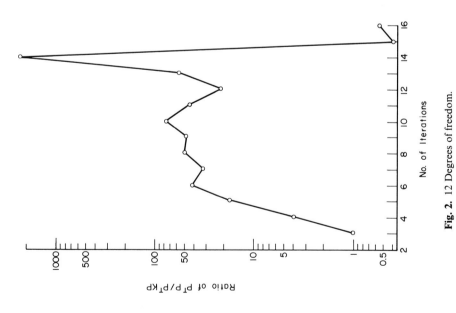

Fig. 2. 12 Degrees of freedom.

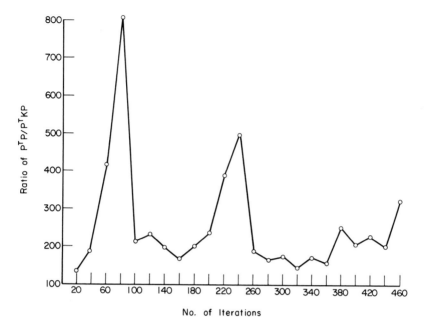

Fig. 4. 2646 Degrees of freedom.

that the stresses converge as quickly as the displacements: The error in deflections when convergence is nearly complete is "smooth." (Private communication, A. Yettram, Brunel University, Uxbridge, Middlesex.)

The final value of $\delta^T\delta/\delta^T K\delta$ is interesting also because, if we assume it takes a value somewhere near its maximum in real situations, it is then related to the Rosanoff conditioning number,

$$C_n = \lambda_{max}/\lambda_{min} \tag{13}$$

For if

$$C_d = \delta^T\delta/\delta^T K\delta \simeq 1/\lambda_{min} \tag{14}$$

it follows from Gershgorin's theorem that $\lambda_{max} < 2N - 1$, where N is the semibandwidth, so that

$$C_n \simeq (2N - 1)\delta^T\delta/\delta^T K\delta \tag{15}$$

Figures 1–4 show the variations of $P^T KP$ and $(R^T R)^{1/2}$ during iteration, for a cantilever with six elements and 12 degrees of freedom. The manner of convergence tends to be abrupt and unexpected. Generally a partially converged, conjugate gradient solution is useless.

7. The Alternating-Direction Approach

The last objection is not trivial. We consider again an application of the Cornell technique to optimization, where the first solution is very approximate, as are the first sensitivity coefficients. A linear optimization follows, and a new iterative solution starts with the existing displacements. Then the iterative solution for the sensitivity coefficients takes the first estimates as starting values. The optimization and the solution itself are made to converge together. This may never become feasible, but the interesting possibility underlines the need for iterative methods that converge at a uniform rate, as did the Gauss–Seidel method. The fact that the conjugate gradient method is not competitive for one load case, let alone for many, is a separate problem. (There seems little hope of any iterative method competing with Gaussian elimination if there are very many load cases.)

The alternating–direction algorithm, used by Rashid [2] over a considerable period, satisfies in principle these new requirements. The authors have done very little research into this technique. In order to lead clearly into certain other techniques, the method is presented in a notation slightly different from that adopted by Rashid. Let X, Y, Z be the forces in the x, y, z directions. Let u, v, w be the deflections. Let R, S, T be the residual forces at any stage:

$$\begin{Bmatrix} R \\ S \\ T \end{Bmatrix} = \begin{Bmatrix} X \\ Y \\ Z \end{Bmatrix} - \begin{bmatrix} K_{xx}^{\bullet} & K_{xy} & K_{xz} \\ K_{yx} & K_{yy} & K_{yz} \\ K_{zx} & K_{zy} & K_{zz} \end{bmatrix} \begin{Bmatrix} u \\ v \\ w \end{Bmatrix} \qquad (16)$$

The objective is to invert only K_{xx}, K_{yy}, and K_{zz}, thus reducing the cost of Gauss reduction by a factor of nine. This is achieved by the matrix equivalent of Gauss–Seidel iteration:

$$u_{i+1} = u_i + \omega K_{xx}^{-1} R(u_i, v_i, w_i)$$

$$v_{i+1} = v_i + \omega K_{yy}^{-1} S(u_{i+1}, v_i, w_i) \qquad (17)$$

$$w_{i+1} = w_i + \omega K_{zz}^{-1} T(u_{i+1}, v_{i+1}, w_i)$$

The overrelaxation factor ω is in the region of 1.3. However, rapid convergence depends entirely on certain accelerating techniques developed by Rashid. It is instructive to visualize the process mechanically: The nodes are locked in the y and z directions, and allowed to settle in the x direction. Then these, and the z deflections, are locked, and the y deflections freed, and so on. The rollers exerting such physical constraints evidently give the technique its name.

The ordinary conditioning criteria are not necessarily the best guide to what has been achieved. For example, if the off-diagonal partitions were zero, then the conditioning would be perfect for Rashid's purposes however ill-conditioned K_{xx}, K_{yy}, and K_{zz} might be, since convergence would be immediate. A better criterion would be

$$\frac{u^T K_{xx} u + v^T K_{yy} v + w^T K_{zz} w}{X u^T + Y v^T + Z w^T} \tag{18}$$

which is easily calculated. It is observed that the off-diagonal partitions are especially large if Poisson's ratio is high.

8. Multivector Iteration

Especially on large computers, the cost of disk transfers is high, and militates strongly against iterative methods. A disadvantage of the alternating-direction approach was considered to be that the transfer of element stiffnesses, or shape functions, was repeated three times per cycle. (It was only recently appreciated that the derivatives at integrating points may be kept permanently on core, and this argument now loses much of its force.)

Thus, perhaps misguidedly, the research on multivector iteration [1] was based on the requirement that the residuals were to be calculated only once per iteration. Thus Rashid's approach was based on Gauss–Seidel, while ours was based on Jacobi iteration. It is known that Jacobi iteration does not, in general, converge unless very small values of ω are used.

It was therefore necessary to calculate R, S, T all using u_i, v_i, w_i: we shall now call $R(u_i v_i w_i)$ for example simply R_i. We return to the ideas of the conjugate gradient approach, but with several directions of choice simultaneously available:

$$\begin{Bmatrix} u \\ v \\ w \end{Bmatrix}_{i+1} = \begin{Bmatrix} u \\ v \\ w \end{Bmatrix}_i + \alpha \begin{Bmatrix} K_{xx}^{-1} R_i \\ 0 \\ 0 \end{Bmatrix} + \beta \begin{Bmatrix} 0 \\ K_{yy}^{-1} S_i \\ 0 \end{Bmatrix} + \gamma \begin{Bmatrix} 0 \\ 0 \\ K_{zz}^{-1} T_i \end{Bmatrix} \tag{19}$$

Thus α, β, γ are three scalars chosen so as to minimize the potential energy. They are properly regarded as generalized deflections, associated with three generalized forces and a 3×3 generalized stiffness matrix.

In truth, this was not the case for which preliminary trials appeared so attractive. Following a technique used by R. W. Clough in 1962, six generalized deflections were used, the other three multiplying u, v, and w themselves and thus allowing them to scale up in the early stages of iteration.

The results were completely unexpected. Initial convergence was remarkably good, as the trials had suggested, but after about 10 iterations there was very little further change, and the residuals remained large. "Laziness factors" exhibited this in an unmistakeable way [1].

In all, 40 algorithms were tried. Overrelaxation factors helped, but rather unpredictably, random numbers did not help, even a change of algorithm during the iteration did not greatly help. However, algorithms were not equally bad. The best appeared to be that using nine directions:

$$
\begin{Bmatrix} K_{xx}^{-1}R_1 \\ 0 \\ 0 \end{Bmatrix}
\begin{Bmatrix} 0 \\ K_{yy}^{-1}S_i \\ 0 \end{Bmatrix}
\begin{Bmatrix} 0 \\ 0 \\ K_{zz}^{-1}T_i \end{Bmatrix}
\begin{Bmatrix} u_i \\ 0 \\ 0 \end{Bmatrix}
\begin{Bmatrix} 0 \\ v_i \\ 0 \end{Bmatrix}
$$

$$
\begin{Bmatrix} 0 \\ 0 \\ w_i \end{Bmatrix}
\begin{Bmatrix} u_{i-1} \\ 0 \\ 0 \end{Bmatrix}
\begin{Bmatrix} 0 \\ v_{i-1} \\ 0 \end{Bmatrix}
\begin{Bmatrix} 0 \\ 0 \\ w_{i-1} \end{Bmatrix}
$$

The details are recorded [1]. Now that the algorithm has failed, the authors feel that the mistake was to base the algorithm on Jacobi iteration rather than Gauss–Seidel.

9. Conclusions

Various equation-solving algorithms have been discussed. The following conclusions are drawn:

1. With the possible exception of Rashid's approach, iterative methods are not yet ready for general use.
2. Gaussian elimination is very effective, particularly with frontal house-keeping, with two or three very small machine language subroutines, and with dynamic arrays transferred to disk without implied DO-loops.
3. Attention should be given to the housekeeping techniques available. In particular, intraelement data should be kept with other element data; much nodal data besides the forces and stiffness can be stored frontally.

Appendix I. The Prefront with Variable Numbers of Degrees of Freedom at Different Nodes

Hellen [9] in his system BERSAFE has modified the prefront algorithm so that it accepts, e.g., 3 degrees of freedom at corner nodes, and 1 degree of freedom at midside nodes. When a new node number is encountered, he first inquires whether it has 1 or 3 degrees of freedom. If it has three, he allocates three destinations to that node number, and whenever it recurs he puts the same three destinations into the destination vector. In general, they are not consecutive destinations.

Appendix II. Avoidance of Zeros within the Front

If, in teaching the front algorithm, one uses rather more complex examples, one finds that empty spaces appear that are not immediately reoccupied. This is embarrassing, because the more intelligent students appreciate immediately that in Gaussian reduction it represents a loss of efficiency.

Empty spaces within the front cannot be avoided, except by shifting, which is again inefficient. Two artifices have been adopted in the latest Swansea prefront:

(i) As the frontwidth changes, the occupied region of the subvector should change immediately. What delays this movement is the existence of long-lived variables near the extremity of the occupied region. During the search for the last appearance of a new node, it is possible to tabulate the new nodes in the current element and their last appearance, and then to give the short-lived nodes the destinations near the extremity of the front. The extra computing is negligible, but the prefront becomes more complicated.

(ii) It is possible to arrange that the front has two ends, thus increasing the chance of releasing any vacancies. The search for destinations starts at the center of the subvector and moves progressively outward in both directions. Again the computing cost is negligible.

References

1. Kan, D. K. Y., "Solution Techniques for Large Finite Element Problems concerning Creep and Temperature Transients in Gas Turbine Blades," Ph.D. Thesis, Univ. of Wales (1971).
2. Rashid, Y. R., "Three Dimensional Analysis of Elastic Solids," *Int. J. Solids Structures* 1311–1331 (1969); 195–207 (1970).
3. Fox, R. L., and Stanton, E. L., "Developments in Structural Analysis by Direct Energy Minimization," *AIAA J.* 1308–1312 (1968).
4. Melosh, R. J., and Bamford, R. M., "Efficient Solution of Load-Deflection Equations," *J. Amer. Soc. Civil Eng. (Struct. Div.)* 661–676 (1969).
5. Irons, B. M., "A Frontal Solution Program for Finite Element Analysis," *Int. J. Numer. Methods Eng.* 5–32 (1970).
6. Cornell, C. A., Reinschmidt, H. F., and Brotchie, J. F., "A Method for the Optimum Design of Structures," *Int. Symp. Digital Comput. Struct. Eng.* Newcastle upon Tyne (July 1966).
7. Irons, B. M., "Roundoff Criteria in Direct Stiffness Solutions," *AIAA J.* 1308–1312 (1968).
8. Boyle, E. F., and Jennings, A., "Accelerating the convergence of elasto-plastic stress analysis," *Int. J. Numer. Methods Eng.*, to be published.
9. Hellen, T. K., Letter concerning BERSAFE, *Int. J. Numer. Methods Eng.* 149 (1971).

FLING—A FORTRAN Language for Interactive Graphics

W. J. Batdorf and S. S. Kapur

LOCKHEED-GEORGIA COMPANY
MARIETTA, GEORGIA

This paper suggests a universal language for computer graphics. It is important that such a language be adopted in order to bring some form of standardization into being for interactive graphics programs. FLING, at present, consists of 16 machine-dependent graphic subroutines. The functions and calling sequences of the various subroutines are explained here in detail. Several example problems are included to illustrate further the use of the various subroutines. FLING has been tested and implemented on three totally different graphic systems: IBM's 360 graphic system with Model 2250 scopes, Univac's 418 with a DEC 340 scope, and Lockheed's MAC-16 computer with an IDI scope. The interface package for the IBM graphic system is included in this report.

1. Introduction

Throughout the aerospace industry, emphasis is being directed toward automating the various design processes with interactive graphic displays. However, this automation is proceeding very slowly and cautiously. Managers are aware that a variety of new graphic hardware is being offered by IBM, CDC, and Univac. They also know that many smaller companies are producing a wide variety of low-cost graphic display devices. Throughout all of this intense activity, there is essentially "zero effort" on the part of the computing industry to standardize a set of FORTRAN commands for computer graphic equipment. For this reason, a "wait and see" attitude has developed.

Sophisticated graphic systems are now operating or will be soon put into operation at Wright-Patterson Air Force Base, NASA, and at several universities. Most of these organizations will begin to develop or modify the manufacturer's software to meet their own particular requirements. Although these efforts to improve existing concepts are certainly praiseworthy, the end result will be an even greater number of graphic software systems. For instance, CUG (Colorado University Graphics) is a system under development at the University of Colorado. Someday it may be of particular concern to the Lockheed-California Company whether or not an application written under CUG can be made to operate on the IBM graphic system.

The problem of incompatibility associated with interactive graphics is quite serious, and it may hinder the development of new programs for the next several years. As with all new developments, industry will solve the problem when it is to their best advantage. Although it may be many more years before the computing industry will admit that a standard language is necessary and can be developed, all knowledgeable persons associated with computer graphics know that this will eventually occur.

At the fifth conference on Electronic Computation sponsored by the American Society of Civil Engineering, the keynote speaker, Professor Steven J. Fenves, took a very concerned attitude regarding the prospects for standardization of software during the next few years:

> It appears clearly that "business as usual" will not work. We simply cannot afford as a profession to continue in the piecemeal fashion that we have used so far. The duplication of effort in producing programs and systems for a bewildering number of organizations, on a hodge-podge of machines, and in any languages we can lay our hands on cannot continue. Not only is it wasting the best resources of the profession to do today's jobs, but it simply does not allow for the replenishment and enrichment that is necessary for the profession's continued success, and even its survival.

It would be presumptuous to assume that FLING would be totally acceptable to the entire computing industry as the standard language for computer graphics, but nevertheless, it is proposed as perhaps an initial standard. Some justifications for this remark will be presented in this paper.

FLING was initially developed in connection with an interactive structural analysis program designed to be machine independent. It is apparent, however, that this language can be used for any computer graphics application program. All of the various manufactuers' software support packages perform essentially the same operations. The associated subroutines could be placed into four categories:

> Initialization and termination subroutines
> Image-generation subroutines
> Buffer-manipulation subroutines
> Attention-related subroutines

A basic set of 16 graphic subroutines has been defined. It is believed that it constitutes a comprehensive set of subroutines for any graphics application program. Three subroutines are used for initialization and termination. In addition to these, there are five image-generation subroutines, four buffer-manipulation subroutines, and four subroutines associated with interrupt processing. The function of each of these subroutines is explained in Section 2 and a variety of example problems are presented in Section 3 to demonstrate the use of the subroutines. If all 16 of these subroutines are implemented on four different computers, and the application program is constructed with calls to these subroutines, the program will operate on each of the computer configurations. The following paragraphs describe how FLING was implemented on three totally different graphic systems—an IBM 360 system with Model 2250 scopes, a Univac 418 with a Digital Equipment 340 scope, and Lockheed's MAC-16 computer interfaced with an IDI (Information Displays Inc.) scope.

If each of the subroutines presented in Section 2 represents a fundamental operation, it should be possible to simulate that operation on either the IBM, CDC, or Univac computer simply by using the manufacturer's existing software. In this sense, the subroutine package has standardized the calling sequence to the various manufacturers' graphic subroutines. Although there is not a simple one-to-one correspondence, this was the approach used to develop the graphics package for the IBM 360 graphics system. This interface package simply performs its necessary functions by calling GSP (IBM's graphic subroutine package) subroutines. In many cases, only five or six instructions are required to perform the task assigned the subroutine.

A Univac 418 computer connected to a DEC (Digital Equipment Corporation) 340 scope is in use at the Lockheed-Georgia Research Laboratory. Most of these subroutines are written in FORTRAN, and bit manipulation was performed by combining the proper integer words to create the bytestream. The interrupt processor was written in a machine language called ART.

FLING has also been thoroughly tested on the MAC-16-IDI graphics system. This is a low-cost graphics system being developed within the Lockheed Corporation. For this computer the subroutine package was written in LEAP by D. M. Smith. LEAP is the assembly language developed for the MAC-16 by the Lockheed Electronics Company.

2. Basic Graphic Subroutines

To write a graphics program, one must do the following:

(a) Initialize the graphic system; i.e., establish communication between the user's program and the graphic console.

(b) Initialize the appropriate display buffers. (A display buffer is an area assigned by the programmer where the graphic orders for displaying images are stored.) A single display consists of one or several buffers. When a buffer is initialized, a unique identification code is automatically assigned to it. This code is used by the programmer to identify the buffer when processing light-pen interrupts.

(c) Create displays by calling various image-generation routines. The graphic orders for the display are stored in the display buffers. Each element within the buffer may, at the option of the user, be assigned a key which is an integer between 1 and 511. This key is used to identify elements within the buffer when they are selected with the light pen.

(d) Perform image modifications or calculations by processing program interrupts. These interrupts can be achieved through the attention-related devices: the light pen, function box, keyboard, and tracking cross. The control for any graphics program is based on the sequence of computer interrupts.

(e) Terminate the use of the display device.

The graphics subroutines can be categorized as follows:

Initialization and Termination Subroutines
    SIGNON
    TERM
    IDISPB
Image-Generation Subroutines
    POINT
    LINE
    TEXT
    DNUM
    CIRCLE
Buffer-Manipulation Subroutines
    ON
    OFF
    RDISPB
    MODPAR

Attention-Related Subroutines
   WAIT
   KEYBRD
   FBOX
   CROSS

The functions and calling sequences for these subroutines are outlined on the following pages.

2.1. INITIALIZATION AND TERMINATION SUBROUTINES

Subroutine SIGNON

The purpose of the SIGNON subroutine is to initialize the graphic system. It must be called at the beginning of every graphics program.

Calling Sequence: CALL SIGNON

Subroutine TERM

Subroutine TERM is used to terminate the graphics portion of the program in a normal manner. This subroutine should be called when the graphics portion of the application program is completed.

Calling Sequence: CALL TERM

Subroutine IDISPB

The IDISPB subroutine is called to initialize a display buffer. This subroutine must be called before a display bytestream is created for the buffer. An identification code for the buffer is automatically assigned by a call to this subroutine, and it is stored as the first word in NDISP. If a light-pen interrupt occurs on the buffer, the I.D. code is returned through WAIT. A light-pen code is used to specify if interrupts on this buffer will be queued. The upper and lower limits of the user's input data are also indicated at this time. The limits specified for the input data are mapped so that they correspond to the upper right and lower left corners of the scope. The type of input data, real or integer, must be indicated. Any coordinate information returned from the scope is given in terms of the latest coordinate system available for the buffer. Images generated for this buffer are scissored at the scope boundaries only if the scissoring option is exercised. Parameters assigned by this call can be modified by a call to MODPAR.

Calling Sequence. CALL IDISPB (NDISP, MODE, ILX, ILY, IUX, IUY, LP, ISC, NSIZE)
   NDISP = the name of the display buffer
   MODE = 1 if the user's coordinate data is in integer mode

MODE = 2 if the user's coordinate data is real

ILX, ILY = integers corresponding to the lower limits of the user's data—corresponds to the lower left corner of the scope

IUX, IUY = integers corresponding to the upper limits of the user's data—corresponds to the upper right corner of the scope

LP = 0 light-pen interrupts will not be queued

LP = 1 light-pen interrupts will be queued for the buffer

ISC = 0 the images are not to be scissored

ISC = 1 the images are to be scissored at the scope boundaries

NSIZE = the total length of the display buffer. This parameter is ignored on the IBM system.

2.2. IMAGE-GENERATION SUBROUTINES

Subroutine POINT

The POINT subroutine generates the display bytestream for one point or a series of points. All image-generation routines create bytestreams in an include or visible status; that is, all elements will be displayed when the buffer is placed in the display queue. Whether the points are to be scissored is determined by the current scissoring option given for the buffer in IDISPB.

Calling Sequence: CALL POINT (NDISP, KEY, X, Y, NPTS)

NDISP = the name of the display buffer

KEY = the key assigned to this set of points. All points generated by a call to this subroutine are assigned the same key

X, Y = an array in user coordinates, indicating where the points are to be displayed

NPTS = the total number of points.

Subroutine LINE

The LINE subroutine generates the display bytestream for a single line or a series of connected lines. Whether the lines are to be scissored is determined by the current scissoring option given for the buffer in IDISPB.

Calling Sequence: CALL LINE (NDISP, KEY, X1, Y1, X2, Y2, NCT)

NDISP = the name of the display buffer

KEY = the key assigned to this set of lines. All lines generated by a call to this subroutine are assigned the same key

X1, Y1 = user coordinates of the beginning point of the first line

X2, Y2 = a pair of arrays in user coordinates defining the end points of the lines

NCT = the total number of lines

Note: If NCT = 3 straight lines will be drawn from:
$$X1, Y1 \text{ to } X2(1), Y2(1)$$
$$X2(1), Y2(1) \text{ to } X2(2), Y2(2)$$
$$X2(2), Y2(2) \text{ to } X2(3), Y2(3)$$

Subroutine TEXT

This subroutine is used to generate the bytestream for characters; for example, light-pen controls. The characters must be packed (one byte per character) and left-justified in the array ITEXT. Whether the characters are scissored is determined by the scissoring option given for the buffer in IDISPB.

Calling Sequence: CALL TEXT (NDISP, KEY, ITEXT, NCH, KCODE, X, Y)

NDISP = the name of the display buffer where the text data will be stored
  KEY = the key associated with the ITEXT data
 ITEXT = a BCD array indicating the character strings to be displayed
  NCH = the total number of characters
KCODE = 1 normal size characters
      = 2 large size characters
  X, Y = the position of the lower left-hand corner of the character string in user coordinates.

Example: CALL TEXT (NDISP, 10, 3HSET, 3, 1, 900, 100)

This call will display SET in normal size at position (900, 100). A key of 10 is returned whenever an interrupt occurs on SET.

Subroutine DNUM

This subroutine generates the bytestream necessary for displaying a single number, fixed or floating point, from core. The floating point number may be displayed in either an E, F, or I format. If the number is too big for F format, it is automatically displayed in E format.

Calling Sequence: CALL DNUM (NDISP, KEY, X, Y, C, ICODE)

NDISP = the display buffer where the bytestream for C will be stored
  KEY = the key assigned to this element
  X, Y = the position in user coordinates where C will be displayed
    C = a fixed or floating point constant or variable
ICODE = 1 the number is displayed in E12.6 format
ICODE = 2 the number is displayed in F12.6 format
ICODE = 3 the number is displayed in I12 format

Subroutine CIRCLE

Subroutine CIRCLE generates the bytestream necessary for a circle of radius R. The IBM and Univac circles are generated by drawing a 24-sided regular polygon, while the circles on the IDI scope are generated by hardware.

Calling Sequence: CALL CIRCLE (NDISP, KEY, CENX, CENY, RADIUS)
 NDISP   = the name of the display buffer
 KEY   = the key assigned to this circle
 (CENX, CENY) = center of the circle in user coordinates
 RADIUS   = radius of the circle in the user coordinate system.

2.3. BUFFER-MANIPULATION SUBROUTINES

Subroutine ON

This subroutine is called either to place a buffer in the display queue or to change the status of a keyed element so that it is visible. At least one element must have been created for the buffer before it is turned "on." When the call refers to an element, the buffer, if not already in the display queue, is automatically queued. If the element or the buffer referred to by the call is already in an "on" status, the call is ignored.

Calling Sequence: CALL ON (NDISP, KEY)
 NDISP = the name of the display buffer
 KEY = 0 the display buffer is placed in the display queue
 KEY > 0 the element with this key is modified so that it can be seen and the buffer is placed in the display queue.

Subroutine OFF

Subroutine OFF is used to place an existing buffer or a keyed element within that buffer in an "off" status. If the call refers to a display buffer, the buffer is removed from the display queue; if the call refers to an element, the beam is blanked while displaying that element. An element not previously keyed cannot be placed in "off" status. If there is more than one element with the same key the call only affects the first element with this key. If the element or the buffer referred to by the call is already in "off" status, the call is ignored.

Calling Sequence: CALL OFF (NDISP, KEY)
 NDISP = the name of the display buffer
 KEY = 0 the display buffer is omitted from the display queue
 KEY > 0 the status of the element with this key in NDISP is modified so that the element cannot be seen. The element is not removed from the display buffer, nor is the display buffer removed from the display queue.

Subroutine RDISPB

This subroutine is used to remove a particular element and all elements that follow it from a display buffer.

Calling Sequence: CALL RDISPB (NDISP, KEY)
NDISP = the name of the display buffer to be reset
KEY   = 0 the entire display buffer is reinitialized, and the bytestream is cleared. In this case, the buffer must be turned off
KEY   > 0 all elements in the bytestream starting from the element with this key to the end of the buffer are erased. The length of the display buffer is modified accordingly.

Subroutine MODPAR

This subroutine is designed to modify the display buffer parameters initially defined in IDISPB. Zooming can conveniently be accomplished by modifying ILX, ILY, IUX, IUY, and then regenerating the displays. If several overlapping images are being displayed that are light-pen sensitive, it may be convenient to disable light-pen interrupts temporarily with MODPAR.

Calling Sequence: CALL MODPAR (NDISP, MODE, ILX, ILY, IUX, IUY, LP, ISC, KODE)
All the parameters except KODE have been previously defined in IDISP. If
KODE = 1 reset MODE, ILX, ILY, IUX, IUY
KODE = 2 reset LP
KODE = 3 reset ISC
KODE = 4 reset all parameters.

2.4. ATTENTION-RELATED SUBROUTINES

Subroutine WAIT

The subroutine WAIT permits the programmer to retrieve information concerning the light-pen interrupts. Various codes can be used which will cause this subroutine to clear or not clear previous interrupts, or to wait or not wait for a light-pen interrupt. Normally, one would want to clear all interrupts and wait. Once an interrupt is processed, it is removed from the queue. The previous interrupt then becomes the latest interrupt. This feature enables a programmer to process stacked interrupts. An interrupt on a particular buffer destroys all information concerning previous interrupts on that buffer, thus, stacked interrupts must be on distinct buffers. One hundred interrupts can be stacked at any one time. After a light-pen interrupt occurs, a two-word integer array is returned.

Calling Sequence: CALL WAIT (INTARY, ICODE)

INTARY (1) = the I.D. code of the display buffer associated with the interrupt. If no interrupt has occurred, a value of zero is returned.

INTARY (2) = the key of the element associated with the interrupt. If no key was assigned then INTARY (2) = 0

ICODE   = 0 Clear the Interrupt Queue (I.Q.) and wait for an interrupt.

ICODE   = 1 Do not clear I.Q. but wait for an interrupt

ICODE   = 2 Clear I.Q.

ICODE   = 3 Do not clear the I.Q. and do not wait for an interrupt.

Subroutine KEYBRD

The KEYBRD subroutine is used to update a display buffer from the keyboard. In addition, the input data is also transferred to the application program via ITEXT. An element not previously keyed cannot be updated. The new length of the bytestream for this element must not exceed the original bytestream length.

Calling Sequence: CALL KEYBRD (NDISP, KEY, ITEXT, NCH, ICODE)

NDISP = the display buffer to be updated

KEY   = the key of the element to be updated. A key must be specified.

ITEXT = a real or integer array returned from KEYBRD. The type of data returned in ITEXT is determined by ICODE

NCH   = the total number of characters that will be accepted from the keyboard. Presently, 72 characters is the maximum per call

ICODE = an integer code indicating what is to be done with the keyboard data. If

ICODE = 1 Keyboard data is a floating point number if E 12.6 format

ICODE = 2 Keyboard data is a floating point number in F 12.6 format

ICODE = 3 Keyboard data is an integer in I12 format

ICODE = 4 Keyboard data is text information.

Example: Dimension NDISP (60), ARY (10).

1. DO 10 I = 1, 10
2. 10 ARY (I) = 6Hbbbbbb
3. Call TEXT (NDISP, 8, ARY, 60, 1, 512, 512)
4. Call KEYBRD (NDISP, 8, C, 6, 1).

Statements 1, 2, 3 display 60 blanks at (512,512) with a key of 8. Statement 4 is a call to KEYBRD to update this element. If the number 3.1416 (6 characters) is typed from the keyboard, it will be displayed at (512,512) and the value of C becomes 3.1416.

Subroutine FBOX

A specific call to FBOX will cause an application program to wait for the depression of a function key. The number of the function key that was depressed is then returned. The interrupts from the function box are not stacked, that is, only the last interrupt is ever returned.

Calling Sequence: CALL FBOX (NB, NLIST, NCT, ICODE)
    NB   = the number of the function key that was depressed
    NLIST = an integer array specifying which function keys are enabled
            for this call
    NCT   = the total number of words in NLIST
    ICODE = 0 clear the previous interrupt, and wait for an interrupt
    ICODE = 1 do not clear the interrupt, and do not wait for an interrupt. If
    NLIST(1) = 2
    NLIST(2) = 6
    NLIST(3) = 10
    NLIST(4) = -18 (NCT = 4)
    Function keys 2, 6, 10–18, are enabled.

Subroutine CROSS

This subroutine is used to display a tracking cross on the screen, return the latest position of the tracking cross, and to remove the tracking cross. Light-pen tracking is automatic, and the tracking cross can be moved with the light pen whenever the cross is being displayed. Interrupts from the tracking cross are not placed in the interrupt queue. Only one tracking cross at a time can be displayed. If the tracking cross is already being displayed, a call to CROSS, with KODE = 1, merely repositions the cross.

Calling Sequence: CALL CROSS (NDISP, X, Y, KODE)
    NDISP = the name of the display buffer. It is used solely to determine the
           user's coordinate system
    X, Y   = user coordinates. Depending on KODE, the cross is either
           displayed at (X, Y) or (X, Y) is returned as the current position
           of the tracking cross
    KODE = 0 the tracking cross is removed from the screen
    KODE = 1 the tracking cross is displayed at position (X, Y)
    KODE = 2 the current position of the tracking cross in user coordinates
           is returned.

3. Example Problems

The following five example problems were prepared to check out and demonstrate the use of the various subroutines. For a person with previous graphic programming experience, the use of the routines is self-evident. For an experienced FORTRAN programmer with no graphic experience, the system can be mastered in less than a week.

Example Problem 1

The first example problem is designed to illustrate the use of the following subroutines: SIGNON, TERM, IDISPB, TEXT, ON, OFF, WAIT, RDISPB.

Problem Statement

Initially, four asterisks (∗), and two light-pen controls, DELETE and RESET, are displayed on the scope (see Fig. 1).

(*a*) A light-pen attention on two points causes a straight line to be drawn between these points.

(*b*) If the light-pen is pointed to DELETE and then to one of the lines, this line is made invisible.

(*c*) A light-pen interrupt on RESET initializes the display.

Fig. 1. Example problem 1 operating on the IBM 360 with an IBM 2250 display unit.

```
100  *EXECUTE EXP1  MAP
200  *COMPILE EXP1 INPUT CARDS OUTPUT TAPE LIST SOURCE
300  *R ExP1
400  C        THIS PROGRAM DISPLAYS FOUR POINTS,DEPENDING ON THE
500  C        ATTENTION CREATED, THE PROGRAM WILL DO THE FOLLOWING.
600  C           1.LIGHT PEN ATTENTION ON A PAIR OF POINTS GENERATES
700  C              THE LINE JOINING THESE POINTS.
800  C           2.LIGHT PEN ATTENTION ON DELETE AND ANY LINE SEGEMNT
900  C              MAKES THAT LINE INVISIBLE.
1000 C           3.LIGHT PEN ATTENTION ON RESET CLEARS DISPLAY
1100          DIMENSION X(4),Y(4),INTARY(4),NDISP1(512),NDISP2(1000)
1200          INTEGER X,Y
1300          DATA X(1),Y(1),X(2),Y(2),X(3),Y(3),X(4),Y(4)
1400         1/300,300,300,600,600,300,600,600/
1500 C     INITIALIZE THE GRAPHIC SYSTEM
1600          CALL SIGNON
1700 C     INITIALIZE A DISPLAY BUFFER FOR DISPLAYING THE
1800 C     FOUR POINTS, THE WORDS DELETE AND RESET
1900          CALL IDISPB(NDISP1,1,0,0,1023,1023,1,0,512)
2000 C     INITIALIZE A DISPLAY BUFFER FOR DISPLAYING
2100 C     LINE SEGMENTS JOINING ANY TWO POINTS
2200          CALL INDISPB(NDISP2,1,0,0,1023,1023,1,0,1000)
2300          KEY1=0
2400 C     GENERATE THE FOUR CORNERS OF THE SQUARE
2500          DO 10 I=1,4
2600       10 CALL TEXT(NDISP1,I,1H*,1,1,X(I),Y(I))
2700 C     GENERATE DELETE AND RESET KEYS
2800          CALL TEXT(NDISP1,5,6HDELETE,6,1,100,800)
2900          CALL TEXT(NDISP1,6,5HRESET,5,1,800,800)
3000          CALL ON(NDISP1,0)
3100 C     KOUNT=FLAG FOR 1ST OR 2ND POINT OF LINE SEGMENT
3200 C     KFLAG=FLAG ASSOCIATED WITH DELETE
3300       15 KOUNT=0
3400 C     MODIFY LIGHT PEN ATTENTION ON LINES
3500          KFLAG=0
3600       16 CALL MODPAR(NDISP2,0,0,0,0,0,KFLAG,0,2)
3700 C     WAIT FOR LIGHT PEN ATTENTION
3800       20 CALL WAIT(INTARY,0)
3900          IF(INTARY(1).NE.NDISP1(1)) GO TO 30
4000          KEY=INTARY(2)
4100          IF (KEY=5) 70,90,100
4200 C     TURN OFF LINE, RESET FLAG
4300       30 IF(KFLAG.EQ.0) GO TO 20
4400          CALL OFF(NDISP2,INTARY(2))
4500          KFLAG=0
4600          GO TO 16
4700 C     CHECK INTERRUPT KEY ON NDISP1
4800       70 IF(KOUNT.EQ.1) GO TO 80
4850 C           INTERRUPT ON FIRST POINT
4900          NH1=X(KEY)
5000          NV1=Y(KEY)
5100          KOUNT=1
5200          GO TO 20
5300 C     INTERRUPT ON 2ND POINT,DRAW LINE
5400       80 KEY1=KEY1+1
5500          NH2=X(KEY)
5600          NV2=Y(KEY)
5700          CALL LINE(NDISP2,KEY1,NH1,NV1,NH2,NV2,1)
5800          CALL ON(NDISP2,0)
5900          KOUNT=0
6000          GO TO 20
6100 C     INTERRUPT ON DELETE, SET FLAG
6200       90 KFLAG=1
6300          GO TO 16
```

```
6400 C       INTERRUPT ON RESET, CLEAR DISPLAY
6500    100  CALL OFF(NDISP2,0)
6600         CALL RDISPB(NDISP2,0)
6700         GO TO 15
6800         END
6900 *S EXP1    9
7000 EXP1         EQU      0000000,    0053000
7100 SSP          EQU      0000233,    0015000
7200 SIGNON       EQU      0000001,    0015000
7300 IDISPB       EQU      0000016,    0015000
7400 TEXT         EQU      0000077,    0015000
7500 ON           EQU      0000322,    0015000
7600 MODPAR       EQU      0000121,    0015000
7700 WAIT         EQU      0000134,    0015000
7800 OFF          EQU      0000340,    0015000
7900 LINE         EQU      0000311,    0015000
8000 RDISPB       EQU      0000344,    0015000
8100 SST          EQU      0000352,    0015000
8200 PROG,   SIZE 003402
```

EXAMPLE PROBLEM 2

Example 2 illustrates the use of the following subroutines: FBOX, CROSS, POINT, DNUM, KEYBRD.

Problem Statement

Five buttons on the function box (1, 2, 3, 4, and 28) are first enabled, and then are used to perform the following functions:

(a) Depressing button 1 causes the tracking cross to be displayed. The cross can be moved around until function button 28 is depressed. At this time, the cross is turned off and a point is displayed at the current position of the cross. This process may be repeated to create other points.

(b) Depressing function key 2 causes an integer multiple of 100.001 to be displayed at the location of the last point created.

(c) When function button 3 is depressed, the computer will assume that an update from the keyboard will be made by the user. The number to be updated is first selected with the light-pen, and the new value is then "typed in" from the keyboard and displayed.

(d) When function button 4 is depressed, the program is terminated.

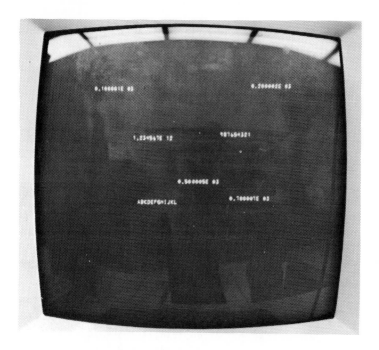

Fig. 2. Example problem 2 operating on the IBM 360 with an IBM 2250 display unit.

```
8300  *COMPILE EXP2  INPUT CARDS OUTPUT TAPE LIST SOURCE
8400  *R EXP2
8500        DIMENSION NDISP1(1000),NDISP2(1000),INTARY(4),NLIST(2)
8600        NLIST(1)=1
8700        NLIST(2)=-4
8800        NHX=100
8900        NVX=100
9000        C=0.0
9100        KEY=0
9200        NSIZE=1000
9300        CALL SIGNON
9400  C     STORE POINTS IN NDISP1
9500  C     STORE NUMBERS IN NDISP2
9600        CALL IDISPB(NDISP1,1,0,0,1023,1023,0,0,NSIZE)
9700        CALL IDISPB(NDISP2,1,0,0,1023,1023,1,0,NSIZE)
9800  C     WAIT FOR AN INTERRUPT FROM FUNCTION BOX
9900      4 CALL FBOX(NB,NLIST,2,0)
10000       GO TO (10,20,30,40),NB
10100 C     DISPLAY CROSS AND CREATE POINTS
10200     10 CALL CROSS(NDISP1,NHX,NVX,1)
10300        CALL FBOX(NB,28,1,0)
10400        CALL CROSS(NDISP1,NHX,NVX,2)
10500        CALL CROSS(NDISP1,NHX,NVX,0)
10600        CALL POINT(NDISP1,0,NHX,NVX,1)
10700        CALL ON(NDISP1,0)
10800        GO TO 4
```

```
10900 C        DISPLAY INTERGER MULTIPLES OF 100,001
11000     20 C=C+100,001
11100           KEY=KEY+1
11200           CALL DNUM(NDISP2,KEY,NHX+20,NVX+20,C,1)
11300           CALL OFF(NDISP1,0)
11400           CALL ON(NDISP2,0)
11500           GO TO 4
11600 C        PROCESS LIGHT PEN INTERRUPT, AND UPDATE NDISP2
11700     30 CALL WAIT(INTARY,0)
11800           KK=INTARY(2)
11900           CALL KEYBRD(NDISP2,KK,A,12,1)
12000           GO TO 4
12100 C        TERMINATE PROGRAM
12200     40 CALL TERM
12300           STOP
12400           END
12500 *S EXP2      $$
12600 EXP2            EQU       0000000,      0053000
12700 $SP             EQU       0000074,      0015000
12800 SIGNON          EQU       0000034,      0015000
12900 IDISPB          EQU       0000051,      0015000
13000 FBOX            EQU       0000110,      0015000
13100 CROSS           EQU       0000124,      0015000
13200 POINT           EQU       0000132,      0015000
13300 ON              EQU       0000205,      0015000
13400 $,+             EQU       0000152,      0015000
13500 DNUM            EQU       0000171,      8015000
13600 OFF             EQU       0000201,      0015000
13700 WAIT            EQU       0000213,      0015000
13800 KEYBRD          EQU       0000227,      0015000
13900 TERM            EQU       0000240,      0015000
14000 $ST             EQU       0000245,      0015000
14100 PROG,    SIZE    004235
```

EXAMPLE PROBLEM 3

The third example was taken from the IBM manual, Form C27-6932-3. This publication describes the graphic subroutine package, GSP.

Problem Statement

The program displays a circle of eight basic size X's with one large X as its center. Function buttons 1, 2, and 3 are also enabled. Depending on the attention created, the program will do the following:

(a) A light-pen attention on the Center X causes labels, NUM1–NUM8, to be displayed adjacent to the outer X's.
(b) A light-pen attention on an outer X generates a circle whose center is the detected X.
(c) A light-pen attention on a circle causes that circle to be invisible.
(d) A light-pen attention on any label makes all the labels invisible.
(e) Programmed function button 1 terminates the graphics program.
(f) Programmed function button 2 makes all circles invisible.
(g) Program function button 3 makes all the circles visible.

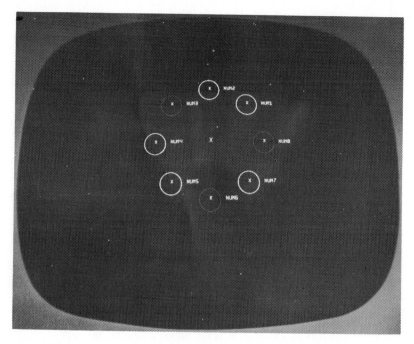

Fig. 3. Example problem 3 operating on the LEC/MAC 16 and the IDI display unit.

```
14200  *COMPILE EXP3   INPUT CARDS OUTPUT TAPE LIST SOURCE
14300  *R EXP3
14400        DIMENSION INTARY(4),NDISP1(1000),NDISP2(1000),NDISP3(1000)
14500        DIMENSION CIRX(8),CIRY(8),LABEL(8),NLIST(3)
14600        REAL LABEL
14700        DATA LABEL(1),LABEL(2),LABEL(3),LABEL(4),LABEL(5),LABEL(6),
14800       1LABEL(7),LABEL(8)/4HNUM1,4HNUM2,4HNUM3,4HNUM4,4HNUM5,
14900       24HNUM6,4HNUM7,4HNUM8/
15000        CALL SIGNON
15100  C     NDISP1 BUFFER FOR X-S
15200  C     NDISP2 BUFFER FOR LABELS
15300  C     NDISP3 BUFFER FOR CIRCLES
15400  C     ALL BUFFERS ARE LIGHT PEN SENSITIVE
15500        CALL IDISPB( NDISP1,2,0,0,4096,4096,1,0,1000)
15600        CALL IDISPB( NDISP2,2,0,0,4096,4096,1,0,1000)
15700        CALL IDISPB( NDISP3,2,0,0,4096,4096,1,0,1000)
15800  C     INITIALIZATION FOR FBOX
15900        NLIST(1)=1
16000        NLIST(2)=-3
16100  C     GENERATE CENTER X IN LARGE CHARACTER MODE
16200        CX=2047,
16300        CY=2047,
16400        CALL TEXT(NDISP1,10,1HX,1,2,CX,CY)
16500  C     GENERATE DISPLAY BUFFERS FOR X-S,LABELS,AND,CIRCLES
16600        R=1200,
16700        C=3,1415926/180,0
```

```
16800       THETA=45.0
16900       DO 20 K=1,8
17000       RADIAN = THETA*FLOAT(K)*C
17100       CIRX(K)=CX+R*COS(RADIAN)
17200       CIRY(K)=CY+R*SIN(RADIAN)
17300       CALL TEXT(NDISP1,K,1HX,1,1,CIRX(K),CIRY(K))
17400       CALL TEXT(NDISP2,K,LABEL(K),4,1,CIRX(K)+300',CIRY(K))
17500    20 CALL CIRCLE(NDISP3,K,CIRX(K),CIRY(K),200.)
17600 C     DISPLAY ALL X=S
17700       CALL ON(NDISP1,0)
17800 C     TURN OFF EACH CIRCLE AND DISPLAY BUFFER
17900       DO 90 I=1,8
18000    90 CALL OFF(NDISP3,I)
18100       CALL ON(NDISP3,0)
18200 C     CLEAR ALL INTERRUPTS
18300   100 CALL WAIT(INTARY,2)
18400       INTARY(1)=0
18500       NB=0
18600 C     WAIT FOR AN INTERRUPT FROM THE LIGHT PEN OR FUNCTION BOX
18700   106 CALL WAIT(INTARY,3)
18800       IF(INTARY(1).NE.0) GO TO 108
18900       CALL FBOX(NB,NLIST,2,1)
19000       IF(NB.EQ.0) GO TO 106
19100 C     AN INTERRUPT WAS RECEIVED FROM THE FUNCTION BOX
19200       GO TO (250,210,200),NB
19300 C     AN INTERRUPT WAS RECEIVED FROM THE LIGHT PEN
19400 C     TEST  WHICH BUFFER  IS ASSOCIATED WITH LIGHT-PEN ATTENTION
19500   108 IF(INTARY(1).EQ.NDISP1(1)) GO TO 110
19600       IF(INTARY(1).EQ.NDISP2(1)) GO TO 300
19700       IF(INTARY(1).EQ.NDISP3(1)) GO TO 350
19800       GO TO 100
19900 C     ATENTION ASSOCIATED WITH NDISP1, CHECK THE KEY
20000   110 IF (INTARY(2)=10)150,120,120
20100 C     DISPLAY ALL LABELS
20200   120 CALL ON(NDISP2,0)
20300       GO TO 100
20400 C     DISPLAY DESIRED CIRCLE
20500   150 KEY=INTARY(2)
20600       CALL ON(NDISP3,KEY)
20700       GO TO 100
20800 C     TURN ON ALL CIRCLES
20900   200 DO 205 I=1,8
21000   205 CALL ON (NDISP3,I)
21100       GO TO 100
21200 C     TURN OFF ALL CIRCLES
21300   210 DO 215 I=1,8
21400   215 CALL OFF(NDISP3,I)
21500       GO TO 100
21600 C     TERMINATE THE PROGRAM
21700   250 CALL TERM
21800       STOP
21900 C     TURN OFF ALL LABELS
22000   300 CALL OFF(NDISP2,0)
22100       GO TO 100
22200 C     TURN OFF THE DESIRED CIRCLE
22300   350 KEY=INTARY(2)
22400       CALL OFF(NDISP3,KEY)
22500       GO TO 100
22600       END
22700 *S EXP3     3
22800 EXP3        EQU       0000000,      0053000
22900 SSP         EQU       0000622,      0015000
23000 SIGNON      EQU       0000001,      0015000
23100 IDISPB      EQU       0000031,      0015000
```

```
23200 TEXT        EQU      0000264,     0015000
23300 $./         EQU      0000113,     0015000
23400 FLOAT       EQU      0000127,     0015000
23500 $.*         EQU      0000176,     0015000
23600 COS         EQU      0000150,     0015000
23700 $.+         EQU      0000256,     0015000
23800 SIN         EQU      0000173,     0015000
23900 CIRCLE      EQU      0000310,     0015000
24000 ON          EQU      0000577,     0015000
24100 OFF         EQU      0000655,     0015000
24200 WAIT        EQU      0000370,     0015000
24300 FBOX        EQU      0000412,     0015000
24400 TERM        EQU      0000632,     0015000
24500 $ST         EQU      0000663,     0015000
24600 PROG,   SIZE 006737
```

EXAMPLE PROBLEM 4

This program allows the user to perform free-hand sketching. Initially, the tracking cross is displayed at location (100, 100), and three light-pen controls, RESET, POS, and DRAW are provided. The program at this time is in a sketch mode and will do the following:

(a) A light-pen strike on RESET will remove any previous sketch and initialize the cross at (100, 100).

(b) A light-pen strike on POS will allow the cross to be repositioned without sketching. A light-pen strike on DRAW at this time will end this mode and cause the program to resume the sketching mode.

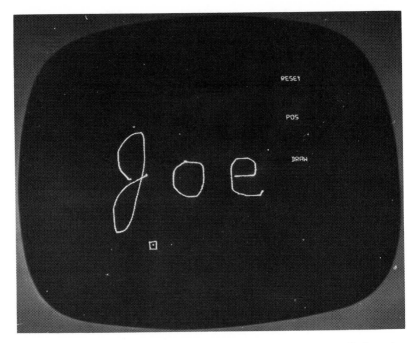

Fig. 4. Example problem 4 operating on the LEC/MAC 16 and the IDI display unit.

```
24700 C      PROGRAM FOR FREE HAND SKETCHING WITH THE LIGHT PEN
24800        DIMENSION NBUF(2000),INT(4)
24900        CALL SIGNON
25000 C      INITIALIZE THE DISPLAY BUFFER
25100      5 CALL IDISPB(NBUF,1,0,0,1023,1023,1,0,2000)
25200 C      CREATE AND DISPLAY THE LIGHT PEN CONTROLS
25300      6 CALL TEXT(NBUF,1,5HRESET,5,2,900,800)
25400        CALL TEXT(NBUF,2,3HPOS,3,2,900,600)
25500        CALL TEXT(NBUF,3,4HDRAW,4,2,900,400)
25600        CALL ON(NBUF,0)
25700        NXL=100
25800        NYL=100
25900 C      CLEAR ALL INTERRUPTS AND DISPLAY CROSS
26000     15 CALL WAIT(INT,2)
26100        CALL CROSS(NBUF,NXL,NYL,1)
26200 C      SET UP LOOP FOR SKETCHING MODE
26300 C      DRAW LINES GREATER THAN ONE UNIT IN LENGTH
26400 C      LINES NOT LIGHT PEN SENSITIVE SINCE LINE KEY = 0
26500     10 CALL WAIT(INT,3)
26600        IF(INT(1).EQ.NBUF(1))GO TO 30
26700        CALL CROSS(NBUF,NX,NY,2)
26800        IF(IABS(NXL-NX)+IABS(NYL-NY).LT.1)GO TO 10
26900        CALL LINE(NBUF,0,NXL,NYL,NX,NY,1)
27000        CALL ON(NBUF,0)
27100        NXL=NX
27200        NYL=NY
27300        GO TO 10
```

```
27400 C      AN INTERRUPT OCCURED ON NBUF PROCESS THIS INTERRUPT
27500    30 KK=INT(2)
27600       GO TO (20,40,10),KK
27700 C      ERASE THE SKETCH AND RESET THE DISPLAY BUFFER
27800    20 CALL OFF(NBUF,0)
27900       CALL RDISPB(NBUF,1)
28000       GO TO 6
28100 C      WAIT FOR AN INTERRUPT BEFORE RESUMING SKETCH MODE
28200    40 CALL WAIT(INT,0)
28300       IF(INT(2).NE.3) GO TO 40
28400       CALL CROSS(NBUF,NXL,NYL,2)
28500       GO TO 15
28600       END
```

EXAMPLE PROBLEM 5

This demonstration program allows a user to play tic-tac-toe with the computer. X's are used for the first player and 0's for the second player. The computer will indicate a win, lose, or draw. An interrupt on FBOX enables three function buttons, which are used as follows:

(a) If the computer is to go first, function button 1 is depressed; otherwise, play is initiated when the light-pen is pointed to a dot in the center of one of the squares (see Fig. 4).
(b) If play is to be restarted, function button 2 is depressed.
(c) If function button 3 is depressed, the graphics program is terminated.

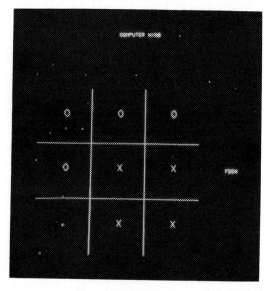

Fig. 5. Example problem 5 operating on the Univac 418 with a DEC 340 display unit.

```
28700 *COMPILE EXP5 INPUT CARDS OUTPUT TAPE LIST SOURCE
28800        DIMENSION SS(9),SP(9),WP(8,3)SQ(9,4)
28900        INTEGER SS,SP,WP,SQ,SV
29000        DIMENSION NLIST(2),RESULT(3,2),INTARY(4)
29100        DIMENSION NDISP(250,2),NDISP1(600),NDISP2(400)
29200        DATA X/1HX/,0/1HO/,BOX/4HFBOX/
29300        DATA WP(1,1), WP(1,2), WP(1,3),WP(2,1),WP(2,2),WP(2,3)
29400        1/1,2,3,4,5,6//WP(3,1),WP(3,2),WP(3,3),WP(4,1),WP(4,2),
29500        2WP(4,3)/7,8,9,1,4,7//WP(5,1),WP(5,2),WP(5,3),WP(6,1),
29600        3WP(6,2),WP(6,3)/2,5,8,3,6,9//WP(7,1),WP(7,2),WP(7,3),
29700        4WP(8,1),WP(8,2),WP(8,3)/1,5,9,3,5,7/
29800        DATA SQ(1,1),SQ(2,1),SQ(3,1),SQ(4,1),SQ(5,1),SQ(6,1)/
29900        11,1,1,2,2,2//SQ(7,1),SQ(8,1),SQ(9,1),SQ(1,2),SQ(2,2),
30000        2SQ(3,2)/3,3,3,4,5,6//SQ(4,2),SQ(5,2),SQ(6,2),SQ(7,2),
30100        3SQ(8,2),SQ(9,2)/4,5,6,4,5,6//SQ(1,3),SQ(2,3),SQ(3,3),
30200        4SQ(4,3),SQ(5,3),SQ(6,3)/7,0,8,0,7,0//SQ(7,3),SQ(8,3),
30300        5SQ(9,3),SQ(1,4),SQ(2,4),SQ(3,4)/8,0,7,0,0,0//SQ(4,4),
30400        6SQ(5,4),SQ(6,4),SQ(7,4),SQ(8,4),SQ(9,4)/0,8,0,0,0,0/
30500 C     NOMENCLATURE
30600 C                  WP = TABLE OF WINNING PATHS
30700 C                  SQ = TABLE OF SQUARES DEFINING THE PATHS
30800 C                       ON WHICH EACH SQUARE LIES
30900 C                  SS = DEFINES THE STATUS OF EACH SQUARE
31000 C                  SP = DEFINES THE STATUS OF EACH PATH
31100 C****     INITIALIZE THE GRAPHIC SYSTEM
31200        CALL SIGNON
31300 C****     INITIALIZE THE DISPLAY BUFFERS
31400        CALL IDISPB(NDISP1,1,0,0,1023,1023,1,1,600)
31500        CALL IDISPB(NDISP2,1,0,0,1023,1023,1,1,400)
31600        DO 10 I=1,2
31700        CALL IDISPB(NDISP(1,I),1,0,0,1024,1024,1,1,250)
31800    10 CONTINUE
31900 C        BLOCK,9POINTS,FBOX KEY AND RESULT OF A GAME,
32000 C****   NEXT 12 STATEMENTS GENERATE THE TICTACTOE BLOCK
32100 C      A KEY=0 IS ASSIGNED TO ELEMENTS FORMING THE BLOCK
32200        CALL LINE(NDISP1,0,100,300,700,300,1)
32300        CALL LINE(NDISP1,0,100,500,700,500,1)
32400        CALL LINE(NDISP1,0,300,100,300,700,1)
32500        CALL LINE(NDISP1,0,500,100,500,700,1)
32600 C        NDISP(1,1) = DISPLAY BUFFER FOR DISPLAYING X-S
32700 C        NDISP(1,2) = DISPLAY BUFFER FOR DISPLAYING O-S
32800 C****   NEXT 10 STATEMENTS CREATE 9 POINTS, THE X-S AND THE O-S
32900        KEY=0
33000        DO 70 I=1,3
33100        NV=200+I
33200        DO 70 J=1,3
33300        NH=200+J
33400        KEY=KEY+1
33500        CALL TEXT(NDISP1,KEY,1H.,,1,1,NH,NV)
33600        CALL TEXT(NDISP(1,1),KEY,X,1,2,NH,NV)
33700        CALL TEXT(NDISP(1,2),KEY,O,1,2,NH,NV)
33800    70 CONTINUE
33900 C****  NEXT 4 STATEMENTS FOR INITIALIZATION AND CREATION OF FBOX KE
34000        NCT=2
34100        NLIST(1)=1
34200        NLIST(2)=-3
34300        CALL TEXT(NDISP1,10,BOX,4,1,800,400)
34400 C****  NEXT 4 STATEMENTS CREATE ELEMENTS NECESSARY FOR DISPLAYING
34500 C      THE OUTCOME OF A GAME
34600        CALL TEXT(NDISP2,1,13HCOMPUTER WINS,13,1,400,900)
34700        CALL TEXT(NDISP2,2,11HPLAYER WINS,11,1,400,900)
34800        CALL TEXT(NDISP2,3,4HDRAW,4,1,400,900)
34900 C****  DISPLAY THE TICTACTOE BLOCK,NINE POINTS AND THE FBOX KEY
35000    82 DO 85 I=1,3
```

```
35100      85 CALL OFF(NDISP2,1)
35200         DO 90 I=1,9
35300         CALL ON(NDISP1,1)
35400         CALL OFF(NDISP(1,1),I)
35500      90 CALL OFF(NDISP(1,2),I)
35600 C****    INITIALIZATION - COMPUTERS STRATEGY
35700     100 DO 110 I=1,9
35800         SS(I)=0
35900         SP(I)=0
36000     110 CONTINUE
36100 C       NOMENCLATURE
36200 C            IFLAG    KEEPS TRACK OF THE TOTAL NUMBER OF MOVES
36300 C            JFLAG    DETERMINES WHETHER COMPUTER OR PLAYER
36400 C                     MADE THE FIRST MOVE
36500 C            KFLAG    KEEPS TRACK OF WHO PLAYED THE LAST MOVE
36600         JFLAG=1
36700         IFLAG=0
36800     200 CALL WAIT(INTARY,0)
36900 C****    INTERRUPT MUST OCCUR ON DISPLAY BUFFER NDISP1
37000         IF(INTARY(1).EQ.NDISP1(1))GO TO 220
37100         GO TO 200
37200     220 KEY=INTARY(2)
37300         IF(KEY-10) 260,240,200
37400 C****      LIGHTPEN ATTENTION OF FBOX
37500 C          FOR COMPUTER TO MOVE FIRST, DEPRESS FUNCTION KEY 1
37600 C          TO RESTART THE GAME, DEPRESS FUNCTION KEY 2
37700 C          TO TERMINATE PROGRAM,DEPRESS FUNCTION KEY 3,
37800     240 CALL FBOX(NB,NLIST,NCT,0)
37900         IF(NB-2) 250,82,650
38000 C****    COMPUTER MOVES FIRST IN CENTER SQUARE
38100     250 JFLAG=1
38200         KEY=5
38300     280 KFLAG=1
38400         GO TO 300
38500 C****    PLAYER MADE HIS MOVE SET KFLAG EQUAL TO -1
38600     260 KFLAG=-1
38700 C****    UPDATE SS AND SP ARRAYS
38800     300 SS(KEY)=KFLAG
38900         DO 310 I=1,4
39000         JJ=SQ(KEY,I)
39100         IF(JJ.EQ.0)GO TO 310
39200         SP(JJ)=SP(JJ)+KFLAG
39300     310 CONTINUE
39400 C****    DISPLAY THE MOVE PLAYED LAST
39500         CALL OFF(NDISP1,KEY)
39600         JJ=(3-JFLAG*KFLAG)/2
39700         CALL ON(NDISP(1,JJ),KEY)
39800 C****    TEST FOR A WIN
39900         DO 350 I=1,8
40000         IF(IABS(SP(I)).EQ.3)GO TO 500
40100     350 CONTINUE
40200 C****    TEST FOR A DRAW
40300         IFLAG=IFLAG+1
40400         IF(IFLAG.EQ.9)GO TO 550
40500 C****    TEST WHO MOVED LAST
40600         IF(KFLAG.EQ.1)GO TO 200
40700 C****    NEXT STATEMENTS DESCRIBE COMPUTER-S STRATEGY
40800 C****    STEP 1
40900 C            IF COMPUTER CAN WIN ON NEXT MOVE, DO SO
41000         DO 400 I=1,8
41100         IF(SP(I).EQ.2)GO TO 450
41200     400 CONTINUE
41300 C****    STEP 2
41400 C        IF PLAYER CAN WIN ON HIS NEXT MOVE, BLOCK HIM
```

```
41500          DO 405 I=1,8
41600          IF(SP(I).EQ.-2)GO TO 450
41700    405 CONTINUE
41800 C****    STEP 3
41900 C          OCCUPY THE SQUARE WITH HIGHEST STRATEGIC VALUE
42000          IF(SS(1).EQ.-1)GO TO 408
42100          IF(SS(3).EQ.-1)GO TO 410
42200          GO TO 418
42300    408 IF(SS(9).EQ.-1)GO TO 412
42400          GO TO 418
42500    410 IF(SS(7).EQ.-1)GO TO 412
42600          GO TO 418
42700    412 DO 415 I=2,8,2
42800          KEY=1
42900          IF(SS(I).EQ.0)GO TO 280
43000    415 CONTINUE
43100    418 SV=0
43200          IS=0
43300          DO 440 I=1,9
43400          IF(SS(I).NE.0)GO TO 440
43500          KV=0
43600          DO 420 J=1,4
43700          JJ=SQ(I,J)
43800          IF(JJ.EQ.0)GO TO 420
43900          KV=KV+1+IABS(SP(JJ))
44000    420 CONTINUE
44100          IF(SV.GT.KV)GO TO 440
44200          SV=KV
44300          IS=I
44400    440 CONTINUE
44500          GO TO 470
44600    450 IP=I
44700          DO 460 I=1,3
44800          IS=WP(IP,I)
44900          IF(SS(IS).EQ.0)GO TO 470
45000    460 CONTINUE
45100    470 KEY=IS
45200          GO TO 280
45300 C****            DISPLAY THE RESULT
45400    500 K=(3-KFLAG)/2
45500          CALL ON(NDISP2,K)
45600          GO TO 200
45700    550 CALL ON(NDISP2,3)
45800          GO TO 200
45900 C****    TERMINATE THE GAME
46000    650 CALL TERM
46100          END
```

EXAMPLE PROBLEM 6

Example 6 illustrates the use of the MODPAR subroutine with respect to zooming images on the scope. This is accomplished by modifying the size of the coordinate window.

Problem Statement

A triangular figure and two light-pen controls, LARGE and SMALL, are displayed on the scope. Continuous interrupts on LARGE increase the relative size of the triangle. Continuous interrupts on SMALL decrease the size of the image being displayed.

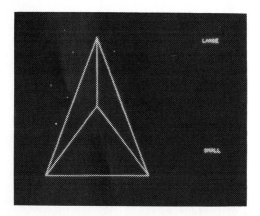

Fig. 6. Example problem 6 operating on the Univac 418 with a DEC 340 display unit.

Fig. 7. Example problem 6 illustrating display zooming features.

```
46200  *COMPILE EXP6 INPUT CARDS OUTPUT TAPE LIST SOURCE
46300  *R EXP6
46400        DIMENSION NDISP1(1000),NDISP2(1000),NX(5),NY(5)
46500        DIMENSION INTARY(4)
46600        INC=100
46700        ILX=0
```

```
46800        ILY=0
46900        IUX=1023
47000        IUY=1023
47100        NX1=200
47200        NY1=200
47300        NX(1)=500
47400        NY(1)=1000
47500        NX(2)=800
47600        NY(2)=200
47700        NX(3)=200
47800        NY(3)=200
47900        NX(4)=500
48000        NY(4)=600
48100        NX(5)=500
48200        NY(5)=1000
48300        CALL SIGNON
48400        CALL IDISPB(NDISP1,1,ILX,ILY,IUX,IUY,0,1,1000)
48500        CALL IDISPB(NDISP2,2,ILX,ILY,IUX,IUY,1,0,1000)
48600        CALL TEXT(NDISP2,1,5HLARGE,5,1,900,800)
48700        CALL TEXT(NDISP2,2,5HSMALL,5,1,900,400)
48800        CALL ON(NDISP2,0)
48900     60 CALL LINE(NDISP1,0,NX1,NY1,NX,NY,5)
49000        CALL LINE(NDISP1,0,NX(4),NY(4),NX(2),NY(2),1)
49100        CALL ON(NDISP1,0)
49200     80 CALL WAIT(INTARY,0)
49300        KEY=INTARY(2)
49400        GO TO (100,200),KEY
49500    100 CALL OFF(NDISP1,0)
49600        CALL RDISPB(NDISP1,0)
49700        ILX=ILX+INC
49800        ILY=ILY+INC
49900        IUX=IUX-INC
50000        IUY=IUY-INC
50100    110 CALL MODPAR(NDISP1,1,ILX,ILY,IUX,IUY,0,0,1)
50200        GO TO 60
50300    200 CALL OFF(NDISP1,0)
50350        CALL RDISPB(NDISP1,0)
50400        ILX=ILX-INC
50500        ILY=ILY-INC
50600        IUX=IUX+INC
50700        IUY=IUY+INC
50800        GO TO 110
50900        END
51000 *S EXP6    6
51100 EXP6        EQU      0000000,      0053000
51200 SSP         EQU      0000263,      0015000
51300 SIGNON      EQU      0000125,      0015000
51400 IDISPB      EQU      0000142,      0015000
51500 TEXT        EQU      0000166,      0015000
51600 ON          EQU      0000242,      0015000
51700 LINE        EQU      0000231,      0015000
51800 WAIT        EQU      0000246,      0015000
51900 OFF         EQU      0000331,      0015000
```

Appendix A. IBM 360 Interface Package

```
52000        SUBROUTINE SIGNON
52100        COMMON/GRASP/NULL,IGSP,ISCOPE,IATL
52200        NULL=-5
52300        CALL INGSP(IGSP,NULL)
52400        CALL INDEV(IGSP,8,ISCOPE)
```

```
52500          CALL CRATL(ISCOPE,IATL)
52600          CALL ENATN(IATL,1,-31,34)
52700          RETURN
52800          END

52900          SUBROUTINE TERM
53000          COMMON/GRASP/NULL,IGSP,ISCOPE,IATL
53100          CALL TMGSP(IGSP)
53200          RETURN
53300          END

53400          SUBROUTINE IDISPB(NDISP,MODE,ILX,ILY,IUX,IUY,LP,ISC,NSIZE)
53500          COMMON/GRASP/NULL,IGSP,ISCOPE,IATL
53600          CALL INGDS(ISCOPE,NDISP)
53700          CALL SGRAM(NDISP,2)
53800          IF(MODE.EQ.1) GO TO 10
53900          CALL SDATM(NDISP,1)
54000          CALL SDATL(NDISP,FLOAT(ILX),FLOAT(ILY),FLOAT(IUX),FLOAT(IUY))
54100          GO TO 20
54200      10  CALL SDATM(NDISP,3)
54300          CALL SDATL(NDISP,ILX,ILY,IUX,IUY)
54400      20  CALL SGDSL(NDISP,0,0,1023,1023,0,0,1023,1023)
54500          IF(ISC.EQ.0)GO TO 30
54600          CALL SSCIS(NDISP,1)
54700          GO TO 40
54800      30  CALL SSCIS(NDISP,3)
54900      40  IF(LP.NE.0) CALL SLPAT(NDISP,1)
55000          RETURN
55100          END
55200          SUBROUTINE POINT(NDISP,KEY,X,Y,NCT)
55300          COMMON/GRASP/NULL,IGSP,ISCOPE,IATL
55400          IF(KEY.LE.0)GO TO 10
55500          CALL PPNT(NDISP,X,Y,KEY,NULL,1,NCT)
55600          GO TO 20
55700      10  CALL PPNT(NDISP,X,Y,NULL,NULL,1,NCT)
55800      20  RETURN
55900          END

56000          SUBROUTINE LINE(NDISP,KEY,X1,Y1,X2,Y2,NCT)
56100          DIMENSION X2(1),Y2(1)
56200          COMMON/GRASP/NULL,IGSP,ISCOPE,IATL
56300          CALL STPOS(NDISP,X1,Y1)
56400          IF(KEY.LE.0)GO TO 10
56500          CALL PLINE(NDISP,X2,Y2,KEY,NULL,1,NCT)
56600          GO TO 20
56700      10  CALL PLINE(NDISP,X2,Y2,NULL,NULL,1,NCT)
56800      20  RETURN
56900          END

57000          SUBROUTINE TEXT(NDISP,KEY,ITEXT,NCH,KCODE,X,Y)
57100          DIMENSION ITEXT(1)
57200          COMMON/GRASP/NULL,IGSP,ISCOPE,IATL
57300          CALL SCHAM(NDISP,KCODE+2)
57400          IF(KEY.LE.0)GO TO 60
57500          CALL PTEXT(NDISP,ITEXT,NCH,KEY,NULL,1,X,Y)
57600          GO TO 80
57700      60  CALL PTEXT(NDISP,ITEXT,NCH,NULL,NULL,1,X,Y)
57800      80  RETURN
57900          END

58000          SUBROUTINE DNUM(NDISP,KEY,X,Y,C,ICODE)
58100          DIMENSION IANS(3),NM(3)
58200          COMMON/GRASP/NULL,IGSP,ISCOPE,IATL
58300          DATA NM/16,14,19/
58400          IF(ICODE.LE.0)GO TO 20
```

```
58500          IF(ICODE.GT.3)GO TO 20
58600          CALL INCORE(C,IANS,NM(ICODE),1,12,6,0)
58700          CALL TEXT(NDISP,KEY,IANS,12,1,X,Y)
58800       20 RETURN
58900          END

59000          SUBROUTINE CIRCLE(NDISP,KEY,CENX,CENY,RADIUS)
59100          DIMENSION X(24),Y(24),IX(24),IY(24)
59200          EQUIVALENCE (X(1),IX(1)),(Y(1),IY(1)),(CX,ICX),(CY,ICY),
59300         1(RAD,IRAD)
59400          CX=CENX
59500          CY=CENY
59600          RAD=RADIUS
59700          MODE=2
59800          IF(IRAD.GE.2**24)GO TO 10
59900          MODE=1
60000          CX=ICX
60100          CY=ICY
60200          RAD=IRAD
60300       10 THETA=0
60400          DTH=3.1415926/12.
60500          DO 20 I=1,24
60600          THETA=THETA+DTH
60700          X(1)=CX+RAD*COS(THETA)
60800       20 Y(1)=CY+RAD*SIN(THETA)
60900          IF(MODE.GT.1)GO TO 40
61000          DO 30 I=1,24
61100          IX(1)=X(1)
61200       30 IY(1)=Y(1)
61300       40 CALL LINE(NDISP,KEY,X(24),Y(24),X,Y,24)
61400          RETURN
61500          END

61600          SUBROUTINE ON(NDISP,KEY)
61700          IF(KEY.LE.0)GO TO 10
61800          CALL INCL(NDISP,KEY)
61900          GO TO 20
62000       10 CALL INCL(NDISP)
62100       20 CALL EXEC(NDISP)
62200          RETURN
62300          END

62400          SUBROUTINE OFF(NDISP,KEY)
62500          IF(KEY.LE.0)GO TO 10
62600          CALL OMIT(NDISP,KEY)
62700          GO TO 20
62800       10 CALL OMIT(NDISP)
62900       20 RETURN
63000          END

63100          SUBROUTINE RDISPB(NDISP,KEY)
63200          IF(KEY.LE.0)GO TO 10
63300          CALL RESET(NDISP,KEY)
63400          GO TO 20
63500       10 CALL RESET(NDISP)
63600       20 RETURN
63700          END

63800          SUBROUTINE MODPAR(NDISP,MODE,ILX,ILY,IUX,IUY,LP,ISC,KODE)
63900          GO TO (100,200,300,100),KODE
64000      100 IF(MODE.EQ.1)GO TO 10
64100          CALL SDATM(NDISP,1)
64200          CALL SDATL(NDISP,FLOAT(ILX),FLOAT(ILY),FLOAT(IUX),FLOAT(IUY)
64300          GO TO 20
```

```
64400        10 CALL SDATM(NDISP,3)
64500           CALL SDATL(NDISP,ILX,ILY,IUX,IUY)
64600        20 CALL SGDSL(NDISP,0,0,1023,1023,0,0,1023,1023)
64700           IF(KODE.EQ.1)GO TO 400
64800       200 JP=2
64900           IF(LP.NE.0) JP=1
65000           CALL SLPAT(NDISP,JP)
65100           IF(KODE.EQ.2)GO TO 400
65200       300 IF(ISC.EQ.0)GO TO 30
65300           CALL SSCIS(NDISP,1)
65400           IF(KODE.EQ.3)GO TO 400
65500        30 CALL SSCIS(NDISP,3)
65600       400 RETURN
65700           END

65800           SUBROUTINE WAIT(INTARY,IWAIT)
65900           DIMENSION KKEY(10),INTARY(4)
66000           COMMON/GRASP/NULL,IGSP,ISCOPE,IATL
66100           INTARY(1)=0
66200           JWAIT=IWAIT/2
66300           IF(IWAIT-2*JWAIT.EQ.0) CALL ENATN(IATL,34)
66400           KWAIT=2-JWAIT
66500           CALL RQATN(IATL,INTCD,KWAIT,KKEY,34)
66600           IF(INTCD.EQ.0) GO TO 100
66700           INTARY(1)=KKEY(1)
66800           INTARY(2)=KKEY(4)
66900       100 RETURN
67000           END

67100           SUBROUTINE KEYBRD(NDISP,KEY,TEXT,NCH,ICODE)
67200 C         DATA IN NDISP MUST HAVE A KEY
67300 C         TEXT STORES FOUR CHARACTERS PER WORD
67400           DIMENSION NDISP(1),TEXT(1),CHAR(3),NM(3)
67500           COMMON/GRASP/NULL,IGSP,ISCOPE,IATL
67600           DATA NM/2,0,5/
67700           CALL CRATL(ISCOPE,ISTLKB)
67800           CALL ENATN(IATLKB,32)
67900           CALL ICURS(NDISP,KEY,NULL,1)
68000           CALL RQATN(IATLKB,INTCD,2,NULL,32)
68100           CALL RCURS(NDISP)
68200           CALL ENATL(IATLKB)
68300           CALL GSPRD(NDISP,CHAR,NCH,1,NULL,KEY)
68400           KWORDS=(NCH-1)/4+1
68500           IF(ICODE.LT.1) GO TO 40
68600           IF(ICODE-4) 10,20,40
68700        10 CALL INCORE(CHAR,TEXT,NM(ICODE),1,12,6,0)
68800           GO TO 40
68900        20 DO 30 I=1,KWORDS
69000        30 TEXT(I)=CHAR(I)
69100        40 RETURN
69200           END

69300           SUBROUTINE CROSS(NDISP,X,Y,KODE)
69400           IF(KODE-1) 10,20,30
69500        10 CALL ENTRK(NDISP)
69600           GO TO 40
69700        20 CALL BGTRK(NDISP,X,Y)
69800           GO TO 40
69900        30 CALL RDTRK(NDISP,X,Y)
70000        40 RETURN
70100           END

70200           SUBROUTINE FBOX(NB,NLIST,NCT,ICODE)
70300           COMMON/GRASP/NULL,IGSP,ISCOPE,IATL
```

```
70400          DIMENSION NLIST(1),LIST(31)
70500          IF(ICODE.EQ.0)CALL ENATN(IATL,1,-31)
70600          K=0
70700          DO 20 I=1,NCT
70800          IF(NLIST(1).LT.0) GO TO 10
70900          K=K+1
71000          LIST(K)=NLIST(I)
71100          GO TO 20
71200    10    J1=-NLIST(I)-NLIST(I-1)
71300          DO 16 KK=1,J1
71400          K=K+1
71500    16    LIST(K)=NLIST(I-1)+KK
71600    20    CONTINUE
71700          DO 40 I=1,K
71800    40    CALL MLITS(ISCOPE,4,LIST(I))
71900          JWAIT=2-ICODE
72000    50    CALL RQATN(IATL,NB,JWAIT,NULL,1,-31)
72100          DO 60 I=1,K
72200          IF(NB.EQ.LIST(I)) GO TO 70
72300    60    CONTINUE
72400          IF(ICODE.EQ.0)GO TO 50
72500          NB=0
72600    70    CALL MLITS(ISCOPE,2)
72700          RETURN
72800          END
```

Acknowledgments

The talents of D. M. Smith, D. R. Chand, and C. M. Bradley are reflected in this report. Their testing and careful implementation of FLING on the MAC-16 computer graphic system is sincerely appreciated.

PART VI

Numerical Methods for a Changing Technology

Trends and Directions in the Applications of Numerical Analysis

Richard H. Gallagher

DEPARTMENT OF STRUCTURAL ENGINEERING
CORNELL UNIVERSITY
ITHACA, NEW YORK

Two major avenues of current development in finite-element analysis are discussed. The first concerns the amenability of existing large-scale general purpose analysis programs to the inclusion of complementary and mixed energy models. The second area relates to the expansion of finite-element analysis to interdisciplinary physical problems, of which structural behavior is a component part.

1. Introduction

Any attempt to give a widely applicable measure of trends and directions in numerical analysis for structural mechanics at this juncture in this volume is foreshadowed by the detailed presentations already given by individual authors. In themselves, these define a large share of current directions. The objective in this paper, therefore, is to discuss just two major avenues of current development not otherwise covered in this book.

The first concerns the utilization of mixed and complementary energy element formulations in existing large-scale computer programs for finite-element analysis. The value of hybrid forms of the mixed approach is amply demonstrated by Pian [1]. The hybrid formulation permits construction of element stiffness matrices which are directly useful in available programs. The same can be said of element formulations derived via complementary energy principles and transformed to stiffness matrix format [2]. It is desirable, nevertheless, that means be available to perform system analysis within the same formulative framework as element analysis. The central question here asks if major revisions are necessary to accomplish this goal with existing programs.

The second avenue of development represents the expansion of finite-element analysis from a topic concerned exclusively with structural mechanics to a tool applicable to a wide range of physical problems. Since this book deals exclusively with structural mechanics, an excursion into nonstructural applications might appear inappropriate. The examples to be cited, however, represent behaviors that are either coupled to structural behavior or are essential to proper determination of structural behavior. These include fluid–structure interaction, thermal analysis, and the study of the flow of bodies of water subjected to wind action and thermal input.

The ability to cope with nonstructural physical problems by use of the finite-element method is felt to be of importance for another reason, quite apart from any connection with the objectives of structural mechanics. These give additional amortization to the extremely large costs of development of the general purpose finite-element program developed initially for structural analysis. Potential savings derive also from the commonality of input and output format and documentation.

2. Alternative Variational Principles

The use herein of the term "alternative variational principles" is intended to designate methods of finite-element analysis which are alternative to the stiffness analysis method. In the strictest sense of variational formulations, the stiffness method corresponds to the principle of minimum potential energy and alternatives take the form of the minimum complementary and Reissner energy principles. The interdisciplinary problems to be discussed subsequently, however, have fostered entirely different approaches to finite-element formulation which are not properly described in the variational context, e.g., the method of weighted residuals with a Galerkin criterion. With such possibilities in mind, the term "alternative variational principles" is retained with the loose definition just ascribed.

It is also necessary, at the outset, to justify the designation of alternative variational principles as major current avenues of development in finite-element analysis. Philosophically, the prevailing imbalance of system analysis procedures, weighted almost exclusively toward the stiffness method, motivates interest in the utilization of the dual (complementary) procedure and in the canonical (mixed) form as well. Duality was emphasized in early matrix analysis work [3, 4] and took the form, on the complementary side, of the matrix force method. For reasons discussed by Gallagher and Dhalla [5], this method advanced negligibly in the period during which the stiffness method grew exponentially in terms of system analysis concepts, element representation, and coverage of various special phenomena (e.g., buckling).

The mixed model concept was first given attention in 1965 [6] and it is fair to say that it also renewed interest in complementary methods. In each case the motivating factor was the simplicity afforded in the treatment of plate flexure. Now, the concern of finite-element practitioners has expanded to the analysis of curved shells, prompting the parallel use of displacement fields in representation of stretching behavior and stress fields for representation of flexure [7], i.e., a mixed model.

Other reasons are often advanced for the growth of finite-element formulations alternative to the stiffness method. Probably the foremost justification arises from the promise of establishing "bounds" upon the solution with use of a single size of idealization. Although it has not been given attention here, methods of estimating solution accuracy must inevitably assume greater importance with the establishment of widely available large-scale computer programs. At this juncture, however, the procedures to be followed in assuring the calculation of error bounds are not only incompletely defined, but the parameters whose bounds are obtained (e.g., strain energy) generally lack structural design significance.

We turn now to the central consideration of this section: the impact of complementary and mixed formulations upon existing stiffness analysis programs. It is convenient to examine first the complementary energy principle, which seeks a stationary value of the following functional (Π_c) as the condition of compatible deformation

$$\Pi_c = \tfrac{1}{2} \int_V \lfloor \sigma \rfloor [E]^{-1} \{\sigma\} \, dV - \int_{S_\Delta} \bar{\Delta} \cdot T \, dS \tag{1}$$

where $\lfloor \sigma \rfloor$ is the stress vector, $[E]^{-1}$ represents the material constitutive law in compliance format, V is the volume of the structure, S_Δ is the portion of the boundary upon which the displacements $\bar{\Delta}$ are prescribed, and T are the corresponding surface tractions. Initial strains are excluded for algebraic simplicity.

Transformation of Eq. (1) into finite-element algebraic relationships requires the selection of self-equilibrating stress fields $\{\sigma\}$ expressed in terms of discrete stress parameters $\{\bar{\sigma}\}$ and "shape functions" which define the field in spatial coordinates. Stress parameters have been employed successfully in some of the recent advances in finite-element complementary analysis [8, 9]. Special advantages are obtained however, if stress functions, rather than stresses, are designated as the discrete parameters because the form of stiffness relationship for flexure corresponds to the complementary stress function form for stretching, and vice-versa (taking due account of the inverse relationship of the constitutive laws). For stretching, the appropriate stress function is the well-known form due to Airy [5]. For bending, it takes a form proposed by Southwell [10]. Stress functions for three-dimensional elasticity are surveyed by Charlwood [11].

Thus, if we designate such stress function parameters by $\{\Phi\}$ and use these in a proper relationship to $\{\sigma\}$ and T in Π_c, we can obtain the discretized form

$$\Pi_c = (\lfloor \Phi \rfloor / 2)[f]\{\Phi\} - \lfloor \Phi \rfloor \{b\} \tag{2}$$

where $[f]$ represents the flexibility matrix and $\{b\}$ represents a vector obtained by appropriate integration of the second term in Eq. (1), where such integration involves only the boundary (S_u) subject to the specified displacements. Application of the first variation yields

$$[f]\{\Phi\} = \{b\} \tag{3}$$

Since the analyst only rarely confronts the problem of specified nonzero boundary displacements, the integral on S_u yields a zero value and one must refer to other considerations to produce the right-hand side of Eq. (3). When the appropriate derivatives of the stress functions are included as nodal parameters this problem is easily handled, since the known edge stresses are given by such derivatives and the effect is to produce a right-hand side vector. One may seek to avoid higher-order elements, however, in which case the stress boundary conditions are examined to yield supplemental equations symbolized here as

$$[q]\{\Phi\} = \{a\} \tag{4}$$

These are constraint equations, tying together certain of the stress function parameters, and must be accounted for before imposition of the first variation, i.e., Eq. (3) is to be modified. The methods of accounting for constraints are taken up in the next section, with the objective of ascertaining how existing stiffness analysis programs can do so.

There are a number of forms of mixed models, but to illustrate the features important to our discussion of the usefulness of existing stiffness analysis programs it is sufficient to deal only with Reissner's variational principle. Here, the functional is given by

$$\Pi_R = -\tfrac{1}{2}\int_V \lfloor \sigma \rfloor [E]^{-1}\{\sigma\}\, dV + \int_V \lfloor \sigma \rfloor \{\varepsilon\}\, dV - \int_{S_\Delta} T(\Delta - \bar{\Delta})\, dS$$
$$- \int_{S_\sigma} \bar{T}\Delta\, dS \tag{5}$$

where $\{\varepsilon\}$ designates the strain field and S_σ is the portion of the surface on which the tractions $\bar{T}$ are prescribed. In discretizing this functional we retain $\{\Phi\}$ as stress parameters and establish assumed displacement fields, expressed in terms of nodal displacements $\{\Delta\}$ which can be related to the strains via imposition of the strain-displacement equations. Upon substitution of both fields in Eq. (5) and the performance of required integrations, the following functional is obtained:

$$\Pi_R = \lfloor \Phi \; \Delta \rfloor \begin{bmatrix} f & q^T \\ q & 0 \end{bmatrix} \begin{Bmatrix} \Phi \\ \Delta \end{Bmatrix} - \lfloor \Phi \; \Delta \rfloor \begin{Bmatrix} b \\ a \end{Bmatrix} \tag{6}$$

and by variation with respect to both Φ and Δ

$$\begin{bmatrix} f & q^T \\ q & 0 \end{bmatrix} \begin{Bmatrix} \Phi \\ \Delta \end{Bmatrix} = \begin{Bmatrix} b \\ a \end{Bmatrix} \tag{7}$$

which may be solved to yield directly both the stress function and displacement parameters.

Two observations must be made with respect to Eq. (7). First, it is apparent that the zero lower right quadrant of the coefficient matrix requires care in the selection of a solution algorithm. Second, the submatrices forming the coefficient matrix have been assigned a symbolism identical to that used above for complementary energy models. The designation of $[f]$ in this manner is apparently appropriate, since it stems from the first term in Π_p which is identical except for sign with the first term of Π_c—the source of the flexibility matrix in Eq. (3). The identification of $[q]$ with the coefficient matrix of the constraint conditions, Eqs. (4), is best established in the discussion of methods of accounting for constraints, to be reviewed next.

Before concluding this discussion of alternative variational principles, it is pertinent to comment briefly about the coordination of the alternatives in a single theoretical framework. Prager [12] has presented such a coordination within the confines of structural mechanics. For more general physical situations, of the type to be described later, a unified theoretical basis was recently given by Sandhu and Pister [13] and, in a different form, by Sewell [14]. These developments demonstrate that variational principles in continuum mechanics have not in all cases existed in well-established classical form prior to their utilization in numerical methods, but have rather emerged in parallel with certain developments. A case in point is the variational principle derived by Herrmann [15] for incompressible materials.

3. Constraint Equation Procedures

Two alternative approaches are commonly adopted in the treatment of constraint condition in finite-element analysis: (1) Use of the constraint equations to solve for certain degrees of freedom in terms of the others and subsequent reduction of the size of the system prior to the imposition of the first variation, or (2) Construction of an augmented functional, where the additional terms are the products of the constraints and corresponding Lagrange multipliers.

The first approach [16] has the advantage of reducing the size of the system of equations to be solved. A disadvantage from the viewpoint of established stiffness analysis programs is the extensive matrix manipulation required. In simplest terms, the reduction can be accomplished by defining the following transformation relationships from Eqs. (4), based on the identification of the degrees of freedom Φ_1 as those to be removed from the full set $\{\Phi\} = \lfloor \Phi_1 \quad \Phi_2 \rfloor^\mathsf{T}$

$$\begin{Bmatrix} \Phi_1 \\ \Phi_2 \end{Bmatrix} = \begin{bmatrix} -q_1^{-1} & q_2 \\ \hline & I \end{bmatrix} \{\Phi_2\} + \begin{bmatrix} -q_1^{-1} \\ \hline 0 \end{bmatrix} \{b\} \tag{8}$$

This transformation is applied to the functional; the reduced form of the functional may then be varied to yield the reduced algebraic equations comparable to Eq. (5).

It should also be noted that the choice of degrees of freedom to be removed is not unique and may in fact involve poor conditioning or even singularity of the matrix to be inverted. This point is discussed at length in Walton and Steeves [17].

The Lagrange multiplier method, conversely, expands the size of the system of equations to be solved but possesses a key advantage with respect to existing stiffness analysis programs as described later. In accordance with this classical concept, each of $1, \ldots, i, \ldots, n$ constraint conditions is multiplied by a parameter λ_i, the Lagrange multiplier. The full set of such products is added to the functional, yielding for the case of Π_c

$$\overline{\Pi}_c = (\lfloor \Phi \rfloor/2)[f]\{\Phi\} - \lfloor \Phi \rfloor\{b\} + \lfloor \lambda \rfloor[q]\{\Phi\} - \lfloor \lambda \rfloor\{a\} \tag{9}$$

where $\overline{\Pi}_c$ is termed the "augmented functional," and by variation with respect to both $\{\Phi\}$ and $\{\lambda\}$, we obtain the algebraic equations

$$\begin{bmatrix} f & q^\mathsf{T} \\ q & 0 \end{bmatrix} \begin{Bmatrix} \Phi \\ \lambda \end{Bmatrix} = \begin{Bmatrix} b \\ a \end{Bmatrix} \tag{10}$$

which have the same form as Eqs. (6) resulting from the Reissner variational principle, with the generalized displacements $\{\Delta\}$ of the latter cast in the role of the Lagrange multipliers, $\{\lambda\}$.

In point of fact, a familiar method of derivation of Reissner's variational principle is to begin with either the principle of potential or complementary energy, relax continuity requirements (represented by the surface integrals), and seek satisfaction of such conditions in at least an average sense by use of constraint conditions with Lagrange multipliers. Thus, in the preceding, the generalized displacements serve this purpose. They pertain to approximate satisfaction of equilibrium conditions along boundaries where tractions are prescribed, where the actual distribution of applied load will not, in general, correspond to the assumed element stress field.

To use existing stiffness programs in the formation of the preceding coefficient matrix, a special element "constraint stiffness" can be visualized of the form

$$\begin{bmatrix} 0 & q^{\mathrm{T}} \\ q & 0 \end{bmatrix} \begin{Bmatrix} \Phi \\ \lambda \end{Bmatrix} = \begin{Bmatrix} 0 \\ b \end{Bmatrix} \tag{11}$$

where now all quantities are meant to apply at the element level; the Lagrange multipliers therefore assume the character of degrees of freedom, with the physical significance of displacements in this example. It is necessary that the program used be able to accommodate nonsquare element relationships; assembly of the system coefficients follows immediately from direct stiffness concepts. The submatrix $[f]$ is found in the usual manner of direct stiffness analysis.

This approach can be generalized. The respective variational principles apply to functionals which consist of the sum of volume and surface integrals. The volume integrals are representable in forms identical to the conventional stiffness matrices and construction of the related coefficient matrix (e.g., $[f]$ in Eq. (10)) follows the procedures of direct stiffness analysis. For surface integrals, on the other hand, special constraint stiffnesses are formed, as described by Eq. (11). Once established, these also admit assembly to form the relevant components of the full coefficient matrix.

4. Interdisciplinary Applications

As a second general area of future trends in the application of numerical analysis we discuss the topic of interdisciplinary problems, defined here as simultaneous analysis for behaviors which are traditionally given independent analytical characterization. The simplification of actual interdisciplinary behavior to independent representation has been a limit imposed historically by analysis capabilities. The removal of these limits is a major contribution of enlarged numerical analysis capabilities, matching that of the inroads made into nonlinear problems. Already a number of practical interdisciplinary problems have been solved. We review three of these in what follows.

The first concerns thermostructural analysis. The practical significance of integrated thermal and structural analysis cannot be overemphasized.

The most desirable situation occurs when the analytical models for transient thermal analysis and structural analysis are in complete correspondence. Any disparity between these representations is a source of error which is amplified in the integration of the transients in time. Furthermore, uncoordinated thermal and elastic analyses are highly inefficient, requiring large costs in the transference of data from the thermal to the thermal stress analysis.

An approach to the formulation of element and system equations for finite-element thermal analysis, parallel to those of structural analysis, was presented by Nickell and Wilson in 1966 [18]. Since then, many contributions have been added, including procedures based on alternate variational principles. Design office applications, at least to steady-state heat conduction, are routine. A detailed review of the state of the art in this field was prepared by the writer [26] in 1971.

It is useful to define the general form of the algebraic equations for finite-element thermal analysis, both for the immediate context and for the later comments on thermal pollution analysis. Such equations are of the form

$$[D]\{\dot{T}\} + [B]\{T\} = \{Q\} \tag{12}$$

where $\{Q\}$ is the vector of (time-dependent) thermal loads at the element joints. $[B]$ is the "conductivity matrix." $[D]$ is the "heat capacitance" matrix. $\{T\}$ is the vector of element joint temperatures. and $\{\dot{T}\}$ is the time derivative of T.

The advantages of this type of formulation of the heat transfer, which are partly surmised from Eq. (12) and a consideration of the character of finite-element structural analysis, are the following:

1. Equations governing the heat transfer throughout a complex system can be constructed automatically, as is done in a finite-element structural analysis, based on the concept of element "conductivity" and "capacitance" matrices.

2. The calculation of the element heat transfer matrices can be accomplished as part of the process of formation of element structural analysis matrices. The problem data, as well as certain computational steps, are common to both.

3. Temperature distributions are identically of the form needed for subsequent structural analysis.

4. The scope of the heat-transfer analysis capability is widened to account for anisotropic materials, with no significant increase of analytical complexity or computational cost.

Coupling of thermal and structural behavior exists and may be accounted for by extension of the preceding, but is rarely of design significance. More important from the standpoint of frequency of occurrence in practical design is the problem of fluid–structure behavior. Such problems exist in the design of ships in floating drydock, in the sloshing of fluids in containers, and in shore and undersea structures.

The fluid–structure interaction problem can be described in matrix algebraic form, at an instant in time, as follows [19]:

$$\begin{bmatrix} M & 0 \\ L^T & N \end{bmatrix} \begin{Bmatrix} \ddot{\Delta} \\ \ddot{p} \end{Bmatrix} + \begin{bmatrix} C & 0 \\ 0 & D \end{bmatrix} \begin{Bmatrix} \dot{\Delta} \\ \dot{p} \end{Bmatrix} + \begin{bmatrix} K & -L \\ 0 & H \end{bmatrix} \begin{Bmatrix} \Delta \\ p \end{Bmatrix} = \begin{Bmatrix} P \\ 0 \end{Bmatrix} \tag{13}$$

where $\lfloor \Delta \quad p \rfloor$ lists the joint displacements and pressures, dots signify derivatives with respect to time, $[M]$, $[C]$, $[K]$, and $\{P\}$ are the conventional structural mass, viscous damping, stiffness, and load matrices, and $[N]$, $[D]$, and $[H]$ are corresponding matrices for fluid pressure. Interaction is governed by the coefficients of the matrix $[L]$. Zienkiewicz and Newton [19] give a thorough development of Eqs. (13) and describe the solution to various specific problems.

As a final illustration of an interdisciplinary problem, we examine the determination of circulation and the temperature state of lakes and reservoirs. The growth of interest in environmental quality has spawned intensive numerical studies in this area. The problems of concern have many sources and forms.

For example, nuclear generating stations seek to draw upon the cooler, deeper lake water and return it to the surface in a heated state. Ecologists claim that an effect of elevating the lake temperature is to increase growth of plant life, which speeds eutrophication, or aging, of the lake. Deleterious effects on aquatic life are also predicted. The significance of these factors in a given set of circumstances clearly depends upon the lake temperature changes, whose prediction involves a combined flow and heat-transfer analysis, with the possible inclusion of the effect of wind action on the surface of the lake.

Analytical studies of this problem, performed with use of finite difference approximations to the governing differential equations, are themselves of very recent origin [20–22]. In each case a one-dimensional (depthwise) model is adopted, and attention is directed to a description that accounts for a stratification of the temperature distribution which features a nearly constant temperature in a layer close to the surface. Equation (12) applies as a solution scheme for this problem except that now, in accordance with the theory of Sundaram and Rehm [22], the matrix $[B]$ must represent a nonlinear dependence upon temperature. The extensive successful use of finite-element analysis in geometric and material nonlinearities indicates that this behavior can be dealt with and that finite-element analysis is appropriate to the problem.

The general lake pollution problem is broader than the prediction of depthwise temperature profiles, and introduces the coupling of thermal and fluid behavior. Sources of lake pollution include waste products of cities, industrial waste, and the natural runoff of agricultural lands. The effects of these pollutants depends largely on where they go in the lake and how long they remain—factors which are dependent on the flow and the temperature distribution. For this, a finite-element model can be constructed by combination of the thermal analysis equation (12) and the flow portion of Eq. (13). For the more general case of fluid flow it is appropriate to introduce flow velocities as joint parameters, designated as $\{u\}$, so that we now find the algebraic equations to be solved to be of the form:

$$
\begin{bmatrix} D & 0 & 0 \\ 0 & J & 0 \\ 0 & 0 & 0 \end{bmatrix} \begin{Bmatrix} \dot{T} \\ \dot{u} \\ \dot{p} \end{Bmatrix} + \begin{bmatrix} B & W^{\mathrm{T}} & 0 \\ W & R & S \\ 0 & 0 & Y \end{bmatrix} \begin{Bmatrix} T \\ u \\ p \end{Bmatrix} = \begin{Bmatrix} Q \\ Z \\ X \end{Bmatrix} \tag{14}
$$

where $[W]$, a function of T and u, gives the coupling between thermal and fluid flow.

A major advantage of the finite-element procedure in the preceding problem is the usefulness of algorithmic tools already established for finite-element analysis in structural mechanics. Formidable difficulties in the solution of the resulting equations lie ahead, however. The basic form of Eq. (14) is unsymmetric, the coefficients are nonlinear functions of the solution parameters T, u, and p, and must be integrated in time. Much theoretical and computer program development, as well as the assimilation of experience in application and correlation with known solutions and test data, lies ahead before this specific potential of finite element interdisciplinary analysis is realized.

5. Concluding Remarks

We have identified two general areas as representative of major trends in development of finite-element forms of numerical analysis.

In the expansion of finite-element formulations to encompass the widest range of alternative variational principles it was suggested that extensions of existing large-scale, stiffness-based computer programs may not be as formidable as might be anticipated. In order to exploit this expanded capability properly, however, a deeper understanding is needed of the practical aspects of conditions under which bounds on solution parameters can be assured. Closer identification of the solution parameters to design objectives is also needed.

With regard to interdisciplinary problems, an examination of practice will disclose that the level of application is currently very low, except for the simplest and most important circumstance, integrated thermostructural analysis. Also, a very great amount of basic theoretical work is yet to be accomplished. Such work will emphasize the establishment of the forms of discretized models and the solution of the resulting time-dependent equations. Relatively little attention will be drawn by questions of individual finite-element representation and formulation, since research already devoted to the latter in structural mechanics is directly applicable to the other technologies.

Motivation for the extension of the finite-element method to a broader range of physical problems might be thought to come from increased computational efficiency, ability to cope with more complex forms of a given class of problem, and savings through a unified general-purpose computer program.

It is difficult to prove that a finite-element approach to a problem which is generally solved by alternative means, e.g., finite differences, is computationally more efficient. Indeed, the heat-transfer studies of Emery and Carson [23] claim that the opposite is true for cases they study. Due to high costs in software development and in other aspects peripheral to computation, the direct computational costs are of less importance.

The ability to deal with complicated load and geometric situations is probably the strongest factor in seeking unified finite-element analysis capabilities. The efficiencies furnished by unified capabilities, however, are remote from realization primarily because organizations responsible for development of existing programs of this type did not envision the broadened applicability when initiating program development. Those programs which are versatile and reliable in time-dependent analysis are most easily extended and transformed to deal with more general physical problems.

The latter consideration touches upon the final point to be commented upon. The foremost requirements, in the current practical utilization of general-purpose finite-element analysis programs relate to reliability and accuracy. A plateau of reliability in numerical analysis of time-dependent problems has not yet been reached; this circumstance was aptly demonstrated by Nickell [24]. The situation is more poorly defined in the solution of nonlinear algebraic equations as occur in plasticity and finite displacement analysis and in fluid flow situations governed by the full Navier–Stokes equations. Procedures for the assessment of solution accuracy as influenced by round-off error are emerging from the realm of mathematical exercises into workable schemes for actual problems. Some progress in this direction is described by Tong [25].

References

1. Pian, T. T., "Hybrid Models." This volume.
2. Fraeijs de Veubeke, B. (ed.), "Upper and Lower Bounds in Matrix Structural Analysis," in *Matrix Methods of Structural Analysis*, Macmillan Co., N.Y., 1964.
3. Langefors, B., "Analysis of Elastic Structures by Matrix Transformation with Special Regard to Semimonocoque Structures," *J. Aeronaut. Sci.* **19**, No. 7, 451–458 (1952).
4. Argyris, J. H., and Kelsey, S., *Energy Theorems and Structural Analysis.* Butterworths, London and Washington, D.C., 1961. (Published serially in Aircraft Eng. Magazine, 1954–1956.)
5. Gallagher, R. H., and Dhalla, A. K., "Direct Flexibility—Finite Element Elastoplastic Analysis," *Proc. Int. Conf. Nucl. Reactor Struct., 1st* Berlin (1971).
6. Herrmann, L. R., "A Bending Analysis for Plates," *Proc. Conf. Matrix Methods Struct. Mech.* AFFDL TR 66-80 (1965).
7. Connor, J., and Will, G., "A Mixed Finite Element Shell Formulation," *Recent Advances in Matrix Methods of Structural Analysis and Design* (Gallagher, R. *et al.*, ed.). Univ. of Alabama Press, Tuscaloosa, Alabama, 1971.
8. Belytschko, T., and Hodge, P. G., "Plane Stress Limit Analysis by Finite Elements," *Proc. ASCE J. Eng. Mech.* **96**, No. EM 6, 931–944 (December 1970).
9. Anderheggen, E., "Finite Element Plate Bending Equilibrium Analysis," *Proc. ASCE J. Eng. Mech. Div.* No. EM 4 (August 1969), 841–858.
10. Fraeijs de Veubeke, B., and Zienkiewicz, O., "Strain Energy Bounds in Finite Element Analysis by Slab Analogy," *J. Strain Anal.* **2**, No. 4 (1967).
11. Charlwood, R. C., "Dual Formulations of Linear Elasticity Using Finite Elements," *Computer-Aided Engineering* (Gladwell, G. L. M., ed.). Univ. of Waterloo, 1971.
12. Prager, W., "Variational Principles of Linear Elastostatics for Discontinuous Displacements, Strains, and Stresses," *Recent Progress in Applied Mechanics: The F. Odqvist Volume*, pp. 463–474. Wiley, New York, 1967.
13. Sandhu, R., and Pister, K., "Variational Principles for Boundary Value and Initial-Boundary Value Problems in Continuum Mechanics," *Int. J. Solids Struct.* **7**, 639–654 (1971).
14. Sewell. M. J., "On Dual Approximation Principles and Optimization in Continuum Mechanics," *Phil. Trans., Roy. Soc. London* **265**, No. 1162, 317–351 (1967).
15. Herrmann, L. R., "Elasticity Equations for Incompressible and Nearly Incompressible Materials by a Variational Theorem," *AIAA J.* **3**, No. 10, 1896–1900 (1964).
16. Greene, B. E., "Application of Generalized Constraints in the Stiffness Method of Structural Analysis," *AIAA J.* **4**, No. 9, 1531–1537 (1966).
17. Walton, W. C., and Steeves, E. C., "A New Matrix Theorem and its Application for Establishing Independent Coordinates for Complex Dynamical Systems with Constraints," NASA TR R-326 (October 1969).
18. Nickell, R. E., and Wilson, E., "Application of the Finite Element Method to Heat Conduction Analysis," *Nucl. Eng. Design* **4**, 276–286 (1966).
19. Zienkiewicz, O. C., and Newton, R. E., "Coupled Vibrations of a Structure Submerged in a Compressible Fluid," *Proc. Symp. Element Tech.*, ISSC, Stuttgart (June 1969).
20. Orlob, G., and Selna, L., "Temperature Variations in Deep Reservoirs," *Proc. ASCE J. Hydrd. Div.* **96**, No. HY 2, 391–410 (1970).
21. Liggett, J., and Choi, Y., "Prediction of Lake Stratification," ASCE Meeting Preprint 1295 (January 1971).
22. Sundaram, J. R., and Rehm, R. G., "Formation and Maintenance of Thermoclines in Temperate Lakes," *AIAA J.* **9**, No. 7, 1322–1329 (1971).

23. Emery, A., and Carson, W., "An Evaluation of the Use of the Finite-Element Method in the Computation of Temperature," *Trans. ASME J. Heat Transfer* 136–145 (1971).
24. Nickell, R. E., "On the Stability of Approximation Operators in Problems of Structural Dynamics," *Int. J. Solids Struct.* **7**, 301–319 (1971).
25. Tong, Pin, "On the Numerical Problems of Finite Element Methods," *Computer Aided Engineering* (Gladwell, G. L. M., ed.). Waterloo Univ., 1971.
26. Gallagher, R. H., "Interpretive Report: Computational Methods in Nuclear Reactor Structural Design for High Temperature Applications," Oak Ridge Nat. Lab. Rep., ORNL-4756, Feb. 1973.

Vehicle Crashworthiness

S. P. Desjardins

DYNAMIC SCIENCE
A DIVISION OF ULTRASYSTEMS, INC.
PHOENIX, ARIZONA

A survey of aircraft crashworthiness is presented. The survivable crash environments for rotary-wing aircraft, light fixed-wing aircraft, and fixed-wing transport aircraft are summarized. General causes of injury as related to the structural response of the aircraft are presented, followed by a more detailed discussion of the highlights of a recent accident study on rotary-wing aircraft. A brief discussion of the structural crashworthiness of light fixed-wing, fixed-wing transport, and rotary-wing aircraft is presented together with areas of potential improvement.

1. Introduction

Human injury and death have always been of personal concern to the injured and his immediate family and friends; however, the increase in injuries and deaths resulting from modern forms of transportation, primarily automobiles and aircraft, have at last reached proportions demanding national concern. In the past, efforts were made to reduce the casualties by accident prevention measures. Although this effort has been fruitful, recent trends have been to accept the fact that some accidents will continue to occur despite these efforts and that vehicles should be designed to provide a degree of protection to the occupant, at least in milder accidents.

It is recognized that an aircraft and, to a lesser extent, an automobile is designed to perform a particular mission and that crashworthiness must take a second place to operational performance. However, it has been shown repeatedly that structural crashworthiness can be achieved simply by having designers make informed choices of alternate approaches during the design stages of the vehicle. In some instances it has been shown that drastic improvements in structural crashworthiness can actually be accompanied by a decrease in cost and weight.

The purpose of this paper is to present some of the principles for providing structural crashworthiness for vehicles. The paper presents an overview of the crash environment, injury causes and patterns, and structural crashworthiness generally applicable to aircraft. The type of information presented is representative of that which can be applied to any vehicle, including automobiles. It is hoped that the information presented will stimulate interest, creative thought, and effort dedicated to decreasing needless injury and fatalities in aircraft and automotive accidents that are potentially survivable.

2. Crash Environment

The aircraft crash environment ranges from the insignificant hard landing to the nonsurvivable. When any vehicle crashes, motion continues until the kinetic energy has been attenuated, primarily through the application of force through distance. The decelerative force is a function of the kinetic energy of the vehicle and the distance through which it moves during deceleration. In the crash of an aircraft with a purely vertical velocity component, the movement is permitted by deformation of both the terrain upon which the aircraft crashes and the structure of the aircraft. If the aircraft crashes on soft soil, considerable soil deformation will occur and the decelerative load will be less by virtue of the distance traveled against the force compacting or moving the soil, as well as against the crushing strength of the fuselage. If the aircraft crashes on a rigid surface such as concrete, the deformation distance will essentially be supplied entirely by the crushing fuselage and result in increased G loading.

If the vehicle crashes with a high longitudinal component of velocity, the longitudinal decelerative loading can depend on many factors. These include friction, plowing and gouging of earth, longitudinal crush strength in the case of barrier or vehicle impact, or local crush strength of the vehicle impacting similar barriers such as trees, posts, and rocks.

Combinations involving longitudinal, vertical, and lateral components of velocity also include longitudinal, vertical, and lateral decelerative loads. Relatively high vertical and vehicle bending loads can also be applied through the process of rapidly changing the direction of the longitudinal velocity component of the vehicle structure, such as occurs when an aircraft with high longitudinal velocity impacts a relatively rigid surface at even a very slight angle. Consequently, consideration must be made for the existence of high vertical deceleration loads and fuselage fracture in aircraft accidents consisting of primarily longitudinal impact velocity components.

The crash environment for helicopters provides a high potential for roll-over because of the vertical location of the center of gravity and because of the turning rotor. Rotor strikes on trees or other obstacles tend to flip the aircraft on its side. This can occur during descent or after initial impact, and it can occur either during a crash or it can cause a crash. Since more than half of the significant survivable accidents of rotary-winged aircraft now involve roll-over, lateral retention and strength and ceiling support strength in the occupied regions are of extreme significance.

Another environment hazard can be created by the main rotor of the helicopter. When the aircraft crashes, the rotor blades deflect downward, and, in addition to striking the ground and rolling the aircraft, the torsional and bending load transmitted to the rotor mast and transmission tends to tear this massive component free and displace it into the occupied sections of the aircraft. In addition, vertical deflection of a rotor at impact may permit the blades to pass through parts of the fuselage, creating an intrusion hazard for the occupants in those regions.

During a typical crash the aircraft structure progressively collapses at its crush strength. The total deformation distance of the structure is a function of the kinetic energy of the aircraft and the structure depth and strength. The decelerative loads transmitted to the occupied sections of the aircraft are thus reduced from those experienced by the contact point of the aircraft. Additional deformation distance reduces the loads experienced by the occupants and occupied portions of the aircraft and increased strength of the crushing section increases the efficiency of the system. The increase in crush strength is limited, of course, by the strength of the fuselage surrounding the occupants and the occupant tolerance.

Survivable accidents are those accidents in which forces transmitted to the occupant through his seat and restraint system do not exceed human tolerance, and in which the fuselage remains sufficiently intact so that it provides a protective shell within which an occupant could survive. Note that "survivable accidents" are a function of present-day fuselage strength. Therefore, injuries and fatalities occurring within this environmental regime need not have occurred if proper protective provisions, such as adequate seating and restraint systems, had been included in the interior design of the vehicle. Increasing the structural crashworthiness of the fuselage would also increase the severity of the defined survivable crash.

Figure 1 illustrates a nonsurvivable rotary-wing aircraft crash. The protective envelope has been totally destroyed during the crash and subsequent roll-over. The seats were torn free and are visible together with crewmen's helmets in the foreground of the picture.

Fig. 1. Example of a nonsurvivable crash.

In an effort to establish the crash environment of a survivable accident, an extensive study of accident records was accomplished. From this, the velocity and acceleration data were estimated and documented in Turnbow *et al.* [1]. The accident cases were limited to those in which one or more of the following conditions applied:

Substantial structural damage
Postcrash fire
Personnel injuries
At least one person survived the crash.

Crashes considered to be nonsurvivable such as midair collisions were not included, since the study was oriented toward determining the environmental conditions of a survivable crash. Sufficient data were gathered to enable a statistical evaluation to be made for vertical and longitudinal velocity changes and floor accelerations for rotary-wing and light fixed-wing aircraft as one group and fixed-wing transport aircraft as another group.

This study showed that, within the accuracy of the analysis, the crash environments of rotary-wing and light fixed-wing aircraft could be considered
identical except in the lateral direction. Figures 2 and 3 are shown as an
example of the results of this study. Figure 2 shows a plot of vertical and
longitudinal velocity change as a function of cumulative frequency of

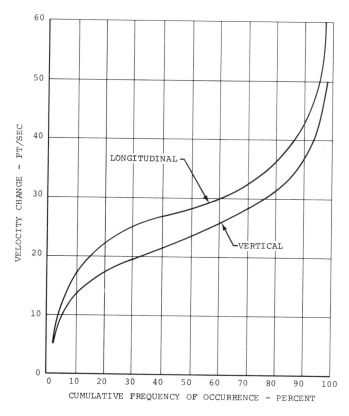

Fig. 2. Velocity changes for survivable rotary- and light fixed-wing aircraft accidents [1].

occurrence and Fig. 3 shows the same relationships for estimated peak
deceleration for rotary-wing and light fixed-wing aircraft.

It will be noted that survivable crashes have relatively mild velocity changes
and deceleration loads. The curves are approximately linear between the 20th
and the 80th percentile, but the tails of the curves vary exponentially.

The 95th percentile survivable crash conditions were selected as reasonable
design environments and are recommended for use in the design of retrofit
components such as seats, restraint systems, and other attached components

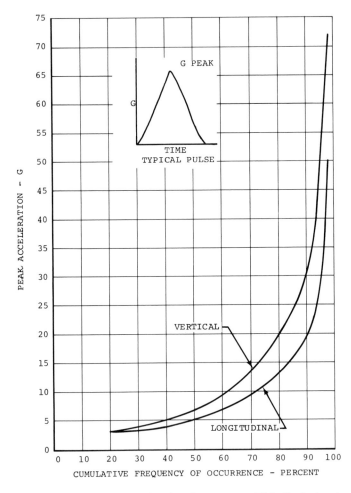

Fig. 3. Distribution of impact accelerations for rotary- and light fixed-wing aircraft.

in existing aircraft fuselages. Design environments for rotary-wing and light fixed-wing aircraft are presented in Table 1 and for large fixed-wing transport aircraft in Table 2. It should be emphasized that increased structural crashworthiness of aircraft fuselages and seating systems would, by definition, increase the severity of a survivable crash and also increase the severity of the 95th percentile design environment.

TABLE 1

SUMMARY OF DESIGN PULSES FOR ROTARY- AND LIGHT FIXED-WING AIRCRAFT[a]

Impact direction	Velocity change (ft sec^{-1})	Peak G	Pulse duration T second
Longitudinal (cockpit)	50	30	0.104
Longitudinal (passenger compartment)	50	24	0.130
Vertical	42	48	0.054
Lateral (fixed-wing)	25	16	0.097
Lateral (rotary-wing)	30	18	0.104

[a] From Turnbow et al. [1].

TABLE 2

SUMMARY OF DESIGN PULSES CORRESPONDING TO THE 95th PERCENTILE ACCIDENT OF FIXED-WING TRANSPORT AIRCRAFT[a]

Impact direction	Velocity change (ft sec^{-1})	Peak G	Pulse duration T second
Longitudinal (cockpit)	64	26	0.153
Longitudinal (cabin)	64	20	0.200
Vertical	35	36	0.060
Lateral (cockpit)	30	20	0.093
Lateral (cabin)	30	16	0.116

[a] From Turnbow et al. [1].

3. Injury Causes and Patterns

Crash injuries are caused by several different mechanisms, including:

Crushing
Impalement
Local body impact
Decelerative loading
Ejection
Postcrash hazards

The crushing injuries caused by fuselage collapse are caused simply by inadequate fuselage strength in the vicinity of the occupant. The crash loads collapse the fuselage and the occupant is crushed within the wreckage. Injuries are extensive and protection can be increased only by redesign of the fuselage itself.

Impalement type injuries are normally the result of partial collapse of structure. If tubular framework or thin structural bracing is designed to buckle inward rather than outward, it can then pass through the occupied area and impale the occupant. Impalement injuries can also be sustained by impact of an improperly restrained occupant with controls or other components of relatively small cross section located within the occupant strike zone. An example of this type of injury is the penetration of a non-energy-absorbing steering column into the chest of an inadequately restrained automobile driver.

Local impact injuries can be sustained in the same manner as impalement type injuries. First, the improperly restrained occupant can impact components within his strike range during high deceleration loading, or the unrestrained occupant can impact whatever structure intersects his path. Head impacts with instrument panels and side walls are typical examples of this type of injury. Another example is head and face impact of commercial airline occupants with the back of the seat located in front of them. This again is due to inadequate upper torso restraint or seats positioned too close to each other.

Another way of receiving these types of impact injuries is from other aircraft components penetrating the occupied section of the aircraft, main rotor blades and dynamic stabilizers on rotary-winged aircraft and inadequately restrained equipment or cargo which is released by decelerative loading during the crash.

Injuries caused by decelerative loading of the occupant can produce local injury of the occupant torso in the area of restraint, skeletal structural failure under the points of restraint, flexure or compressive failure of the spine, and massive hemorrhage resulting from ruptured blood vessels or massive organs being torn from their suspension. Human tolerance to decelerative loads is increased through proper restraint system and seat design. Increasing the contact area of the restraint system, maintaining its position over skeletal strong points, and providing energy-absorbing systems which permit controlled movement thus reducing the decelerative loads are methods which can be used to reduce the frequency and severity of this type of injury.

Ejection injuries usually fall into the general categories already mentioned, depending of course on the particular situation. Occupants are sometimes ejected from aircraft and impact the ground, buildings, or other vehicles. This results in either excessive overall decelerative loads or local impact injuries. On occasion, the occupant is thrown from the vehicle and then passed over, resulting in crushing injuries. Contact with jagged metal in passing through fractured fuselage structure can result in dismemberment which, together with massive lacerations (which in themselves can be fatal), can result in death due to loss of blood or shock.

If occupants survive the initial impact environment and are incapacitated due either to concussion, physical injury such as broken arms and legs, or entrapment in the interior of the vehicle, postcrash hazards such as fire and water can cause injury or death. In any case, very little time is available in typical aircraft crashes between the cessation of impact environment and the onset of fire, so that rapid egress must be made to escape this hazard.

A study was conducted by Mr. Joseph L. Haley, Jr., U.S. Army Board for Aviation Accident Research, Fort Rucker, Alabama [2]. Accidents of three different kinds of Army helicopters which occurred in 1967, 1968, and 1969 were studied to establish type of accident, accident kinematics, structural performance of the aircraft, and injury patterns. The types of helicopters studied were the Light Observation Helicopter (LOH), the Utility Helicopter (UH), and the Cargo Helicopter (CH). The LOH is 4-place, the UH is 13-place, and the CH is 36-place; all are turbine-powered vehicles.

It was shown in this study that 158 out of 2546 accidents were nonsurvivable. These 158 nonsurvivable accidents, 6.2% of the total, produced 59.8% of the fatalities. However, 40.2% of the fatalities occurred in survivable accidents, clearly indicating the need for improvement in crashworthiness of these aircraft. These data are summarized in Table 3.

TABLE 3
SUMMARY OF SURVIVABLE VERSUS NONSURVIVABLE ACCIDENT DATA

	Survivable	Nonsurvivable	Total
Accidents	2,388	158	2,546
Occupants	10,599	735	11,334
Total fatalities	439	655	1,094
Percentage	40.2	59.8	

The data summarized in Table 4 show that 5% of the occupants drowned, 29.9% died from thermal causes, 4% died from unknown causes, and 61.1% died from impact trauma, including 22.6% percent from head impact. Furthermore, the study showed that, of 11,334 occupants aboard these crashed aircraft, 1094 (9.6% of the total) received fatal injuries and 2699 (23.8% of the total) received nonfatal injuries. These data show that 33.4% of all occupants involved in the accidents studied received injuries.

TABLE 4
SUMMARY OF FATAL INJURY DATA

Cause	Percentage
Unknown	4.0
Drowning	5.0
Thermal	29.9
Impact trauma	61.1
Head impact	*22.6*

Among the 2699 injured, 774 received injuries to the head and/or face. In 592 of these cases, the head injury was the primary injury and the only injury in 275 cases. There were 175 fatalities (16% of the total) attributed to head and/or face injuries. These data illustrate the inability of present restraint systems to keep the head from impacting sidewalls or components within the strike range, or the nonusage of existing restraint systems.

Back injury was frequently sustained by occupants of rotary-wing aircraft; according to the study, 525 back injuries were sustained. These injuries were made up of 180 (34.4%) compression failures, 146 (27.8%) strain/sprain injuries, and 199 (37.9%) contusions, lacerations, or unknown.

A more detailed study of 1449 of the 2546 accidents revealed that the CH is much more hazardous than the UH and the LOH. The percentage of fatalities to total number of personnel aboard was 5.6 for the LOH, 9.1 for the UH, and 22.2 for the CH. Percentage of thermal to total fatalities for survivable and nonsurvivable accidents was 11.1 for the LOH, 32.8 for the UH, and 43.5 for the CH. Percentage of thermal to total fatalities for survivable accidents was 8 for the LOH, 40.2 for the UH, and 73.5 for the CH.

Figure 4, a summary of the percentage of occupants receiving various injuries, shows that one of the most significant causes of injury is nonuse of restraint systems, particularly in the CH. It further shows that crushing and entrapment are about comparable for UH and CH, but about four times more frequent than for LOH. It shows that about six times more injuries are caused by nonuse of restraint systems in UH than in LOH and about 29 times as many for CH.

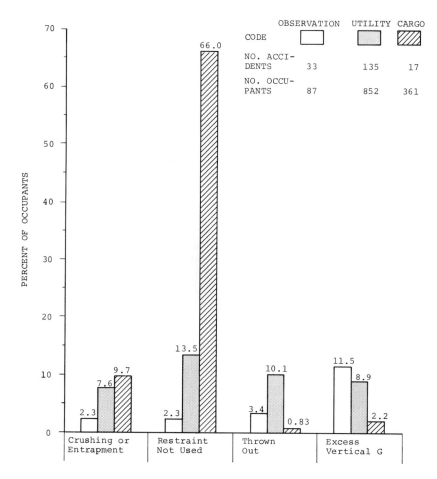

Fig. 4. Occupant injury experience, January 1967 through December 1969 [2].

Very few occupants are injured as a result of being thrown out of the LOH and the CH, but a significant number occur in the UH. Excessive decelerative loads are about five times more prevalent in LOH than in CH and about four times more prevalent than in UH. The data further show that, in the accidents investigated, LOH transmissions did not displace sufficiently to penetrate the occupiable area. In addition, postcrash fires occurred in only 9% of the 33 LOH crashes, whereas 33% of the UH crashes included post-crash fire and 82% of the CH included postcrash fire.

The most serious deficiency in the LOH fuselage appeared to be the tendency for fracture between the cockpit and troop seats in high-speed water impacts.

Utility helicopter transmission displacement was noted in one out of four accidents and main rotor blade penetration of the occupied section of the aircraft occurred in one out of eight accidents. Transmission and main rotor blade penetration of occupied areas occurred in 44 of 135 accidents evaluated.

Figure 5 illustrates transmission displacement into the cabin of a UH. Troop seats which are located directly in front of the transmission support structure are now under the transmission, illustrating the hazard of inadequate transmission support.

Fig. 5. Example of transmission penetration of cabin.

An example of cockpit damage due to main rotor strike is shown in Fig. 6. As can be seen, this impact resulted in destruction of the entire forward and upper cockpit structure; however, the accident was classified as survivable because of the intact cabin. Penetration of livable space by the transmission or main rotor blades resulted in 15 fatalities and 22 injuries for 135 UH accidents studied in detail.

Displaced transmissions in UH are not only a missile hazard but also create a postcrash fire hazard by rupturing hydraulic and lubricating oil lines and electrical wiring. Some troops in UH expressed fear of being crushed or trapped by the transmission as an excuse for not wearing their lap belts. Since the large side doors on these helicopters are usually lost during impact, the majority of the cases shown in Fig. 4 were thrown out during roll-over impacts because of no restraint.

Fig. 6. Example of main rotor penetration into cockpit.

The study of CH showed that 66% of the personnel were unrestrained during the crashes studied, revealing that occupants of these aircraft simply are not using the restraint systems provided. Part of the reason for this is that the belts with which the troop seats are equipped are simply not convenient to use. Although this is not valid justification, incorporation of more convenient systems could increase the usage.

Transmission and main rotor blades penetrated the occupied volume of the CH at about the same frequency as the UH and were the cause of five fatalities. Fire occurred in 14 CH, the fuselage rolled in 8, and sideward impact of the fuselage occurred in 4 of the 17 accidents studied in detail. The relatively high number of fuselage penetrations by main rotor blades or transmissions may be one of the causes for the high percentage of postcrash fires in CH as numerous lubricating, hydraulic oil, and electrical lines are located in the path of these components.

4. Crash Survival

4.1. General

Crash survival requires that the vehicle occupant both survive the impact and be able to evacuate the vehicle before the postcrash environment hazards such as fire and water become intolerable. In turn, impact and postcrash hazard survival depend on several vehicle features, including the following:

Impact Survival
>   Crashworthy structure
>   Adequate restraint
>   Noninjurious environment
>   Crash force attenuation

Postcrash Hazard Survival
>   Fire protection
>   Ease of egress

In order of priority, maintenance of living space for the occupant and thus the requirement for a crashworthy structure must be considered first. If the occupant is killed by collapsing structure, reduction of impact forces on the occupant and elimination of fires are unimportant. Consequently, the structure of the aircraft must be designed to provide a protective envelope around the occupants throughout the crash sequence as a first requirement.

Next in order of priority is that the occupant be restrained in a manner that eliminates serious or fatal injury due to secondary impact with surrounding structure or by loose objects. The restraint system and the interior of the vehicle must be designed such that occupant movement within the compartment is arrested before impact occurs with controls or other potential injurious objects located within the occupied volume.

The crash force, of course, must be limited to within human tolerance as a function of the restraint system in use. This can be accomplished either by design of the structure or a design combination of structure, restraint, and seating systems. Since airframe collapse occurs when loads exceed the strength of the structure and since modern aircraft and automotive structural strengths normally fall within human deceleration tolerance limits when adequate restraint is used, the postcrash condition of the vehicle can be used as an indicator of potential survivability. Relative intactness of structure or portions of structure implies survivable deceleration loads were imposed and that occupant survival could be expected, dependent on the other variables mentioned previously.

Solutions to the postcrash hazards threat include the incorporation of crashworthy fuel systems such as presently included in the UH-1D/H aircraft and improvement of structural crashworthiness so that massive damage to fuel and oil lines and tanks is minimized. Other design provisions can also protect the occupant for a limited period of time or aid in his rapid egress to help reduce the hazard toll. These provisions include fire-resistant clothing, gloves, helmets, and face shields which provide the wearer with additional protection during emergency egress. Design of the aircraft to minimize the number of disabling injuries which can be sustained by the occupant will increase his mobility after the crash and reduce casualties from fire and water. In addition, proper design of escape paths through planned placement of seats, design of restraint system releases, and the placement of exits fall into the general area of structural design for improving the probability of occupant survival.

4.2. STRUCTURAL CRASHWORTHINESS

The capability of a vehicle structure to provide a protective envelope during an accident depends to a great extent on its strength, mass distribution, total kinetic energy at the time of impact, and the distance over which the kinetic energy is dissipated. Combinations of kinetic energy and stopping distance exist which result in total disintegration of the vehicle, e.g., an aircraft flying into a mountainside as illustrated by the barrier crash test shown in Fig. 7. Combinations also exist where the decelerative forces are very small, e.g., those encountered during a well-executed belly landing of an aircraft on a prepared surface or in accidents where the kinetic energy is small. Obviously, crashworthiness is of little significance under these crash condition extremes. It is in the region of moderate to severe crash conditions that crashworthiness plays its useful role. These conditions encompass survivable accidents. Various aspects of improving chances for occupant survival in typical aircraft types are included in the following discussion.

(a) Light Fixed-Wing Aircraft

Light aircraft (General Aviation category; under 12,500 lb) have the highest accident rate and the highest frequency of serious injuries. Typically, in these aircraft the occupants are located far forward and have very little structure between the nose of the aircraft and the cockpit. The strength of the structure is inadequate to make efficient use of the distance available, and the strength of the fuselage is insufficient to maintain a protective envelope for the occupant. Furthermore, restraint systems in this type of aircraft are normally totally inadequate. A stall at any significant height can result in an almost vertical drop to the ground and usually results in fatalities.

What can be accomplished through designing for crashworthiness is illustrated by the design of a light aircraft for a specific agricultural use: aerial applicators for crops. In the early 1960s Aviation Safety Engineering and Research (now Dynamic Science) prepared a list of recommendations to improve crash survival of light aircraft. Since aerial application of sprays and powders to crops is extremely hazardous, the recommendations were first applied to these aircraft. The recommendations were as follows:

1. Locate the *cockpit*/cabin as *far aft as possible* in the fuselage and provide a large amount of energy-absorbing structure ahead of the occupants.

Fig. 7. Example of high energy barrier impact.

2. Design the *cockpit/cabin area* as the *strongest part of the fuselage* ("island of safety") in order to maintain the occupant's environmental integrity until the energy-absorbing action of surrounding structures is exhausted in progressive collapse.
3. Locate all *heavy components below and forward* of the cockpit/cabin to prevent crushing of the occupiable area by inertial loads.

4. Provide as much *space* as practicable *between* the *engine and cockpit* to allow aft displacement of the engine without affecting the integrity of the instrument panel or surrounding structure.

5. Locate the *fuel tanks as far away as* practicable from the *cockpit/cabin* and the engine to prevent ignition of spilled fuel and to increase escape time when postcrash fire does occur.

6. Design a *sturdy, smooth keel to reduce* abrupt decelerations resulting from *plowing effects* in low angle accidents and during heavy, flat impacts on soft terrain.

7. Design *roll-over structure or cabin roof* with sufficient strength to provide crash protection regardless of the direction of impact.

8. Provide the occupants with *seat belts, shoulder harness, and inertia reels* of sufficient strength to resist failure up to the point of complete cockpit/cabin collapse.

Figure 8 is a front quarter view of a specific aircraft designed with these recommendations in mind. The aft placement of the cockpit is apparent as is the low forward placement of the wings.

Fig. 8. Front quarter view of aircraft.

The revised nose section places more and stronger structure between the occupant and the probable point of impact and thus provides greater energy absorption ahead of the cockpit cabin area, reducing the loads imposed on the occupant and his protective envelope. Also, aft displacement of the engine will not force its immediate intrusion into the cockpit.

Placing as much weight as possible below and forward of the occupant, instead of behind reduces the loading on the protective envelope in which the occupant is located. This reduces the chance of cabin crushing from inertial loads. The sturdy, smooth keel reduces the scoop-forming tendency of buckled nose sections and decreases the tendency for gouging or plowing of earth during the crash. The result is a decrease in the decelerative loading associated with acceleration of masses of soil.

Increased strength of the cockpit area, of course, increases the ability of the structure to maintain protective envelopes around the occupants and the tubular structure over the cockpit provides structure capable of withstanding roll-over loads.

When adequate restraint systems are used in these aircraft, occupants have survived all but the most severe impacts. Figure 9 shows the postaccident

Fig. 9. Damage to forward fuselage section (right side view).

configuration of the forward section of the aircraft that was previously shown in Fig. 8. The aircraft stalled at about 200 feet and impacted the ground in a nose-down dive orientation [3]. In most other aircraft, this would have been a nonsurvivable accident; however, the occupant of this aircraft survived with moderate injuries. Notice the progressive deformation of the forward structure including the nose and the leading wing edges and that the cockpit is intact.

In a study of the old and new generation aerial applicator crash data, it was shown that the new generation aircraft had a better record. Where one out of four occupants of old generation aircraft received serious or fatal injuries, one out of fifteen were seriously or fatally injured in crashes of the new generation aircraft [4]. These data are summarized in Table 5.

TABLE 5

Injury Experience in Aerial Application Accidents, 1960–1961 (Taken from Reference 4)

Year	Acci-dents	Occu-pants	Injuries				Total injury (%)
			Fatal	Percentage	Serious	Percentage	
Old generation							
1960	235	235	32	13.6	23	9.8	23.4
1961	225	226	34	15.0	27	11.9	27.0
Total	460	461	66	14.3 Av.	50	10.8 Av.	25.2 Av.
New generation							
1960	34	4	0	0.0	3	8.8	8.8
1961	40	40	1	2.5	1	2.5	5.0
Total	74	74	1	1.4 Av.	4	5.4 Av.	6.8 Av.

The large mass located over the occupied section that is typical of high-winged aircraft can create a hazard when inertial forces drive the wing downward during a crash. This effect is emphasized in twin engine aircraft where the weight of the wings, fuel, and engines located on the wings may amount to half the gross weight of the aircraft [5]. In addition, the lower fuselage structure is generally not as strong as that in low-wing aircraft. This, of course, is due to the displacement of the wing spars and supporting reinforcement structure to the top of the fuselage. As a result, cockpit and cabin penetration and general collapse are more likely to occur in crashes of high-wing than in crashes of low-wing aircraft under similar impact conditions. Notice that the placement of the wings on the aerial applicator complies with the requirement for locating heavy components low and forward (Recommendation 3 on the list just presented).

The principles just discussed are directly applicable to the light fixed-wing category of aircraft; however, it should be apparent that these features are also generally applicable to most vehicles.

(b) *Fixed-Wing Transport Aircraft*

Fixed-wing transport aircraft with pressurized, cylindrically shaped fuselages inherently contain the desirable strength characteristics associated with maintenance of protective envelopes. These aircraft usually crash with a relatively low sink rate and a relatively high longitudinal velocity component. Because of the rapid direction changes required during nose-down impact with the ground, or during a slide when the aircraft body comes in contact with ground discontinuities such as hillocks, severe bending loads can be imposed on the relatively long fuselage. Typical crashes of this sort result in fuselage fracture immediately forward and/or immediately aft of the wing reinforcement sections. A typical fuselage fracture is shown in Fig. 10.

Fig. 10. Typical fuselage fracture.

Crash energy in an impact into a wall or a hill must be sustained by the nose of the aircraft, and the force must be transmitted down the body of the aircraft. The overall result is that, from the standpoint of impact survival, the tail section is the safest place of the aircraft in which to sit. Since the greatest weight of the aircraft is concentrated in the wing, it can be expected that the section of fuselage forward of the wings will have to absorb most of the crash energy. The logical recommendation would be to place all occupants in the fuselage section aft of the wings, but practical and economical considerations prohibit this. As a partial solution, however, occupants should not be placed over those sections of the aircraft in which fracture can be expected.

Some factors which would increase the structural crashworthiness of this aircraft category are as follows:

Placing occupants in areas other than those that are expected to fracture in the event of a crash.

Placing occupants in areas away from strong underfloor structure connected to the belly of the aircraft.

Increasing the space between seats.

Adding upper torso restraints or installing aft-facing seats.
Increasing seat and tie-down strengths.
Increasing the ease of egress.
Reinforcing the fuselage under wings in high-winged aircraft.

Placing occupants in areas other than immediately in front of and immediately behind the wing would minimize the hazards associated with fracture such as loss of seat retention resulting in occupant crushing or forcible contact with the jagged metal formed by the fracture. Galleys, closets, and lavatories could be placed in these areas instead of occupants.

The wing spars and reinforcements extend from the floor of the aircraft to the belly in some smaller low-winged aircraft. In a crash, maximum forces felt by the belly of the aircraft can be transferred without reduction directly to the floor of the aircraft and thus to the seat of the occupant located in this area. Placement of the occupant seats in areas where structural deformation can take place between the belly and the floor will thus reduce the impact environment which must be attenuated by the seat or sustained by the occupant. Figure 11 presents a sketch of the cross section of the 707 and 747

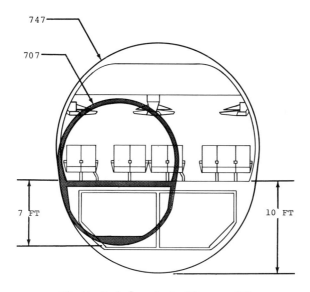

Fig. 11. Underfloor depth: 747 versus 707.

superimposed. The deformation distance is increased from around 7 feet to 10 feet. This trend also applies, of course, to the length of the aircraft and indicates that, with respect to impact trauma, the larger aircraft is the more crashworthy. Fuselage integrity, however, remains to be evaluated.

Increasing the space between the seats would reduce the injurious environment by placing the forward seat back out of the strike zone of the rear seat occupant. This would reduce the number of face and head impacts during longitudinal deceleration.

The severity of the environment which the occupant can tolerate would be increased substantially by the addition of upper torso restraint harnesses. This harness not only distributes the load over a larger body area, but also restrains the occupant from jackknifing over the lap belt. This, of course, would reduce the number of head strikes, neck injuries, torso injuries, and back injuries resulting from jackknifing and impacting either the seat in front or the side of the aircraft. This same effect, of course, could be accomplished more efficiently by use of an aft-facing passenger seat as the back of the seat would then support the body. Aft-facing seats maximize contact area, thereby decreasing the injury potential and increasing the severity of the environment that would be tolerable to the occupant.

Increasing the strength of the seat and tie-downs is needed to ensure maintenance of restraint. This would not only increase protection to the individual seat occupant but to the seat occupants forward of the freed seat. Increasing the space between seats, of course, would also improve the ease of egress and permit a faster and easier exit from the aircraft after impact.

Reinforcing the fuselage section strength under wings in high fixed-wing aircraft would increase the capability to maintain a protective shell for occupants. To achieve the degree of protection required, however, might require both reinforcement of the fuselage and a breakaway wing concept to reduce the load. Permitting large portions of the wings to break away at a certain vertical decelerative load would reduce the force the supporting structure would be required to carry, thus easing the fuselage strength requirement.

(c) *Rotary-Wing Aircraft*

The vertical take-off and landing (VTOL) capability of rotary-wing aircraft provides them with a unique usage flexibility. Consequently, rotary-wing aircraft are used on many varied missions including heavy lift or crane applications, troop transport, observation, supply, and rescue. The variations in use result in many differently constructed aircraft. These aircraft typically are equipped with large, loosely connected doorways and windows and very little overhead support. This is particularly true for utility-type aircraft. Furthermore, the location of the main rotor and mast makes it efficient to locate heavy components such as transmissions and engines overhead.

Some factors which would increase the structural crashworthiness of rotary-wing aircraft are as follows:

Transferral of mass from the top of the fuselage to the cockpit or cabin floor.

Localized strengthening of supporting structure and attachments at locations of large concentrations of mass attached to upper structure.

Design of cockpit and cabin structure to increase elastic energy absorption or provide for plastic energy absorption at loads less than the general collapse load of the primary cabin structure.

Increasing the strength of the roof-supporting structure.

Use of stronger doors and more reliable sidewalls in the fuselage.

Use of energy-absorbing structure in the subfloor and side wall structure.

Use of energy-absorbing landing gear.

Use of energy-absorbing seats.

Use of improved restraint systems.

Transfer of mass from the top of the fuselage to the cockpit or cabin floor will reduce the inertial load on the fuselage and thus help reduce the strength requirement necessary for the fuselage to maintain a protective envelope around the occupant. Obviously, all of the mass cannot be removed; however, depending on the design of the aircraft, as much as possible should be displaced to the floor.

Localized strengthening at locations of large concentrations of mass should be accomplished to assure adequate retention. This would include transmissions and/or engine tie-down supports.

Fig. 12. Example of fuselage crushing due to deceleration of massive overhead structure.

Figure 12 demonstrates the crushing damage that can be caused by decelerative loads of massive overhead objects on supporting fuselage structure.

Because of the high sink rate associated with helicopter crashes, fuselages must be capable of withstanding large vertical decelerations. Cockpit and cabin structures, therefore, that can absorb elastic and/or plastic energy at loads less than the protective envelope strength will increase survivability.

Again, because of the relatively large sink-rate crash velocity and tendency for side impact of helicopters, as much energy-absorbing material as possible should be incorporated into the subfloor and side-wall structure. The energy-absorbing material must be deformable at loads below the general cabin collapse load in order to maintain the protective envelope around the occupant while reducing the loads transmitted to the seats.

This energy-absorbing structure must have the properties required to carry its operational loads, but should be capable of plastic deformation when loaded in the direction of its thickness. One concept that has been proposed as an example of such a structural member is shown in Fig. 13. The concept shown is a beam that plastically deforms by first shearing rivets and then deforming the beam webs in the die formed in the beam base.

As stated previously, the roof is inadequately supported in many helicopters primarily because of the large side openings provided in the fuselage. The supporting structure should be sufficient to withstand the decelerative loads imposed without extensive structural collapse which would decrease the possibility of maintaining the protective envelope.

Use of energy-absorbing landing gear would reduce the loads transmitted to the fuselage section and thus reduce the structural requirements. Stronger doors and sidewalls would help support the roof structure, provide a structure within which energy-absorbing material can be placed, and provide the side walls for the protective envelope in which the occupant is restrained. This latter consideration is important, since quite frequently an inadequately restrained occupant protrudes from the aircraft after loss of doors and receives serious or lethal injuries as a result.

Energy-absorbing seats and improved restraint systems would lower the loads transmitted to the seat occupant and provide him more adequate restraint in all directions including the lateral direction.

(d) *Structural Dynamics Computer Simulation*

Mathematical models have been developed and programmed on computers for studying the effects of variables on structural crashworthiness. One such program was developed by Dynamic Science under contract to the U.S. Army Air Mobility Research and Development Laboratory (USAAMRDL) and was used to study the UH-1 helicopter [6]. The program developed was

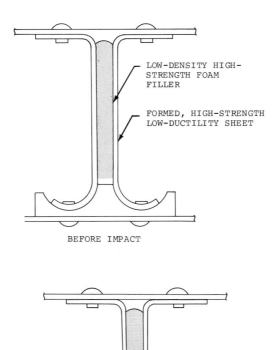

LOW-DENSITY HIGH-
STRENGTH FOAM
FILLER

FORMED, HIGH-STRENGTH
LOW-DUCTILITY SHEET

BEFORE IMPACT

AFTER IMPACT

Fig. 13. Cap and web combination beam design with potential energy-absorbing capability [1].

capable of calculating the dynamic response of the helicopter airframe subjected to vertical crash loading. The model was a nonlinear lumped mass model of 23 degrees of freedom composed of 14 lumped masses connected by 31 springs as shown in Fig. 14. The solution program generated both plotted and tabulated data for 126 different variables.

After development, the model was employed to simulate the response of the UH-1D helicopter to vertical loading. The resulting theoretical data were checked against empirical data measured in a full-scale drop test of this aircraft. After the program was adjusted by use of the empirical data, final simulations for this aircraft were calculated and served as a base line for a parametric study in which the load-limiting characteristics of various portions of the aircraft were evaluated.

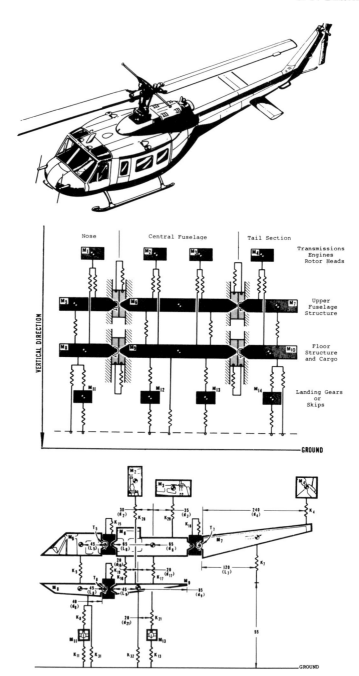

Fig. 14. Crash dynamics simulation model for UH-1D/H helicopter [6].

The results of the parametric study showed that a significant reduction in both floor and transmission accelerations could be achieved by increasing the strength of the landing gear system and redesigning the fuselage belly to provide additional depth of crushable material. Figure 15 shows the percentage of reduction in floor deceleration for improving the landing gear, the fuselage belly structure, and both simultaneously. It can be seen that incorporation of both modifications are predicted to result in a 65% reduction in floor deceleration.

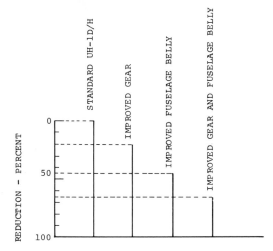

Fig. 15. Floor acceleration comparisons (30 ft sec^{-1} vertical impact) for UH-1D/H [6].

It is apparent that the development and use of accurate models can result in the development of optimum criteria for the design of vehicles without the requirement for extensive testing. Care must be taken, however, in the choice of the model, depending on the desired use of the program. Very rigorous analytical models may require such extensive running time that they would not be economically advantageous for use as parametric analysis tools. On the other hand, less rigorous models designed for parametric analysis would be unsuitable for detailed prediction of structural response.

5. Conclusions

Conclusions that can be drawn include the following points:

1. Major injuries and fatalities occurring in accidents rated as survivable are indicative of the need for increased emphasis on designing for structural crashworthiness.

2. Vehicles can be designed to provide crash protection to the occupant at little or no penalty to the weight and cost of the vehicle.

3. Vehicle crashworthiness can be improved by providing the design principles to designers and by emphasizing the importance of this aspect of design responsibility. Decisions and design choices can then be influenced and in cases where two or more comparable choices are possible, the choice in favor of improved crashworthiness can be made.

4. Overall vehicle structural crashworthiness requires design for crash loads with an objective of maintaining a protective envelope around the occupant and reducing the crash loads transmitted to him by controlled deformation of surrounding structure.

5. Seating and restraint systems should be designed to provide adequate restraint in all loading directions and to minimize decelerative loading of the occupant.

6. Seating and restraint systems should have the strength required to remain in place until the surrounding structure collapses.

7. Some analytical procedures are available and some are being developed which can be used to evaluate and optimize structures and subsystems for occupant survival. Additional tools are needed to enable overall systems to be optimized for both structural crashworthiness and mission performance.

8. Creative innovation is needed to develop structural design components and concepts that can provide the stiffness and strength needed to perform the primary design function, but that will efficiently and progressively deform with fracture during crash loading.

References

1. Turnbow, J. W. *et al.*, "Crash Survival Design Guide," Dynamic Science, USAAVLABS Tech. Rep. 70-22, U.S. Army Aviation Mater. Lab.,* Fort Eustis, Virginia (1971) (latest edition to be published).
2. Haley, J. L., "Analysis of Existing Helicopter Structures to Determine Direct Impact Survival Problems," U.S. Army Board for Aviation Accident Res., Fort Rucker, Alabama (1971).
3. "Crash Injury Investigation, S-2B Snow Aerial Applicator Aircraft Accident," Aviation Crash Injury Res., Phoenix, Arizona (October 18, 1961).
4. Bruggink, G. M. *et al.*, "Injury Reduction Trends in Agricultural Aviation," *Aerospace Med.* **35**, No. 5 (1969).
5. "Crash Survival Investigation Textbook," Dynamic Sci., A Division of Marshall Ind., The "AvSER" Facility, 1800 W. Deer Valley Drive, Phoenix, Arizona (October 1968).
6. Gatlin, C. I. *et al.*, "Analysis of Helicopter Structural Crashworthiness," Volumes I and II, USAAVLABS Tech. Reps. 70-71A and 70-71B. U.S. Army Air Mobility Res. and Develop. Lab., Fort Eustis, Virginia, AD 880 680 and AD 880 678 (January 1971).

* Now U.S. Army Air Mobility Res. and Develop. Lab.

Computational Fracture Mechanics

J. R. Rice and D. M. Tracey

BROWN UNIVERSITY
PROVIDENCE, RHODE ISLAND

Some areas of fracture mechanics which are being developed through computational stress analysis methods are surveyed. These include the numerical determination of elastic stress–intensity factors, the elastic–plastic analysis of near crack deformation fields, three-dimensional analysis of cracked bodies, and the description of fracture mechanisms on the microscale.

In addition, finite-element procedures are presented for the accurate numerical determination of elastic–plastic fields in the immediate vicinity of a crack tip. These are based on asymptotic studies of crack tip singularities in plastic materials, the results of which are summarized here and further extended for the nonhardening case. A new finite-element is presented which allows the requisite crack tip opening and associated $1/r$ shear strain singularity for this case, but with strictly nonsingular dilatation. This is employed in the elastic–perfectly-plastic solutions for small-scale plane strain yielding at a crack tip, and for yielding from small scale to limit load conditions in a circumferentially cracked round bar. Resulting numerical solutions are shown to be in excellent accord with analytical predictions, and parameters of the near tip field of interest in developing a fracture criterion are discussed.

1. Introduction

Current fracture mechanics research is focused in two principal directions: the development of phenomenological explanations of crack extension behaviors, and the description of micromechanical processes of material separation on the microscale. Both have come to rely strongly on computational methods of stress analysis.

In the first, the goal is to correlate crack extension behavior in subcritical growth by fatigue or stress corrosion, or in critical growth due to an overload, in terms of parameters from analytical solutions which characterize the near tip stress field. Elastic fracture mechanics is a case in point: When the crack extension behavior of interest is accompanied by a small crack-tip plastic zone, in comparison to crack depth and uncracked dimensions of a flawed specimen or structure, the correlation is in terms of the elastic stress–intensity factor. This is the coefficient of the inverse square root crack tip singularity in an elastic stress field. It serves to characterize the influence of applied loads and flaw geometry on the near tip field for such small scale yielding conditions, even though the predicted elastic stress field is wrong in detail within the plastic region.

Hence the analytical problem in elastic fracture mechanics is to determine the stress–intensity factor. Several numerical methods have been developed for this, including boundary collocation, numerical solution of integral equations, and finite elements. There is now a substantial literature which we shall review briefly here.

Plasticity effects limit this approach, and there is much current work on attempting to define and make use of parameters from elastic–plastic solutions which might similarly characterize the near crack tip field. This regime must be understood not only to deal with flawed structures failing under large scale yielding conditions, but also to allow fracture test results on small (and hence often fully plastic) precracked laboratory specimens to be accurately interpreted for assessing the safety of a flawed structure under nominally elastic conditions. Analysis in this elastic–plastic range is based principally on finite-element methods. These must, however, reveal sufficient detail on a fine scale at the crack tip, and for this reason it is necessary to take special precautions in the design of near tip finite elements.

Our approach is based on using asymptotic studies of elastic–plastic crack tip singularities as a guide to the development of displacement assumptions within elements. Previous investigations of this type are reviewed, and a new finite-element is described which allows the $1/r$ shear-strain singularity appropriate to the nonhardening idealization. Application of this element to the plane-strain small-scale yielding problem and to the circumferentially cracked round bar problem leads to highly accurate descriptions of the near tip field, and these may be useful in a phenomenological assessment of counterparts to the elastic stress–intensity factor in correlating fracture behavior.

There is, however, no single parameter which can uniquely characterize the near crack tip field in the large-scale yielding range, especially when prior stable crack advance under increasing load must be considered. Hence studies of fracture on the microscale are of significance not only for basic under-standing and as guides to alloy design, but also for suggesting suitable crack extension criteria to employ in flaw stress analysis and test correlations at the macroscopic level. Very much remains to be done in clarifying the mechanics of separation processes on the microscale, and in merging models at this level with macroscopic crack stress analysis for fracture prediction. We discuss the work to date in these areas and point out some of the challenging computational problems of plastic deformation, finite strain, and instability which appear at the microstructural level.

Our paper is divided into sections on the numerical determination of elastic stress–intensity factors in two-dimensional problems; crack tip plasticity, singular finite-element formulations, and results; three-dimensional crack problems, especially surface flaws; and fracture mechanics problems on the microscale. For a general background on analytical aspects of the subject, the reader may wish to consult the review papers by Paris and Sih [1], Rice [2], and McClintock [3].

2. Numerical Determination of Elastic Stress Intensity Factors (Two-Dimensional Problems)

The stress field at the tip of a sharp crack in an isotropic, linear, elastic material under loading conditions symmetric about the crack surface (Mode I) contains a stress singularity of the form

$$\sigma_{rr} + \sigma_{\theta\theta} \rightarrow 2K(2\pi r)^{-1/2}\cos(\theta/2)$$

$$\sigma_{zz} \rightarrow 2vK(2\pi r)^{-1/2}\cos(\theta/2) \qquad (1)$$

$$\sigma_{\theta\theta} - \sigma_{rr} + 2i\sigma_{r\theta} \rightarrow iK(2\pi r)^{-1/2}\sin(\theta)e^{i\theta/2}$$

where (r, θ, z) is a cylindrical polar system with origin lying at the point of interest along the crack front, with the z direction parallel to the crack tip, and with $\theta = \pm\pi$ on the crack surfaces (see Fig. 1). Here i is the unit imag-inary number, and K is the stress–intensity factor. This same stress distribu-tion with $\sigma_{zz} = 0$ applies to thickness averages in the simplest two-dimensional theory of generalized plane stress. The intensity factor is the parameter on which elastic fracture mechanics is based, and hence there is considerable interest in its numerical determination. We review some numerical methods for determining K here for two-dimensional problems of plane strain and

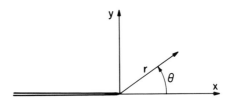

Fig. 1. Coordinates for description of near tip stress states.

generalized plane stress. Three-dimensional problems are discussed in a subsequent section.

The numerical methods may be divided broadly into those based on analytical representations of solutions (principally through analytic function theory) and those based on finite-element methods. Some of the former are limited as to the class of problems which may be handled, whereas the usual accuracy problems near singularities arise with the latter, and must be circumvented.

2.1. Boundary Collocation

Muskhelishvili's stress functions take the form [2]

$$\phi'(\zeta) = \zeta^{-1/2} f(\zeta) + g(\zeta) \tag{2}$$

$$\psi'(\zeta) = -\phi'(\zeta) - \zeta \phi''(\zeta) + \zeta^{-1/2} f(\zeta) - g(\zeta)$$

for a single straight Mode I crack penetrating in from the boundary of a body, where $\zeta = re^{i\theta} = x + iy$ and where the functions f and g are analytic everywhere within the body, including points along the crack line. For an internal Mode I crack of length l, the same form applies with $\zeta^{-1/2} f(\zeta)$ replaced by $\zeta^{-1/2}(\zeta + l)^{-1/2} F(\zeta)$, where $F(\zeta)$ is again analytic everywhere within the body, including the crack line. It is rigorously true that f (or F) and g have expansions of the form

$$f = \sum_0^\infty a_n \zeta^n, \qquad g = \sum_0^\infty b_n \zeta^n \tag{3}$$

where a_0 is expressible as $(8\pi)^{-1/2} K$, which converge in a neighborhood of the crack tip up to a radius equal at least to that of the nearest portion of external boundary, or of some other singularity of the problem.

The boundary collocation method as employed by Gross et al. [4–6] for edge cracks, and in the modified form by Kobayashi et al. [7] for an internal crack, in rectangular specimens adopts truncated power series in ζ for f and g. These are assumed to apply everywhere within the body, and the coefficients are chosen to match imposed stress conditions at discrete points of the external boundary. Commonly an excess number of collocation points are chosen so that an overdetermined system is obtained which is then solved in the sense of obtaining a least square minimization of the total error over the discrete points.

The method is attractive because it automatically satisfies traction-free boundary conditions on the crack surfaces. There does remain, however, a question in need of resolution as to the limitations set by the limited radius of convergence of complete power series for f and g.

2.2. Approximate Conformal Mapping

Another general method that has been used to obtain crack solutions is that of approximate conformal mapping, which may be applied to cracks emanating from holes in infinite bodies or to edge cracks in simply connected bodies. The technique involves finding accurate polynomial approximations to the mapping function which transforms the physical cracked domain into a circular region. The motivation to mapping is the fact that if a map of the form of a polynomial or ratio of polynomials is available, the stress functions expressed in terms of the auxiliary plane complex spatial variable can be obtained exactly by solving a finite system of equations. Bowie [8–10] treated the problem of an isolated circular hole with radial cracks and edge-notched strips using this method. Kaminskij [11] considered the case of isolated elliptical holes weakened by edge cracks. The stress–intensity factor may be unambiguously defined if the approximate polynomial mapping is chosen to keep the crack tip sharp (as may be done, whereas other types of corners must be rounded by such an approximation).

Bowie and Neal [12] have used a hybrid mapping, collocation technique to treat the doubly connected circular disk with an internal crack. The procedure is to choose a simple function which maps a circle and its exterior onto the crack and its exterior. Truncated series for the stress functions in the auxiliary plane are chosen with the coefficients determined by collocating on the mapped external boundary. Along with stress and displacement, force and moment boundary values are used in the collocation.

2.3. Integral Equations

A crack may be represented as a continuous distribution of dislocations. Rice [2] has outlined a method whereby the solution for an isolated dislocation in a body may be used to generate a singular integral equation governing the crack problem, and has shown how the equation is reduced to a regular Fredholm equation which may be solved numerically in a straightforward fashion. Related methods have been employed by Grief and Sanders [13] for a stringer reinforcement on a cracked plate and by Bueckner [14] for edge cracks.

More generally, solutions to plane elasticity problems may be given in the form of integrals, taken around the boundary, of fundamental singularities times unknown weighting functions, with the resulting singular integral equations being solved numerically. Cruse [15] and Cruse and Van Buren [16] have made effective use of this method for the numerical solution of three-dimensional elastic-crack problems; no results seem yet to be available on the application of the method to plane problems. Tirosh [17] has solved some crack problems for Mode III (antiplane strain) deformation through a related technique, but with an important modification which assures crack tip accuracy: For the fundamental singularity he chooses the field of an isolated dislocation in an infinite body with a semi-infinite crack. This automatically leaves the crack surfaces traction free, and one has to deal only with an integral equation along the external portion of the boundary.

2.4. FINITE-ELEMENT METHODS

The usual failure of numerical methods near singularities such as a crack tip requires either the use of special crack-tip finite-elements which embed the inverse square root singularity or, if standard elements with polynomial interpolation functions are used, the use of indirect procedures such as extrapolation to the tip or energy release methods.

Chan et al. [18] discussed extrapolation methods of determining K from constant stress triangular elements. The procedure is to plot the product of $r^{1/2}$ with some stress component (say $\sigma_{\theta\theta}$), as a function of distance along some ray emanating from the tip, and to extrapolate this as a smooth curve to the tip so as to estimate K. The result is, of course, generally quite different from the value which would be estimated based on stresses in the elements nearest the tip. Alternatively, the exptrapolation may be based on a product of $r^{-1/2}$ with a displacement, making use of the known displacements associated with the stress singularity (e.g. Rice [2])

$$u_x + iu_y = K/2G(r/2\pi)^{1/2}(\kappa - \cos\theta)e^{i\theta/2} \tag{4}$$

Here G is the shear modulus, and $\kappa = 3 - 4v$ for plane strain, or $(3 - v)/(1 + v)$ for plane stress, where v is Poisson's ratio. Chan et al. reported an agreement within 4–5% of known solutions for K, when approximately 2000 degrees of freedom were used, by using extrapolations based on $\sigma_{\theta\theta}$ directly ahead of the tip or on u_y along the crack surface.

Alternatively, K can be determined from a calculation of the decrease in potential energy of a body due to an increase in crack length, with applied loads and displacement constraints remaining fixed. For plane strain the relation between K and the energy release rate dP/dl is given by

$$dP/dl = -(1 - v^2)K^2/E \tag{5}$$

Hayes [19] used this method in treating a variety of configurations with simple triangular elements to obtain about 5% accuracy when 1000 degrees of freedom were allowed. His program was written to change the crack length automatically after a solution, by successively canceling reaction forces on nodes ahead of the crack. His use of an overrelaxation equation solver made this an efficient scheme, since the master stiffness matrix is only slightly perturbed in the process.

A related method discussed by Chan et al. [18] calculates the energy release rate without the necessity of actually re-solving a new problem for a slightly extended crack. This is done through Rice's J integral [20]:

$$J = \int_\Gamma [(W - \sigma_{xx}\, \partial u_x/\partial x - \sigma_{xy}\, \partial u_y/\partial x)\, dy + (\sigma_{yx}\, \partial u_x/\partial x + \sigma_{yy}\, \partial u_y/\partial x)\, dx] \quad (6)$$

Here W is the elastic strain-energy density ($=\frac{1}{2}\sigma_{ij}\varepsilon_{ij}$ for a linear material) and the path Γ on which the integral is taken is an arbitrarily chosen contour beginning at any point on the lower crack surface of Fig. 1, encircling the tip, and ending at any point on the upper crack surface. The integral has a value which is independent of the particular path chosen; there is no restriction that the material be *linear* elastic, but instead only that its stress–strain relations be consistent with the existence of a strain-energy function (i.e., $\sigma_{ij}\, d\varepsilon_{ij} = dW$, an exact differential). The physical interpretation of J is as the energy release rate [2, 20], and hence in the case of a linear elastic material

$$J = (1 - v^2)K^2/E \quad (7)$$

Chan et al. chose an integration path Γ for the integral lying far from the tip and coinciding with the boundary of their edge-cracked specimen. Resulting values of K, obtained through evaluating the integral from the triangular finite-element solution, were reported to have an accuracy essentially similar to that of the extrapolation method.

Other methods of solving for K have been discussed by Barone and Robinson [21] and Rice [22]. These methods use elastic reciprocity properties to formulate new boundary-value problems whose solution leads to a determination of K, but through calculations which do not require numerical accuracy in the near tip region.

The alternative to the preceding methods is that of directly embedding the elastic singular term in the displacement assumption for the near tip finite elements. Wilson [23] developed an axisymmetric ring element of circular cross section centered at the crack tip for investigation of the circumferentially cracked round bar under torsion. The stiffness of the element was formed by integrating the strain-energy density of the dominant $r^{-1/2}$ singularity over the circular cross section of the element. Conventional triangular ring elements covered the remainder of the mesh. The undetermined parameters in the formulation were K and the displacements of nodes not lying on the crack-tip element boundary. The same procedure is applicable to tension, and Hilton and Hutchinson [24] have followed a similar procedure in obtaining deformation plasticity solutions for cracks under in-plane deformation. Tracey [25] used a mesh composed of isosceles trapezoidal-shaped elements focused into the crack tip. The elements nearest to the tip had a $r^{1/2}$ variation of displacement specified, while the adjacent elements were treated as ordinary isoparametric elements. The near-tip interpolation function was designed to guarantee interelement displacement continuity. More details of this will be given in the section presenting our elastic–plastic numerical results.

3. Crack Tip Plasticity

As we have noted, studies on crack-tip plasticity are important for extension of the phenomenological fracture mechanics approach to the large-scale yielding range, and also for setting boundary conditions on models of microscale separation mechanisms at the crack tip. In both cases, rather detailed descriptions of stress and deformation on a size scale that is small compared to overall plastic region dimensions seem to be required. Lee and Kobayashi [27], Marcal and King [28], Swedlow and co-workers [29, 30], Wells [31], and others have presented finite-element solutions for yielding near cracks or sharp-tipped notches, and much has been learned from these concerning the growth and shape of the plastic region and transitional behavior from the elastic to fully plastic ranges.

However, we think it unrealistic to expect that standard finite-element methods will give the detailed results desired in the near-tip region, as will be more apparent with the ensuing discussion. For this reason, our own computational work has relied heavily on a merging of computer methods with what is known from asymptotic studies of crack tip singularities in plastic materials. Here we refer specifically to the papers of Cherepanov [32], Hutchinson [33, 34], Rice [2, 20, 35], and Rice and Rosengren [36], which have elucidated the structure of plane-strain and plane-stress singularities at crack tips, both for materials idealized as nonhardening and for power law strain hardening materials [i.e., (stress) $\propto$ (strain)^N in the plastic range]. Indeed, approximate small-scale yielding solutions have been given [20, 33, 34, 36] for these cases on the basis of a deformation plasticity formulation and the J integral. Previous papers by Hilton and Hutchinson [24] and Levy et al. [37] have made similar use of the asymptotic studies in finite-element analyses. The first of these introduced a circular near-tip element in which the dominant power-law-hardening singularity strain and displacement distribution is assumed, but with an unknown amplitude. Levy *et al.* introduced a different type of singular element, which we shall describe later; our work is a continuation and refinement of that method.

Nearly all the results of the asymptotic studies and computational treatments have been for stationary cracks, and one of the major unresolved problems of the field is in clarifying the elastic–plastic mechanics of quasi-static crack advance. McClintock [3, 38] and Rice [2] have discussed this type of problem, for which the history-dependent nature of plastic stress–strain relations and the feature of crack advance into previously deformed material lead to near-tip strain distributions which are very different from those for stationary cracks. For example, Rice [20] has shown that a $1/r$ shear strain singularity results in the regions above and below the tip of a *stationary* plane strain crack in a perfectly plastic material, whereas McClintock [38] showed that nonsingular strains result in the case of a continuously advancing crack in a *rigid*-perfectly-plastic material under increasing imposed displacements at its boundary, and Rice [2] showed that a logarithmic strain singularity resulted at the tip for conditions of steady-state crack advance in an *elastic*-perfectly-plastic material. This is an area in which much remains to be done toward developing computational accuracy paralleling that now attainable for the stationary crack case, and analyses of this type are of obvious importance for a correlation of fracture tests in which substantial stable crack extension precedes the running crack instability (e.g., cracks in thin, ductile sheets). On the other hand, the stationary crack model alone seems appropriate for the abruptly initiated fractures which frequently result under conditions of plane strain constraint at the crack tip.

3.1. CRACK-TIP STRESS FIELD

The numerical solutions that we report in the next section are for stationary cracks under plane strain (or nearly so) conditions at the tip, and the material is idealized as isotropic and elastic–perfectly-plastic (of the Mises type). Large geometry change effects on the form of governing equations are neglected, although we shall consider these in the section on micromechanics. For these cases, Rice [2, 20] has given approximate arguments for validity of the stress state of the Prandtl slip line field (Fig. 2) in providing the limiting stress state as $r \to 0$ at the crack tip for cases of contained plastic yielding. Also, Hutchinson [34] and Rice and Rosengren [36] have noted that this field is the limit, as the hardening exponent N approaches zero, of their dominant singularity solutions for crack tip stresses. However, it cannot be expected that in this limit the "dominant singularity" does in fact dominate, as has been learned from the Mode III case, and indeed the Hutchinson–

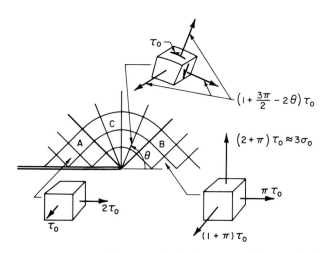

Fig. 2. Prandtl field as the limiting stress distribution as $r \to 0$, for contained plain strain yielding of a non-hardening material.

Rice–Rosengren equations have no unique solution for the strain distribution when $N = 0$.

Thus we reexamine the nature of the near crack tip field here both as motivation for our singular finite element formulation and as an extension of prior asymptotic studies of this type. In view of the boundedness of stresses in a nonhardening material it may be assumed that $r \, \partial\sigma_{ij}/\partial r \to 0$ as $r \to 0$, and hence the two stress equilibrium equations in polar coordinates take the form

$$\sigma_{rr} - \sigma_{\theta\theta} + \partial\sigma_{r\theta}/\partial\theta = 0, \qquad 2\sigma_{r\theta} + \partial\sigma_{\theta\theta}/\partial\theta = 0 \tag{8}$$

for the angular variation of the stress state at $r = 0$. The Mises yield condition $s_{ij}s_{ij} = 2\tau_0^2$, where s_{ij} is the stress deviator and τ_0 the yield stress in shear, and may be rewritten as

$$(\sigma_{rr} - \sigma_{\theta\theta})^2/4 + \sigma_{r\theta}^2 + (\sigma_{rr} + \sigma_{\theta\theta} - 2\sigma_{zz})^2/12 = \tau_0^2 \tag{9}$$

This is satisfied in any angular sector at the tip which is in the plastic state, and after differentiation with respect to θ and use of the equilibrium equations to simplify the result, we obtain (for $r = 0$)

$$6[\partial\sigma_{r\theta}/\partial\theta][\partial(\sigma_{rr} + \sigma_{\theta\theta})/\partial\theta] = \partial[(\sigma_{rr} + \sigma_{\theta\theta} - 2\sigma_{zz})^2]/\partial\theta \tag{10}$$

By the flow rule for plastic strain increments,

$$\dot{\varepsilon}_{ij}^p = (\dot{\varepsilon}_{kl}^p\dot{\varepsilon}_{kl}^p/2)^{1/2}s_{ij}/\tau_0. \tag{11}$$

Thus, if there is a strain singularity at the tip in the angular sector under consideration, we must have $s_{zz} = 0$ there because conditions of plane strain prohibit a singularity in ε^p_{zz}. This means that

$$\sigma_{zz} = (\sigma_{rr} + \sigma_{\theta\theta})/2 \tag{12}$$

at $r = 0$ in such a sector. We may also note that this last equation would be valid for a rigid–plastic material, and that it would be approached as a limit for an elastic–plastic material subjected to monotonically increasing plastic deformations. Hence Eq. (12) is strictly valid in an angular sector where there is a strain singularity, and it would seem essentially correct even in angular sectors for which the plastic strains are nonsingular, although the argument cannot in general be made rigorous in sectors of the latter type.

With Eq. (12) the right side of Eq. (10) vanishes, and hence the stress state at $r = 0$ satisfies either

$$\text{(a)} \quad \partial\sigma_{r\theta}/\partial\theta = 0, \quad \text{or} \quad \text{(b)} \quad \partial(\sigma_{rr} + \sigma_{\theta\theta})/\partial\theta = 0. \tag{13}$$

In sectors for which (a) holds, the equilibrium equations and yield condition lead to stress states of the form

$$\sigma_{r\theta} = \pm\tau_0, \quad \sigma_{rr} = \sigma_{\theta\theta} = \sigma_{zz} = \text{const} \pm 2\tau_0\theta \quad \text{in (a) sectors} \tag{14}$$

Stress fields of this type which apply over a finite range of r are known as "centered fans" in slip line theory, and these appear above and below the crack tip in the Prandtl field of Fig. 2. In those sectors where (b) holds, one finds, as a consequence of the equilibrium equations, that

$$\sigma_{xx}, \quad \sigma_{xy}, \quad \sigma_{yy}, \quad \sigma_{zz} \quad \text{are independent of } \theta \text{ in (b) sectors.} \tag{15}$$

That is, the stress components when referred to cartesian coordinates are constant at the tip in (b) sectors. These are known as "constant state" regions in slip line theory, and occur directly ahead of the tip and in regions adjacent to the crack surfaces for the Prandtl field.

Hence, to the extent that Eq. (12) holds, the stress distribution surrounding the crack tip over all angular ranges for which there is plastic yielding must be made up of centered fan sectors of type (a) and constant stress sectors of type (b). This holds for stationary as well as advancing cracks. If the entire angular range surrounding the crack tip is yielding, then the only possible stress distribution of this type, corresponding to a continuous stress variation with θ (as would be expected for problems of contained plastic yielding) and to symmetrical Mode I loading conditions, is that given by the Prandtl field of Fig. 2. Indeed, our computational results and those of Levy et al. [37] support this assertion of the Prandtl field as the limiting stress distribution as $r \to 0$ at least for problems of small scale contained yielding. However, it cannot be asserted a priori that Eqs. (13)–(15) govern in nonsingular plastic sectors.

Also, fully plastic flow fields of limit analysis may involve a variety of near tip stress distributions [2, 3], depending on overall geometry of the cracked specimen, and these may involve discontinuous stress distributions and nonyielding sectors, as well as discontinuous deformation concentration into shear bands emanating from the tip.

In comparison, Rice's [20] approximate treatment was based on simplifying the yield condition to the statement that the maximum in-plane shear stress is constant. This is the same as deleting the last term on the left in Eq. (9), so that Eq. (10) has zero on the right and thus Eqs. (13) and their consequences apply in *all* plastic sectors, whether singular or not. Within this approximation, all plastic sectors therefore consist either of centered fan or of constant stress regions. In the most general case, there may also be nonyielding sectors at the tip, although if the entire angular range is yielding and the stresses are continuous, then only the Prandtl field of Fig. 2 may result at $r = 0$. Note that when the Prandtl field is present the maximum tensile stress directly ahead of the tip is approximately $3\sigma_0$, where σ_0 is the tensile yield stress.

3.2. DISPLACEMENTS AND STRAIN SINGULARITY

We have remarked that near tip stress states, at least in singular regions, are familiar from slip line theory and indeed, for the approximation of the yield condition by the maximum in-plane shear criterion, slip line theory applies in all yielding regions. As Rice noted [20], a feature of a centered fan slip-line field at a stationary crack tip is that there is a nonunique displacement at the tip in the sense that a different displacement vector (u_x, u_y) results at $r = 0$ for each different ray of the fan along which the tip is approached. That is, the displacements at $r = 0$ vary with θ in the fan and hence there is a discrete opening displacement of the crack surfaces at the tip. Radial and circumferential lines are zero extension rate directions so that ε_{rr} and $\varepsilon_{\theta\theta}$ are nonsingular, the singular deformation consisting of a pure shear $\gamma_{r\theta}$ which becomes infinite as r^{-1}. Singularities result where slip lines focus to a point and thus the strain components are, in general, nonsingular in the constant stress sectors with a unique displacement resulting at $r = 0$ as the tip is approached through these sectors. There is also, however, a possibility of a sliding displacement discontinuity emanating from the tip along a slip line, and these frequently occur in limit flow fields [3]. The features of r^{-1} strain singularities, tip opening displacements, and lines of displacement discontinuity at limit load seem to be general features of crack tip fields in nonhardening materials, in the sense that they are also familiar from the Mode III case [3] and from the two-dimensional plane stress case [33, 34].

To study the near tip field in plane strain without recourse to its approximate representation in terms of slip lines, consider the polar coordinate strain components

$$\varepsilon_{rr} = \cos \theta \; \partial u_x/\partial r + \sin \theta \; \partial u_y/\partial r$$

$$\varepsilon_{\theta\theta} = r^{-1}(-\sin \theta \; \partial u_x/\partial \theta + \cos \theta \; \partial u_y/\partial \theta) \tag{16}$$

$$\gamma_{r\theta} = r^{-1}(\cos \theta \; \partial u_x/\partial \theta + \sin \theta \; \partial u_y/\partial \theta) - \sin \theta \; \partial u_x/\partial r + \cos \theta \; \partial u_y/\partial r$$

where it is convenient for this discussion to write the strain-displacement gradient relations as we have in terms of Cartesian displacements. We shall consider that the displacements may vary with θ at $r = 0$ and examine the restrictions placed on this variation by the flow rule. One does however expect bounded displacement components at the tip and thus it appears reasonable to assume that $r \; \partial u_i/\partial r \to 0$ as $r \to 0$.

Any plastic-strain singularity must conform to the incompressibility condition and since elastic strains are bounded we must therefore have $(\varepsilon_{rr} + \varepsilon_{\theta\theta})$ bounded at the tip. This means that $r(\varepsilon_{rr} + \varepsilon_{\theta\theta}) \to 0$ as $r \to 0$ and thus Eqs. (16) require that the displacements at $r = 0$ satisfy the constraint

$$\sin \theta \; \partial u_x^0/\partial \theta = \cos \theta \; \partial u_y^0/\partial \theta, \tag{17}$$

where we use the notation $u_i^0(\theta)$ for $u_i(r, \theta)$ at $r = 0$. This same restriction applies also to displacement rates $\dot{u}_i^0$. We therefore have $r\varepsilon_{rr}$ and $r\varepsilon_{\theta\theta}$ both going to zero at the tip. On the other hand, it is seen from Eq. (16) that as $r \to 0$

$$r\gamma_{r\theta} \to \cos \theta \; \partial u_x^0/d\theta + \sin \theta \; \partial u_y^0/\partial \theta = (\cos \theta)^{-1} \; \partial u_x^0/\partial \theta$$

$$= (\sin \theta)^{-1} \; \partial u_y^0/\partial \theta \tag{18}$$

where the last two versions on the right follow from Eq. (17). Hence $\gamma_{r\theta}$ (or $\dot{\gamma}_{r\theta}$) exhibits a singularity of strength r^{-1} in any sector for which the crack tip displacements (or rates $\dot{u}_i^0$) vary with θ.

Recalling our previous study of the crack tip stress state, in which singular sectors were shown to be either of the centered fan [Eq. (14)] or constant stress [Eq. (15)] type, we see from the flow rule [Eq. (11)] that the stress state in a fan sector at a stationary crack tip is consistent with a singularity dominated by the polar shear rate $\dot{\gamma}_{r\theta}$, and the crack tip displacement rates may indeed vary with θ in such sectors. On the other hand, such a singularity violates the flow rule and hence is inadmissible in a constant stress sector. Crack tip displacement rates therefore cannot vary with θ in such sectors, with one exception: If the constant stress state of the sector is such that for some angle θ within it $\sigma_{r\theta} = \pm\tau_0$, then it is possible to have a discontinuity in $\dot{u}_i^0$ at that angle with $\partial \dot{u}_i^0/\partial \theta$ vanishing elsewhere. In this case the jump version of Eq. (17) applies across the discontinuity. It is in fact possible to view such a discontinuity as a centered fan sector with a vanishing angular range, since $\sigma_{r\theta} = \pm\tau_0$ at this angle.

It is of interest to note that while equilibrium considerations required singular sectors to be either of the fan or constant stress type, flow rule considerations show that only the former can in fact exhibit the $1/r$ singularity. Neighboring sectors *may* be of the constant stress type but this is not necessarily so, except for the maximum in-plane shear approximation to the yield condition.

Our finite-element formulation incorporates the preceding features of the near-tip field, in that we choose displacement assumptions which permit a variation with θ at the tip subject to the constraint of Eq. (17). This allows a direct calculation of the crack tip opening displacement, of the angular strength of the strain singularity, and of the tip stress state. Detail at this level is of course unattainable from conventional finite-element formulations.

For reporting our results we shall use the standard notation [2, 20] for the strain singularity, writing Eq. (18) in the form

$$\gamma_{r\theta} \to \frac{\tau_0}{G} \frac{R(\theta)}{r} \tag{19}$$

where τ_0/G is the initial yield strain in shear, and where the angular strength of the strain singularity is written so that $R(\theta)$ may be interpreted as an *approximate* measure of the linear extent of the plastically strained region at angle θ. From the preceding discussion, R (or more precisely, $\dot{R}$) is nonzero only in fan sectors. The function R is related to the displacement components, for by comparing Eqs. (18) and (19) we see that

$$\partial u_x{}^0/\partial\theta = (\tau_0/G)R\cos\theta, \qquad \partial u_y{}^0/\partial\theta = (\tau_0/G)R\sin\theta \tag{20}$$

Also, by integrating the last of these, the opening displacement between the upper and lower crack surfaces at the tip is

$$\delta_t = 2(\tau_0/G)\int_0^\pi R(\theta)\sin\theta\,d\theta \tag{21}$$

When the Prandtl field applies, the limits are the angular range $\pi/4$ to $3\pi/4$ of the fan sector.

If one adopts a deformation theory of plasticity, which models the material as if it were nonlinear elastic, then the J integral remains path independent for contours Γ passing through the plastic region provided that the appropriate W is used. Taking the Prandtl field as the near tip stress state and shrinking Γ to the tip, Rice [20] showed that J could be expressed as

$$J = (2\tau^2{}_0/G)\int_{\pi/4}^{3\pi/4} R(\theta)[\cos\theta + (1 + 3\pi/2 - 2\theta)\sin\theta]\,d\theta. \tag{22}$$

For small-scale yielding, J retains its linear elastic value of Eq. (7), and this serves as the basis for some approximations.

4. Singular Finite-Element Formulation and Results

Finite-element incremental elastic–plastic solutions to the small-scale-yielding plane-strain problem and the large-scale-yielding of a circumferentially cracked tension bar are described in this section. As has been emphasized, the goal is to obtain reliable results at the crack tip singularity as well as globally. These nonlinear problems are linearized by specifying the load in small, finite steps and solving for the resulting deformation at each step by the tangent modulus approach. Within each increment an iterative scheme is adopted which allows convergence to the best representative constitutive matrices of yielded elements for the increment. The scheme is outlined here for the isotropic, perfectly plastic Mises material idealization; it could also serve as the basis for treatment of more general cases.

4.1. ITERATIVE PROCEDURE

To illustrate the procedure consider $\mathbf{s}^0$ as the deviatoric stress vector of an element at the beginning of a particular load increment. To remain general let $\mathbf{s}^0$ lie inside the yield surface. Estimate that the strain increment due to the current load increment is parallel to that of the solution for the previous load increment, and scaled according to increment size. Considering this estimate to be entirely elastic, calculate the corresponding fictitious final stress $\mathbf{s}^2$ $(= \mathbf{s}^0 + 2G\Delta\mathbf{e}^{\text{est}}$, where G is the elastic shear modulus and $\Delta\mathbf{e}^{\text{est}}$ is the deviatoric part of the strain increment estimate). Figure 3 is a π-plane

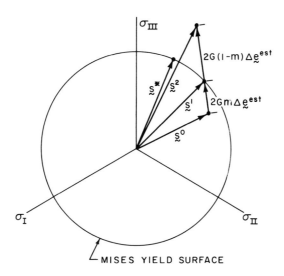

Fig. 3. π-plane view of stress states of an element during a load increment.

projection of the stress vectors and the Mises yield surface, which appears as a
circle of radius $\sqrt{2}\tau_0$ in this plane. The material would yield at s^1. For this
case of transition from elastic to elastic–plastic behavior within an increment
the partial-stiffness approach of Marcal and King [28] is followed. The
matrix $\mathbf{D}$ which relates deviatoric stress increment to deviatoric strain
increment is divided into elastic and elastic–plastic portions according to the
ratio $m = |s^1 - s^0|/|s^2 - s^0|$:

$$\mathbf{D} = m\mathbf{D}^{\text{el}} + (1 - m)\mathbf{D}^{\text{el–pl}} \tag{21}$$

$\mathbf{D}^{\text{el}}$ is the diagonal matrix $2G\mathbf{I}$ and $\mathbf{D}^{\text{el–pl}}$ is equal to $2G(\mathbf{I} - \mathbf{n}\mathbf{n}^{\text{T}})$, where $\mathbf{n}$
is the unit vector $s/\sqrt{2}\tau_0$ normal to the yield surface at the stress state $\mathbf{s}$.

Previous analyses [27, 28] have used the unit normal vector $\mathbf{n}^1$ at the initial
yield state s^1 to define $\mathbf{D}^{\text{el–pl}}$ for the increment. This corresponds to a simple
Euler integration of the actual constitutive "rate" equation over the elastic–
plastic portion of the increment. The large-load step sizes which are necessary
for computational economy for some problems make this an unacceptable
approximation due to the associated large stress increments at each step.
Hayes [19] specified $\mathbf{D}^{\text{el–pl}}$ for the increment by using the unit vector in the
direction of the estimated average stress vector predicted by the Marcal–
King procedure,

$$s^1 + G(\mathbf{I} - \mathbf{n}^1\mathbf{n}^{1\text{T}})(1 - m)\,\Delta\mathbf{e}^{\text{est}}$$

In the present scheme we use the unit vector $\bar{\mathbf{n}}$ in the direction of the
average of s^1 and s^2,

$$\bar{\mathbf{n}} = \frac{s^1 + s^2}{|s^1 + s^2|} = \frac{s^1 + G(1 - m)\,\Delta\mathbf{e}^{\text{est}}}{|s^1 + G(1 - m)\,\Delta\mathbf{e}^{\text{est}}|} \tag{22}$$

because this has the remarkable feature that the corresponding $\mathbf{D}^{\text{el–pl}}$ transforms $(1 - m)\,\Delta\mathbf{e}^{\text{est}}$ to a stress increment which, when added to s^1, results in
a final stress s^* which precisely meets the yield criterion. The strain estimates
are found to converge quite rapidly using this procedure so that the approximation within an increment is essentially the use of a secant between the
initial and final stress states to define the yield surface during the increment.

To prove that the stress state s^* lies on the yield surface when $\bar{\mathbf{n}}$ is used to
determine the plastic flow during the increment, we must prove that

$$s^{*\text{T}}s^* - s^{1\text{T}}s^1 = 0, \quad \text{or} \quad (s^* + s^1)^{\text{T}}(s^* - s^1) = 0 \tag{23}$$

Equation (23) is proved by demonstrating the orthogonality of the vectors
$(s^* + s^1)$ and $(s^* - s^1)$. The elastic–plastic constitutive relation based on $\bar{\mathbf{n}}$
sets $(s^* - s^1)$ normal to $\bar{\mathbf{n}}$,

$$(s^* - s^1) = 2G(\mathbf{I} - \mathbf{n}\mathbf{n}^{\text{T}})(1 - m)\,\Delta\mathbf{e}^{\text{est}}. \tag{24}$$

Adding $2\mathbf{s}^1$ to both sides of Eq. (24) and recognizing that $2\mathbf{s}^1 + 2G(1-m)$ $\Delta\mathbf{e}^{est}$ is parallel to $\bar{\mathbf{n}}$ by definition, we see that $(\mathbf{s}^* + \mathbf{s}^1)$ is parallel to $\bar{\mathbf{n}}$, and hence Eq. (23) is satisfied.

4.2. Element Design

Three types of finite elements were used in the analysis. For elastic solutions the $r^{-1/2}$ singular element [25] was used nearest the crack tip with arbitrary quadrilateral four-node isoparametric elements over the remainder of the configuration. To study plastic effects at the crack tip, a new singular element was designed, similar to that of Levy et al. [37], which has a $1/r$ shear strain singularity (with a bounded dilatational strain) and a uniform strain as admissible deformations.

The singularity elements have the shape of isosceles triangles and are focused along radial lines into the crack tip. However, they are treated as degenerate isosceles trapezoids in the sense that four nodes are assigned to the elements, one at each vertex, even though two of the nodes coincide at the crack tip. Levy et al. [37] introduced this coincident node technique to study the crack tip displacement variation. Contrary to their procedure, however, the coincident nodes were here constrained to move as a single point in obtaining the elastic response of the cracked body, since the nonunique crack tip displacement is a plasticity effect.

The variation of stress and hence constitutive relation in the plastic case within elements was accounted for in the following approximate manner. Each near-tip element was viewed as the composite of three subelements, each extending one-third of the height of the element. The area average strain of an individual sub-element was used in evaluating the stress state and constitutive matrix representative of the subelement. The three subelement stiffnesses were then formed and added to obtain the total element stiffness matrix. For the adjoining isoparametric elements the midpoint strain was judged adequate to calculate the stress representative of the entire element.

To obtain elastic–plastic solutions the procedure was to specify the $r^{-1/2}$ element just up to the load necessary to yield one of the subelements. Thereupon the r^{-1} element was used with its associated nonunique crack tip displacement capability. Clearly the elastic singularity implies yielding under infinitesimal load so that there is some error involved in the plastic solution by specifying the $r^{-1/2}$ near tip strain distribution up to finite loads. Actually for the size element used at the tip this error should be very small. In the round bar problem the near tip element extended a distance $\frac{1}{72}$ of the crack length, a distance at which the singularity is expected to dominate. Using the area average strain basis the first subelement yields when the strain at $\frac{3}{16} \times \frac{1}{72} \times$ crack length satisfies yield. The neglect of plasticity in the analysis until this load level should have a minor effect on the solution away from the tip at this load and as loading proceeds, the crack tip solution should show little evidence of this numerical transition procedure.

The interpolation functions used are most easily described in terms of the natural coordinates of the elements. Taking the element edges as coordinate lines $\xi = 0, 1$ and $\eta = 0, 1$ with the nodes I, J, K, L at the intersections we have the following correspondence with the physical coordinates:

$$\mathbf{x} = \mathbf{x}^I(1 - \xi)\eta + \mathbf{x}^J(1 - \xi)(1 - \eta) + \mathbf{x}^K\xi(1 - \eta) + \mathbf{x}^L\xi\eta \qquad (25)$$

Equation (25) may be thought of as a mapping of the physical region onto a unit square in the (ξ, η) plane. For instance Fig. 4 illustrates the map of a near-tip element of angular extent 2α and height s_0. Notice that the edge

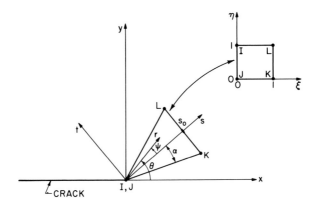

Fig. 4. Typical near crack tip element.

$\xi = 0$ with two distinct nodes I and J maps onto one point—the crack tip—in the (x, y) plane so that in this case $\mathbf{x}^I = \mathbf{x}^J$. The inverse map of this particular element in terms of the local Cartesian coordinates (s, t) and local polar coordinates (r, ψ) is

$$\xi = s/s_0, \qquad \eta = (\tan \psi/\tan \alpha + 1)/2 \qquad (26)$$

The elastic singularity element has the interpolation function

$$\mathbf{u} = \mathbf{u}^{IJ}(1 - \sqrt{\xi}) + \mathbf{u}^K(1 - \eta)\sqrt{\xi} + \mathbf{u}^L\eta\sqrt{\xi} \qquad (27)$$

The unique displacement of the crack tip nodes I and J is denoted by $\mathbf{u}^{IJ}$. From Eq. (26) we see that this displacement distribution corresponds to the expected form of $\sqrt{r}$ times a smooth function of angle. Along the edges $\eta = 0, 1$ displacement is a two-parameter function of $\xi(a + b\sqrt{\xi})$ so that there is displacement compatibility across them. Along the $\xi = 1$ edge the displacement is linearly interpolated between $\mathbf{u}^K$ and $\mathbf{u}^L$ so that there is complete inter-element compatibility when four node isoparametric elements are joined there. When the nodes are denoted as K, L, M, N the latter element has the form

$$\mathbf{u} = \mathbf{u}^L(1 - \xi)\eta + \mathbf{u}^K(1 - \xi)(1 - \eta) + \mathbf{u}^M\xi(1 - \eta) + \mathbf{u}^N\xi\eta \qquad (28)$$

The plastic singularity element interpolation function was derived through consideration of displacement distributions which correspond to a $1/r$ shear-strain singularity and importantly also a bounded dilatational strain. Levy *et al.* [37], in treating the small-scale-yielding problem, used a bilinear polar coordinate interpolation function for near tip pie-shaped elements which allowed a $1/r$ singularity in $\varepsilon_{\theta\theta}$ as well as $\varepsilon_{r\theta}$, leaving it to the numerical solution to choose a bounded $\varepsilon_{\theta\theta}$. However, it is impossible with their element to make the $1/r$ part of $\varepsilon_{\theta\theta}$ vanish for all θ. Here the dilatation boundedness condition Eq. (17) is precisely satisfied throughout the near tip elements. In the notation of Fig. 4 the condition met by the displacement components $u_s(\xi, \eta)$ and $u_t(\xi, \eta)$ is

$$\partial u_t(0, \eta)/\partial \eta = \tan \psi \, \partial u_s(0, \eta)/\partial \eta \qquad (29)$$

The $1/r$ shear singularity results for any assumed displacement function which allows a crack tip displacement variation, as in the general discussion of the last section.

Only the displacements $(u_s{}^I, u_s{}^I)$ and $(u_s{}^J, u_t{}^J)$ of the nodes at $(0, 1)$ and $(0, 0)$ enter in the displacement distribution along $\xi = 0$ in the present formulation. Hence these degrees of freedom completely determine the strength of the shear singularity within the element as seen from the first of Eqs. (20) rewritten in present notation,

$$R(\psi) = \frac{G}{\tau_0 \cos \psi} \frac{\partial u_s(0, \eta)}{\partial \eta} \frac{d\eta}{d\psi} \qquad (30)$$

With no attempt to enforce continuity of $R(\psi)$ at element boundaries $u_s(0, \eta)$ was chosen linear in η so that, to first order in ψ, $R(\psi)$ would be constant within an element,

$$u_s(0, \eta) = u_s{}^J + (u_s{}^I - u_s{}^J)\eta \qquad (31)$$

Equations (29) and (31) then establish $u_t(0, \eta)$ to within a constant which in turn is determined from the end conditions $u_t(0, 1) = u_t{}^I$ and $u_t(0, 0) = u_t{}^J$. We find the distribution

$$u_t(0, \eta) = u_t{}^{IJ} + \tan \alpha(u_s{}^I - u_s{}^J)\eta(\eta - 1) \qquad (32)$$

which has the displacement components $u_t{}^I$ and $u_t{}^J$ equal to each other and commonly called $u_t{}^{IJ}$. This constraint $u_t{}^I = u_t{}^J$ is consistent with the intuitive feeling that the shear singularity is governed mostly by a u_s displacement variation. The exact expression for $R(\psi)$ is

$$R(\psi) = \frac{G}{\tau_0 \cos^3 \psi} \frac{u_s^I - u_s^J}{2 \tan \alpha} \tag{33}$$

This is recognized as a first-order finite difference approximation to $R(\psi)$ expressed either as we have earlier or as [20] $G/\tau_0[\partial u_r(0, \psi)/\partial \psi - u_\psi(0, \psi)]$ when the displacement u_s is transformed to polar coordinates and the equation is linearized in its angular dependence.

A bilinear displacement variation throughout the element due to displacement of nodes K and L was specified. Also the distributions (31) and (32) were weighted in the ξ direction by the factor $(1 - \xi)$. The complete interpolation function for the shear singularity element is then given by

$$u_s = \frac{u_s^I + u_s^J}{2}(1 - \xi) + [u_s^L\eta + u_s^K(1 - \eta)]\xi + \frac{u_s^I - u_s^J}{2}(2\eta - 1)(1 - \xi)$$
$$\tag{34}$$
$$u_t = u_t^{IJ}(1 - \xi) + [u_t^L\eta + u_t^K(1 - \eta)]\xi + \tan \alpha(u_s^I - u_s^J)\eta(\eta - 1)(1 - \xi)$$

If the two coincident nodes move as one, so that $u_s^I = u_s^J$, the element is nothing more than the conventional constant strain triangle. Hence the possibility of strain free rigid body motion of the element is present. The interelement displacement compatibility condition is satisfied since displacement varies linearly on all interelement edges.

A 9-point numerical integration of the element stiffness matrices was performed. The integration stations were at $\xi, \eta = \frac{1}{6}, \frac{1}{2}, \frac{5}{6}$ and each station was weighted by $\frac{1}{9}$ of the area of the element. In the fan regions expected at the crack tip there is no angular variation in deviatoric stress state when reference is to a polar coordinate system. Thus to enhance the accuracy of using subelement area average strains to evaluate the stress of the subelement, the stresses and strains of the near tip elements, which follow from Eq. (34), were referred to polar coordinates.

4.3. COMPUTING DETAILS

A version of the general-purpose finite-element program MARC4 was used in this work. Calculations were done on the Brown University IBM 360/67 in double precision arithmetic. The master stiffness equations were solved by direct elimination. The shear singularity element formulation required double precision; single precision calculations resulted in very erratic strain distributions. Experience with other formulations in single precision using isoparametric or polar elements (and the $r^{-1/2}$ singular element in the elastic case) gave no similar direct hint of arithmetic precision difficulties, establishing the shear singularity formulation as particularly sensitive to computational mode.

4.4. SMALL SCALE YIELDING RESULTS

The problem under consideration is the plane strain contained yielding of an elastic–plastic plane with a semi-infinite edge crack under the boundary condition that the singular field of the elastic solution, Eqs. (1) and (4), is asymptotically approached as $r \to \infty$. This boundary layer formulation was proposed by Rice [20] for the analysis of sharply cracked bodies with a crack tip plastic zone which is small compared to significant configuration dimensions. The finite-element model involved a finite region about the crack with a near-tip element dimension chosen small compared to region size so that the outer boundary could effectively be considered at infinity. The displacement field (4) was imposed at the nodes on the outer edge with K as the loading parameter. Taking advantage of symmetry, only the upper half of the region ($y \geqslant 0$, using the coordinates of Fig. 1) was treated. Ahead of the crack on $y = 0$ the displacement component u_y and the shear traction were zero.

The mesh was composed of four rings of $7.5°$ focussed isosceles trapezoids followed by eight rings of $15°$ elements making a total of 192 elements and 229 nodes. The nodes described arcs of radius

$$r = 0, 0.5, 1, 1.625, 1.5^2, 2^2, \ldots, 5.5^2$$

The nodes on $r = 2.25$ not common to the adjacent $7.5°$ and $15°$ elements were constrained to maintain interelement compatibility.

The plastic solution was obtained by specifying successive increments in K equal to 25% of K_0—the stress intensity factor which causes the first subelement to yield. At each load increment the solution was the result of three iterations on the representative element constitutive matrices. Loading ceased when elements of the fifth ring yielded so that the extent of the plastic zone was always small compared to the outer radius.

From the exact elastic distribution (1) we find that initial yield occurs at an angle of $\cos^{-1}[(1 - 2v)^2/3] \approx 87°$, for $v = 0.3$. Furthermore, K_0 for yielding a radius r_y can be determined from $K_0/\sigma_0(2\pi r_y)^{1/2} = 1.10$ for this Poisson ratio. In this problem subelement yielding was based on the subelement midpoint stress so that r_y was $\frac{1}{12}$. The element between $82.5°$ and $90°$ yielded first; the midpoint angle being $86.25°$ indicates excellent finite-element agreement with the theoretical value. The finite element yield load parameter $K_0/\sigma_0(2\pi r_y)^{1/2}$ was 1.07—less than 3% deviation from theory. The angular near tip stress variation was also in excellent agreement.

The crack tip opening displacement [twice the y component of displacement of the node at $(0, \pi)$] made dimensionless by the similarity parameter $K^2/E\sigma_0$ is plotted in Fig. 5 as a function of K/K_0. From dimensional considerations $\delta_t/(K^2/E\sigma_0)$ is constant but because of the numerical procedure the value varies with K/K_0 until the near tip plastic field is established. A

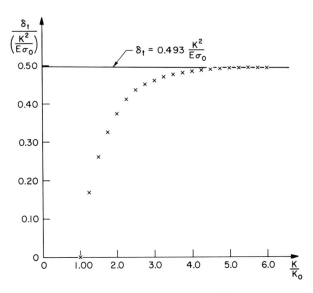

Fig. 5. Dimensionless crack tip opening displacement $\delta_t/(K^2/E\sigma_0)$ versus loading parameter K/K_0 for small-scale-yielding problem.

value of 0.493 was achieved at $K/K_0 = 4.75$ and it did not change for the remainder of the loading so that we conclude that for small-scale yielding

$$\delta_t = 0.493 \, K^2/E\sigma_0 \tag{35}$$

Levy *et al.* [37] found a factor of 0.425 from their incremental plasticity finite element results. The current estimate is thought more accurate since their polar element involved a dilatational singularity along with the expected shear singularity. With the physically less precise deformation theory of plasticity the J integral can be used to estimate the factor : Assuming an $R(\theta)$ symmetrical about $\theta = 90°$ Rice [20] predicted a factor of 0.613. Using $R(\theta)$ from the nonhardening limit of the power law hardening singularity Rice and Johnson [39] showed that the resulting value was 0.717.

The crack tip stress field as represented by the stresses of the twenty-four subelements nearest the crack tip approaches a distribution similar to the

Prandtl field that was discussed in the last section. The progression of $\sigma_{r\theta}(\theta)/\sigma_0$ from the elastic distribution ($K/K_0 = 1$) to the fully developed distribution ($K/K_0 > 4.5$) which is plotted in Fig. 6 is typical of all the stress components. The most obvious connection between the solution and the Prandtl field is that both have distinct fans in the range $45° < \theta < 135°$.

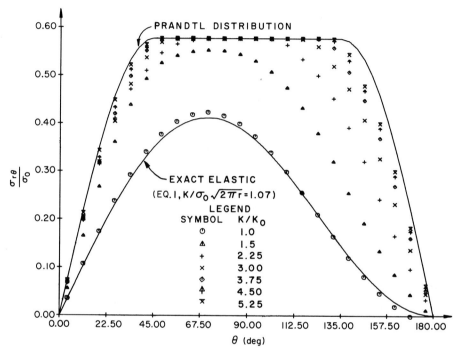

Fig. 6. Crack tip shear stress distribution at various load steps for the small scale yielding problem.

Also the stress $\sigma_{\theta\theta}$ of the subelement in the range $0 < \theta < 7.5°$ reaches the value $2.96\sigma_0$ at $K/K_0 = 5.25$ which certainly is in excellent agreement with the Prandtl value of 2.97 for $\sigma_{\theta\theta}(0,0)$. Two important assumptions used in deriving the Prandtl field were that yielding completely surrounds the crack tip and that the out-of-plane deviatoric stress s_{zz} vanishes at all values of θ. Neither condition was met in the finite element solution. The two elements between $165°$ and $180°$ remain elastic throughout the loading and $s_{zz} = 0$ only in the fan. Hence it is not surprising that the stress distributions of the Prandtl constant state region were not realized in detail. Yet from the fan results this problem does indeed show the value of using analytical work to guide in the design of numerical procedure.

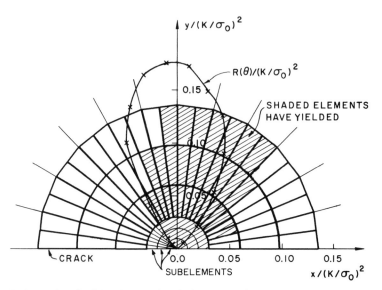

Fig. 7. Strength $R(\theta)$ of the $1/r$ shear singularity and plastic zone extent in terms of similarity coordinates $(x, y)/(K/\sigma_0)^2$, for small scale yielding.

Fig. 7 shows the near tip mesh, yielded elements and the strength $R(\theta)$ of the shear strain singularity, from the solution at $K/K_0 = 5.25$, all referred to the similarity coordinates $(x, y)/(K/\sigma_0)^2$. The contour defined by the yielded elements does not precisely define the elastic–plastic boundary due to stress variation within elements. By interpolation between load steps we find that $r_{P,\max}$, the maximum linear extent of the elastic–plastic boundary, is $0.152(K/\sigma_0)^2$ and this is at $\theta = 71°$. The boundary crosses the $\theta = 0$ line at a radius $r_{P,0} = 0.041(K/\sigma_0)^2$. The strength $R(\theta)$ was determined from the crack tip nodal displacements in accord with Eq. (33) (with $\psi = 0$). The function peaks in the range $90°$–$97.5°$ with a value

$$R_{\max} = 0.177(K/\sigma_0)^2 \qquad (36)$$

In comparison, Levy et al. [37] found factors of 0.157, 0.036, and 0.155 for $r_{P,\max}, r_{P,0}$, and $R_{\max}$, respectively. The vanishingly small value of R outside of the angular range $45° < \theta < 135°$ clearly defines this range as the fan region active in the blunting of the crack tip.

4.5. CIRCUMFERENTIALLY CRACKED ROUND BAR

The axisymmetric round bar considered has a circumferential crack penetrating its outer surface to a depth of $\frac{1}{2}$ the bar radius and a length of $4D$, where D is the bar diameter. The boundary condition was that the ends of the bar move uniformly in the axial direction with zero shear tractions. The length was sufficient to have a uniform axial stress state at the ends during the entire loading sequence. A mesh with 384 nodes and 340 axisymmetric ring elements was used to represent the upper half of the bar which is naturally described in terms of a (ρ, z, ϕ) cylindrical coordinate system with the centerline coinciding with the z axis and the crack along $D/4 \leqslant \rho \leqslant D/2$ in each meridional plane $\phi = $ const. Cross sections of the elements near the crack tip were focussed isosceles trapezoids, near the ends rectangles, and joining the two groups were arbitrary quadrilaterals. As viewed on a meridional plane there were 13 rings of $7.5°$ trapezoids encircling the crack tip, as for the previous small scale yielding solution. Introducing an (r, θ) polar coordinate system in this plane (with the crack along $\theta = \pm \pi$) the nodes of the trapezoids describe arcs of radius

$$r = (D/288)(0, 1, 1.5^2, 2^2, 2.5^2, 3^2, 4^2, 5^2, 6^2, 48, 60, 72, 91, 120)$$

The stress intensity factor for this geometry as a function of net section stress σ_{net} and diameter D was found by Bueckner [40] to be within one percent of $0.240\sigma_{net}(\pi D)^{1/2}$. The nodal displacements at $r/D = \frac{1}{288}$ and σ_{net} from the elastic solution were used in conjunction with the theoretical plane strain near tip field (4) to estimate K; however, nodes within $37.5°$ of the crack were not considered for, in this range, $\varepsilon_{\phi\phi}$ was of the same order of magnitude as the in-plane strains. A simple average of the discrete estimates of K results in a factor of 0.244 which is within 2% of Bueckner's solution. The stress σ_{zz} of the subelement between $0 < \theta < 7.5°$ corresponds to 0.247 which is 3% from Bueckner.

The accuracy of the partial-stiffness treatment, Eq. (21), of elements making the elastic to elastic–plastic transition within a load increment and the rate of convergence of the mean yield surface normal technique are greatly affected by the size of the load increment. A successful procedure in terms of convergence rate involved regulating the load increment so that only elements within 10% of yield would yield during an increment, and also allowing three iterations for each increment. For small scale yield the sufficient load step sizes were prejudged by assuming that the elastic regions responded proportionally to load up to yield. The displacement of the end of the bar was increased to 38.2 times the end displacement $(u_z^{end})_0$ at first yield in 42 increments in the following manner

2 steps of $0.1(u_z^{end})_0$,   6 of 0.2,   3 of 0.3,   3 of 0.4,   2 of 0.5,   6 of 0.7,

9 of 1.0,   7 of 1.5,   4 of 2.25

The load-deflection curve, σ_{net}/σ_0 versus $Eu_z^{end}/\sigma_0 D$, is presented in Fig. 8. The end displacement was increased until the limiting elastic–plastic zone

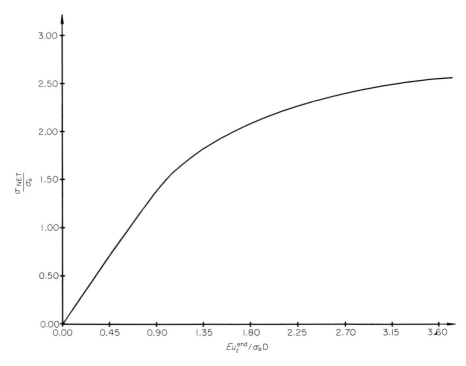

Fig. 8. Load deflection curve for round bar.

was achieved. The corresponding limit stress for this Mises idealization is $\sigma_{net} = 2.56\sigma_0$. In comparison, Shield [41] found a limit pressure of $5.69\tau_0$ for axisymmetric frictionless indentation of an elastically rigid–perfectly-plastic Tresca half-space. While the flow was confined to a radius of 1.58 times the punch radius in Shield's problem, the present finite-element result is that yielding spreads to include the outside surface of the bar at limit load. Figure 9 shows the yielded regions at different stages of loading. If the plasticity had been confined to the bar interior Shield's limit load could serve to establish bounds to the Mises limit load for the crack depth chosen by invoking corollaries to the limit theorems of plasticity.

When the Mises and Tresca materials are assigned identical shear limits the ratio of the finite-element limit stress to Shield's is $4.43/5.69 = 0.78$; when matched in tension the ratio is $2.56/2.85 = 0.90$. Figure 10 is a plot of the two net section stress distributions, σ_{zz}/σ_0 versus $4\rho/D$ at $z = 0$, at limit load. The larger Tresca stresses over most of the section can be explained in terms of the factor of 0.9 between the Mises and Tresca limit loads. The Tresca curve monotonically decreases from a centerline value of $3.60\sigma_0$ to $2.57\sigma_0$ while the Mises curve increases from $2.10\sigma_0$ to $3.05\sigma_0$ at the crack tip (or

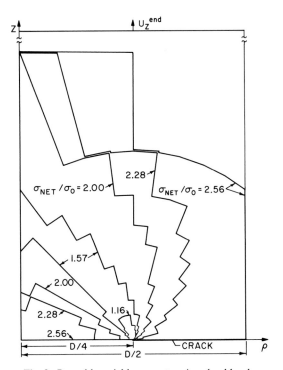

Fig. 9. Round bar yield zones at various load levels.

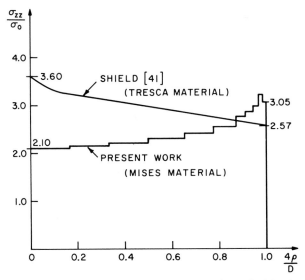

Fig. 10. Round bar net section stress distribution at limit load.

punch surface in Shield's context) when averaging the near tip subelement values of 2.94, 2.98, and 3.23.

The crack tip small scale yielding solution was essentially that of the previous asymptotic problem. As explained, the normalized crack tip opening displacement $\delta_t/(K^2/E\sigma_0)$ increases from value zero to a characteristic small scale yielding value over a small initial load range due to numerical procedure. Also, once the plastic zone extends to an appreciable fraction of crack length $\delta_t/(K^2/E\sigma_0)$ dramatically increases with K signaling the beginning of "large scale" yielding. The value of $\delta_t/(K^2/E\sigma_0)$ increased to within 10% of the asymptotic solution value 0.493 at the eighth load increment corresponding to $\sigma_{net}/\sigma_0 = 0.37$ and the plasticity was confined to the first ring of elements about the crack tip. The value was 10% higher than the asymptotic solution at $\sigma_{net}/\sigma_0 = 0.85$ and at this state plasticity was confined to a radius of $D/32$. This plastic zone size could be used to set a rough upper limit to the applicability of linear fracture mechanics treatment of crack tip plasticity.

The function $R(\theta)$ in the small-scale yielding range differed slightly from the distribution for the asymptotic problem in that the function peaked between 82.5° and 90°; in the former solution R_{max} was in the range 90° to 97.5°. This

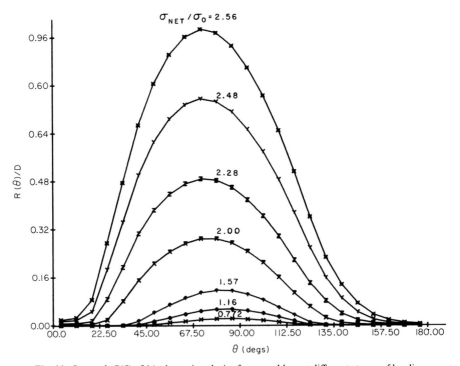

Fig. 11. Strength $R(\theta)$ of $1/r$ shear singularity for round bar at different stages of loading.

is most likely the influence of the centerline which is felt due to the relatively coarse mesh of the present problem. In Fig. 11 $R(\theta)/D$ is plotted for various load states in the large scale yielding range. In conjunction with the elastic–plastic zones of Fig. 9, one can see that R serves as a reasonable estimate to the extent of the plastic zone at angles which are within the $1/r$ shear fan while the plastic zone at the angle remains interior to the specimen boundaries. As loading progresses the fan region extends from the Prandtl range of $45° < \theta < 135°$ to the larger range of $15° < \theta < 157.5°$. This may, however, reflect a failure of the numerical solution to accurately meet the stress-free crack surface boundary condition in the innermost element as fully plastic conditions are reached.

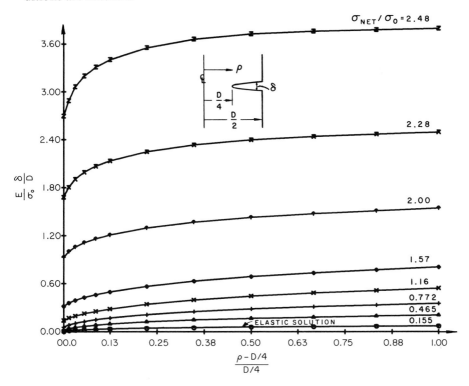

Fig. 12. Round bar crack profile at various load levels.

Figure 12 is a plot of the opening displacement δ of the neighboring points on the crack surfaces, as a function of distance from the crack tip, for various net stress levels. It is clear from these curves that, throughout the large-scale-yielding range, the crack tip opening displacement, δ_t ($=\delta$ at $\rho = D/4$) is very significant even when compared to the flank opening δ at $\rho = D/2$.

These results may be helpful in correlating large scale yield fractures of this geometry since if fracture is controlled by a critical crack tip state, it is quite likely reflected in the attainment of a critical crack opening displacement, at least for cases such as the present for which the stress triaxiality does not vary appreciably from small scale to general yielding conditions.

5. Three-Dimensional Problems

Both structural applications of fracture mechanics and the interpretation of experimental results require advances in three-dimensional stress analysis. Work to date has been limited, and has focused on the stress analysis of part-through-the-thickness surface cracks in the walls of plate or shell structures, as representative of flaw types in applications, and on the transition from plane strain to plane stresslike constraint near the tip of a straight through-the-thickness crack in a plate. This last problem is, of course, important to the interpretation of fracture test results which are typically obtained on plate specimens precracked in this way. It is also of interest for the informa-tion it sheds on the actual three-dimensional aspects of what is commonly treated as a two-dimensional problem.

5.1. THROUGH CRACK IN A PLATE

The straight, through-the-thickness crack in a plate has been studied by Aryes [42] using finite difference methods, by Cruse [15] and Cruse and Van Buren [16] using a numerical solution of singular integral equations over the specimen boundary, and by Levy et al. [43] using finite-element methods. Only Ayres gave results for this problem in the plastic range, but he made no special provisions beyond mesh refinement for attaining near tip accuracy. Levy et al. employed a singular element similar to that of their earlier plane strain study [37] and to that of the last section, with layers of polar arrays of the elements being stacked through the plate thickness.

Figure 13 is replotted from Cruse's [15] results on a compact ($2h \times 2h \times h$, where h is plate thickness) fracture test specimen containing an edge crack of length h which is wedged open by end forces. Lines of constant value for the parameter $\alpha \, [= \sigma_{zz}/v(\sigma_{xx} + \sigma_{yy})]$ are shown for half the plane of material directly ahead of the crack. The parameter is called the "degree of plane strain" since such conditions correspond to $\alpha = 1$. One sees that plane strain conditions are indeed approached at the crack tip, although the fall off to-ward a plane stress state ($\alpha = 0$) is quite rapid. Similar results were found by Levy et al. [43], who studied a circular plate of six thicknesses in radius with a through crack having its tip at the plate center. For boundary conditions they imposed the stresses σ_{rr} and $\sigma_{r\theta}$ of the charactristic $r^{-1/2}$ singularity appropriate to the two-dimensional-plane stress theory. Figure 14 shows their

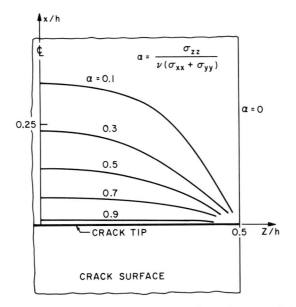

Fig. 13. Results showing the transition from plane strain to plane stress behavior near the tip of a through-the-thickness crack in an elastic plate, from Cruse [15].

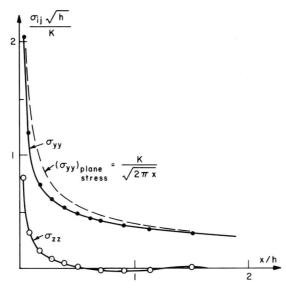

Fig. 14. Results demonstrating the rapid transition to a plane stress condition away from the crack tip in a through-the-thickness cracked elastic plate, from Levy et al. [43].

results for σ_{yy} and σ_{zz} on the line in the plate middle surface directly ahead of the crack. The dashed line shows the two-dimensional plane stress result for σ_{yy}. Again, the rapid approach to a plane stress state may be noted; σ_{zz} is negligible even in the middle surface beyond a distance of about a half-thickness. The last two figures tend to make plausible the rather large ratio of plate thickness to plastic zone size found necessary to assure plane strain conditions within the plastic region at fracture [44].

5.2. SURFACE CRACKS

Several different computational approaches have also been taken for problems of part-through-the-thickness surface cracks. For example, Kobayashi and Moss [45] and Smith and Alavi [46] have employed alternating methods in three-dimensional elasticity, based on the Boussinesq solution for a half space and on that for removal of tractions from an embedded circular or elliptical shaped crack in an infinite body, to estimate K for circular arc and semielliptical surface cracks in plates. The papers by Ayres [42] and Levy et al. [43] employing finite difference and isoparametric finite-element formulations, respectively, have discussed the elastic and elastic–plastic fields near a semielliptical surface crack in a plate.

Also, Rice and Levy [47] have developed a model for problems of long surface cracks (in comparison to plate thickness) which reduces these to problems in the two-dimensional theory of plane stress and bending for a plate containing a line spring which represents the part-cracked section. Their original work reduced the problem to two coupled integral equations, solved numerically, for the force and moment transmitted across the line spring. However [48], the model has been extended to cracks in shells, and a finite-element formulation has been developed for incorporation in existing two-dimensional plate and shell programs. Results for K have been given for surface cracks of various shapes in plates, and for axial and circumferential semielliptical cracks in the wall of a cylindrical tube. Their model shows promise of extension to the plastic range, and to application within shell analysis programs to a variety of surface crack locations in pressure vessels.

6. Micromechanics and Development of Fracture Criteria

Here we discuss the use of computational methods in the description of fracture processes on the microscale, and in the merging of such studies with elastic–plastic analyses at the continuum level so as to develop rational fracture criteria. The area is not yet very much studied, and hence our emphasis will be in part on pointing out what we consider to be opportunities for productive use of computational stress analysis methods for problems of this type. These include the effects of plastic flow, finite strains, and deformation instabilities.

6.1. DUCTILE FRACTURE MECHANISMS

Fracture mechanisms in structural metals, apart from low-temperature cleavage in steels, generally involve substantial plastic flow on the microscale. This arises through the formation of small voids, typically by the decohesion or cracking of hard inclusion particles, which undergo large ductile expansion until final separation results from coalescence of arrays of these voids. That is, fracture arises as a kinematic result of large plastic flow. This is so even for materials such as the high-strength steels and aluminum alloys which may, under plane strain conditions, show macroscopically brittle crack advance with plastic zone sizes in the millimeter range. On a scale of, say, 5 to 100 μm the resulting fracture surfaces show evidence of great ductility with local strains on the order of unity. McClintock [3, 38, 49] has discussed this fracture mechanism in detail. He and Rice and Tracey [50] have applied continuum plasticity solutions for cavity expansion as models for hole growth.

In general, however, the modeling of void growth should include a treatment of finite shape changes, interactions between neighboring voids, and the possibly unstable coalescence of neighboring voids or void arrays. This necessarily involves numerical formulations for large deformations, of the type presented, for example, by Hibbit *et al.* [51] and Needleman [52]. In fact, Needleman's paper contains a finite-element solution for the large ductile expansion of a periodic array of initially cylindrical holes in a power law hardening material. His procedure was based on a form for the incremental constitutive law at finite strain proposed by Budiansky [53]. This is consistent with a variational principle for the rate problem, as in the small distortion theory, and the finite-element method was based directly on it. In contrast, Hibbitt *et al.* propose a constitutive law in which the Jaumann stress rate is employed, rather than Budiansky's time derivative of a stress measure referred to convected coordinates, and no variational principle is then applicable. The formulation begins instead directly from the principle of virtual work and the resulting statement of equilibrium is differentiated to derive the governing finite-element equations of the rate problem.

Computations of the type done by Needleman also serve to predict the slight dilatation which should appear in macroscopic plastic constitutive laws as a consequence of void growth. Berg [54] has suggested that such dilatational constitutive laws could be used as a basis for stress analysis procedures which include ductile fracture initiation as a *consequence* of a proper stress analysis, rather than as an ad hoc supplement to such an analysis.

Here the idea is that such constitutive laws, which already include the volume change due to void growth, may also ultimately permit localization in a band as representative of unstable void coalescence on the microscale. This may occur when the hardening in an increment of deformation is just balanced by the softening due to the increased porosity through void growth, provided also that the kinematic condition is met of existence of a plane of zero extension rates. The types of problems for which such an approach may be valuable (e.g., metal-forming processes) typically involve extensive plastic flow and will require a numerical formulation appropriate to finite strain. Very little has been done to date. Indeed, even the classic problem of ductile fracture initiation in the necked region of a round tensile bar is unsolved at present.

The mechanics of void initiation from inclusions has also received little attention, although the problem of determining the stress state in and near a nonyielding inclusion in a ductile matrix is certainly within the capability of existing analysis methods. Huang [55] has presented such a study for a circular cylindrical inclusion in a Ramberg–Osgood power hardening material through a method of Fourier expansion and finite differences which, unfortunately, does not seem to offer the possibility of general application.

6.2. Fracture Mechanisms at a Crack Tip

The elastic–plastic crack stress analyses discussed elsewhere in this paper were based on conventional small strain-analysis procedures, in that effects of geometry changes on the governing equations were neglected. This is obviously incorrect within a distance from the tip comparable in size to the predicted opening displacement. Analysis at such a scale is important since ductile fracture mechanisms are operative in this very near tip region where large strains occur.

Rice and Johnson [39] have shown how the solutions based on neglect of geometry changes may be employed to set boundary conditions on a localized analysis of the large crack tip deformations for the nonhardening model. In this case it is important that the distribution $R(\theta)$ of the strength of the strain singularity be known (as, for example, in Figs. 7 and 11), for from it the crack tip velocity field is computed and this is the boundary condition for the local large-strain analysis. The analysis is based on the application of slip line theory to the near tip region, and McClintock [3, 56] has similarly discussed large geometry changes at the tip.

When the tip is drawn as progessively blunted by increasing load as in Fig. 15, the constant stress regions A and B as in the Prandtl field of Fig. 2 remain, but they are separated by a fan of straight slip lines C which is no

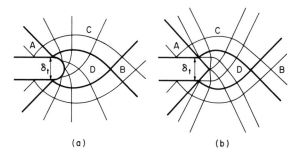

Fig. 15. Two possible solutions to the slip line analysis of the blunting of an initially sharp-tipped crack.

longer centered. Instead it feeds into a spiral region D ahead of the tip of a size roughly comparable to the crack opening. We have seen that δ_t is of the order of an initial yield strain times the maximum linear dimension of the plastic zone. Representative numerical values for a wide range of low and high strength structural metals are between 0.003 and 0.03 times the zone size. Hence, in Fig. 15 we see a minute fraction of the total plastic region and, indeed, the blunted tip would appear essentially as a point when viewed on the size scale of the plastic zone. For this reason, Rice and Johnson suggested that crack tip blunting could be studied for the contained plastic yielding range by applying rigid–plastic theory locally to region D, using the fact that straight slip lines of the fan transmit a uniform velocity along their length and hence that radial velocity, as a function of angle, should differ negligibly from the result for the solution which neglects geometry change effects. That is, the radial displacement rate $\dot{u}_r{}^0(\theta)$ from solutions of the type discussed earlier is taken as the normal velocity, as a function of slip line angle, imposed from the fan region along the boundary between C and D. From this it is straightforward to numerically calculate, in order, the velocity field in terms of slip line coordinates, the movement of the crack tip, the resulting physical coordinates of points of the slip line field in D, and hence the entire local strain and displacement solution for crack tip blunting. Rice and Johnson describe a computational scheme for doing this and give some representative results.

The greatest difficulty with such solutions is illustrated in Fig. 15: As McClintock [3] has emphasized, solutions for an initially sharp crack are nonunique. It is possible to find a solution involving smooth blunting as in (a); it is also possible to find solutions in which the crack tip retains sharp corners of singular strain rate as in (b). Indeed, there appears to be nothing in continuum plasticity to enable a choice. McClintock suggests from observations of fracture surfaces that the latter is the more realistic picture.

Strain-hardening effects cannot be properly included in the slip line analysis of blunting. Nor can the interactive effects of growing voids. Hence, it would seem desirable to apply some of the numerical methods for finite strain to the crack tip blunting case. The problem is not straightforward, however, due to the extremely small size of the large strain region in comparison to plastic zone dimensions (or, equivalently, to the great strain gradients involved), and to the nonuniqueness as illustrated by Fig. 15.

As we have noted earlier, very little work has been done on cracks which advance in a quasi-static fashion under increasing load. Of course, near tip accuracy is paramount in this case as always when fracture prediction is a goal of the numerical solution. Perhaps a finite-element treatment could be based on a focused mesh which moves relative to the material in each increment. Also, it would seem necessary that some plausible model of crack advance at the microstructural level be analyzed in parallel with the continuum calculations in this case, for otherwise the increment in crack length accompanying a given increment in load is not determined. McClintock [3, 38] has suggested that a decohering layer ahead of the crack may provide a proper model. The layer is imagined to represent a region of material in which void growth is already in its unstable stages, so that a falling stress versus separation-distance relation applies as a boundary condition on the continuum plasticity problem. The amount of crack advance due to a load increment in this formulation would correspond to that length ahead of the current crack tip across which zero load is transmitted.

Additional computational problems arise with subcritical crack growth by stress corrosion and fatigue. For the first of these, the mechanical features of the near tip state could be determined through any program suitable for quasi-static crack advance. In the case of fatigue, important computational problems include the determination of the cyclic deformation states near the tip, its progressive blunting and sharpening, and the role of interference of previously deformed material with crack closure at its tip.

7. Conclusion

Numerical procedures for accurate determination of elastic stress intensity factors for the general two-dimensional crack problem were reviewed. The elastic–perfectly-plastic crack tip deformation state was investigated through a finite element treatment which was designed to allow the $1/r$ shear singularity and associated crack tip opening displacement predicted from a detailed asymptotic study. The importance of basing crack tip numerical procedure on analytical results was emphasized when accuracy sufficient to develop fracture criteria is required.

The small scale yielding problem was modelled and expressions for crack tip opening displacement, shear singularity amplitude, and plastic zone extent were presented.

A finite-element solution to the large scale yielding of the circumferentially cracked round tension bar was presented. The global solution as reflected through the resulting limit load, net section stress distribution, yield zones, and crack profile and the local crack tip solution were discussed as relevant to fracture testing and prediction.

The three-dimensional aspects of flawed structures were discussed and numerical treatments of the subject were reviewed. Finally, ductile fracture mechanisms and specifically crack tip fracture on the microscale were discussed.

Acknowledgment

We wish to acknowledge financial support from the National Aeronautics and Space Administration under Grant NGL-40-002-080 to Brown University.

References

1. Paris, P. C., and Sih, G. C., "Stress Analysis of Cracks," in *Fracture Toughness Testing and Its Applications*, pp. 30–76. ASTM-STP-381. Philadelphia, Pennsylvania, 1965.
2. Rice, J. R., "Mathematical Analysis in the Mechanics of Fracture," in *Fracture: An Advanced Treatise* (Liebowitz, H., ed.), Vol. 2, pp. 191–311. Academic Press, New York, 1968.
3. McClintock, F. A., "Plasticity Aspects of Fracture," in *Fracture: An Advanced Treatise* (Liebowitz, H., ed.), Vol. 3, pp. 47–225. Academic Press, New York, 1971.
4. Gross, B., Srawley, J. E., and Brown, W. F., "Stress Intensity Factors for a Single Edge Notch Tensile Specimen by Boundary Collocation of a Stress Function," NASA Tech. Note D-2395 (1964).
5. Gross, B., and Srawley, J. E., "Stress Intensity Factors for Single Edge Notch Specimens in Bending or Combined Bending and Tension," NASA Tech. Note D-2603 (1965).
6. Gross, B., and Srawley, J. E., "Stress Intensity Factors for Three Point Bend Specimens by Boundary Collocation," NASA Tech. Note D-3092 (1965).
7. Kobayashi, A. S., Cherepy, R. B., and Kinsel, W. C., "A Numerical Procedure for Estimating the Stress Intensity Factor of a Crack in a Finite Plate," *J. Basic Eng.* **86**, 681–684 (1964).
8. Bowie, O. L., "Analysis of an Infinite Plate Containing Radial Cracks Originating from the Boundary of an Internal Circular Hole," *J. Math. Phys.* **35**, 60–71 (1956).
9. Bowie, O. L., "Rectangular Tensile Sheet with Symmetric Edge Cracks," *J. Appl. Mech.* **31**, 208–212 (1964).
10. Bowie, O. L., "Analysis of Edge Notches in a Semi-Infinite Region," *J. Math. Phys.* **45**, 356–366 (1966).
11. Kaminskij, A. A., "Critical Loads in Regions Weakened by Holes with Cracks," in *Kontsentratsija Naprjazhenij*, Vol. 1, pp. 130–136. Naukovaja Dumka, Kiev, 1965.

12. Bowie, O. L., and Neal, D. M., "A Modified Mapping-Collocation Technique for Accurate Calculation of Stress Intensity Factors," *Int. J. Fracture Mech.* **6**, 199–206 (1970).
13. Greif, R., and Sanders, J. L., "The Effect of a Stringer on the Stress in a Cracked Sheet," *J. Appl. Mech.* **32**, 59–66 (1965).
14. Bueckner, H. F., "Some Stress Singularities and Their Computation by Means of Integral Equations," in *Boundary Value Problems in Differential Equations* (Langer, R., ed.), pp. 215–230. Univ. of Wisconsin Press, Madison, Wisconsin, 1960.
15. Cruse, T. A., "Lateral Constraint in a Cracked Three Dimensional Elastic Body," *Int. J. Fracture Mech.* **6**, 326–328 (1970).
16. Cruse, T. A., and Buren Van, W., "Three Dimensional Elastic Stress Analysis of a Fracture Specimen with an Edge Crack," *Int. J. Fracture Mech.* **7**, 1–15 (1971).
17. Tirosh, J., "A Direct Numerical Method for Stress and Stress Intensity Factor in Arbitrary, Cracked, Elastic Bars Under Torsion and Longitudinal Shear," *J. Appl. Mech.* **37**, 971–976 (1970).
18. Chan, S. K., Tuba, I. S., and Wilson, W. K., "On the Finite Element Method in Linear Fracture Mechanics," *Eng. Fracture Mech.* **2**, 1–17 (1970).
19. Hayes, D. J., "Some Applications of Elastic Plastic Analysis to Fracture Mechanics," Ph.D. Thesis, Imperial College of Sci. and Technol. (1970).
20. Rice, J. R., "A Path Independent Integral and the Approximate Analysis of Strain Concentration by Notches and Cracks," *J. Appl. Mech.* **35**, 379–386 (1968).
21. Barone, M. A., and Robinson, A. R., "Approximate Determination of Stresses near Notches and Corners in Elastic Media by an Integral Equation Method," Univ. of Illinois Civil Eng. Stud., Struct. Res. Ser. No. 374 (1971).
22. Rice, J. R., "Some Remarks on Elastic Crack Tip Stress Fields," *Int. J. Solids Structures*, **8**, 751–758 (1972).
23. Wilson, W. K., "On Combined Mode Fracture Mechanics," Westinghouse Res. Lab., Res. Rep. 69-1E7-FMECH-R1, Pittsburgh, Pennsylvania (1969).
24. Hilton, P., and Hutchinson, J. W., "Plastic Intensity Factors for Cracked Plates," *Eng. Fracture Mech.* (in press).
25. Tracey, D. M., "Finite Elements for Determination of Crack Tip Elastic Stress Intensity Factors," *Eng. Fracture Mech.* **3**, 255–265 (1971).
26. Wilson, W. K., Plane Strain Crack Toughness Testing of High Strength Metallic Materials, ASTM-STP-410, pp. 75–76, Philadelphia, Pennsylvania, 1969.
27. Lee, C. H., and Kobayashi, S., "Elasto-plastic Analysis of Plane Strain and Axisymmetric Flat Punch Indentation by the Finite Element Method," *Int. J. Mech. Sci.* **12**, 349–370 (1970).
28. Marcal, P. V., and King, I. P., "Elastic-Plastic Analysis of Two-Dimensional Stress Systems by the Finite Element Method," *Int. J. Mech. Sci.* **9**, 143–155 (1967).
29. Swedlow, J. L., Yang, A. H., and Williams, M. L., "Elasto-Plastic Stresses and Strains in a Cracked Plate," in *Proc. Int. Conf. Fracture, 1st, Sendai, 1965* (Yokobori, T., et al., eds.), Vol. I, pp. 259–282. Jap. Soc. for Strength and Fracture of Mater., Tokyo, 1966.
30. Swedlow, J. L., "Elasto-Plastic Cracked Plates in Plane Strain," *Int. J. Fracture Mech.* **5**, 33–44 (1969).
31. Wells, A. A., "Crack Opening Displacements from Elastic-Plastic Analyses of Externally Notched Tension Bars," *Eng. Fracture Mech.* **1**, 399–410 (1969).
32. Cherepanov, G. P., "Crack Propagation in Continuous Media," *Appl. Math. Mech.* (*PMM*) **31**, 476–488 (1967).
33. Hutchinson, J. W., "Singular Behavior at the End of a Tensile Crack in a Hardening Material," *J. Mech. Phys. Solids* **16**, 13–31 (1968).
34. Hutchinson, J. W., "Plastic Stress and Strain Fields at a Crack Tip," *J. Mech. Phys. Solids* **16**, 337–347 (1968).

35. Rice, J. R., "The Mechanics of Crack Tip Deformation and Extension by Fatigue," in *Fatigue Crack Propagation*, ASTM-STP-415, pp. 247–309 (1967).
36. Rice, J. R., and Rosengren, G. F., "Plane Strain Deformation Near a Crack Tip in a Power Law Hardening Material," *J. Mech. Phys. Solids* **16**, 1–12 (1968).
37. Levy, N., Marcal, P. V., Ostergren, W. J., and Rice, J. R., "Small Scale Yielding Near a Crack in Plane Strain: A Finite Element Analysis," *Int. J. Fracture Mech.* **7**, 143–156 (1971).
38. McClintock, F. A., "Local Criteria for Ductile Fracture," *Int. J. Fracture Mech.* **4**, 101–130 (1968).
39. Rice, J. R., and Johnson, M. A., "The Role of Large Crack Tip Geometry Changes in Plane Strain Fracture," in *Inelastic Behavior of Solids* (Kanninen, M. F., *et al.*, eds.), pp. 641–672. McGraw-Hill, New York, 1970.
40. Bueckner, H. P., "Coefficients for Computation of the Stress Intensity Factor K_I for a Notched Round Bar," in *Fracture Toughness Testing and Its Applications*, ASTM-STP-381, pp. 82–83, Philadelphia, Pennsylvania, 1965.
41. Shield, R. T., "On the Plastic Flow of Metals Under Conditions of Axial Symmetry," *Proc. Roy. Soc. Ser. A* **233**, 267–287 (1955).
42. Ayres, D. J., "A Numerical Procedure for Calculating Stress and Deformation Near a Slit in a Three Dimensional Elastic-Plastic Solid," *Eng. Fracture Mech.* **2**, 87–106 (1970).
43. Levy, N. J., Marcal, P. V., and Rice, J. R., "Progress in Three-Dimensional Elastic-Plastic Stress Analysis for Fracture Mechanics," *Nucl. Eng. Design* (in press).
44. Srawley, J. E., and Brown, W. F., "Fracture Toughness Testing," in *Fracture Toughness Testing and Its Applications*, ASTM-STP-381, pp. 133–193. Philadelphia, Pennsylvania, 1965.
45. Kobayashi, A. S., and Moss, W. L., "Stress Intensity Magnification Factors for Surface-Flawed Tension Plate and Notched Round Tension Bar," *Proc. Int. Conf. Fracture, 2nd*, Brighton, England (1968).
46. Smith, F. W., and Alavi, M. J., "Stress Intensity Factors for a Part Circular Surface Flaw," in *Pressure Vessel Technology, Part II, Materials and Fabrication*, pp. 793–800. ASME, New York, 1969.
47. Rice, J. R., and Levy, N., "The Part Through Surface Crack in an Elastic Plate," *J. Appl. Mech.* (in press) (ASME Paper No. 71-APM-20, 1971).
48. Levy, N., and Rice, J. R., "Surface Cracks in Elastic Plates and Shells," Brown Univ. Tech. Rept. NGL 40-002-080/6 to NASA, 1971.
49. McClintock, F. A., "A Criterion for Ductile Fracture by the Growth of Holes," *J. Appl. Mech.* **35**, 363–371 (1968).
50. Rice, J. R., and Tracey, D. M., "On the Ductile Enlargement of Voids in Triaxial Stress Fields," *J. Mech. Phys. Solids* **17**, 201–217 (1969).
51. Hibbitt, H. D., Marcal, P. V., and Rice, J. R., "A Finite Element Formulation for Problems of Large Strain and Large Displacement," *Int. J. Solids Struct.* **6**, 1069–1086 (1970).
52. Needleman, A., "Void Growth in an Elastic-Plastic Medium," *J. Appl. Mech.*, **39**, 964–970 (1972).
53. Budiansky, B., "Remarks on Solid and Structural Mechanics," *Problems of Hydrodynamics and Continuum Mechanics*, pp. 77–83. Soc. for Ind. and Appl. Math., Philadelphia, Pennsylvania, 1969.
54. Berg, C. A., "Plastic Dilation and Void Interaction," in *Inelastic Behavior of Solids* (Kanninen, M. F., *et al.*, eds.), pp. 171–210. McGraw-Hill, New York, 1970.
55. Huang, W.-C., "Theoretical Study of Stress Concentrations at Circular Holes and Inclusions in Strain Hardening Materials," Harvard Univ. Rep. SM-45 (1970).
56. McClintock, F. A., "Crack Growth in Fully Plastic Grooved Tensile Specimens," in *Physics of Strength and Plasticity* (Argon, A. S., ed.), pp. 307–326. M.I.T. Press, Cambridge, Massachusetts, 1969.

Biomechanics

George Bugliarello

COLLEGE OF ENGINEERING
UNIVERSITY OF ILLINOIS AT CHICAGO CIRCLE
CHICAGO, ILLINOIS

After a brief outline of the role of mechanics in living systems and a definition of the multiple tasks of biomechanics, some of the key mechanical processes occurring in living systems are categorized and illustrated by examples. The need for numerical solutions in the study of such processes is discussed, with examples of the state of the art in three areas: microcirculation, lung elasticity and the wall of arterioles. Biomechanics differs in a number of important respects from the mechanics of man-made objects (artimechanics) and the mechanics of inorganic nature (physimechanics). Development of more effective numerical approaches, particularly for dealing with extremely small dimensions and with flow processes, is necessary.

1. Mechanics in Living Systems

A living organism is an enormously complex system in which the mechanical, chemical, thermal, electrical, and radiation aspects are intimately interwoven and act synergistically. An essential step toward understanding the system as a whole, and toward the ability to combine man-made components rationally with living components, is to study each aspect of the system per se.

The basic building block of the organism, the cell, is in itself a very complex system. It carries out exchanges with its surroundings by means of

625

fluid-mechanics processes as well as other processes. Fluid processes govern the distribution of gases and nutrients to the interior of the cell and the removal of the wastes of the metabolic process from its interior. The limit of effectiveness achievable in the internal exchange processes governs the maximum attainable size of the cell, leaving the aggregation of cells as the only avenue for the formation of larger organisms. Such an aggregation takes various forms—from simple colonies in which all cells perform essentially the same function, to highly specialized organisms, which depend again on flow processes to bring signals, nutrients, and other substances to the various cells, and to remove from the cells the products of their metabolism. Finally, the exchanges between the organism as a whole and its environment also occur largely through fluid processes, and so does the propulsion of many organisms.

If flows are a key physiological vehicle of the living process—and hence a major portion of biomechanics must be concerned with them—other problems of a mechanical nature, ranging from the mechanics of the structures that provide containment and support for the organism and its components to the lubrication of moving joints, are also of paramount importance.

2. The Tasks of Biomechanics

Since time immemorial, man has endeavored not only to understand nature, but also to imitate it in the artifacts it has constructed. The column—particularly the greek column with its esthetic evocations of the tree—and the pathetic wings of the tailor of Ulm [1] are but two examples. In more recent times, the ambitions and skills of man have extended to the implanting of man-made parts in the living organisms. Dental prostheses, the oldest implants, are actually a few hundred years old. Artificial internal organs, such as heart pacemakers, artificial heart valves, and femoral pins, are contemporary, and symbolize what is possibly the most ambitious program of all times underaken by technology—the fight against mortality.

Thus, the tasks of biomechanics are multiple and vast:

1. Consider those aspects of the structure, function and behavior of a living system that can be described in mechanical terms.
2. Guide the mechanical design of machines (prostheses, tools, and other devices) that complement or assist the living system, or interact with it in any other way.
3. Extract from the analysis of the mechanical aspects of living systems information useful for the understanding and design of man-made systems (bionics [2, 3]).

In basic terms, regardless of tasks, biomechanics is concerned with the entire range of problems of mechanics: the development of constitutive equations, their embodiment in suitable field equations, and the solution of such equations. Furthermore, it is concerned with problems of optimization and of the interaction between mechanical systems and other systems.

Since the task of understanding the mechanics of living systems not only is the necessary premise to the implanting of man-made organs and to the technological imitation of living processes, but also, at this moment, confronts the mechanician with the most unusual and difficult challenges, it will constitute the main focus of the present paper.

3. A Function-Oriented Taxonomy of the Mechanics of Living Systems

A taxonomy of the mechanics aspects of living systems presents considerable difficulties, in view of the complex nature of most biological materials, the intimate interplay of fluid- and solid-mechanics aspects, and the multiple functions and modes of operation of biological organisms and their components. Fortunately, taxonomic perfectionism, if ever a possible or even a useful enterprise, is not required for the purpose of presenting an overview of the role that numerical computations can play in biomechanics. What is sufficient for such a purpose, rather, is a classification that will indicate— accepting overlaps, some vagueness and ambiguity—the major mechanical processes or functions encountered in living systems. The following classification, although not comprehensive or even entirely consistent, endeavors to do so, and lists *some* examples for each category. It groups the principal mechanical processes in five categories: flows, containment, support, material handling, and sensing.

3.1. FLOW PROCESSES

The flow processes in living systems are of a large variety. *Cellular flow processes*, occurring in the interior of the cell, are still far from being completely understood, either in the essential nature of the mechanism causing the flow or in the characteristics of the flow. This is so even in passive cellular elements such as erythrocytes (red blood cells), where flow occurs primarily as a result of the deformation that the cell undergoes while traveling in the circulatory system. Particularly complex are flows occurring in the small channels, a few angstroms in diameter, that characterize cellular membranes, and in which the classical mechanics formulation must give way to quantum hydrodynamics approaches.

In general, from the mechanics viewpoint, the most significant characteristics of cellular flows are two: (1) the problem of understanding and describing a process in which mechanics is closely coupled with biochemical energetics and (2) the theoretical and experimental problems generated by the very small dimensions of the domain within which these flows occur. The same problems occur in the case of intercellular flows—the flows in the spaces between the cells.

Circulatory and conveyance processes present a large number of cases. Understandably, the two processes that thus far have received the greatest attention from the mechanics viewpoint are the blood circulatory system and the respiratory system. More recently, there have been some studies of the flow in the ureters [4], in the gastrointestinal system, and in the crystalline portion of the eye (the convection currents that provide an exchange of oxygen between the outer surface of the eye and the crystalline itself).

For analysis purposes, the blood circulation can be divided into two portions: the macrocirculation, whose primary function is the conveyance of blood from the heart to the tissues in the peripheral beds and back, and the microcirculation, which continues the conveyance process in the peripheral beds and is the seat of exchange processes between the bloodstream and the peripheral tissues. The respiratory system also can be analyzed in terms of its macroportion and its microportion (the processes occurring in the lung alveoli).

The mechanic characteristics of these circulatory and conveyance systems are extremely varied and present problems which are specific to each system or portion of the system. It is fair to say that, at this moment, only few portions of some of the systems have been analyzed from a mechanics viewpoint, and that numerical or analytic studies have been even more limited.

Active propulsion mechanisms are mechanisms whereby propulsion is generated by an action by the organism or one of its parts. These mechanisms present a great challenge to the mechanician, both for the intrinsic elegance and complexity of the mechanics problems involved and because of the opportunities that they offer for technological innovation, that is, for the transfer to the technological domain of solutions evolved by the living organisms. From the purpose of a mechanics analysis, the mechanisms can be divided into two categories: those internal to the organism and those external to it, having as a purpose the propulsion of the organism as a whole.

The mechanisms internal to the organism include lumped-parameter positive displacement mechanisms, such as the heart, and distributed parameter systems, such as the veins or ureters. A mechanism which, in its arrangement and small scale, finds no counterpart in technological devices is the ciliate movement that propels mucus in the bronchial channels.

External active propulsion mechanisms are as varied as the internal ones, ranging from the swimming of fish to the flying of birds to the propulsion of spermatozoa (the latter being a process closely related in its mechanics and energetics to ciliate movements). The propulsion of amebas is significant as an example of an active external propulsion produced by an internal propulsion mechanism—the flow of cytoplasm in the interior of the ameba itself.

Passive propulsion mechanisms are the result of specific configurations of biological organisms, or their components, which allow them to be conveyed by external agents. They are exemplified by windborn seeds or by the motion of erythrocytes. Very little, at this moment, is known of the hydro or aerodynamics of these motions.

Exchange and mixing processes constitute an extremely important category, ranging from the flows in and around leaves to the processes in blood capillaries and in the alveoli of the lungs, to the processes in the stomach and the intestines. The scope for analysis here is extremely large, with very few aspects investigated thus far to any depth—primarily the flow in the capillaries and that in the alveoli.

Lubrication processes play an essential role in making possible the movement of support structures of the organism, and the functioning of other moving components of living systems. They include the sinovial processes in the articulations and the movement of pleural membranes. In general, the processes are highly sophisticated and only partly approachable with conventional lubrication theory.

Many *sensory and communication processes* in living organisms rely on flow mechanisms for emission or reception of signals—from sound emission to hearing to heat sensing. The hearing of humans, and sound, have been investigated for a long time, but there are many aspects of the performance of sensory organs, from the otolyths in more primitive hearing systems to the tympans in the legs of insects, which demand investigation and hold technological promise [2]. So does also the transmission of sound and vibration and, for that matter, of impact, through different regions, organs, or tissues. The mechanics of impact transmission is clearly of major significance in the design of effective protection against impact accidents, from car accidents to explosions.

3.2. CONTAINMENT PROCESSES

The cells, the flow channels inside the organism, as well as the entire organism itself, need to be contained. A large variety of arrangements provide for this need, ranging from membranes to vessel walls to skin. Possibly the most complex membranes are those of cells, since they contain mechanisms for the active transport of essential substances to and from the cell. The visceral pleura is an example of an organ-containing membrane, a membrane, furthermore, with the constraint that it facilitate the movement of its outer surface. The pleura is largely a thick layer of collagen, in many regions running in parallel bands, and of irregular elastin fibers. Its outer surface is moist and slippery to facilitate its movement, while the inner surface is loosely connected to the limiting membrane of the lung lobules beneath.

Skin is a composite material, multilayered and multicellular in cross section. Surfaces that contain the fluids moving inside the organism include blood vessels, the intestinal tract, the stomach, etc. Organs such as the gastrointestinal tract or the respiratory system represent internalization of what originally were external surfaces, and are thus different from containment organs such as blood vessels, that from the beginning were of an internal nature. In the blood vessels themselves, there are profound differences in structure according to the location and function of the vessel.

3.3. SUPPORT AND PROTECTION

In higher animals, the support and protection functions are carried either exoskeletons or endoskeletons. Exoskeletons range from the inflexible outer shield of the turtle to the articulated shell of the lobster. Endoskeletons also assume many configurations. Their parts include multipurpose bone structures such as the legs, which provide both support and propulsion; the arms; the spine, which acquires very different characteristics according to the nature of the animal's support—whether it is supported completely by its environment, as in the case of fish, or it is supported by the contact of its own bones with the environment, as in the case of animals moving on legs. In birds, the endoskeleton structure includes the wing bones, with their dual requirements of strength and light weight.

The mechanics problems in each of these support and protection structures are of a double nature: (1) the static problem connected with the ability of the structure to support the weight of the animal or organ, and (2) the dynamic problem of making the movement of the animal possible, compatibly with the exigencies of the support structure. Interesting examples have been proposed of possible application of the exoskeleton concept to the design of machinery [5].

In the case of support structures in the vegetable domain, exemplified by trunks, stalks, etc., roots, the problem is primarily a static one. The above ground structures, however, are generally very flexible and exhibit daring examples of cantilever arrangements [6].

The support and protection structures either enclose or are surrounded by material that can be generically called tissue. The mechanical properties of this material are important to a complete specification of the response characteristics of a living organism, to an analysis of its locomotion, etc.

Given the extreme heterogeneity of the composition of tissue, and its non-uniform geometric shapes, the establishment of constitutive equations and the solution of field equations present enormous difficulties. Thus, at present, most of the mechanical models available for tissue are of a very simplified nature, based on empirical experimentation.

3.4. MATERIAL HANDLING AND PROCESSING

The handling and mechanical processing of food materials is essential to the survival of many living organisms. The phenomena involved range from oral systems with teeth, to the burrowing and mudsucking systems of some lower organisms. Although the mechanics of teeth and of the oral apparatus present a challenge of first magnitude, and enormous sums are being spent yearly throughout the world for dental care, at this moment this area has not received sufficient attention by the mechanician.

The challenge in the case of teeth arises from a variety of factors, ranging from the structure of the tooth as a composite material, to the mode of support of the tooth, which is essentially suspended in its cavity by ligaments, to the mechanics of the comminution process, to the rheological properties of the food being processed, and to the analysis of combined mechanical and thermal stresses to which teeth are subjected.

3.5. OTHER PROCESSES AND FUNCTIONS

It would be beyond the scope of this paper to continue at great length—as one could—a list of processes and functions. However, it is important to note that the processes and functions reviewed apply also to prostheses, the artifacts intended to fulfill the function of a natural process or organ. Here, the essential problem is one of design—how to integrate successfully with the operation of the biological organism the operation of an artifact, which possesses, in general, entirely different mechanical and chemical character-istics. The complexity of the problem is exemplified by the present dental prostheses—as yet far from perfect—and by the frequent failures of bone implants. In the design of prostheses, such as artificial limbs or artificial hearts, which need to be powered, a major challenge is offered by the in-triguing possibility of transferring to the prostheses muscular power from some other region of the body.

4. Characteristics of Biomechanics Problems

Although the characteristics of the mechanics problems in the situations outlined in the previous section vary from case to case, it is possible to single out some characteristics that are common to essentially all biological situations.

In the first place, the geometries are generally far more complicated than those of man-made objects. For example, in the circulatory system, the conducts (vessels) are never uniform. They are tapered from the regions of higher flow to those of lower flow, have a large number of curves and bifurcations (so a steady-state regime is reached very seldom in practice), and only in first approximation can be considered axisymmetric.

Secondly, the materials are very frequently of great rheological complexity. Blood, skin, vessel walls, cytoplasm, and bone itself are all examples of materials whose rheological properties change with physiological or pathological conditions, are nonlinear and of extremely difficult experimental determination (being, e.g., markedly different when tested on cadavers and in living organisms).

Third, the phenomena are generally time dependent—from the pulsation of blood to the locomotion process to the batch processing of the gastrointestinal system to the periodicities of protoplasmic motion.

Fourth, the dimensions in which the essential characteristics and factors of many biomechanical phenomena manifest themselves are extremely small, pushing to the limits the validity of continuum theories. This is not to say that the microscopic scale is unimportant in other domains of mechanics, but simply that frequently, in such domains, very statisfactory mechanics theories can be developed at the macroscopic level, or on the assumption of a homogeneous microscopic structure. The distinguishing characteristic of biomechanics phenomena is the generally greater heterogeneity of the microscopic or cellular structure.

A fifth characteristic is the intimate and often inseparable combination of several systems that from their very inception have grown together and that functionally cannot be analyzed in isolation. At this moment, no artifact produced by our technology even approaches this characteristic of living organism. The situation in the case of living systems is exemplified by the mechanics of the lung, which, as discussed later, cannot be analyzed without considering simultaneously the characteristics of the circulatory and airway systems that supply it. Thus, while in the mechanics of man-made devices or of inorganic nature, it is frequently adequate to analyze interacting systems on an individual basis and then to describe their interaction only on a lumped-parameter basis (or, if on a distributed parameter basis, by considering only limited regions of interaction), this is often not the case in the study of mechanical process in organic nature.

For instance, gravity waves in a man-made channel can be analyzed with only limited reference to the rheological characteristic of the bottom and walls (or, rather, such rheological characteristics are usually susceptible to simple description). On the other hand, the propagation of pressure waves in the circulatory system is strongly influenced by the complex rheological nature of the vessel wall. In a system such as the lung, the mechanics of the wall of the blood capillaries cannot be separated from that of the alveolar wall in which the capillaries are embedded. (In certain cases, however, even in biomechanics it is possible to acquire a sufficient degree of approximation in the description of a process by neglecting some of the interactions.)

In brief, in biomechanics there is a far greater need for what could be called *synmechanics approaches*, that is, approaches in which several systems must be analyzed simultaneously. Clearly, these approaches are unthinkable without the power of a computer, and hence the recourse to numerical solutions. The particular characteristics of the problems requiring such solutions is that they involve interactions of mechanical processes composed of a large number of elements, each of a complex rheological nature. Thus, one must deal either with continuum idealizations or, possibly more effectively, with stochastic simulations capable of reproducing the time-dependent nature of the biological process and the heterogeneity of the materials involved in it.

5. The Role of Numerical Solutions in Biomechanics

Numerical solutions are needed in biomechanics for essentially the same reasons as in other domains of application of mechanics—the difficulties in obtaining analytical solutions, or the need to simulate complex systems with a large number of variables.

In many biological situations, because of the characteristics outlined in the previous sections, analytic solutions are of much greater difficulty than in the traditional inorganic domains of mechanics, thus accentuating the need for numerical solutions.

Another important reason that heightens the need for mathematical, and, hence, essentially numerical solutions, is the difficulty in obtaining experimental information in situations where the dimensions of the domain to be studied are too small for our instrumentation to reach or where non-destructive tests are necessary for clinical or other purposes. For example, no instrumentation presently exists that can measure the flow inside cells or inside blood capillaries, to the degree of detail necessary to achieve an understanding of the characteristics of the flow processes and to relate them to the physiological process.

Nondestructive testing is exemplified by the determination of the rheological characteristics of the arterial walls for diagnostic purposes, in order to assess their degree of arteriosclerosis and rigidity. Obviously there are measurements that need to be atraumatic, and at present can best be performed indirectly, by measuring some other parameter and relating that parameter, by means of a mathematical model, to the desired variable.

Another reason for the use of numerical studies is the construction of models to assess the performance of a physiological system under abnormal conditions—for example, the dynamics of ambulation in a reduced gravity field or the performance of an artificial heart valve. Clearly under such conditions a numerical simulation becomes necessary, and needs to be complemented, whenever possible, by experiments (or vice-versa).

6. State of the Art Examples of Numerical Solutions in Biomechanics

The nature of some biomechanical problems and the scope and challenges of the pertinent numerical computations can perhaps best be conveyed by specific examples. The first example will consider the problem of the microcirculation; the second, the problem of lung elasticity; and the third, the problem of the arteriole wall.

6.1. MICROCIRCULATION

The microcirculation is the "business end" of the circulatory system [7]. It offers a classical example of a situation in which great experimental difficulties are encountered in obtaining detailed information about the flow process, on account of:

1. The very small dimensions of the components involved. In humans, the arterioles which convey the blood to the capillaries range in diameter from 100 to 7 μm, the capillaries—where the gaseous exchanges take place between blood and the surrounding tissue—have a diameter around 7 μm, and the erythrocytes are biconcave, deformable disks, 7–1 μm in diameter and about 2 μm thick.

2. The nonuniform geometry, which offers extreme difficulties to analytic solutions.

3. The time-dependent nature of the processes which arise from two factors: (*i*) the changes, still largely unexplored, in the rheological characteristics of the erythrocytes produces by the mass transfer processes (the exchange O_2 of and CO) in the capillaries; (*ii*) the pulsating nature of the flow.

The currently available numerical solutions have been restricted to flow in the capillaries, that is, in the vessels of smaller diameter, and have dealt with idealized flow field configurations and only with simplified time-dependent processes.

Figure 1 shows an example of solutions (streamlines) obtained by assuming rigid impermeable boundaries and steady state creeping flow of a newtonian

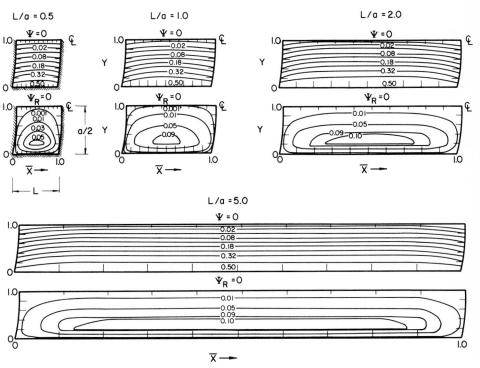

Fig. 1. Streamlines in the plasma between erythrocytes in a simplified model of the capillary [8].

plasma [8]. Figure 2 shows the corresponding shear stresses applied to the capillary by the plasma circulating in the gaps between the erythrocytes. These solutions have been extended to a sinusoidal time variation of the velocity of the erythrocytes [9] and are being extended to the assumption of deformable erythrocytes, treated as membrane-bounded fluid regions of higher viscosity. The crudity of these approaches, complex as they may already appear in terms of the state of the art of numerical solutions, is evident if one considers the real situation, which involves a vessel wall both deformable and highly porous, a vessel that is neither uniform in cross section nor straight, the presence of entrance phenomena, and complex rheological characteristics of the erythrocytes.

No numerical solutions are currently available concerning the flow in the arterioles, particularly in the higher ranges where multiple erythrocytes span the cross section, but the ratio of vessel diameter to erythrocyte diameter is not high enough to enable the flow to be considered homogeneous. The only information available for this portion of the microcirculation is either of an experimental nature, or of a simplified, analytical nature.

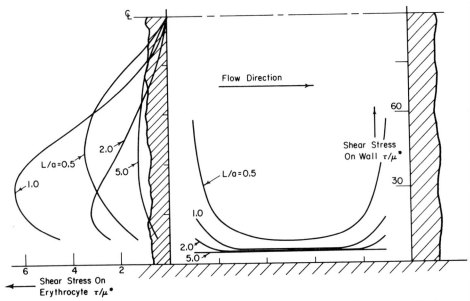

Fig. 2. Shear stresses on capillary wall and erythrocyte due to the motion of plasma in the gaps between erythrocytes [8].

Information about the mechanics and hydrodynamical properties of the erythrocytes, which represent the key elements and the raison d'étre of the circulation, is also extremely limited. Figure 3 shows sample results of a finite-element solution for the flow around two-dimensional erythrocyte shapes [10]. Except for the capillary simulations mentioned earlier, no detailed numerical information is available for three-dimensional shapes, or for the deformation of erythrocytes—shaped bodies of prescribed rheological characteristics under the influence of shear fields or of neighboring erythrocytes.

6.2. Lung Elasticity

Figure 4 is a schematic diagram of a pulmonary region showing the alveoli and surrounding walls. The walls of the alveoli contain blood capillaries, surrounded by collagen and elastin fibers that connect to other walls or structures. The healthy human lung contains approximately 4×10^8 alveoli, whose interconnected walls form a continuous elastic network. The structure of the meshwork is not homogeneous. The airways (alveolar ducts) leading to the alveoli, which contain 25 % of the total lung gas volume, are cylindric in shape. In each respiratory unit formed by the alveolar duct, and associated alveoli, there are location-dependent gradients of the amount of elastin, collagen, and smooth muscle. The characteristics of the three

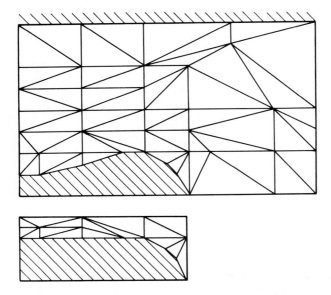

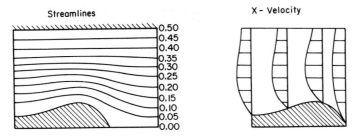

Streamlines   X - Velocity

Fig. 3. Fine-element solution for creeping flow around a two-dimensional erythrocyte shape [10].

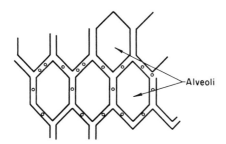

Alveoli

Fig. 4. Idealized cross section of a region of the lung meshwork.

systems of branching tubes—the pulmonary arteries, the pulmonary veins and the airways—which come together in the lung are also different from those of the lung meshwork. By being intimately connected to it, they are likely to influence the deformation of the meshwork, although no sufficient information is available on this point.

As summarized in a recent review [11], state of the arts biomechanics characterizations of the healthy lung are based on the assumption of a single degree of freedom system, with elastic behavior governed by a lung compliance C, so that the pressure difference ΔP between the airway opening and the pleural surface enveloping the lung is expressed by a compliance term $\Delta V/C$ (where V is the lung volume), by a term expressing the flow resistance of the system of multiple-branch pathways, and by an inertance term, function of the lung volume acceleration. Under conditions of no flow or very small flow, the resistance and reactance terms vanish or become very small, so that ΔP becomes approximately equal to $\Delta V/C$—a relation used to determine the elastic properties of the lung meshwork.

Clearly, this unidimensional approach is far too unsophisticated to provide but the grossest physiological and anatomical information. Thus, in very recent years, attempts have been made to consider the lung as an interconnected network [12], to provide continuum formulations and to develop finite element approaches, concurrently with experimental techniques for measuring the coefficient of the element stiffness matrix [11]. Even these approaches, however, continue to be limited when compared to the complexity of the problem to be solved.

6.3. Wall of Arterioles

The entire circulatory system is lined by a layer of endothelial cells. In the arterioles, the endothelial layer is sheathed by a single layer of muscle cells, which allow the arterioles to contract or expand. Figure 5 shows the cross-

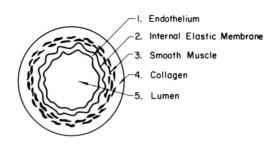

Fig. 5. Cross section of a partially constricted frog arteriole ([13], from an electron micrograph).

section of a partially constricted frog arteriole; the figure was constructed from an electron micrograph [13]. The endothelium is structurally very soft, and hence has a very limited load-carrying capacity. The internal elastic membrane is believed to have an instantaneous modulus of elasticity, similar to that of a relaxed smooth muscle [14]; on the other hand, when smooth muscle is constricted, its modulus can become up to an order of magnitude greater than that of the membrane.

Smooth muscle creeps, but collagen fibers do not. The behavior of the composite material formed by these two elements has been represented by Peterson [15] with the model shown in Fig. 6, which assumes that smooth muscle behaves as a Kelvin–Voigt material and is connected in series to collagen. Gerstenkorn *et al* [13] have challenged the validity of this assumption. Changes in the circumference of the arteriole are affected by tensile forces which must be exerted through the collagen, since muscle cells do not attach to one other. Collagen could exert tensile forces without contribution by the smooth muscle, only when the muscle is relaxed with little visco-elastic response.

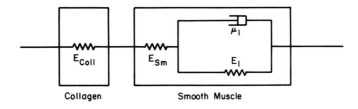

Fig. 6. Kelvin–Voigt model of smooth muscle [15].

A direct stiffness method using finite-element analysis has been applied by Wiederhielm *et al.* [16] to the analysis of the structural response of relaxed and constricted arterioles. Different material properties were assigned to different triangular elements, by using as anatomical components of the wall smooth muscle cells (considered elastic in the relaxed state and viscoelastic in the contracted state) and two types of collagen fibers—one type recticular, circumferentially oriented fibers with a high modulus of elasticity, and the other axially oriented fibers in the exterior portion of the blood vessel, with a low modulus of elasticity. Uniform material properties, states of uniform stress and strain, and isotropic material properties were assumed for each element (although, if known, orthotropic material proper-ties could have been assumed). The wall thickness of approximately 10 μm was subdivided in 14 layers of triangular elements. The numerical results yielded the deformation and the stress histories of the wall under load.

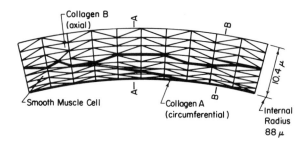

Fig. 7. Discrete element idealization of the wall of an arteriole [16].

A more recent analysis by Simon *et al.* [17], applied to a larger vessel, the aorta, has demonstrated the importance of utilizing finite deformation analysis, rather than modified linear theories, when precise states of stress and deformation are desired. For instance, the Lamé cylinder formula was shown to underestimate the maximum tangential stress by as much as 80%, and to fail to reveal the significant tangential stress gradients that are present also in thin wall vessels. The study shows that only by viewing the anisotropic response of the aorta to finite deformations, it is possible to correctly identify the necessary material properties.

7. Conclusions: Artimechanics, Physimechanics, and Biomechanics

The foregoing examples have attempted to indicate some of the distinguishing characteristics of mechanics phenomena in living systems, and the importance of numerical methods in providing descriptions of such phenomena. Because of the power of the direct stiffness method in dealing with nonhomogeneous materials, numerical solutions of finite elements offer, at this moment, considerable promise in a multitude of problems concerned with the mechanics of tissues and of support or containment materials, such as bone or membranes. On the other hand, achievement of a realistic description of flow processes continues to present extreme difficulties, and so does, as a consequence, the description of complex organs or regions, such as the lung mesh, that combine several different flow systems and containment structures.

Although basic laws and procedures of mechanics do not change with domain of application (in the sense that in every domain one is concerned with the development of constitutive equations, their embodiment in field equations and their integration over a given region), the characteristics of the mechanics problems vary greatly according to the domain considered. In this sense, it appears useful to differentiate among what can be called *physimechanics*, or the mechanics of inorganic nature, *artimechanics*, or the mechanics of man-made objects (artifacts) and *biomechanics*, or the mechanics of living nature. Biomechanics is concerned with all living systems; physimechanics with physical phenomena or materials such as winds, waves, soils, rocks, and earthquakes; and artimechanics with objects such as beams, tubes, containers, hulls, and wings, and with man-made materials such as cement, steel, or plastics. (Man-shaped natural materials such as wood or leather are part of both the biomechanics and the artimechanics domain.)

In each of these domains, significant differences with regard to geometry, time dependency, material complexity, scale of the phenomena and synmechanics effects, require the development of approaches particular to that domain. Until now, the greatest development has occurred in the mechanics of the man-made, because of the generally greater simplicity that at the present state of our technology, each of these five factors has in that domain. Most of the current developments in biomechanics have been based on the extension of techniques developed in the artimechanics domain. Boundary conditions have been idealized, material behavior has been described by simple constitutive equations, small scale phenomena have been approached in terms of continuum theory and so on. While our approaches to the mechanics of the man-made involve idealizations also, the idealizations that we have thus far been forced to use in the description of biomechanics phenomena have been generally far more drastic.

Hence, the greatest challenge is offered to the mechanician and to the numerical methods specialist in developing logical and computational approaches that can enhance our power to (1) analyze living systems, (2) utilize in our technology the lessons learned from such an analysis, and (3) move from today's prostheses to truly symbiotic man–machine systems.

References

1. Eyth, M., *Der Schneider von Ulm*. Duetsche Verlags-Anstalt, Stuttgart and Berlin (n.d.).
2. Bugliarello, G., "Engineering Implications of Biological Flow Processes," in *Chemical Engineering in Medicine* (Leonard, F., ed.) Vol. 62. Amer. Inst. Chem. Engrs., 46–64, 1966.
3. Bugliarello, G., Wong, A. K. C., Hung, T. K., Weissman, M. H., Ahuja, A. S., and Stark, K. P., "Hydro-Bionics: A Perspective on Technological Implications of Biological Flows," in *Advances in Hydrosciences* (Chow, T. C., ed.).
4. Hinman, F., Jr., (ed.), *Hydrodynamics of Micturition*. Thomas, Springfield, Illinois, 1971.
5. Bootzin, E., Galbavy, A., Melow, G., and Muffley, H., "Weapons Command Biomechanics Investigations," in *Bioengineering, an Engineering View* (Bugliarello, G., ed.), San Francisco Press, San Francisco, California, 1968.
6. Hertel, H., *Struktur Form Bewegung*. Krausskopf Verlag, Mainz, 1963.
7. Bugliarello, G., "Some Aspects of the Biomechanics of the Microcirculation, Anno 1969," in *Developments in Mechanics* (Weiss, H. J., Young, D. F., Riley, W. F., and Rogge, T. R., eds.), Vol. 5, pp. 921–962. Iowa State Univ. Press, Ames, Iowa, 1969.

8. Bugliarello, G., and Hsiao, G. C. C., "Mathematical Model of the Flow in the Axial Plasmatic Gaps of the Smaller Vessels," *Biorheology* **7**, 5–36 (1970).
9. Hung, T. K., Weissman, M. H., and Bugliarello, G., "A Numerical Model for Oscillatory Flow and Oxygen Transfer in the Axial Plasmatic Gaps of Capillaries. I Two-Dimensional Disk-Like Rigid Erythrocytes," *Proc. Int. Conf. Hemorheol.*, *2nd* Heidelberg (July 1969) (in press).
10. Bair, R., and Bugliarello, G., "Finite Elements Solution of Flow Around a Two-Dimensional Discoid," Presented, Fluid Mechanics Session, February, 1969, Nat. Meeting Amer. Soc. of Civil Eng. (unpublished).
11. Lee, G. C., and Hoppin, F. G., Jr., "Lung Elasticity," in *The Foundations Objectives of Biomechanics* (Fung, Y. C., ed.) (in press).
12. Mead, J., Takashima, T., and Leith, D., "Stress Distribution in Lungs," *J. Appl. Physiol.* **28**, No. 5, 596–608 (1970).
13. Gerstenkorn, G. F., Kobayashi, A. S., Weiderhielm, C. A., and Rushmer, R. F., "Structural Analysis of an Arteriole by the Direct Stiffness Method," *ASME Trans. J. Eng. Ind.* #**66-HuF6**, 1–6 (1966).
14. Bergel, D. C., "The Static Elastic Properties of the Arterial Wall," *J. Physiol.* **156**, 445–457 (1961).
15. Peterson, L. H., Jensen, R. E., and Parnell, Jr., "Mechanical Properties of Arteries in Vivo," *Circ. Res.* **8**, 622–639 (1960).
16. Wiederhielm, C. A., Kobayashi, A. S., Stromberg, D. D., and Woo, S. L. Y., "Structural Response of Relaxed and Constricted Arterioles," *J. Biomech.* **1**, 259–270 (1968).
17. Simon, B. R., Kobayashi, A. S., Strandness, D. E., and Wiederhielm, C. A., "Large Deformation Analysis of the Arterial Cross-Section," *Trans. ASME* Paper 70-WA/BHF-15.

The Computer in Ship Structure Design

H. A. Kamel, D. Liu, and E. I. White

UNIVERSITY OF ARIZONA
TUCSON, ARIZONA
AMERICAN BUREAU OF SHIPPING
NEW YORK, NEW YORK
STANDARD OIL OF CALIFORNIA

1. Introduction

The area of ship structure design and construction is undergoing continuous change. New configurations as well as previously unattempted vessel sizes are constantly emerging. The use of computers in engineering analysis has been developing and gaining popularity over the last 25 years. The computer hardware is and has been in a continuous state of evolution. Basic computer software has been available for a long time for general scientific and engineering use. Specialized-application software packages designed for the solution of problems in particular engineering disciplines have also been made available during the last decade. Computer software in general and engineering software in particular have also been in a continuous state of change. On the one hand, the software is dependent on the available computer hardware and on the other hand, engineers are becoming increasingly aware of the capabilities of digital computers and their possible application to facilitate engineering analysis and design.

With this turbulent picture in mind, where everything is continually changing and where all items are interdependent, it becomes a formidable task to try to do any crystal ball gazing as was suggested as one of the purposes of this paper. However, we shall attempt to look at the present state of the art and project to the near future. In this attempt a certain amount of bias is unavoidable, so the picture will be distorted.

We shall also take the opportunity to describe some techniques that are already in use, and some that, we believe, may have an important bearing on what is to come. We ask the indulgence of the reader if some of our prophecies are overpowered by the hard facts of life, or if some important current developments in the field are overlooked.

In our approach to the subject we shall try hard to look at ideas with their ultimate usefulness in mind. Investigations of academic nature are useful only in the initial stages of development, where techniques are proved or disproved. In engineering we worry not only about the accuracy of methods in the solution of simple classical problems but also on the reliability of the methods when applied to complex, real-life situations. We are also concerned about the cost and time involved as well as meeting deadlines. It is only when all of the considerations are taken into account that we can talk of engineering analysis.

2. Current and Future Developments in Computer Hardware

The current trends in computer developments from the engineering application viewpoint seem to point in a number of directions that may be summarized as follows:

1. Computer speeds are increasing.

2. Computers with larger core memories are available, including such options as the fast, extended-core memory.

3. Instruction repertoires are becoming more sophisticated. As an example, single operations involving the simultaneous processing of a large number of data are becoming available.

4. Hardware and software enables time sharing and interactive terminals.

5. A very important development has been the marketing of mini- and medium-sized computers that are relatively inexpensive and capable of performing small computations quite efficiently. The low cost of acquiring such units make them particularly attractive for interactive analysis.

6. A great leap forward, from the engineering point of view, is the introduction and marketing of interactive graphics terminals. They can be used either independently together with a mini- or medium-sized computer; or the system may be connected to a large installation. The former arrangement is less expensive to acquire and less problematic in its operation.

3. Current and Future Developments in Structural Analysis Software

A number of large-scale, general-purpose, finite-element programs are now available to the engineering community. Of those, one may mention NASTRAN, ASKA, STARDYNE, STRUDL, and DAISY. These computer codes represent the first generation of engineering analysis programs that were developed with a large, fast computer in mind. The second generation of structural analysis programs will have to take into account the availability of improved man–machine communication through the use of interactive graphics terminals [1, 2].

An engineering applications package should be written with the following objectives in mind.

1. Current engineering problems should be solved more efficiently and more satisfactorily. The emphasis should be clearly on real-life engineering problems rather than those of academic interest. Efforts should be directed toward handling complex structures easily and efficiently.

2. New advances in hardware design must be utilized to the greatest advantage. Periodically, one should be able to pause and reassess progress so far achieved and reevaluate the goals toward which the software development is directed.

3. The ideal engineering application program will use as input an engineering language similar to that which engineers use to communicate among themselves. Input will not be only in the form of numerical data on cards or through teletypes, but through the most convenient choice of several possibilities such as numerical data, tags and labels, or graphic input through a light pen or wired tablets.

4. In the program all possible time-consuming operations that can be performed by the computer will be automated. The user will retain the right to veto any of the automatically generated data and to adjust it to what his physical instinct and practical experience dictates.

5. The program will allow a multishot approach to solving engineering problems with feedback and easy interfacing between the different solutions rather than a single shot aimed at obtaining all the required results in a single large computation.

4. Relationship between the Computer and the Engineer

The digital computer is a recently acquired engineering tool. As all tools it should be regarded as an extension to the user's personality and an amplifier of his capabilities. Computers should be programmed to reduce the amount of work required for the solution of an engineering problem and to make possible the solution of problems that were thus far impossible. The engineer should be retrained to recognize problems that may be conveniently solved and change the emphasis on available methods and techniques. The present relationship between the digital computer and the engineering user leaves a lot to be desired. Far from always being the helpful tool, the computer often enslaves the user. A constant effort is still required to make this relationship more satisfactory.

The importance of the communication problem between man and machine can be clearly appreciated if we study typical expenses and effort encountered in the full scale-analysis of a large vessel. If the total project costs, say, $40,000, $2000 of this amount is required to perform the major analysis on the computer and another $2000 for a series of local analyses to obtain a more detailed stress picture. The remaining expenses are solely to pay for manpower and computer time involved in generating the data and the subsequent evaluation of the results. If such a project may last over a period of 3 months, perhaps only 4 hours of error-free computation are employed to reach the numerical results.

From these figures it is clear that the major problem, which manifests itself in the form of expenditure and time waste, is that of communication. The authors believe that the main effort in the future will be directed toward man–machine interaction. The current use of interactive computer terminals is directed mainly toward solving small problems rather than handling large amounts of data.

5. Future Trends in Ship Structure Analysis

Before embarking on the description of specific techniques that show promise for the future, it may be appropriate to venture into the formulation of some broad prophecies as to how engineering computer analysis will develop in the next decade.

1. The ease of communication between computer and user will enable more intelligent use of the computer as an engineering tool.

2. The more intelligent use of the computer will result in more economical analysis methods based on the solution of a large number of small size problems, feedback, and follow-through capabilities at a greatly reduced cost.

3. In the battle between mini-computers versus time-sharing based on a large computer installation, the mini- and medium-sized computer are at a distinct advantage in engineering applications. The large computers will be mainly used for "number crunching" and will tend to serve as a slave to the

mini-computer in such a setup. Also, the decade will see the growing use of mini-computers as independent units.

4. Graphics display terminals will play an extremely important role in engineering analysis. Since these terminals are more economical when used with mini-computers, many small and medium-sized organizations will prefer such smaller systems for everyday use rather than sharing a large installation. Larger installations will be useful for scientific work and as a backup unit for such small installations.

5. Engineering analysis programs will acquire a more interdisciplinary character. It will be, for example, a safe assumption that the same program package will compute the hydrodynamic and inertia forces on a ship at the same time as the subsequent stress analysis.

6. Computer programs will be used more often in the design stage rather than as a computational check on the final design produced.

7. Large, bulky, general-purpose programs will give way to more intelligent interactive program packages, where the main emphasis will be on engineer–computer interaction.

6. Problems and Useful Techniques in Interactive Analysis

In operating interactively it is ideal to reduce the amount of time required for each computational step to a minimum. This helps to keep the attention of the user, and allows him to interfere between the different steps, thus exercising his personal judgment in order to speed up the computation. Whether the user is interacting with a large computer through a satellite computer, or using an independent off-line mini-computer, it is of great importance to reduce the amount of data handled. Operation with an interactive terminal, or computer, will greatly facilitate model generation and interpretation of results. The interactive nature of the system will also encourage the interfacing or nesting of various solutions to achieve the final results.

A new research program at the University of Arizona was initiated in 1971. It is designed to investigate the possibilities of using interactive graphics to speed up the process of analysis. We shall devote the rest of the paper to describing techniques that show promise applicable to ship structure analysis and lend themselves to interactive analysis. We shall also describe techniques specifically designed to handle the following problems:

1. Creating a more accurate local solution out of a less detailed overall solution.
2. The concept of substructures, and the related but more general procedure of the reduced substructure.

3. Successive problem-size reduction, or condensation aimed at speeding up the solution.
4. Saving of storage space in a small computer. (Sparsely populated matrix representation.)
5. Iterative solution techniques suited to interactive operation.
6. Automatic data generation and interpretation of results.
7. Data banks, or a library of optimally designed, precomputed structural modules.

Owing to constraints of space, reference will be made often to other sources for further information.

7. The Local Analysis Procedure

Local analysis is the process of obtaining a more accurate local solution based on a previously conducted overall analysis. Paulling [3] refers to it as "zooming," an appropriate and catching name.

The major problem encountered in such an analysis is that of extracting the relevant data for the overall solution for use as boundary conditions in the local model. One may try to match the grid eactly at the boundary (see Fig. 1) so that displacement results from the major analysis may be directly used in the local model. This process may produce awkward and unsatisfactory grids.

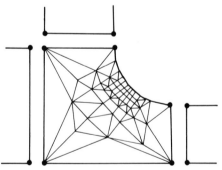

Fig. 1. Local analysis of bracket with matching boundary points.

Another alternative is to generate the local grid for the local solution independently of the overall solution, taking care, however, that the local model possesses nodes corresponding to those of the major analysis (Fig. 2). It would be erroneous to allow the intermediate or excessive boundary nodes of the local solution to deflect freely; therefore, some boundary conditions must be assumed. One possibility is to use force resultants from

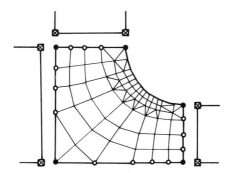

Fig. 2. Local analysis of bracket with excessive boundary points.

the major solution. On the other hand, one may use a suitable interpolation function to determine the deflections of these intermediate points from the known deflections of the overall analysis. The interpolation procedure may result in some disturbance of the static boundary conditions at the interface. This error, however, is of a minor nature.

A generalization of this procedure is that of the reduced substructure. In this approach, the overall equilibrium at the boundary is satisfied and the local analysis is actually incorporated in the major analysis. It will be described in the next section.

8. The Reduced Substructure Technique (RESS)

One of the methods that has been under extensive investigation may be regarded as a generalization of the substructure technique and at the same time of the local analysis, or zooming method. Since 1970, it has found many applications and has proved to be of great use.

Figure 3 shows a local area of a complex structure which is to be modeled fairly accurately without increasing the number of degrees of freedom of the main analysis. The boundary of the structural model in question is to be connected to the rest of the structure at a number of points shown by the full dots in the figure. The remaining points on the boundary, shown by the hollow circles, are necessary for the ease of generation of the local mesh. A process of automatic boundary interpolation is employed by which the displacements of the intermediate boundary nodes, called the dependent nodes, can be directly related to the master or the external nodes. The internal nodes are left free to move. This technique will produce what we term a "reduced substructure" (RESS).

The necessary mathematical formulation for generating a RESS structural module will now be obtained if we subdivide both the deflection matrix **r**

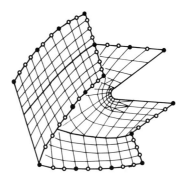

Fig. 3. Use of the reduced substructure technique in ship structural analysis.

and the load matrix **R** in three segments each so that

$$
\mathbf{r} = \begin{bmatrix} \mathbf{r}_e \\ \mathbf{r}_d \\ \mathbf{r}_i \end{bmatrix}, \qquad \mathbf{R} = \begin{bmatrix} \mathbf{R}_e \\ \mathbf{R}_d \\ \mathbf{R}_i \end{bmatrix} \tag{1}
$$

whereby the subscript "e" stands for external (master), the subscript "d" denotes dependent boundary points and the subscript "i" denotes internal or free points.

The stiffness relationship in partitioned form may be obtained by subdividing the master stiffness matrix correspondingly. Then

$$
\begin{bmatrix} \mathbf{K}_{ee} & \mathbf{K}_{ed} & \mathbf{K}_{ei} \\ \mathbf{K}_{de} & \mathbf{K}_{dd} & \mathbf{K}_{di} \\ \mathbf{K}_{ie} & \mathbf{K}_{id} & \mathbf{K}_{ii} \end{bmatrix} \begin{bmatrix} \mathbf{r}_e \\ \mathbf{r}_d \\ \mathbf{r}_i \end{bmatrix} = \begin{bmatrix} \mathbf{R}_e \\ \mathbf{R}_d \\ \mathbf{R}_i \end{bmatrix} \tag{2}
$$

Equation (2) may be written as three separate equations

$$
\mathbf{K}_{ee}\mathbf{r}_e + \mathbf{K}_{ed}\mathbf{r}_d + \mathbf{K}_{ei}\mathbf{r}_i = \mathbf{R}_e \tag{3}
$$

$$
\mathbf{K}_{de}\mathbf{r}_e + \mathbf{K}_{dd}\mathbf{r}_d + \mathbf{K}_{di}\mathbf{r}_i = \mathbf{R}_d \tag{4}
$$

$$
\mathbf{K}_{ie}\mathbf{r}_e + \mathbf{K}_{id}\mathbf{r}_d + \mathbf{K}_{ii}\mathbf{r}_i = \mathbf{R}_i \tag{5}
$$

The assumed relationship between the dependent and external nodes may be represented by

$$
\mathbf{r}_d = \mathbf{T}_{d,e}\mathbf{r}_e \tag{6}
$$

whereby $\mathbf{T}_{d,e}$ is automatically generated.

By substituting Eq. (6) in Eq. (5), one may solve for $\mathbf{r}_i$ in terms of the internal loads $\mathbf{R}_i$ and the external master displacements $\mathbf{r}_e$.

$$
\mathbf{r}_i = \mathbf{K}_{ii}^{-1}\mathbf{R}_i - \mathbf{K}_{ii}^{-1}(\mathbf{K}_{ie} + \mathbf{K}_{id}\mathbf{T}_{de})\mathbf{r}_e = \mathbf{K}_{ii}^{-1}\mathbf{R}_i + \mathbf{T}_{i,e}\mathbf{r}_e \tag{7}
$$

where

$$T_{i,e} = -K_{ii}^{-1}(K_{ie} + K_{id}T_{de})$$ (8)

It is now required to obtain a set of reduced loads R_e* to be transferred to the main structure as representative of the total load vector R on the RESS module. It is difficult to obtain R_e* in terms of the internal loads. However, one may approach the problem from an engineering viewpoint through the introduction of a reasonable assumption. If the internal loads did not exist, then the matrix $T_{i,e}$ given by Eq. (8) yields completely the relationship between internal displacements and external displacements. It will be assumed that the effect of local loading on that relationship is small so that, *for the purpose of load reduction alone*, one may assume the simple relationship,

$$r_i = T_{i,e}r_e$$ (9)

Using Eqs. (6) and (9) in the total displacement vector r given by Eq. (1) one obtains

$$r = \begin{bmatrix} r_e \\ T_{de}r_e \\ T_{ie}r_e \end{bmatrix} = \begin{bmatrix} I \\ T_{de} \\ T_{ie} \end{bmatrix} r_e = T_e r_e$$ (10)

where T_e is the transformation matrix yielding the complete displacement vector r from the external displacements r_e alone. It may be interesting to note that both $T_{d,e}$ and $T_{i,e}$ satisfy the rigid-body movement criterion so that, should r_e represent a rigid body movement of the RESS module, then no internal stresses and strains will be produced.

Using energy principles, the work done by the reduced external forces over the external displacements must be equal to the total work done by the complex force system over the total displacement vector, hence

$$r_e{}^t R_e* = r^t R = r_e{}^t T_e{}^t R$$ (11)

Since this is valid for any set of external displacements r_e we deduce that the reduced forces are to be computed from the expression

$$R_e* = T_e{}^t R$$ (12)

and since the matrix T_e satisfies the rigid-body displacement criterion it is clear that the reduced force R_e* will represent correct static equivalent forces to the total force vector R. Should our RESS deform such that the internal displacements r_i due to the master displacements r_e will be dominant over the local R_i effects then

$$R_e* = R_e + T_{de}^t R_d + T_{ie} R_i$$ (13)

should be sufficiently accurate.

It is also now possible to derive a reduced stiffness matrix $\mathbf{K}_{ee}^*$ for the RESS module using the original stiffness matrix together with Eq. (10) in order to obtain

$$\mathbf{R}_e^* = \mathbf{T}_e{}^t\mathbf{Kr} = \mathbf{T}_e{}^t\mathbf{KT}_e\mathbf{r}_e = \mathbf{K}_{ee}^*\mathbf{r}_e \qquad (14)$$

where

$$\mathbf{K}_{ee}^* = \mathbf{T}_e^t\mathbf{KT}_e \qquad (15)$$

The best method to compute the reduced stiffness $\mathbf{K}_{ee}^*$ is to substitute the appropriate relationships in Eq. (3) directly and take account of the repeated occurrence of certain matrix expressions in order to reduce the computation time.

The computation of a reduced stiffness matrix and a set of generalized external forces will then be used, together with similarly computed RESS modules, and/or a collection of finite elements and simple loads, in order to produce an overall structural model which may be solved for displacements in the ordinary manner. A local analysis must be subsequently performed on the RESS module. Using the previously obtained $\mathbf{T}_{i,e}$ as well as the detailed internal loads $\mathbf{R}_i$, the internal displacements $\mathbf{r}_i$ may be obtained from Eq.(7). The total displacement vector is given by Eq. (10). Subsequently, the detailed stress picture may be computed. From a number of numerical investigations as well as from application to several full-sized ship analyses it has been concluded that internal stresses generated within RESS modules are reliable. The stiffness representation of such modules, as in the case of tanker web-frame brackets (see Fig. 4) is definitely superior to that obtained by simple finite elements. The method has also many other attractive features, such as greater economy with a relatively large number of degrees of freedom as well as the availability of intermediate checks before each module is assembled to the large-scale model. It is hoped that more details of the method and more examples will follow in a later publication.

9. Automatic Data Generation and Interpretation of Results

From previous discussions, it is apparent that the problem of generation of the mathematical model required for the analysis is very demanding in manpower, computer time, and schedule time. We firmly believe that a continuing effort has to be exerted in this direction so that the problem can be solved more satisfactorily. Kamel and Eisenstein [4] should be consulted for details of preliminary work done in this area at the University of Arizona. At the moment, a PDP-15 interactive graphics computer is in the process of being acquired by the University for stimulating research and development in this area.

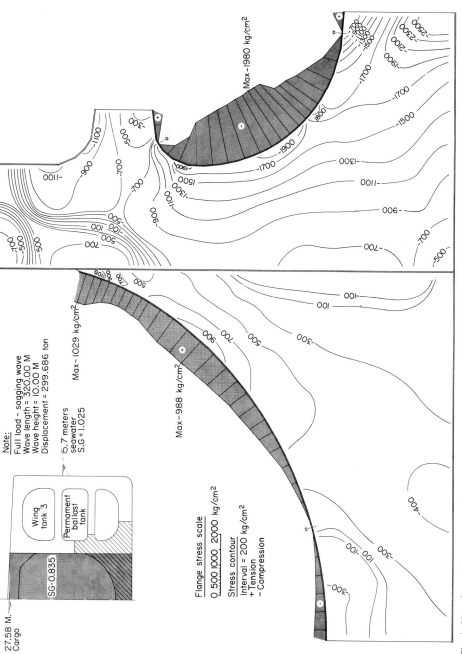

Fig. 4. Use of two-dimensional RESS modules in web-frame bracket analysis. (a) Maximum principal stress contours and flange stresses. (b) Frame 74.

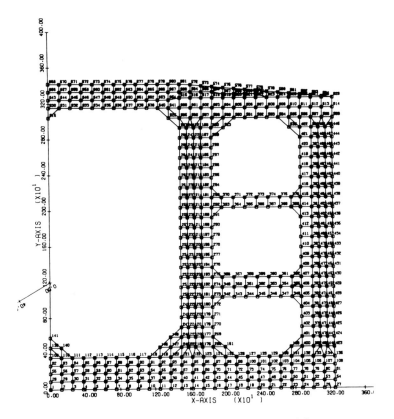

Fig. 5. Automatically generated mesh of a tanker web frame.

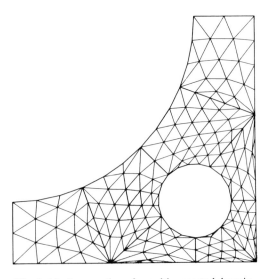

Fig. 6. Mesh generation of a multiconnected domain.

To give an idea, however, of what directions are being followed, Fig. 5 represents an automatically generated mesh for a web frame using MEG, a FORTRAN IV automatic mesh generator designed for complex structures. Figure 6 shows results obtained from another experimental mesh generator, AMEGO, that is more suited to membrane and solids problems. It can be seen from this figure that the resulting mesh, although generated with great ease, is locally unsatisfactory. A solution that approaches the optimum more uniformly can be obtained in a direct manner using an interactive graphics computer such as the PDP-15. Figure 7 shows a RESS module generated using MEG. Other mesh generators are available that are specifically

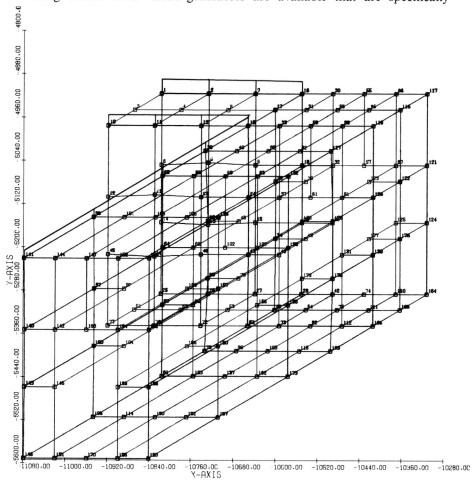

Fig. 7. Use of a three-dimensional RESS module in the analysis of the hatch corner region in a container ship.

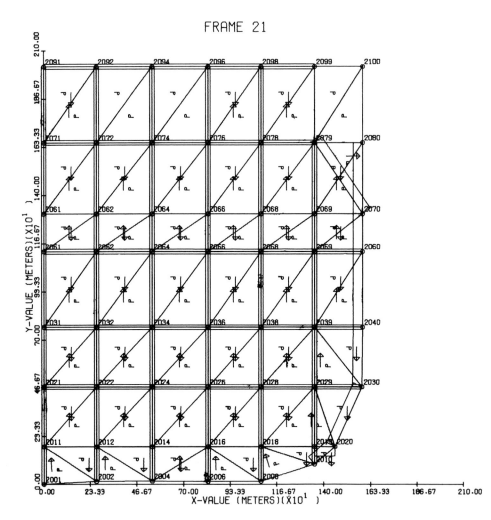

Fig. 8. Automatically generated transverse bulkhead used in the overall analysis of a contained ship.

designed to handle ship structures. Figure 8 gives part of a model automatically generated by EXAM, a program successfully developed and applied to the overall analysis of supertankers [5] and open-deck vessels.

10. Data Banks

The idea of having precomputed structural modules which may be assembled using a simple scalar factor to denote thickness adjustments may be greatly helpful in speeding up and rationalizing the design process in ship structures. The idea has been used by Gellatly [6] in structural optimization, and may be used to considerable advantage in conjunction with the previously described RESS modules in ship structural analysis. With such a capability, one could build up an ideal team relationship between a small interactive computer and a large system that may be programmed to provide the small computer with the major part of the data required to perform the analysis, or even the design. The data precomputed by the large system may take the form of stiffness coefficients, dependent on the thickness parameter, a series of stress constants denoting maximum stresses encountered within the RESS module due to external stimuli, as well as the necessary information to reduce detailed loading to a set of statically or kinematically equivalent corner loads. With some stretch of the imagination we may envisage a standardized library of such modules that may be made available to ship designers everywhere and may be helpful in the preliminary and detailed design stages. Undoubtedly, however, freedom must be left to the designer to develop and compute the properties of alternative modules that he may employ.

11. Multistep Solutions

As mentioned before, the solution of a complex problem may be achieved through the solution of a large number of smaller ones, using perhaps a minicomputer. An ideal system should allow for automatic zooming by which the local analysis may be performed subsequent to a major analysis without the loss of any of the data produced by the major analysis in the process. The program should allow for the merging of several small solutions in order to produce a larger structure. It should also permit, as will be described later, an efficient interative solution technique based on the condensation of a large model to a smaller one using intelligently chosen automated initial guesses and then expanding the results back to the original full size for further local improvement. We believe that such a system combined with data banks, graphic representation of results, and automatic model generation could well be ideal for ship structural design using a small interactive computer backed up either on-line or off-line by a larger system.

The interactive nature of a small computer, together with the informality of approach, and immediate accessability may indeed alter our concept altogether of how equations may be solved. Continuous iterations in order to improve the solution may run independently of and in parallel with the modeling procedure. The stiffness coefficients may be continuously improved, the number of degrees of freedom expanded, the number of finite elements increased, thickness readjusted, without an actual pause ever occurring.

The process of modeling and analysis may become one, dynamic, ever-changing process. Idle time spent on the graphics terminal will be fully utilized by the system to do its "number crunching" quietly behind the scenes.

12. Dynamic Modeling Using Compatible Finite Elements of Different Order

Certain families of finite elements, such as the TRIM3, TRIM6 and TRIM10 element family (see Fig. 9a) are based on displacement functions of different order. It is known that the lower-order elements, such as the TRIM3 give poor results in areas of rapid stress change, yet, in the extreme

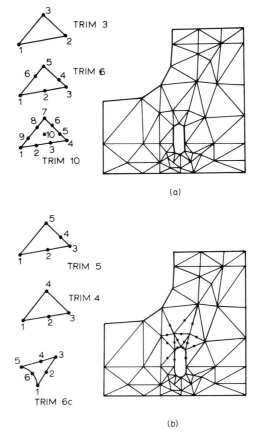

Fig. 9. Dynamic modeling using complete polynomial elements of different order. (a) TRIM 3 model of a bracket with opening; (b) bracket modeled using interface elements.

case of a constant stress field, it yields the exact solution with a minimum number of elements. Higher-order elements such as the TRIM6, and, even more so, the TRIM10 are well suited to areas of high stress gradients. From the point of view of computation time, the TRIM3 elements are superior, since their properties are determined more rapidly, and the bandwidth of the equations is considerably less than the more sophisticated types. The members of the element family are incompatible as they stand, and so arguments arise as to which is the better choice for a particular problem.

It would be possible to employ the various element types in the same solution, while enforcing compatibility along the common boundaries by developing special interface elements of which we give an example.

Starting with a TRIM6 element, we define its node displacement as

$$\mathbf{p}_{(12,1)} = \{\mathbf{p}_1 \quad \mathbf{p}_2 \quad \mathbf{p}_3 \quad \mathbf{p}_4 \quad \mathbf{p}_5 \quad \mathbf{p}_6\} \tag{16}$$

where $\mathbf{p}_i$ gives the two displacement components of node i.

The corresponding node force vector is given by

$$\mathbf{P}_{(12,1)} = \{\mathbf{P}_1 \quad \mathbf{P}_2 \quad \mathbf{P}_3 \quad \mathbf{P}_4 \quad \mathbf{P}_5 \quad \mathbf{P}_6\} \tag{17}$$

where $\mathbf{P}_i$ gives the load on node i.

The $\mathbf{k}$ matrix is of size $(12, 12)$ and can be computed in the standard manner [7, 8].

We now derive an interfacing element of the type TRIM5, in which the side 5, 1 remains straight, whereas the deflection along the other two sides assume a second-order distribution from the rigid TRIM6 element by assuming the relationship

$$\begin{bmatrix} \mathbf{p}_1 \\ \mathbf{p}_2 \\ \mathbf{p}_3 \\ \mathbf{p}_4 \\ \mathbf{p}_5 \\ \mathbf{p}_6 \end{bmatrix} = \begin{bmatrix} \mathbf{I}_2 & 0 & 0 & 0 & 0 \\ 0 & \mathbf{I}_2 & 0 & 0 & 0 \\ 0 & 0 & \mathbf{I}_2 & 0 & 0 \\ 0 & 0 & 0 & \mathbf{I}_2 & 0 \\ 0 & 0 & 0 & 0 & \mathbf{I}_2 \\ \alpha\mathbf{I}_2 & 0 & 0 & 0 & (1-\alpha)\mathbf{I}_2 \end{bmatrix} \begin{bmatrix} \mathbf{p}_1^* \\ \mathbf{p}_2^* \\ \mathbf{p}_3^* \\ \mathbf{p}_4^* \\ \mathbf{p}_5^* \end{bmatrix} \tag{18}$$

or

$$\mathbf{p} = \mathbf{T}_{p,q}\mathbf{p}^* \tag{19}$$

where $\mathbf{p}^*$ is the displacement vector of the TRIM5 element, and α is a scalar, less than unity, and depends on the position of point 6 along side 1, 5. Here also $\mathbf{I}_2$ is a unit matrix of order 2, and $\mathbf{0}$ is a (2×2) zero matrix.

It is easy to see that the stiffness matrix $\mathbf{k}^*$ of the TRIM5 element can be derived from the stiffness matrix $\mathbf{k}$ of the TRIM6 element using the relationship

$$\underbrace{\mathbf{k}^*}_{(10,10)} = \mathbf{T}_{p,q}^t \mathbf{k} \mathbf{T}_{p,q} \tag{20}$$

and the loads

$$\mathbf{P}^* = \mathbf{T}_{p,q}^t \mathbf{P} \tag{21}$$

It is also possible to generate a TRIM4 interface element based on the TRIM6. Starting with the TRIM10 element, we can derive various interfacing elements with combinations of first-, second-, and third-order displacement variations along their boundaries.

In a typical interactive operation, the model is formed using triangular elements, regardless of their type. From the first results, based on an interative solution, areas of stress concentration will be apparent, and the user will introduce additional nodes on the appropriate boundaries (see Fig. 9b). The computer should then automatically revise the element types, and replace the old coefficients with a new one, expand the current displacement vector, adding interpolated displacement values at the intermediate points, and continue with the iterative solution.

13. The Problem of Storage in a Small Computer

Should one choose to operate with a small computer the situation becomes distinctly different from that when using a large installation. In order to solve the largest possible problem with the small machine, the limited configuration must be used to the utmost advantage. This section and some of the following ones will deal with particular problems and suggested techniques aimed at achieving this goal. The other alternative or parallel approach would be to solve a complex problem through a series of smaller solutions in which the interface from one problem to the next is fully automated; and optimum use of the fast communication, possible with a mini-computer, is made.

One particular problem that has to be handled is that of the limited amount of available storage space in the core and backing storage of the small unit. It is of great importance to reduce the storage requirements and for that reason a hard look at the nature of the major space-consuming item, the assembled stiffness matrix $\mathbf{K}$ is necessary so that it may be fitted economically within the available storage.

For a two-dimensional membrane problem modelled using simple quadrilateral elements in which each node possesses two degrees of freedom, the number of nonzero entries in any one row of the stiffness matrix is 9 or 10 regardless of the band width of the system of equations. As the size of the problem increases, it becomes distinctly wasteful to store the rest of the nonzero elements. The only reason such space may be required is that the remaining zero elements will be filled up during a direct elimination process. A banded solution approach takes advantage of the sparsely populated nature of the matrix only partially. Should we choose to use an iterative method of solution such as the Gauss–Seidel scheme one is able then to retain the original format of the master stiffness matrix. The major disadvantage of such a method is the undetermined rate of convergence and the possible excessive computational time due to a large number of iterations, it becomes necessary to investigate various techniques aimed at speeding up the convergence. Kamel and Lambert [9] should be consulted for early research in this area; their methods have also been described for obtaining solutions for eigenvalue problems (vibration, buckling) using sparsely populated matrix representation as well as a simple method for reducing the number of iterations by a factor of two. In further sections of this paper, we will describe various other techniques that may be used to obtain faster convergence based on sparsely populated representation.

14. Method of Additional Constraints (MAC)

The equations of equilibrium of the nodes for a finite-element displacement model is given by

$$\mathbf{Kr} = \mathbf{R} \tag{22}$$

where $\mathbf{r}$ is the displacement vector, $\mathbf{R}$ is the force vector, and $\mathbf{K}$ is the master stiffness matrix.

Instead of solving for the complete set of displacements $\mathbf{r}$, we choose to solve for an alternate smaller vector $\mathbf{q}$ from which the $\mathbf{r}$ can be derived using a relationship of the form

$$\mathbf{r} = \mathbf{T}_{r,q}\mathbf{q} \tag{23}$$

where $\mathbf{T}_{r,q}$ is an interpolation matrix which may be automatically generated. Using energy principles, the corresponding generalized force vector $\mathbf{Q}$ may be obtained from the relationship

$$\mathbf{Q} = \mathbf{T}_{r,q}^t\mathbf{R} \tag{24}$$

Now it is also possible to obtain a reduced stiffness matrix $\mathbf{K}_Q$

$$\mathbf{Q} = \mathbf{T}_{r,q}^t\mathbf{R} = \mathbf{T}_{r,q}^t\mathbf{Kr} = \mathbf{T}_{r,q}^t\mathbf{KT}_{r,q}\mathbf{q} = \mathbf{K}_Q\mathbf{q} \tag{25}$$

where

$$\mathbf{K}_Q = \mathbf{T}_{r,q}^t\mathbf{KT}_{r,q} \tag{26}$$

The solution for the vector $\mathbf{q}$ in terms of the generalized vector $\mathbf{Q}$ should converge faster due to the smaller size of the problem. After the reduced displacements $\mathbf{q}$ have been obtained, one may transform them back into an initial $\mathbf{r}$ vector using Eq. (23). With the new approximate $\mathbf{r}$ vector as a starting point, the iterative procedure is continued for the original problem until convergence. A series of pilot computations have been conducted to evaluate the desirability of such a procedure.

In Fig. 10 a system of five nodes connected to a series of six springs each of unit stiffness is shown. The displacements of the five nodes are represented by the vector $\mathbf{r}$. A set of three fictitious points are chosen, the first, point I, is situated midway between nodes 1 and 2, the second, point II, coincides with node 3, and the third, point III, is midway between nodes 4 and 5. The displacements of the three points, I, II, and III are given by the vector $\mathbf{q}$. A simple linear interpolation was established in which the displacement of point 1 was interpolated from that of point I, point 2 interpolated from points I and II, the displacement of point 3 was taken to be identical to that of II, etc. The resulting relationship is given by

$$
\begin{bmatrix} \mathbf{r}_1 \\ \mathbf{r}_2 \\ \mathbf{r}_3 \\ \mathbf{r}_4 \\ \mathbf{r}_5 \end{bmatrix} = \begin{bmatrix} \frac{2}{3} & 0 & 0 \\ \frac{2}{3} & \frac{1}{3} & 0 \\ 0 & 1 & 0 \\ 0 & \frac{1}{3} & \frac{2}{3} \\ 0 & 0 & \frac{2}{3} \end{bmatrix} \cdot \begin{bmatrix} \mathbf{q}I \\ \mathbf{q}II \\ \mathbf{q}III \end{bmatrix} \tag{27}
$$

Three different loading cases were tried out. Case (a) was that of a unit load imposed at point 3. Solving the five degrees of freedom problem so that the displacements are accurate to within 0.0001 required 28 iterations. Using MAC, the three degrees of freedom system required 11 iterations, the expanded five degrees of freedom system converged immediately (one iteration). In order to investigate the effect of the loading pattern on the solution, a series of equal loads were imposed at each node [loading case (b)]. The solution using the Gauss–Seidel method directly on the five degrees of freedom system required 32 iterations. Using MAC, in conjunction with the Gauss–Seidel method, the reduced system converged in 13 iterations and the fully expanded solution converged immediately. A third loading case was used in which the loads varied linearly [loading case (c)]. Using Gauss–Seidel on the full system required 29 iterations, MAC required 12 iterations in a three degrees of freedom system and nine for the fully expanded configuration. A larger problem with a total number of 10 degrees of freedom was attempted using the same method. The loading was chosen to have an unfavorable distribution. The Gauss–Seidel solution required 90 iterations.

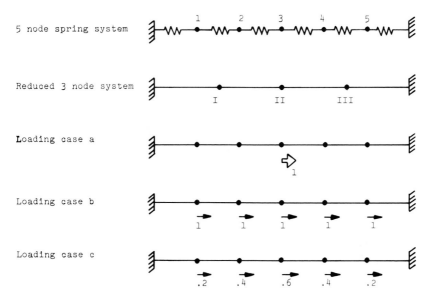

Fig. 10. Example for the method of additional constraints. (a) Five-node spring system; (b) reduced three-node system; (c) loading case a; (d) loading case b; (e) loading case c.

A reduction of the system to five degrees of freedom resulted in convergence after 42 iterations. The subsequent iterations for the full system required 41 iterations.

From these figures it seems that, provided the assumptions are reasonable, the method results in a definite improvement in the total computation time required. The main difficulty would seem to be in the establishment of the reduction matrix $T_{r,q}$. We have reason to believe that such a process can be automated using a mini-computer and that an experienced user may employ it to great advantage.

The previously described process need not be applied only once. A series of nested solutions may enhance the power of the method greatly. For example, a 3000 degrees of freedom system may be transformed in a 1000 degrees of freedom system that may again be transformed in a 200 degrees of freedom system, then an 80 and then a 30. The solution of the 30 degrees of freedom system will speed up that of the 80 which will speed up that of the 200 degrees of freedom system, etc. until the final 3000 degrees of freedom system is solved. The necessary assumptions to reduce one problem to a smaller one may be a challenging and interesting task which would encourage individual judgment and keep up the interest of the user while the computer is performing its operations. In the case of a ship, with let us say 40 transverse frame stations, the assumption of plane cross-sections remaining plane, applied to each frame station independently, will reduce the number of degrees of freedom to three per frame in the case of a symmetric or anti-symmetric loading. Warping of each cross section, in the case of open-deck vessels, may be allowed by using a simple additional warping function. Subsequently, the system may be reduced further by choosing every second or third frame as a master frame and interpolating the freedoms between these frames. In this manner, a quick estimate of the behavior of the ship may be obtained. More refined solutions are possible by removing constraints gradually until the final detailed solution is achieved, or even more economic-ally by releasing the constraints only at local areas where more accurate stress values may be required.

15. Variations on the Gauss–Seidel Iterative Technique

One of the most widely used and most stable techniques for the solution of a set of simultaneous linear equations is the Gauss–Seidel method. The application of the method to the solution of the displacement equations

$$\mathbf{Kr} = \mathbf{R} \qquad (28)$$

converges after a number of passes. In each pass every equation of the set,

$$\sum_j K_{ij} r_j = R_i \qquad (29)$$

is taken in turn and used to obtain a better value for the corresponding displacement r_i

$$\sum_j K_{ij} r_j - K_{ii} r_i + K_{ii} r_i = R_i$$

which we can use to solve for r_i

$$r_i = \frac{R_i - (\sum K_{ij} r_j - K_{ii} r_i)}{K_{ii}} = \frac{R_i - \sum_{j \neq i} K_{ij} r_j}{K_{ii}} \qquad (30)$$

This method will be taken as the basis for a number of generalizations which are being currently investigated as possible useful techniques to be used with an interactive small computer. These methods will be described now.

15.1. Gauss–Seidel iteration Used in Conjunction with a Partitioned Matrix (GSP)

Partitioning of the deflection, load, and stiffness matrices serves to group the various quantities in physically meaningful units and enables the pro-gram to use the interaction pattern between the various groups more effectively. Any submatrix of the partitioned matrices need not always be stored as a fully populated matrix.

A typical equation corresponding to $\mathbf{r}_i$ is a direct generalization of Eq. (29) and is given by

$$\sum \mathbf{K}_{ij}\mathbf{r}_j = \mathbf{R}_i \tag{31}$$

Proceeding in the same manner as before,

$$\sum \mathbf{K}_{ij}\mathbf{r}_j - \mathbf{K}_{ii}\mathbf{r}_i + \mathbf{K}_{ii}\mathbf{r}_i = \mathbf{R}_i$$

which then may be solved for the complete group of displacements $\mathbf{r}_i$ simultaneously rather than for one displacement at a time. The result is given by the following matrix equation

$$\mathbf{r}_i = \mathbf{K}_{ii}^{-1}[\mathbf{R}_i - (\mathbf{K}_{ij}\mathbf{r}_j - \mathbf{K}_{ii}\mathbf{r}_i)] = \mathbf{K}_{ii}^{-1}\left(\mathbf{R}_i - \sum_{j \neq i} \mathbf{K}_{ij}\mathbf{r}_j\right) \tag{32}$$

From the description of the method, it appears to be a cross between direct and iterative solutions. The limiting case of one partitioning corresponds to a direct solution of the equations whereas, on the other hand, a nonpartitioned matrix gives the usual Gauss–Seidel scheme. One expects a faster convergence involving a smaller number of iterations, each iteration, however, taking more time than that of the original method. With prudent planning of the storage space a definite advantage can be gained over a direct solution. One may also perhaps retain the inverse of the diagonal submatrices $\mathbf{K}_{ii}$ once they are formed so that the inverses need only be computed once.

15.2. THE GAUSS–SEIDEL SCHEME INVOLVING TWO EQUATIONS (GS2)

Under this heading, we denote an iterative method in which a pair of displacements r_i and r_j are improved within one step using a simultaneous solution of the two corresponding equations. This operation will be carried only if the cross coefficient K_{ij} exists. In this manner, the sparsely populated nature of the matrix is used to an advantage. The two equations are of the form

$$K_{i,1}r_1 + \cdots + K_{i,i}r_i + \cdots + K_{i,j}r_j + \cdots + K_{in}r_n = R_i$$
$$K_{j,1}r_1 + \cdots + K_{j,i}r_i + \cdots + K_{j,j}r_j + \cdots + K_{jn}r_n = R_j \tag{33}$$

which may be written as

$$K_{ii}r_i + K_{ij}r_j = R_i - \sum_{\substack{p \neq i \\ p \neq j}} K_{ip}r_p$$
$$K_{ji}r_i + K_{jj}r_j = R_j - \sum_{\substack{p \neq i \\ p \neq j}} K_{jp}r_p \tag{34}$$

Equations (34) may be solved for r_i and r_j simultaneously to give

$$r_i = \frac{K_{jj}(R_i - \sum_{p \neq i, p \neq j} K_{ip} r_p) - K_{ij}(R_j - \sum_{p \neq i, p \neq j} K_{jp} r_p)}{(K_{ii} K_{jj} - K_{ij}^2)}$$

$$r_j = \frac{-K_{ji}(R_i - \sum_{p \neq i, p \neq j} K_{ip} r_p) + K_{ii}(R_j - \sum_{p \neq i, p \neq j} K_{jp} r_p)}{(K_{ii} K_{jj} - K_{ij}^2)}$$

(35)

The process is repeated until convergence.

This procedure may be extended further by taking three or more displacement values simultaneously, in which case we end up with a higher-level scheme. However, we doubt if an advantage is to be gained from that, since the arbitrary sparsely populated nature of the matrix cannot be utilized as effectively.

A pilot study made for the purpose of comparing the GS2 with the conventional Gauss–Seidel procedure resulted in convergence using fewer iterations. It is clear from the results that the GS2 procedure is more stable, and the monotonic nature of the convergence is a reassuring factor when used in conjunction with an interactive terminal. For a five degrees of freedom system, the conventional Gauss–Seidel solution needed 31 iterations. The GS2 method required 12. For a 10 degrees of freedom system, the number of iterations were 43 compared to 19. The net result was a moderate reduction in the total solution time. However, further investigation is necessary before definite conclusions can be reached.

15.3. The GS2 for Partitioned Matrices (GSP2)

The two previously described techniques may be used together to provide simultaneous improvements for two sets of deflections provided an interaction exists. The partitioned matrices will be rearranged for convenience to become

$$\begin{bmatrix} K_{ii} & K_{ij} & K_{im} \\ K_{ji} & K_{jj} & K_{jm} \\ K_{mi} & K_{mj} & K_{mm} \end{bmatrix} \begin{bmatrix} r_i \\ r_j \\ r_m \end{bmatrix} = \begin{bmatrix} R_i \\ R_j \\ R_m \end{bmatrix}$$

(36)

The first two equations may be rewritten as

$$K_{ii} r_i + K_{ij} r_j = R_i - K_{im} r_m \qquad K_{ji} r_i + K_{jj} r_j = R_j - K_{jm} r_m$$

(37)

and may be solved simultaneously for r_i and r_j to provide a generalized form of Eq. (35).

$$r_i = (K_{ii} - K_{ij} K_{jj}^{-1} K_{ji})^{-1} [(R_i - K_{im} r_m) - K_{ij} K_{jj}^{-1} (R_j - K_{jm} r_m)]$$

$$r_j = (K_{jj} - K_{ji} K_{ii}^{-1} K_{ij})^{-1} [(R_j - K_{jm} r_m) - K_{ji} K_{ii}^{-1} (R_i - K_{im} r_m)]$$

(38)

It appears, therefore, that we have presented a wide spectrum of solution techniques that vary from being direct to fully iterative. It should be interesting to investigate the efficiency and desirability of such techniques in a computer-controlled environment (large system) and in the user-controlled environment (small interactive system).

16. Conclusion

In this paper we have presented some predictions regarding the future use of computers in the area of ship structural analysis, in particular, and that of the applications to the finite-element analysis in general. Some useful techniques are described, of which the reduced substructure method has already found application in the analysis of tankers and open-deck vessels. Other techniques are suggested for use in conjunction with an interactive graphics system, some relate to modeling and variations on the finite-element theory.

The remaining methods represent early feasibility studies aimed at improvement in currently used iterative solution techniques. Automatic mesh generation is briefly discussed with a few illustrative examples.

Acknowledgments

Some of the research presented in this paper has been generously supported by the American Bureau of Shipping, New York; Airesearch Corporation, Phoenix; Lockheed Missiles and Space Company, Sunnyvale, California; and Standard Oil of California. The University of Arizona Computer Center has contributed valuable advice and help.

The authors would like to thank the numerous contributors to this effort, particularly Mr. Donald Couch of Standard Oil of California, Mr. Michael W. McCabe, graduate assistant at the University of Arizona, and Mr. Anastasios Peratikos of the American Bureau of Shipping. The authors would also like to thank Miss Janette Schmerbeck for typing the manuscript, and Mr. Robert G. Dowdy for the preparation of the drawings.

Appendix A. Nomenclature

r	Deflection matrix
R	Load matrix
K	Stiffness matrix
T_{de}	Boundary displacement interpolation matrix
T_{ie}	Displacement transformation matrix
T_e	Complete displacement transformation matrix
e	Subscript denoting master or external nodes
d	Subscript denoting dependent boundary nodes
i	Subscript denoting internal, or free, nodes

$\mathbf{R}_c^*$ Reduced load matrix
$\mathbf{K}_{ee}^*$ Reduced stiffness matrix
$\mathbf{q}$ Generalized displacement vector
$\mathbf{Q}$ Generalized force vector
$\mathbf{T}_{r,q}$ Generalized displacement transformation matrix
$\mathbf{p}, \mathbf{p}^*$ Element corner displacements
$\mathbf{P}, \mathbf{P}^*$ Element corner forces
$\mathbf{k}, \mathbf{k}^*$ Element stiffness matrices

References

1. Batdorf, W. J., Kapur, S. S., and Sayer, R. B., "The Role of Computer Graphics in the Structural Design Process," *Proc. Conf. Matrix Methods Struct. Mech. 2nd* Wright-Patterson AFB, Ohio, AFFDL-TR68150 (October 1968).
2. Prince, David M., *Interactive Graphics For Computer-Aided Design.* Addison-Wesley, Reading, Massachusetts, 1971.
3. Paulling, J. R., and Payer, H. G., "Hull-Deckhouse Interaction by Finite Element Calculations," *Trans. Soc. Naval Architects Marine Eng.* **76**, 281–308 (1968).
4. Kamel, H. A., and Eisenstein, H. K., "Automatic Mesh Generation in Two and Three Dimensional Interconnected Domains," Paper presented at the *Symp. High Speed Comput. Elastic Struct.* IUTAM, Liege, Belgium, August 23–28 (1970).
5. Kamel, H. A., Birchler, W., Liu, D., McKinley, J. W., and Reid, W., "An Automated Approach to Ship Structure Analysis," *Trans. Soc. Naval Architects Marine Eng.* **77**, 233–268 (1969).
6. Gellatly, R. A., "Optimization and Materials Selection," Paper presented at the *Symp. Comput. Finite Element Anal. Ship Struct.* at the Univ. of Arizona, Tucson, Arizona, March 29–April 2 (1971).
7. Fraeijs De Veubeke, B., "Displacement and Equilibrium Models in the Finite Element Method," in *Stress Analysis* (Zienkiewicz, O. C., and Holister, G. S., eds.), Chapter 9. Wiley, New York, 1965.
8. Hansteen, O. E., "Analysis of Stress Distribution in Shear Walls by the Finite Element Displacement Method," *Int. Kongr. über Anwendungen der Math. Ingenieurwissenschaften*, Weimar (1967).
9. Kamel, H. A., and Lambert, R. L., "Solution of Structural Eigenvalue Problems using Sparsely Populated Matrices," Paper presented at the *Conf. Comput. Oriented Anal. Shell Structures*, Lockheed Palo Alto Research Laboratory, California (August 1970).

Author Index

Numbers in parentheses are reference numbers and indicate that an author's work is referred to although his name is not cited in the text. Numbers in italics show the page on which the complete reference is listed.

A

Abu-Sitta, S. H., 391(2, 15), 392(2), 394, 396 (2), *401*, *402*
Adelman, H. M., 297, 330(15), *336*
Ahmad, S., 14(7, 17-19), 22(7), 24(7), 33, 39(7), *40*, 43(2, 3), *57*, 67(19), *77*
Ahuja, A. S., 626(3), *641*
Alavi, M. J., 616, *623*
Alberts, A. A., 491(11), 492(10, 11), *495*
Allwood, R. J., *77*
Almond, J. C., 82(16), *118*
Almroth, B. O., 294, 301, 304(16), 311, 327, 328, *335*, *336*, 396, *401*
Alway, G. G., 98(48), *120*
Ames, W. F., 267, 271(3), 274, 278, 279(3), 283, *286*
Anderheggen, E., 545(9), *554*
Anderson, D. G. M., 294(6), *335*
Anderson, R. G., 14(19), *40*
Ang, A. H.-S., 395(21), *402*
Arakawa, A., 283, *289*
Araldsen, P. O., 418(2), *436*
Argyris, J. H., 14(10, 13, 14), 22(13), *40*, 79(3), 112(1, 2, 82), 114(83), *118*, *121*, 125(3, 4), 129(7), 131(8), *148*, 349, *352*, 544(4), *554*

B

Atluri, S., 75(37), *78*
Au, T., 459(11), 463(11), 467(11), *477*
Ayres, D. J., 614, 616, *623*

B

Bair, R., 636(10), 637(10), *642*
Balmer, H. A., 355(16, 18), *376*
Bamford, R. M., 86(42), 90(42), 98(42), *119*, 500(4), *511*
Barlow, J., 98(50), *120*
Baron, M. L., 267(6), 270(28, 31), 280(28), *286*, *287*, *288*
Barone, M. A., 591, *622*
Barr, L. K., 268, *287*
Barton, D., 459(6), *477*
Batdorf, W. J., 114(76), *121*, 645(1), *668*
Bauer, F. L., 102(63), *120*
Bayer, R., 82(19), *118*
Bell, K., 129(6), *149*
Belytschko, T., 545(8), *554*
Bemer, R. W., 81(12), 82(12), *118*
Benzley, S. E., 342(2), *350*
Berg, B. A., 297(13), *336*
Berg, C. A., 617, *623*
Bergel, D. C., 639(14), *642*